CMS Books in Mathematics

Ouvrages de mathématiques de la SMC

CMS Books in Mathematics is a collection of graduate level textbooks and monographs published in cooperation with the Canadian Mathematical Society since 1999. This series offers authors the joint advantage of publishing with a major mathematical society and with a leading academic publishing company. The series is edited by Karl Dilcher and Keith Taylor. The more than 20 titles already in print cover such diverse areas as combinatorics, number theory, optimization, dynamical systems and the history of mathematics, to name a few. Books in this series will appeal to all mathematicians, whether pure or applied, whether students or researchers.

More information about this series at http://www.springer.com/series/4318

David M. Jackson · Iain Moffatt

An Introduction to Quantum and Vassiliev Knot Invariants

Springer

David M. Jackson
Faculty of Mathematics
University of Waterloo
Waterloo, ON, Canada

Iain Moffatt
Department of Mathematics
Royal Holloway, University of London
Egham, UK

ISSN 1613-5237 ISSN 2197-4152 (electronic)
CMS Books in Mathematics
ISBN 978-3-030-05212-6 ISBN 978-3-030-05213-3 (eBook)
https://doi.org/10.1007/978-3-030-05213-3

Library of Congress Control Number: 2018963043

Mathematics Subject Classification (2010): 57M27, 57M25, 17B37

This Springer imprint is published by the registered company Springer Nature Switzerland AG
The registered company address is: Gewerbestrasse 11, 6330 Cham, Switzerland

Preface

This book grew from a series of discussions between its authors on the combinatorics of Vassiliev and quantum knot invariants. We were fascinated both by the area itself, and by the rich interaction between the combinatorial and algebraic ideas that supported it. Moreover, we recognised that the area perfectly exhibited many of the important stages that are passed through in the development of a mathematical theory. Such stages may so easily remain hidden when a reader is presented with a complete and polished theory, yet experiencing the thinking behind the transitions from one stage to another is vital for the aspiring mathematician.

Three key themes stood out in these discussions.

- The first theme is the *constructive approach*, which is fundamental to mathematics in general, and to algebraic combinatorics in particular. It is an approach in which the preliminary step to handling and understanding a difficult object (for us, a *knot*) is to derive a description of it as a *combinatorial structure*. This can open the way to decompose the structure into 'elementary pieces' that may then be described in algebraic terms. We may then hope to work backwards to associate an *algebraic structure* with the original object.
- The second theme is how one mathematical object may be transformed into one more amenable to further investigation, while still ensuring that key information (for us, the *isotopy class of a knot*) in the former is retained in the latter.
- The third theme is the use of mathematical techniques for working effectively within the algebras that arise in these processes (e.g. *ribbon Hopf algebras*). In doing so the intention is, of course, to elicit the key information (e.g. *the isotopy class of a knot*) that has been retained in the more amenable object, once the association with the original object (e.g. *a knot*) has been established.

Given that the area of knot theory perfectly illustrates the use of these mathematical processes that are often hidden from students, we felt that an introduction to this theory, and particularly to its interactions between algebra, combinatorics and topology, could serve as an excellent step in the mathematical development of a student. This thinking led to perhaps an unusual expository aim in writing this book. Rather than writing a text that serves primarily as an introduction to quantum knot theory (although we certainly hope that we have provided a worthy introduction to that theory), we wanted to reveal mathematical approaches available for engaging with a difficult mathematical object. Accordingly, we have adopted the following expository principles:

1. to focus on the interplay between combinatorics, algebra, and topology;
2. to take a workmanlike approach to the abstract material, and to the acquiring of a facility with, for example, calculations within the algebras that arise;
3. to write a book to be read, rather than to serve primarily as a reference volume;
4. to use exercises to aid learning rather than to test;
5. to write a book that will serve a broad spectrum of aspiring mathematicians, and not solely those who wish to pursue knot theory;
6. to offer an aspiring knot theorist a grounding in the fundamentals of Vassiliev and quantum knot invariants, thereby enabling a reader to successfully tackle the more specialised literature;
7. and, finally, to keep the volume to a manageable length.

Writing to these principles has, of course, affected what content and which topics we have included or omitted. For example, an expert reader will undoubtedly notice that we have chosen not to give a full development of the theory of quantum groups or monodromy solutions of the Knizhnik–Zamolodchikov equations, nor do we discuss Chern–Simons theory. While these topics are beautiful and central to the area, a proper exposition of them takes us too far from our guiding expository principles. For similar reasons, in a few places we decided to quote results without proof. We also felt we could safely omit these topics since they are served by other excellent texts in the area, such as [37, 90, 144]. In fact, the present volume resulted from paring down a much longer volume. This was a very painful process, particularly when it involved excising wonderful pieces of mathematics, but we believe that the resulting text better serves the broader mathematical community.

In writing this book, our emphasis has been on the use and the flow of mathematical ideas. We have tailored our exposition to an advanced undergraduate or beginning graduate student, who may intend to pursue a research area other than topology or to a researcher from another field who wishes to have a rapid introduction to the area. In this volume, the reader will see versatile general results, such as isomorphism theorems, and how they may be used in practice; will develop an appreciation of how various types of algebras arise, and how the introduction of algebraic objects is driven by application; will see how a profound theory can be built based on the tools acquired in undergraduate studies; and will develop a proficiency in a variety of algebraic, combinatorial, and topological techniques that can be applied in areas beyond the scope of this book.

Finally, we hope that the reader will be captivated by the beauty of the area and fascinated by its algebraic combinatorial nature, as indeed we were.

In preparing this book, we have benefited from and enjoyed conversations with and support from many people. It is impossible to thank each person by name, although special thanks are due to P. Etingov, S. Garoufalidis, S. Lando, and

M. Perry for valuable discussions at various stages. Special thanks are due to our spouses Jennifer and Vikki for their continuing support and patience through this project.

Waterloo, ON, Canada
Egham, UK
October 2018

David M. Jackson
Iain Moffatt

Contents

Part III Vassiliev Invariants

Part IV The Kontsevich Invariant

Frequently Used Notation

$\langle D \rangle$	Kauffman bracket (p. 43)
∇_a	Operator on Jacobi diagrams (p. 298)
$\mathsf{1T}$	$1\mathsf{T}$-relation (p. 202)
$\mathsf{4T}$	$4\mathsf{T}$-relation (pp. 202, 294)
$\mathscr{A}$, $\mathscr{A}_m$	Vector spaces of Jacobi diagrams modulo the STU-relation (p. 223)
$\mathscr{A}^c$, $\mathscr{A}^c_m$	Vector space of chord diagrams modulo $4\mathsf{T}$-relation (p. 215)
$\bar{\mathscr{A}}^c \bar{\mathscr{A}}^c_m$	Vector spaces of chord diagrams modulo the $1\mathsf{T}$- and $4\mathsf{T}$-relation (p. 203)
$\mathscr{A}(X)$, $\mathscr{A}_m(X)$	Vector spaces of Jacobi diagrams on X modulo the STU-relation (p. 295)
$\mathscr{A}^c(X)$, $\mathscr{A}^c_m(X)$	Vector spaces of chord diagrams on X modulo the $4\mathsf{T}$-relation (p. 294)
$\widehat{\mathscr{A}}(X)$	Vector space of all infinite formal linear combinations of Jacobi diagrams modulo the STU-relation (p. 296)
$\widehat{\mathscr{A}}^c(X)$	Vector space of all infinite formal linear combinations of chord diagrams modulo the $4\mathsf{T}$-relation (p. 296)
AS	AS-relation (p. 225)
$(\mathfrak{A}, \mathbf{v})$	Ribbon Hopf algebra (p. 138)
$\mathfrak{B}_n$	Braid group (p. 53)
$\mathsf{BI} - \mathsf{BIII}$	Braid moves (p. 51)
$\mathscr{C}$, $\mathscr{C}_m$	Vector spaces of chord diagrams (p. 186)
$\mathscr{C}(X)$, $\mathscr{C}_m(X)$	Vector spaces of chord diagrams on X (p. 293)
$C(L)$	Alexander-Conway polynomial (p. 34)
$\mathscr{D}$, $\mathscr{D}_m$	Vector spaces of Jacobi diagrams (p. 221)
$\mathscr{D}(X)$, $\mathscr{D}_m(X)$	Vector spaces of Jacobi diagrams on X (p. 295)
FRI	A framed Reidemeister move (p. 40)
$\mathsf{FT}_0 - \mathsf{FT}_7$	Framed Turaev moves (p. 118)
IHX	IHX-relation (p. 225)
$J(L)$	Jones polynomial (p. 31)

$\mathbb{K}$	Ring (p. 341)
$\mathscr{K}$, $\mathscr{K}_m$	Vector spaces of knots (pp. 167, 171)
$\mathscr{K}^f$, $\mathscr{K}^f_m$	Vector spaces of framed knots (p. 212)
$\dot{\mathscr{K}}$, $\dot{\mathscr{K}}_m$	Vector spaces of singular knots (pp. 181, 182)
$\dot{\mathscr{K}}^f$, $\dot{\mathscr{K}}^f_m$	Vector spaces of framed singular knots (p. 213)
MI, MII	Markov moves (p. 57)
$P(L)$	HOMFLYPT polynomial (p. 33)
$\mathsf{qFT}_0 - \mathsf{qFT}_5$	Turaev moves for framed q-tangles (p. 288)
$Q^{\mathfrak{A}}$	Reshetikhin-Turaev invariant (p. 150)
$Q_{(\mathsf{R},\alpha,\beta)}$	Operator invariant (p. 109)
$Q_{(V,W,\mathsf{R},\vec{u},\overleftarrow{u},\vec{n},\overleftarrow{n})}$	Operator invariant (p. 119)
$Q^f_{(\mathsf{R},\alpha,\beta)}$	Framed operator invariant (p. 121)
$Q^f_{(V,W,\mathsf{R},\vec{u},\overleftarrow{u},\vec{n},\overleftarrow{n})}$	Framed operator invariant (p. 118)
R	(Universal) R-matrix (pp. 69, 132)
$\mathsf{R} \in \widehat{\mathscr{A}^c_2}(\uparrow\uparrow)$	Special element $\exp(\frac{1}{2}H)$ of $\widehat{\mathscr{A}^c_2}(\uparrow\uparrow)$ (p. 297)
RI – RIII	Reidemeister moves (p. 13)
$\vec{\mathsf{RI}} - \vec{\mathsf{RIII}}$	Oriented Reidemeister moves (p. 20)
$\mathfrak{sl}_2$, $\mathfrak{sl}_n$	The Lie algebras (p. 140)
STU	STU-relation (p. 221)
$\mathsf{T}_0 - \mathsf{T}_7$	Turaev moves (p. 98)
$T_{\mathfrak{g}}$	Universal $\mathfrak{g}$ weight system (p. 260)
$U(\mathfrak{g})$	Universal enveloping algebra (p. 141)
$U_h(\mathfrak{sl}_2)$	Quantized universal enveloping algebra (p. 143)
$U_q(\mathfrak{sl}_2)$	q-analogue $U(\mathfrak{sl}_2)$ (p. 145)
$\mathscr{V}$, $\mathscr{V}_m$	Vector spaces of Vassiliev invariants (p. 171)
$\mathscr{V}^f$, $\mathscr{V}^f_m$	Vector spaces of framed Vassiliev invariants (p. 212)
$\mathscr{W}$, $\mathscr{W}_m$	Vector spaces of weight systems (p. 200)
$\mathscr{W}^f$, $\mathscr{W}^f_m$	Vector spaces of framed weight systems (p. 214)
$W_{\mathfrak{g},\rho}$	Lie algebra weight system (p. 266)
$W_{\mathfrak{sl}_2}$	$\mathfrak{sl}_2$ weight system (p. 276)
$\check{Z}(T)$	Kontsevich invariant (p. 318)
$\check{Z}^u(T)$	Unframed Kontsevich invariant (p. 324)
$\rho_{\mathsf{R},\mu}(L)$	The knot invariant (p. 84)
Φ	Associator (p. 305)
$\omega(D)$	Writhe (p. 42)

Introduction

The problem we are interested in is the classical problem in knot theory of deciding whether two knots are equivalent. We start intuitively by thinking of a knot as the familiar configuration that results from looping a length of string around itself and threading an end through the loops so formed, as shown below:

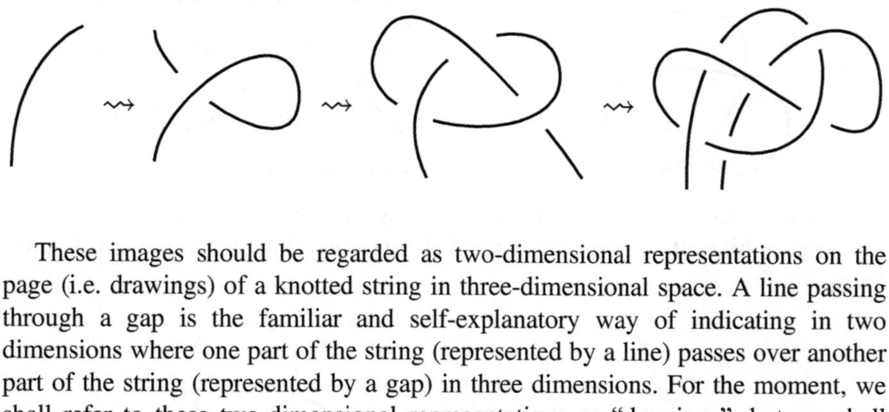

These images should be regarded as two-dimensional representations on the page (i.e. drawings) of a knotted string in three-dimensional space. A line passing through a gap is the familiar and self-explanatory way of indicating in two dimensions where one part of the string (represented by a line) passes over another part of the string (represented by a gap) in three dimensions. For the moment, we shall refer to these two-dimensional representations as "drawings", but we shall soon formalise such representations as "knot diagrams".

One such knot K is said to be *equivalent* to another such knot K' if K may be transformed into K' by simply shifting around parts of the string. We are not allowed to cut the string or to glue pieces of it together. It is an easy matter to start at one end of K and slide this end along the length of the string to produce a straight length of string, and then to loop this around itself, threading an end through the loops to produce K'. Clearly, the theory of such knots is trivial since all such knots are equivalent for we can always reverse what we did to knot the string!

The situation is entirely different if, once the knot is formed in the string, the ends of the string are inseparably sealed together. Thus, instead of a knotted line, there is a knotted circle. An example of this is given by

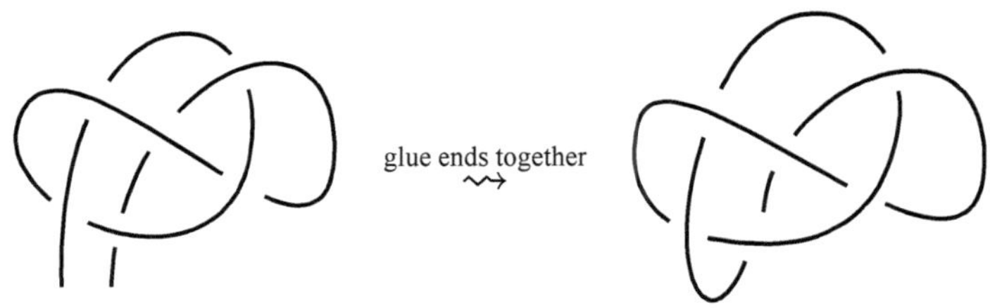

For two such knots K and K', there is no longer the possibility of unknotting K and then retying it as K' since there are no loose ends, as there were for the case of a knotted line.

Intuitively, we regard two knots K and K' as being *equivalent* if we can move the circles of string around in space until the knots look the same. (As we shall see, this idea is formalised by requiring an orientation preserving homeomorphism of $\mathbb{R}^3$ that takes one knot to the other.) We are interested in the question of whether two knots are equivalent or whether they are not.

By forming them out of a piece of string, for example, the reader will quickly convince himself or herself that the following two drawings represent the same knot (i.e. we can deform one so it looks like the other).

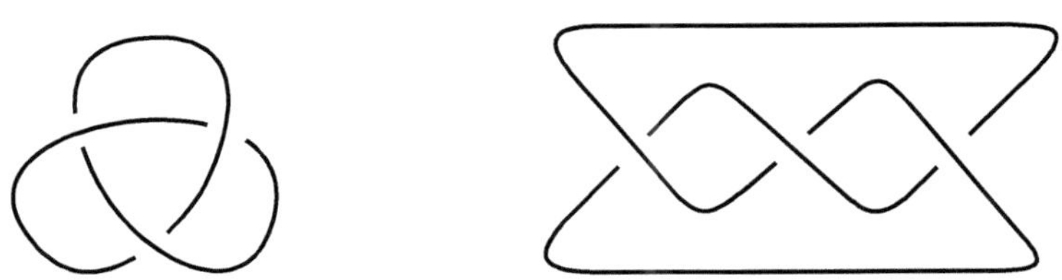

Also by forming them out of a piece of string, say, the reader should also be able to convince himself or herself that the following two drawings represent different knots.

Although the reader will be convinced that these two knots cannot be changed into one another, it is at least theoretically possible that there is some way to do it using a trick that he or she did not spot. Thus, we need a way to *prove* that no such trick exists and that the knots can never be changed into one another. (In Exercise 2.26, we shall see that these two knots are indeed different.)

Deciding whether two drawings represent different knots is a hard problem. As an illustration of the difficulty, the following two knots, called the *Perko Pair*, were thought to be distinct until the 1970s when Perko [148] showed that they are not.

Exercise 1 Verify that the two drawings above do indeed represent the same knot.

Deciding whether knots are inequivalent requires the use of a knot invariant. A *knot invariant* is a function θ from knots to some set (typically a set of numbers, polynomials, or groups) with the property that if K and K' are knots, then

$\theta(K) \neq \theta(K')$ implies that K and K' are different knots.

The intent, of course, is to find a set S (e.g., the set of complex numbers, or a set of polynomials) such that testing for $\theta(K) \neq \theta(K')$ in S is easier than testing that $K \neq K'$ in the set of all knots. Constructing computable and effective knot invariants is difficult. In this book, we describe the construction of two related families of knot invariants called Quantum Invariants and Vassiliev invariants.

It is instructive to reflect upon progress in knot theory during the last two hundred and fifty years. The first appearance of knots in a mathematical context was in a paper by Vandermonde in 1771, followed by some notes on knots by Gauss in 1794 that were little more than collections of drawings. Fifty years later the first systematic attempts to study and classify knots, due in the most part to Tait, Kirkman, and Little, although substantial, were thwarted by the absence of a suitable mathematical framework to support them. Quite simply, although these early proponents of knot theory were often able to show that *two knots were the same*, they were often unable to show that *two knots were different*. This changed with the introduction of algebraic topology in the early twentieth century, and mathematicians such as Dehn, Seifert, Reidemeister, Alexander, and Wirtinger were attracted to the subject. Connections between knots and the topology of 3-manifolds were discovered.

In the 1980s, a revolution in knot theory was sparked through the work of Jones. This resulted in a new class of knot invariants called *Quantum Invariants*, and connections were established with statistical mechanics, quantum physics, quantum groups, and Lie algebras. Drinfel'd, Jones, and Kontsevich each received the Fields Medal, the highest award in mathematics, for work that contributed to knot theory. It is this family of knot invariants that we focus upon here. (For a more detailed history of knot theory, we refer the reader to [153].)

We don't mention Vassiliev invariants.

Before plunging into the mathematics, we briefly describe the structure of this book.

Part I provides an overview of *Basic Knot Theory*. Its purpose is to familiarise the reader with the basic concepts of knot theory that are used in the rest of the book.

Part II describes the construction of a family of knot invariants called *Quantum Invariants*. In brief, the idea behind the material in this part is to construct invariants by slicing a knot diagram up into elementary pieces, such as crossings or critical points, then to form a knot invariant by associating linear maps with these pieces. We shall see that Lie algebras give rise to knot invariants *via* the theory of quantum groups.

Part III describes a very different approach to knot invariants. The idea taken in this part is to study a whole class (or space) of knot invariants rather than individual knot invariants. The invariants in this class are called *Vassiliev Invariants*. We shall see that Lie algebras give rise to Vassiliev invariants through the use of chord diagrams and weight systems.

In Part IV, we bring together the two approaches, showing that they are unified by a single, very powerful knot invariant called *The Kontsevich Invariant*. The invariants arising from Lie algebras in the two ways described in Part II and Part III have dramatically different constructions and philosophies. The culmination of this book, in Part IV, will be the fact that these two constructions result in the same knot invariants. A reader will gain an understanding of how the very different looking theories of Quantum Invariants and Vassiliev invariants fit together as one theory. On the route to this culmination, the reader will see much beautiful mathematics used in a sophisticated and creative way.

Figure 1 shows the chapter dependencies. This book does not have to be read linearly, and, by following the routes described in the second, third, and fourth columns in Fig. 1, it can be used to provide standalone introductions to its three main topics of Quantum Invariants, Vassiliev invariants, and the Kontsevich invariant.

Chapters 1–3 provide the foundation in basic knot theory required for this text. They form a prerequisite for all later chapters. However, a reader who has previously studied some knot theory is likely to be familiar with this material and so may safely skip all of Part I.

A path through the book that focusses only on *Quantum Invariants* is given by either following Chaps. 1–9, or, as a more direct route, a reader can skip Chaps. 4–6. (It is not necessary to read Chaps. 4–6 to follow Chaps. 7–9, but we do recommend reading them.) For a fuller picture of Quantum Invariants, a reader following this route is encouraged to jump to Chaps. 15–17, and then to Sect. 18.2 where the quantum invariant narrative culminates.

A path focussing only on *Vassiliev invariants* is given by jumping to Chaps. 10–14, and making a brief diversion to Sect. 8.4.1 for background on Lie algebras. For a fuller picture of Vassiliev invariants, a reader following this route may then jump to Chaps. 15–17, then to Sect. 18.1 where the Vassiliev invariant narrative culminates.

A path through the book focussing only on the *Kontsevich invariant* is given by reading Sect. 7.1 on tangles, Chap. 13 on Jacobi diagrams, then following Chaps. 15–17.

Despite these options for omitting various topics, we believe the text is best read linearly and encourage a reader to do so.

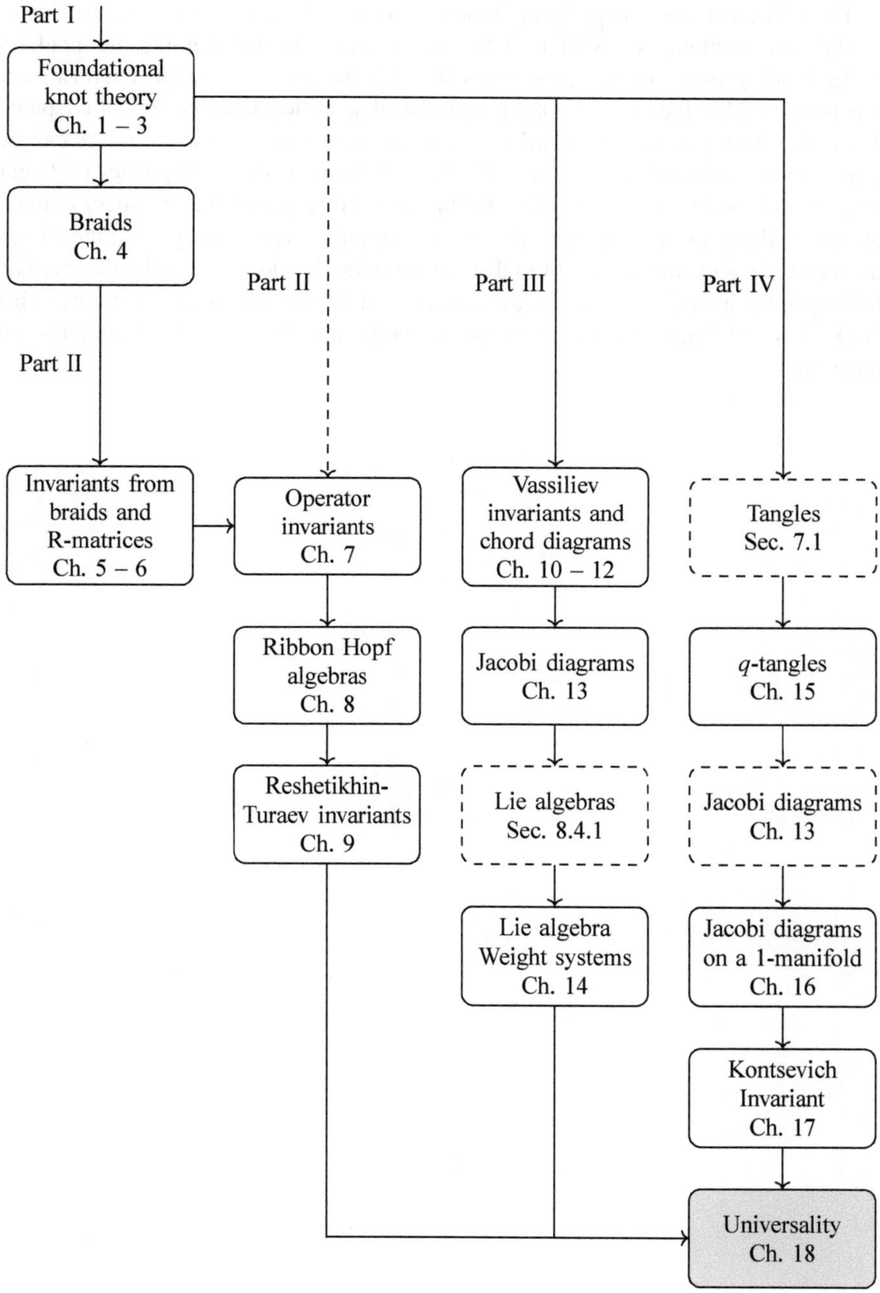

Fig. 1 Chapter dependencies

Exercises can be found throughout the book. These exercises are formative rather than summative, written with the intention of developing the reader's understanding as he or she progresses through the book. As such, exercises are positioned within the main chapter text, rather than collected at the end of chapters. They also form part of the narrative. A collection of hints for the exercises can be found towards the end of the book. Appendix A contains the prerequisites on linear algebra and module theory required for the text. The remaining appendices contain details of algebraic computations that it is important to work through, but have been moved to the appendix so as not to distract from the development of the theory. The bibliography includes all the mathematical sources we consulted to prepare this book. A list of frequently used notation can be found after the table of contents, on page xiii.

Part I
Basic Knot Theory

Chapter 1
Knots

1.1 Knots and Equivalence

We assume that the reader is familiar with basic topology but, for convenience, we recall that a *homeomorphism* is a continuous map with a continuous inverse, and that a function is an *embedding* if it is a homeomorphism onto its image. All the topologies we consider here are those inherited from the standard topology of $\mathbb{R}^n$. Note that throughout the book, we work in the piecewise-linear setting, rather than the smooth one. In particular all embeddings are piecewise-linear.

The physical idea of a knot, that was stated in the introduction as a "knotted circle of string", has a mathematical idealisation encapsulated by the following definition.

Definition 1.1 (Knot). A *knot* is an (piecewise-linear) embedding of the (unit) circle $\mathbb{S}^1$ into $\mathbb{R}^3$.

Fig. 1.1 Some examples of knots

Drawings of a few knots are given in Fig. 1.1 together with their standard symbols. The knot 0_1 is called the *unknot*, 3_1 is the *trefoil*, and 4_1 is the *figure-of-eight knot*. The symbols such as 3_1 used here refer to the names of the knots in the standard Rolfsen tables [158].

D. M. Jackson and I. Moffatt, *An Introduction to Quantum and Vassiliev Knot Invariants*, CMS Books in Mathematics, https://doi.org/10.1007/978-3-030-05213-3_1

Above we provided some examples of knots by providing drawings of them. Knots can also be described by writing down a function from $\mathbb{S}^1$ to $\mathbb{R}^3$. For example, if we describe the points of $\mathbb{S}^1$ by polar coordinates $(1, t)$ for $0 \leq t < 2\pi$, then

$$f : \mathbb{S}^1 \to \mathbb{R}^3 : t \mapsto ((2 + \cos(3t)) \cos(2t), (2 + \cos(3t)) \sin(2t), \sin(3t))$$

provides the parameterised plot of the trefoil knot shown below.

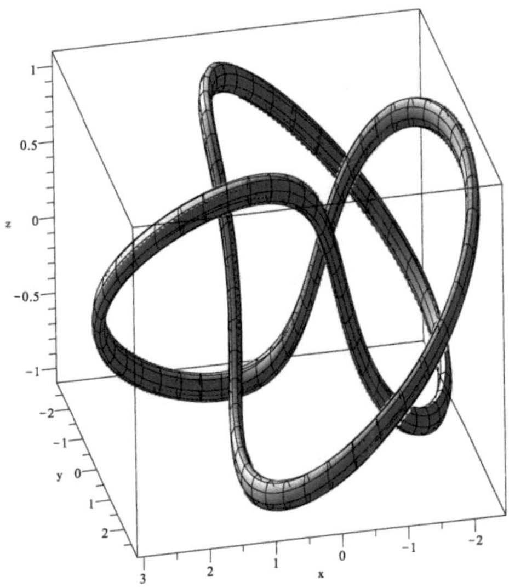

In practice, it is usually best not to specify or work with knots as parameterised curves, but rather to specify them by drawings as we did previously. Nevertheless, it is reassuring to know that knots can be described as parameterised curves in $\mathbb{R}^3$.

Returning to the physical situation of a "knotted circle of string" for a moment, once we have one knotted circle of string, we can of course take a second length of string, and knot with itself and possibly also with the knot we already have, and fuse its end together. This would result in two knotted circles of string, which may be interweaved with each other in such a way that they may or may not be pulled apart. This can be repeated with a third length of string, then a fourth and so on to get any number of knotted circles of string. This gives the concept of a link, which is idealised mathematically by the following definition.

Definition 1.2 (Link). A *link* is a disjoint (piecewise-linear) embedding of n copies of $\mathbb{S}^1$ into $\mathbb{R}^3$. Each copy of $\mathbb{S}^1$ in the link is called a *component* of the link.

In the following drawings of three links

the first link is called the *Hopf link*, and the second the *Borromean rings*.

Note that no embedded circles in a link can intersect and that a knot is precisely a one-component link. More properly, knot theory is the study of links, and it is common practice in the area to prefix a term by "knot" when it should really be prefixed by "link". Usually, but not exclusively, the term "knot invariant", for example, is used to refer to invariants of links as well as knots. This standard abuse of terminology within the discipline should be kept in mind when reading any material about knot theory.

Knots and links are considered up to a relation called *ambient isotopy*. This relation coincides with the intuitive notion of deforming a knotted circle in three-dimensional space without cutting it or gluing parts of it together. The formal definition is the following.

Definition 1.3 (Ambient isotopy of links). Two knots (or links) K and K' are said to be *(ambient) isotopic* if there is a family of homeomorphisms $h_t : \mathbb{R}^3 \to \mathbb{R}^3$, $t \in [0, 1]$, such that $h_1(K) = K'$, $h_0 = \mathrm{id}_{\mathbb{R}^3}$ and $(x, t) \mapsto (h_t(x), t)$ defines a homeomorphism from $\mathbb{R}^3 \times [0, 1]$ to itself.

The idea behind this definition is that two knots are equivalent if $\mathbb{R}^3$ can be continuously deformed in a way that takes one knot to the other.

We shall often write "isotopy" for "ambient isotopy", but it should be understood that all isotopies considered here are ambient isotopies.

Definition 1.4 (Equivalence of links). Two knots or links are *equal* or *equivalent* if they are isotopic. An equivalence class of the set of knots or links modulo isotopy is called an *isotopy class*.

Example 1.5. Consider the following knots $K_1, \ldots, K_4$.

K_1 K_2 K_3 K_4

The knots K_2 and K_1 may be shown to be equivalent by the following sequence of deformations

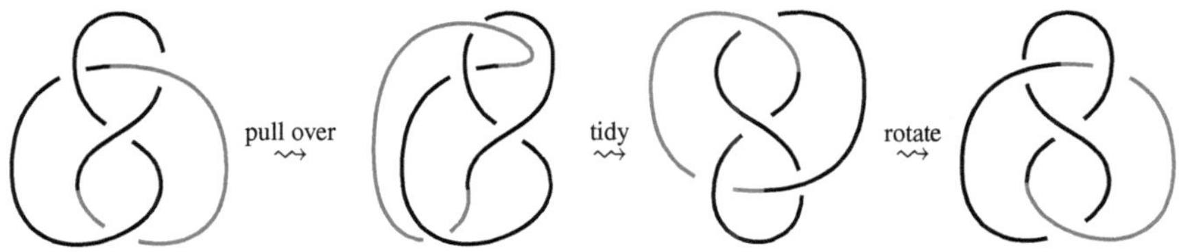

However, the knots K_3 and K_4 are not equivalent (See Exercise 2.26).

Exercise 1.6. Let $p_1, \ldots, p_n, \in \mathbb{Z}$. The *pretzel link* $\ell(p_1, \ldots, p_n)$ is the link defined by the drawing

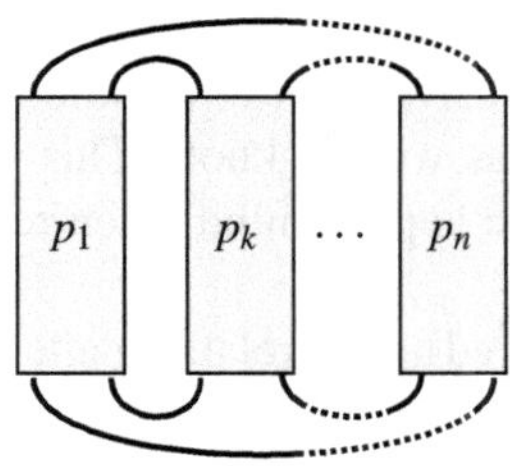

where 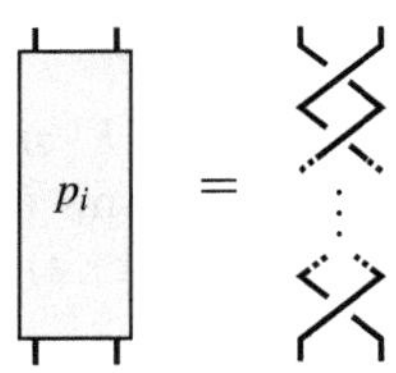 if $p_i \geq 0$, and 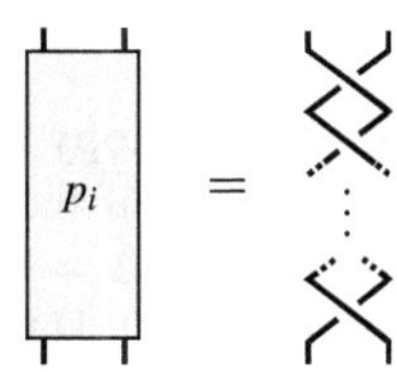 if $p_i < 0$, and

diagrams on the right-hand sides each have p_i crossings. For example,

$\ell(2, 3, -5, 4) =$ 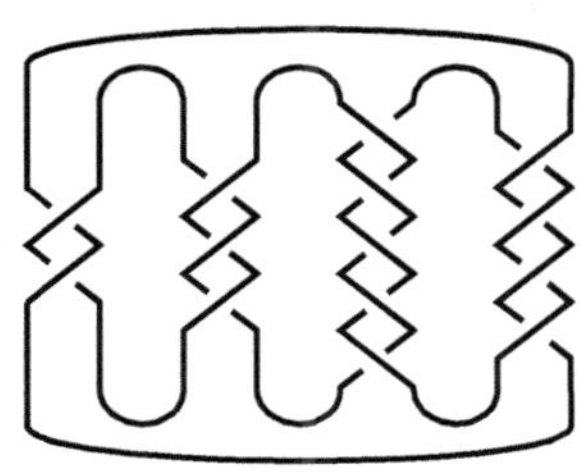

Find a necessary and sufficient condition in terms of the values $n, p_1, \ldots, p_n$ for a pretzel link to be a knot. Show that $\ell(p_1, \ldots, p_n) = \ell(p_2, \ldots p_n, p_1)$.

There is an inherent difficulty in showing that knots are not equivalent. On the one hand, we can demonstrate that two knots or links are equivalent by describing a sequence of deformations taking one to another, as in the first part of Example 1.5. On the other hand, being unable to find such a sequence of deformations does not necessarily imply that such a sequence cannot exist—perhaps we did not look hard enough for one. Instead, we need to prove that no sequence of deformations can exist.

This requires the use of a knot invariant. That is to say, we need to find a (sufficiently discriminating) knot invariant whose evaluations on the two knots are different. We shall discuss this topic in Chap. 2.

With minor abuse of language, it is usual in knot theory to use the term "knot" to mean (i) an embedding of $\mathbb{S}^1$ into $\mathbb{R}^3$, (ii) the image of an embedding under ambient isotopy, or (iii) an equivalence class of embeddings under ambient isotopy, and to do similarly for the term "link". This should not cause any difficulties since any ambiguity is resolved by the context.

Here (and in much of knot theory literature) we shall avoid pathologies and consider only knots and links that correspond closely to the intuitive notion of a knotted string. To do this, we need the following definition.

Definition 1.7 (Tame link). A link is said to be *tame* if it is ambient isotopic to a set of simple closed polygons in $\mathbb{R}^3$ whose sides are straight line segments; otherwise it is said to be *wild*.

The tameness condition excludes infinite limiting processes and ensures that we are studying knots and links that conform to the intuitive idea of a knot or link. Wild knots, on the other hand, may have bizarre and counter-intuitive properties and, although the existence of such knots has been known since the 1940s, remarkably little is known about them. We shall be concerned exclusively with tame knots and links and, henceforth, whenever we use the term "knot" or "link" we always mean a tame one.

Links whose images are polygons will be referred to as *polygonal links* and, since we are restricting attention to tame links and consider links up to isotopy, we can and shall assume that links are polygonal, and therefore amenable to the techniques of piecewise-linear topology. (For background on piecewise-linear topology, see, for example [162].) It will be seen that this provides a combinatorial formulation of ambient isotopy.

Our convention throughout is to draw links, link diagrams and other knotted objects as smooth curves. Nevertheless, the reader should regard all of these smooth curves as a polygonal curves whose vertices and edges are "so small that the curves appear smooth" (as is the case with any pixellated image on a computer screen). This is recorded for reference purposes by the following.

Convention 1.8. Links, link diagrams and other knotted objects are drawn here as smooth curves but are to be regarded as polygonal curves.

The advantage of realising links as polygonal links, and so stepping into the world of piecewise-linear topology, is that we can consider them as combinatorial objects. In particular, we can obtain a combinatorial reformulation of ambient isotopy as follows, where $\partial(T)$ denotes the boundary of the solid triangle T. Such a triangle is represented here by shading its interior.

Definition 1.9 (Δ- and Δ^{-1}-moves). Let L be a polygonal link, let p be a straight line segment of L, and let T be a solid triangle in $\mathbb{R}^3$ such that $L \cap T = p$ and $\partial(T) = p \cup q \cup r$. Then a *$\Delta$-move* is a move which replaces the edge p of L with $q \cup r$ of T. The inverse of this process is called a *Δ^{-1}-move*.

Thus, the Δ- and Δ^{-1}-moves in $\mathbb{R}^3$ are defined by

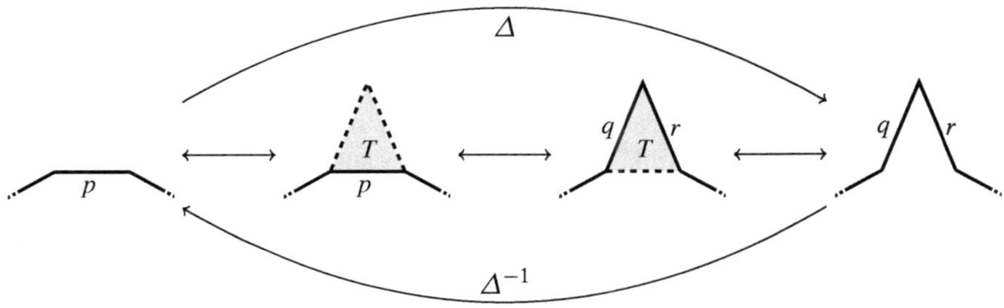

Since T is solid, no straight line segment of L may pass through it. Thus, in other words, the Δ-move replaces the edge p by $q \cup r$ provided no edge of the link passes through the solid triangle T that is used in the construction.

The following theorem gives equivalent formulations of knot equivalence. We exclude its proof and refer the interested reader to [31] for one.

Theorem 1.10. *Let L and L' be (tame) links in $\mathbb{R}^3$. Then the following are equivalent:*

1. *L and L' are ambient isotopic.*
2. *There is an orientation preserving piecewise-linear homeomorphism $h : \mathbb{R}^3 \to \mathbb{R}^3$ such that $h(L) = L'$.*
3. *There is a finite sequence of Δ- and Δ^{-1}-moves taking (a polygonal representative of) L to (a polygonal representative of) L'.*

1.2 Knot and Link Diagrams

Instead of working with knots and links as objects in $\mathbb{R}^3$, we can work with drawings of them on the plane $\mathbb{R}^2$ (in effect we have been doing so above!). We can do this by considering the images of knots and links in $\mathbb{R}^3$ under sufficiently nice (they are termed *regular*) projections from $\mathbb{R}^3$ onto $\mathbb{R}^2$. It is convenient for some arguments to fix a *rectangular Cartesian coordinate system* in the ambient space $\mathbb{R}^3$ and to project onto a fixed plane (typically the plane $z = 0$ in which case the projection maps (x, y, z) to $(x, y, 0)$). We adopt the following convention.

Convention 1.11. The rectangular Cartesian coordinate system, $Oxyz$, is chosen such that all links lie on the same side of the plane of projection.

There may be points of a knot that are mapped by projection to the same point in the plane $z = 0$. Such a point is called a *multiple point*. To make sure that our diagrams do not hide structure of the link, or suggest structure that is not there, we need to insist that all multiple points are "nice" in the following sense.

Definition 1.12 (Regular projection). A projection $p : \mathbb{R}^3 \to \mathbb{R}^2$ of a (polygonal) link L is said to be *regular* if

1. all multiple points of the projection are double points,
2. there are only finitely many double points,
3. no double point contains the image of a vertex.

Here a *vertex* is a point where two edges meet in the piecewise-linear representation of the knot.

A permissible multiple point of regular projection is shown in Fig. 1.2a while some impermissible multiple points are shown in (b), (c) and (d) of the same figure. The intersections in (c) and (d) contain the image of a vertex, and the multiple point in (b) is not a double point.

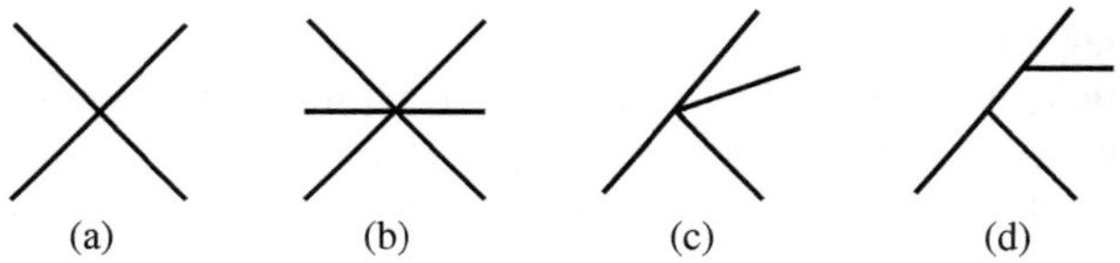

Fig. 1.2 Permissible (**a**) and impermissible (**b**, **c**, **d**) multiple points of a regular projection

Theorem 1.13. *Every link admits a regular projection.*

We exclude this proof and refer the reader to [46]. The idea behind the proof is that if the projection of a link is non-regular, then the link may be suitably adjusted by rotation in $\mathbb{R}^3$ or by making small deformations of the link using Δ-moves so that the projection becomes regular.

A projection of a link does not contain sufficient information to recover the original link since the distinction between over- and under-crossings is lost under projecting. However, if the over- and under-crossing information is attached to every double point of a regular projection then clearly the original link can be recovered up to isotopy. A regular projection of a link with this additional crossing structure is called a *link diagram*. A more formal definition is the following.

Definition 1.14 (Diagram). A *knot or link diagram* is the image of a regular projection of a knot or a link on which an under-/over-crossing structure has been assigned to each double point. The crossing structure is indicated by line breaks in the diagram with the unbroken strand sitting "above" the broken one.

Recall from Convention 1.8 that, for clarity, our convention throughout is to draw links, link diagrams and other knotted objects as smooth curves. Figure 1.3a, b shows (a regular neighbourhood of) the figure-of-eight knot and a corresponding diagram, and Fig. 1.3c, d shows (a regular neighbourhood of) a two-component link and a corresponding (link) diagram.

Sometimes properties of links are best defined through their diagrams. For example, the knot diagrams of 3_1 and 4_1 (but not of 10_{157}) in Fig. 1.1 have the property

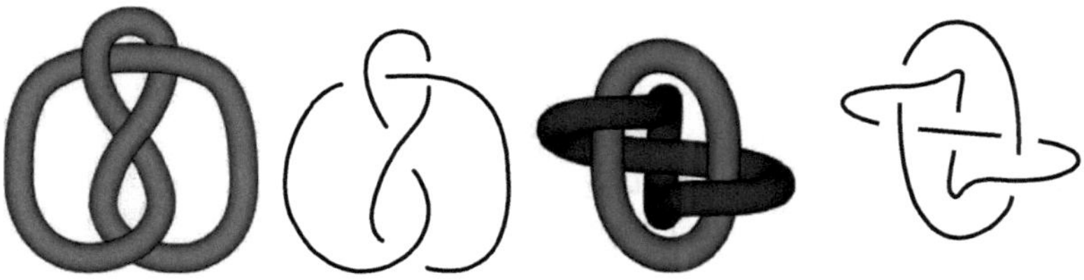

(a) (b) (c) (d)

Fig. 1.3 A knot (**a**) and link (**c**) (drawn using [169]) and their corresponding diagrams (**b**, **d**)

that the crossings alternate between over- and under-crossings in a tour along the component. Such a knot diagram is said to be an *alternating diagram*. Similarly, a link diagram is alternating if in a tour along each component the crossings alternate between over- and under-crossings. An *alternating knot or link* is one that admits an alternating diagram.

Note that there is a subtlety in this definition for it states that a knot or link is alternating if it has *an* alternating diagram. An alternating knot will also have infinitely many diagrams that are not alternating. Thus, if we are given a non-alternating knot diagram, it may or may not represent an alternating knot.

Exercise 1.15. Show that the following non-alternating diagram represents an alternating knot.

While we can prove that a knot or link is alternating by exhibiting an alternating diagram for it, showing that a knot or link is not alternating is a hard problem since we need to prove that none of its infinitely many diagrams is alternating. Figure 1.4 shows the non-alternating knot 8_{20} from the knot tables [158].

Fig. 1.4 The non-alternating knot 8_{20}

Diagrams also enable us to find easy descriptions of some classes of knots and links.

Definition 1.16 (Unknot and unlink). An *unknot* is a knot which is the boundary of an embedded disc $D^2 \subset \mathbb{R}^3$. Similarly, the n-component *unlink* is a link that is the boundary of n disjoint embedded discs in $\mathbb{R}^3$. A knot or link is said to be *trivial* if it is an unknot or unlink.

For example,

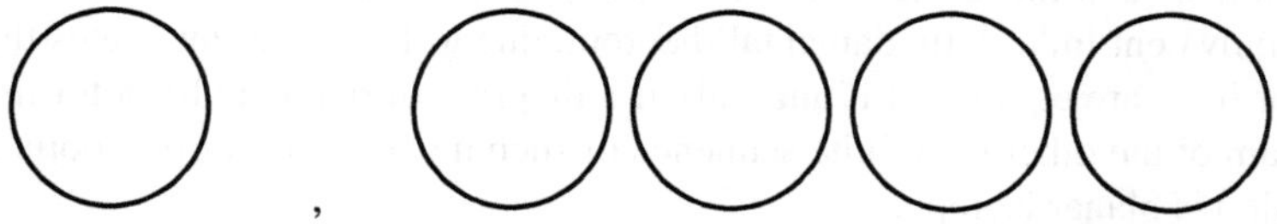

unknot, four component unlink.

An equivalent definition is provided by the following.

Definition 1.17 (Unknot and unlink). A link is an *unlink* if it admits a diagram with no crossings. A one-component unlink is called an *unknot*.

We note that a link is an unlink if and only if it has a diagram with no crossings. However, not every diagram of the unlink has no crossings. For example, below are two diagrams of a three-component unlink

Exercise 1.18. Verify that the following knot is equivalent to the unknot.

Exercise 1.19. A *crossing change* at a crossing in a link diagram is the local change 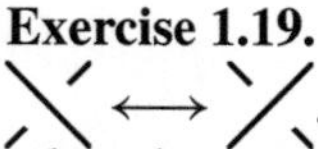. (So a crossing is changed into one of the other type, and there is no other change to the diagram. See Convention 1.21 below.) Crossing changes do not preserve the isotopy class of a link. Prove that every knot diagram can be changed into a diagram of the unknot by crossing changes. Hence, prove that any link diagram can be changed into the diagram of the unlink by crossing changes.

1.3 The Reidemeister Moves

Many different diagrams arise from equivalent links. Thus, to study links through their diagrams it is necessary first to understand how diagrams corresponding to equivalent links are related to each other. This is seen through the *Reidemeister moves*. Each such move makes a local change in a diagram to produce a diagram of an equivalent link. A fundamental theorem, due to Reidemeister, states that two knots or links are equivalent if and only if a diagram of one may be deformed into a diagram of the other by a finite sequence of such moves. The starting point in this discussion is planar isotopy.

An intuitive way to think about planar isotopy is to think of the link diagram as being drawn on a rubber sheet rather than on paper. Then we can change how the diagram looks by stretching and compressing the rubber sheet it is drawn on. In any such deformation, the crossing structure remains unchanged. Planar isotopy can be formally defined as follows.

Definition 1.20 (Planar isotopy). Two polygonal link diagrams are said to be related by *planar isotopy* if they are related by a finite sequence of the moves (where the angles and lengths may vary) of the form

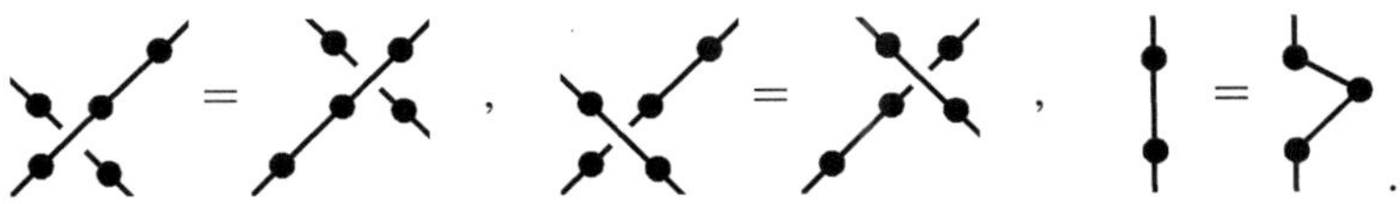

These are called the *relations of planar isotopy*. The point at which two polygonal edges meet is called a *vertex*.

Implicit in these figures (and diagrams for other moves) is that the move acts only on the neighbourhood of interest and that the rest of the diagram is unchanged by the move. We therefore adopt the following convention for displaying local operations or sequences of local operations on link diagrams, as well as on other types of diagram.

Convention 1.21 (Local operations on diagrams). We generally display a neighbourhood of the relevant diagram and leave the reader to recall that the rest of the diagram, which lies outside this neighbourhood, is implicitly present and unchanged. Moreover, when an operation or move on a link diagram or on another type of diagram is specified by a local change, it is to be understood that the diagrams are identical outside of the shown region.

As an example, the following diagrams are related by planar isotopy.

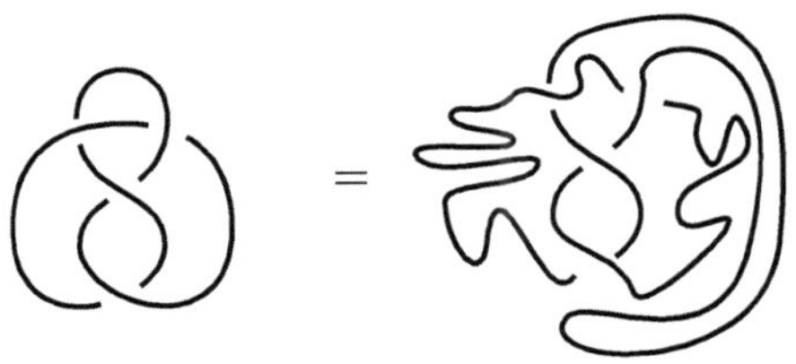

The Reidemeister moves, other than planar isotopy (which is sometimes denoted by R0), are denoted by RI, RII and RIII and are given next in a definition that invokes Convention 1.21.

Definition 1.22 (Reidemeister moves). The *Reidemeister moves* consist of planar isotopy and the following three local changes in a diagram.

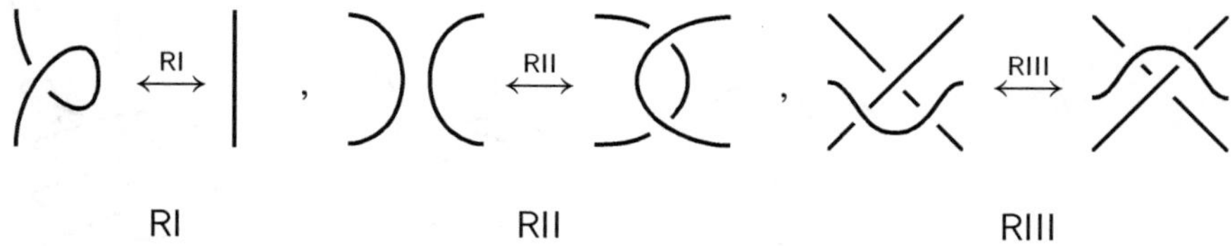

The diagrams are identical outside of the local region shown.

Diagrammatic moves having very similar forms to the Reidemeister moves can be obtained from them. For example, variant forms of RI and RIII are obtained by means of RI, RII and RIII as follows:

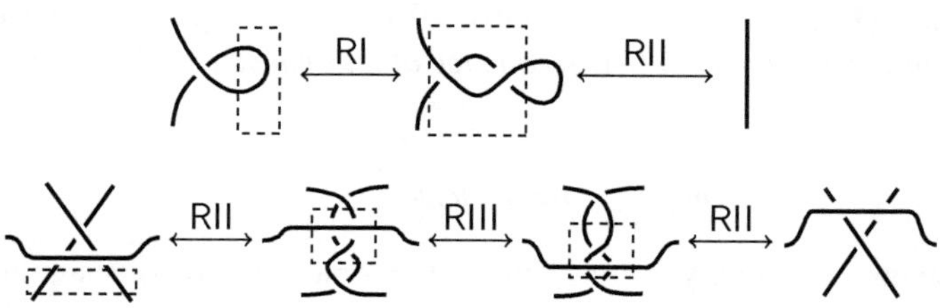

(1.1)

where a dashed box encloses the argument of the Reidemeister move to its immediate right. Although these two moves are consequences of the other Reidemeister moves, it is nevertheless useful (and reassuring) to have these variant forms listed explicitly at our disposal.

Example 1.23. An instance of the use of a sequence of Reidemeister moves is given in Fig. 1.5. The diagrams on the extreme left of the chain of Reidemeister moves are therefore equivalent diagrams. They are, in fact, both diagrams of the trefoil.

Exercise 1.24. Find a sequence of Reidemeister moves relating the following two diagrams:  and 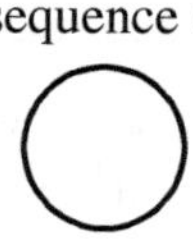 .

We emphasise that, here, the Reidemeister moves include planar isotopy. This convention holds for all of the variants of the Reidemeister moves we see in this book, even when it is not explicitly mentioned.

Convention 1.25 (Reidemeister moves). The term "Reidemeister moves" includes planar isotopy, as well as RI, RII and RIII.

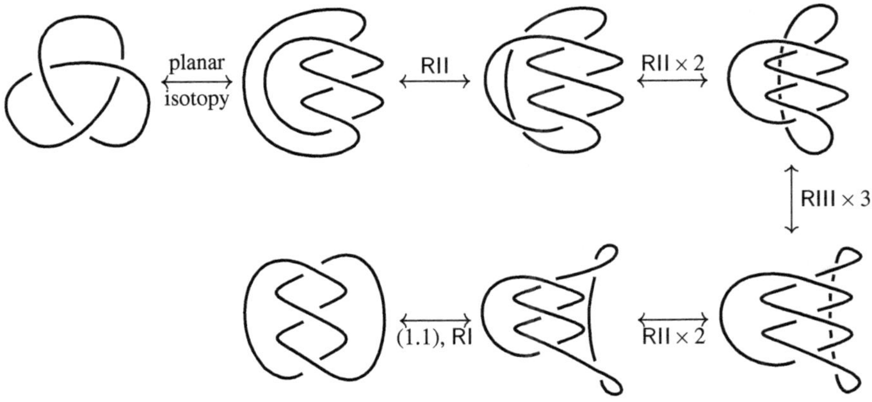

Fig. 1.5 Example of the use of the three Reidemeister moves

The next theorem provides a combinatorial description of knot and link equivalence in terms of diagrams.

Theorem 1.26 (Reidemeister's Theorem). *Two links are equivalent if and only if their diagrams are related by a finite sequence of Reidemeister moves. That is*

$$\frac{\{links\}}{isotopy} \cong \frac{\{diagrams\}}{Reidemeister\ moves}.$$

The following preliminary exercise on the Reidemeister moves is significant for it shows that it may be necessary to make a diagram "more complicated" first by increasing the number of crossings.

Exercise 1.27. Find a sequence of Reidemeister moves that changes the knot in Exercise 1.18 to a diagram of the unknot with no crossings. The knot diagram has 10 crossings. Prove that in any sequence of Reidemeister moves changing it to a diagram with no crossings must include a knot diagram with more than 10 crossings.

Reidemeister's Theorem implies the following.

> The study of knots or links up to isotopy is equivalent to the study of knot or link diagrams modulo the Reidemeister moves.

This is a transformative statement, the first of many that we will see in this book. It enables us to pass from one class of object to another that will permit new techniques to be applied.

1.4 A Proof of Reidemeister's Theorem

To prove Reidemeister's Theorem, it is necessary to understand how to move between links in $\mathbb{R}^3$ and link diagrams in $\mathbb{R}^2$. In particular, we need to understand how a

Δ-move on a link in $\mathbb{R}^3$ changes the link diagram in $\mathbb{R}^2$, and how a Reidemeister move applied to a link diagram changes the corresponding link in $\mathbb{R}^3$ in terms of Δ-moves. It will be convenient to invoke the coordinatisation convention stated in Convention 1.11.

By way of motivation, let L be a polygonal link and D its diagram. It is immediately clear that L can be recovered from D by "lifting" the link above the plane of projection. The lifting process is shown in Fig. 1.6a (on part of the link), where D is the diagram in the plane $z = 0$, and the link L is the object specified by the solid lines, and the dotted lines indicate projections of vertices to the plane $z = 0$.

The application of a single Δ-move is shown in Fig. 1.6. This introduces the vertex C in the plane $z = 2$, and joins it to A and B to form a solid triangle $[ABC]$ that is not pierced by an edge of L. The projection of the result onto the plane $z = 0$ is also shown. It is seen that the image c of C is such that triangle [abc] has no edges passing through it as a solid triangle.

In Fig. (c), the edge $[AB]$ is deleted, together with its image $[ab]$ in the plane $z = 0$. This is the diagram D'. The link specified by the solid lines in and between the planes above $z = 0$ is the link L'.

We call the result of a Δ-move on an embedding of a polygonal link L in $\mathbb{R}^3$ a δ-move, on the projection D.

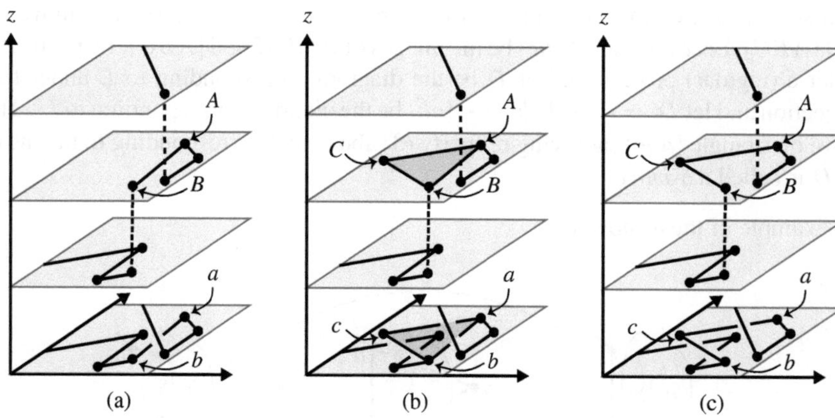

Fig. 1.6 Moving between links and their diagrams and the action of a Δ-move on the diagram

Now suppose, in addition, that L' is a link with a diagram D' such that L' is obtained from L by exactly one Δ-move which replaces an edge $[AB]$ of L with edges $[AC]$ and $[CB]$ of the embedded solid triangle $[ABC]$. An example is shown in the plane $z = 2$ across the links shown in (a), (b) and (c) above. Here, (a) at $z = 2$ shows the edge [AB] of the link L, (b) at $z = 2$ shows the solid triangle [ABC], and (c) at $z = 2$ shows the edges $[AC] \cup [CB]$ of the resulting link L'.

Consider the diagram of $L \cup [ABC]$ (*i.e.* the projection of $L \cup [ABC]$ with crossings assigned to the double points). This diagram consists of the diagram D together with a triangle $[abc] \subset \mathbb{R}^2$ corresponding to $[ABC] \subset \mathbb{R}^3$. The diagram $D \cup [abc]$ has three important properties:

1. D can be recovered from $D \cup [abc]$ by deleting the edges $[ac]$ and $[cb]$;
2. D' can be recovered from $D \cup [abc]$ by deleting the edge $[ab]$;
3. Paths of edges of D which intersect the interior of $[abc]$ will lie completely "above" the triangle $[abc]$ or will lie completely "below" or "above and below" the triangle $[abc]$ in the sense indicated by Fig. 1.7a, b, c, respectively.

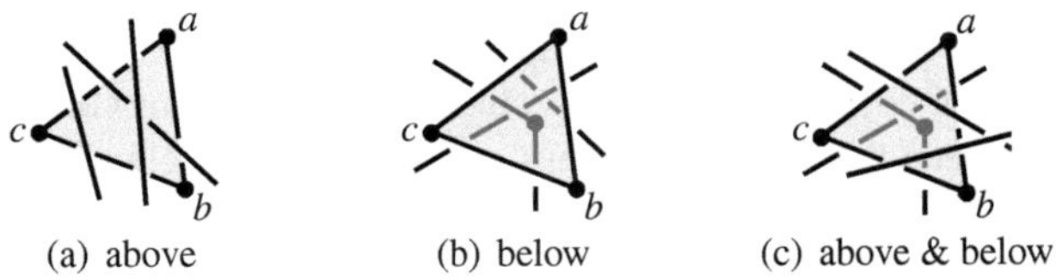

Fig. 1.7 Paths of edges of D lie either above triangle T or below it

A δ-move is the move on diagrams that corresponds to a Δ-move on links.

Definition 1.28 **(δ-move).** Let L be a link that contains an edge $[AB]$ and let $[ABC]$ be a solid triangle embedded in $\mathbb{R}^3$. Let L' be the link obtained by a Δ-move on $L \cup [ABC]$. Let a, b, c and $[abc]$ be the images of A, B, C and $[ABC]$, respectively, under a (regular) projection. Let D be the diagram corresponding to L under this projection and let $D' = (D \cup [abc]) - [ab]$ be the diagram corresponding to L'. The move on triangle $[abc]$ satisfying property (3) above and corresponding to the move on D is called a *δ-move*.

An example of the δ-move is

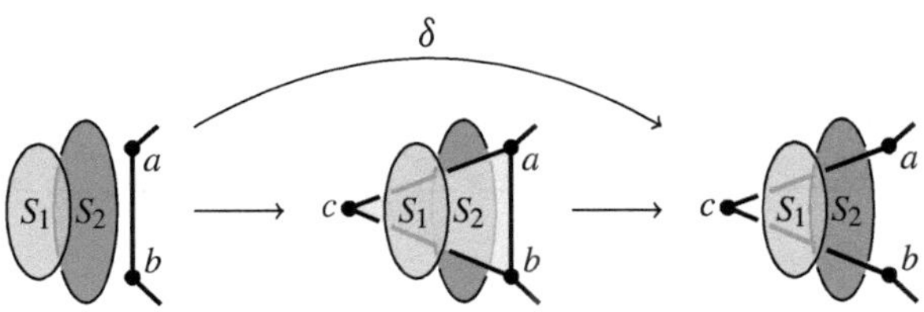

where S_1 and S_2 are sets of edges entirely above or beneath the solid triangle [abc].

We summarise this discussion in the statement of the following result.

Lemma 1.29.

1. *If two links are related by a Δ-move, then their diagrams are related by a δ-move.*
2. *If two link diagrams are related by a δ-move, then, up to isotopy, they determine two links which are related by a Δ-move.*

We can now prove Reidemeister's Theorem. For clarity, we split it into two lemmas corresponding to the "if" and "only if" parts.

In terms of polygons, RI, RII and RIII can be described as

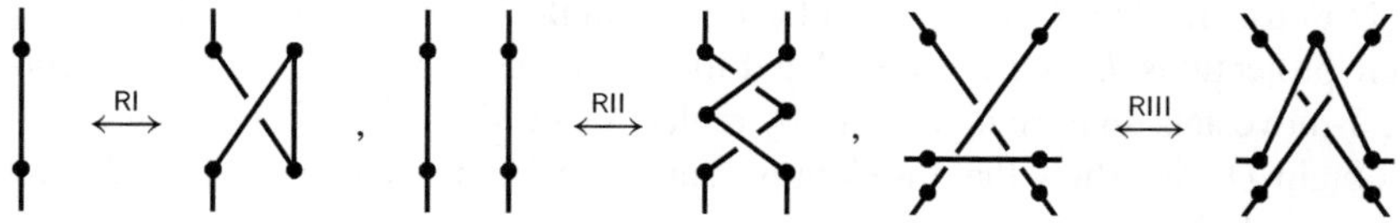

For the remainder of this Section, we shall assume all Reidemeister moves are in this polygonal form.

Lemma 1.30. *Let D and D′ be two link diagrams which are related to each other by a finite sequence of Reidemeister moves and planar isotopies. Then the corresponding links L and L′ in* $\mathbb{R}^3$ *are equivalent.*

Proof. It is enough to prove the lemma for the case when D and D' are related by the application of a single Reidemeister move or by the application of a single planar isotopy move shown in Definition 1.20. The relation which moves a crossing past a vertex clearly results in an isotopy of the link. The remaining moves can be obtained as a finite sequence of δ-moves (or their inverses) as follows:

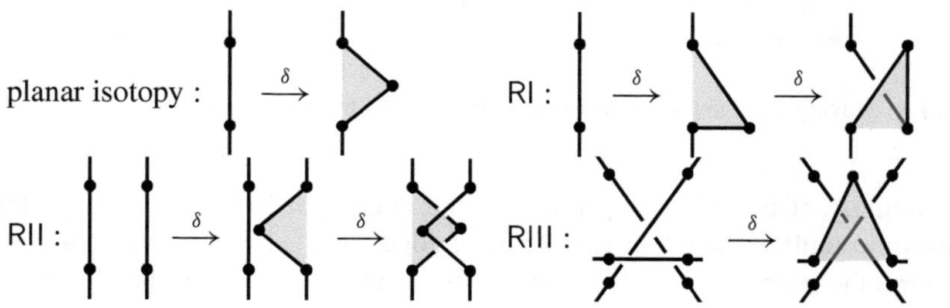

The result then follows by Lemma 1.29. □

The next lemma is a restatement of the "only if" part of Reidemeister's Theorem, and it is in this form that we shall prove this theorem.

Lemma 1.31. *Let L and L′ be equivalent links with diagrams D and D′, respectively. Then D and D′ are related through a finite sequence of Reidemeister moves and planar isotopies.*

Proof. Since L and L' are tame, we may assume they are polygonal. Then as L and L' are equivalent, they are related through a finite sequence of Δ- and Δ^{-1}-moves. It is clearly enough to prove the lemma for the case where L and L' are related by a single Δ-move or Δ^{-1}-move. In fact, since the Reidemeister moves are invertible, it is enough to prove the lemma for the case where L and L' are related by a single

Δ-move. This follows from the fact that if a Δ-move gives rise to a sequence of Reidemeister moves, then the inverse of that Δ-move will give rise to the inverse sequence of Reidemeister moves.

Now assume that L and L' are related by a single Δ-move. Then, by Lemma 1.29, their projections D and D' are related to each other by a single δ-move. Suppose the δ-move acts on a triangle $[abc]$ by replacing an edge $[ab]$ of D with $[bc] \cup [ca]$ to obtain D'. We show that this δ-move can be realised a sequence of Reidemeister moves and planar isotopies.

First there is a small technical issue to overcome. Consider the arc of D defined by the vertex sequence $(dabe)$ (so that this arc contains the edge $[ab]$ and its two incident edges). Either the edge $[da]$ of D intersects the interior of the triangle $[abc]$ in the diagram in $\mathbb{R}^2$ or it does not. If it does then apply a δ-move to a triangle $[adf]$, where f is some point in the exterior of $[abc]$. This is indicated in Fig. 1.8. In performing this δ-move on D, we have replaced a path $(dabe)$ with the path $(dfabe)$ where $[fa]$ does not intersect the interior of $[abc]$. The δ-move is easily seen to be an application of the RI-move or its mirror image.

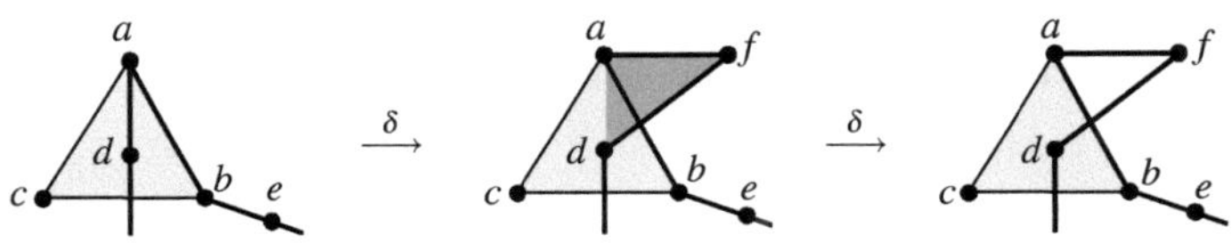

Fig. 1.8 Moving an edge $[ad]$ away from the interior of a triangle by δ-moves

Similarly, if the edge $[be]$ intersects the interior of $[abc]$, we may perform a δ-move to replace the edge with a pair of edges $[eg] \cup [gb]$ so that $[gb]$ does not intersect the interior of $[abc]$. Moreover, this δ-move can be described as a RI-move or its mirror image.

We may now assume, without loss of generality, that the edges of the diagram D which are incident with the edge $[ab]$ do not intersect the interior of the triangle $[abc]$, since if this were not the case we may use the above argument to isotope the link L to obtain a link L'' whose diagram D'' does indeed have this property. Moreover, D and D'' are related by Reidemeister moves, recalling from (1.1) that the mirror image of the RI-move is a consequence of the Reidemeister moves.

Assume that D has this property. Divide the triangle $[abc]$ into small triangles $T_i, i = 1, \ldots, m$ such that for each i, $T_i \cap D$ is empty, or contains part of an edge of D, or contains exactly one vertex of D, or contains exactly one crossing of D. An example that contains each of these possibilities is

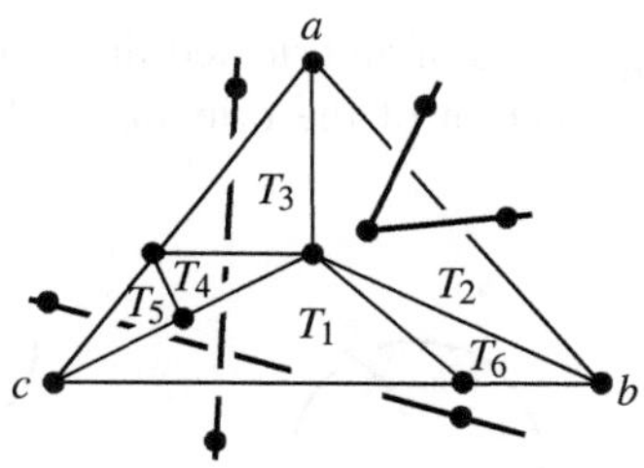

It is clear that such a sub-division exists. Obviously, the result of the δ-move on the triangle $[abc]$ can be obtained as the result of a sequence of δ-moves on the small triangles T_i. Therefore, it remains to show that each of the δ-moves on the small triangles can be realised as a sequence of Reidemeister moves.

To do this, we begin by observing that if D does not intersect the interior of any T_i then the δ-move on T_i is a planar isotopy, and if D does intersect the interior of a T_i then it must do so in one of the ways shown in Fig. 1.9. It is then just a matter of verifying that each of the δ-moves can be described as a sequence of Reidemeister moves and planar isotopies. This completes the proof of the lemma. □

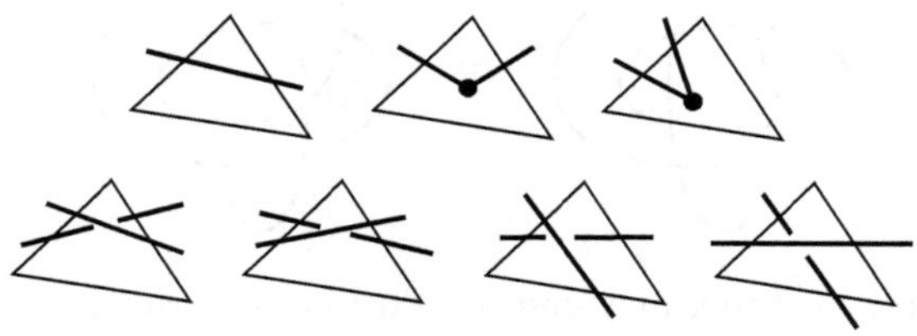

Fig. 1.9 The ways in which D can intersect a triangle T_i

Proof (Reidemeister's Theorem). The theorem follows from Lemma 1.31. □

Exercise 1.32. Verify that each, or at least some, of the δ-moves associated with Fig. 1.9 can be described as a sequence of Reidemeister moves and planar isotopies, as asserted at the end of the above proof.

1.5 Oriented Links

To this point, attention has been devoted to unoriented links, which are disjoint embeddings of unoriented circles in $\mathbb{R}^3$. However, for reasons that will come clear, we shall work mainly with oriented links which are disjoint embeddings of oriented circles in $\mathbb{R}^3$, so that each component of the link has a preferred direction around it.

Definition 1.33 (Oriented link). A knot or link in $\mathbb{R}^3$ is *oriented* if each of its components is oriented. Similarly, a knot or link diagram is *oriented* if each closed curve is oriented.

The orientation of the components of an oriented link is recorded on link diagrams by arrows pointing in the direction of the orientation. Examples of oriented link diagrams are

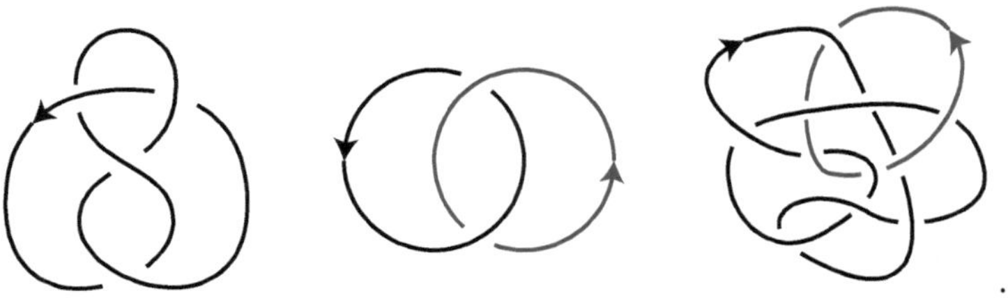

As with knots and links, oriented knots and links are considered up to ambient isotopy. We note that the isotopy preserves the orientation of the components. It should be observed that if an oriented link L' is constructed from a link L by reversing the orientation of one of its components, then L and L' may or may not be equivalent. For example, it can be shown (see [179]) that

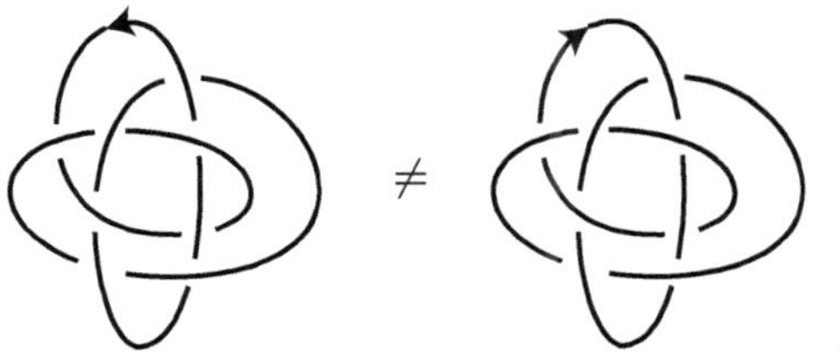

A unique unoriented link may constructed from an oriented link by forgetting the orientations of the components. Conversely, an oriented link with m components gives rise to a collection of 2^m oriented links by assigning all possible orientations to its components. However, in general, we do not know how many of the 2^m resulting links in this collection will be inequivalent.

The following definition and theorem provide the version of Reidemeister's theorem (Theorem 1.26) for oriented links.

Definition 1.34 (Oriented Reidemeister moves). The *oriented Reidemeister moves* consist of planar isotopy and the following three local changes in a diagram.

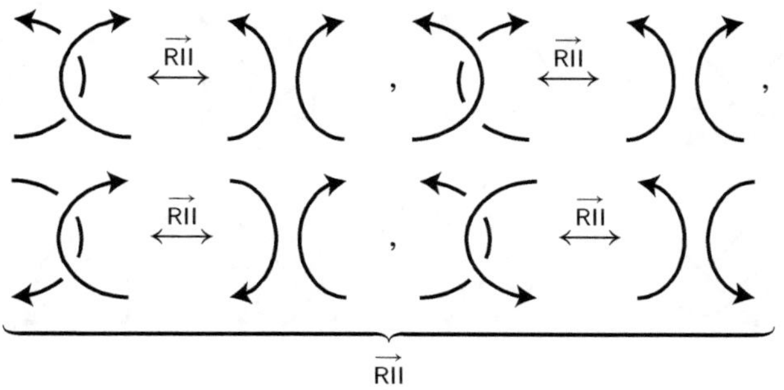

As with the unoriented case, considering oriented links up to isotopy is equivalent to considering oriented link diagrams up to the oriented Reidemeister moves.

Theorem 1.35. *Two oriented links are equivalent if and only if their diagrams are related by a finite sequence of oriented Reidemeister moves. That is*

$$\frac{\{\textit{oriented links}\}}{\textit{isotopy}} \cong \frac{\{\textit{oriented diagrams}\}}{\textit{oriented Reidemeister moves}}.$$

Proof. Throughout the proof, let S denote the set of moves which arise from the oriented Reidemeister moves in Definition 1.22 by considering all possible orientations of the arcs. The set S has *fourteen* elements, *two* arising from RI, *four* $(= 2^2)$ arising from RII and *eight* $(= 2^3)$ arising from RIII.

The only aspect in which oriented links and link diagrams differ from unoriented links and link diagrams is that each component has a specified orientation. Therefore, by Theorem 1.26, two oriented link diagrams correspond to isotopic oriented links if and only if they are related through a finite sequence of moves in the set S together with planar isotopy. It remains to show that the oriented Reidemeister moves from Definition 1.34 generate the set S. There are three cases.

<u>Case 1</u>: One of the oriented RI-moves is exactly that in S formed by orienting the unoriented RI-move. The other move in S is obtained by orienting the unoriented RI-move is obtained as follows:

<u>Case 2</u>: The four oriented RII-moves are exactly the four moves in S arising from the unoriented RII-move.

<u>Case 3</u>: We need to show that the eight moves in S arising from the unoriented RIII-move are all consequences of the oriented Reidemeister moves in Definition 1.34. These are reduced to four by observing that the RIII-move is symmetric with respect to rotation through 180°. Of these four, one move is contained in the definition of the oriented Reidemeister moves, and the other three are consequences as follows:

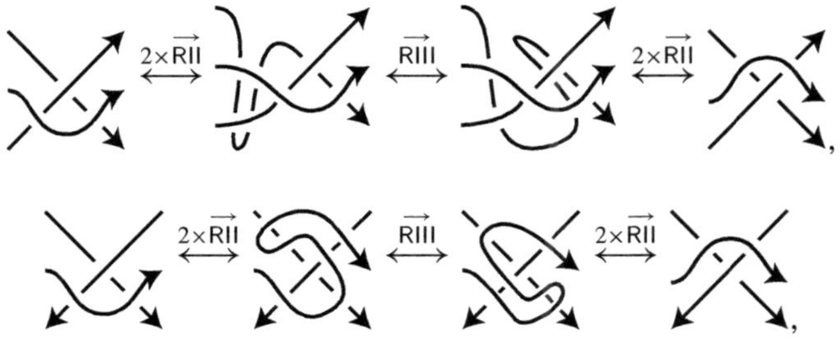

(1.2)

And making use of the move we have just shown, it follows that

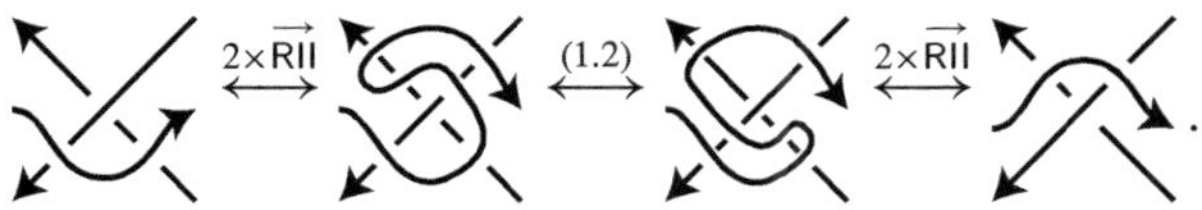

□

We shall in general refer to both oriented and unoriented diagrams simply as diagrams, and similarly for the Reidemeister moves, since the intended meaning will be resolved immediately by context. At times, we shall not specify whether a link is oriented or unoriented and, again, this should not cause any confusion. It is also worth noting that the Reidemeister moves of Definition 1.22 can still be applied to oriented links and that oriented link diagrams related by these moves are equivalent.

Remark 1.36. The set of oriented Reidemeister moves given in Definition 1.34 includes four different RII-moves. In fact, the first two forms of the RII-move (those in which all the arcs point upwards) given in that list can be deduced from the remaining five moves (see [151]). Thus, the list

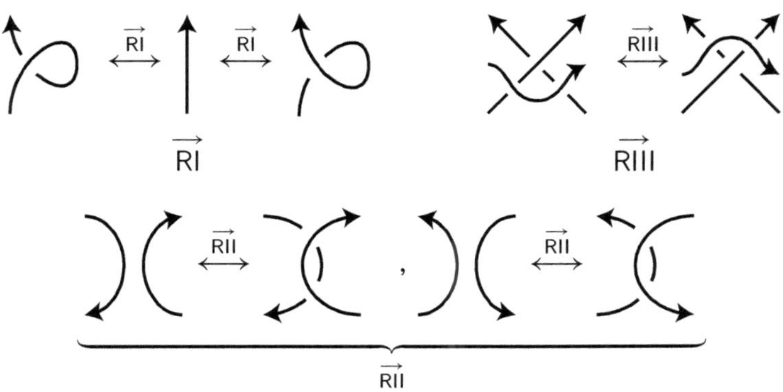

constitutes a complete set of oriented Reidemeister moves. Here, however, we use the longer list of moves since some of the later algebraic and topological constructions become more transparent if we include them.

Chapter 2
Knot and Link Invariants

2.1 The Idea of an Invariant

A fundamental problem of knot theory is determining if two knots, given by their diagrams, say, are equivalent or not. This is, in general, a hard problem. Diagrams for the same knot can look remarkably different. For example, consider the knot diagram

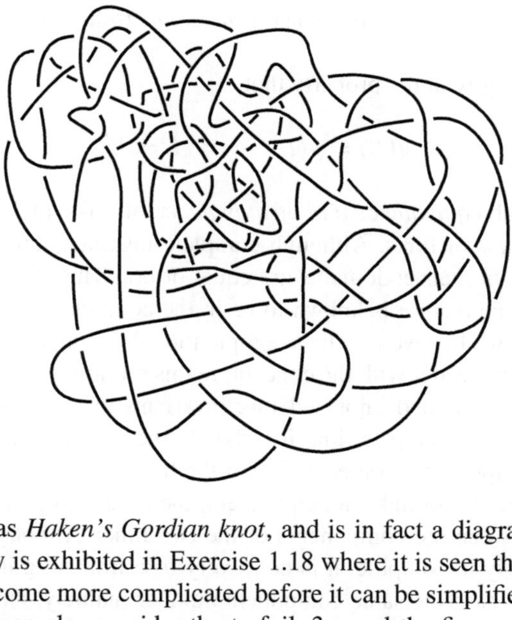

which is known as *Haken's Gordian knot*, and is in fact a diagram of the unknot. Another difficulty is exhibited in Exercise 1.18 where it is seen that a knot diagram might need to become more complicated before it can be simplified.

As a further example, consider the trefoil, 3_1, and the figure-of-eight knot, 4_1, shown in Fig. 1.1. The reader will quickly decide that they are not equivalent, but it is necessary to be able to assert this with certitude. For example, if we were to try every sequence of 1000 Reidemeister moves to transform one to the other, and failed, it might be that a sequence of 1001 Reidemeister moves would succeed! Clearly, tools

D. M. Jackson and I. Moffatt, *An Introduction to Quantum and Vassiliev Knot Invariants*, CMS Books in Mathematics, https://doi.org/10.1007/978-3-030-05213-3_2

of greater sophistication are required for deciding whether two knots or links are different, and this leads to the notion of a link invariant.

Definition 2.1 (Link invariant). A *link invariant* (or *knot invariant*) f is a function on the set of isotopy classes of links: that is,

$$f\colon \frac{\{\text{links}\}}{\text{isotopy}} \to S, \tag{2.1}$$

where S is a set.

The codomain S of a knot invariant is usually selected to be one in which the problem of testing the equality of two of its elements is easier than testing the equivalence of two links (that is, testing whether they belong to the same isotopy class).

A link invariant has the property that if two links, L and L', are equivalent then $f(L) = f(L')$. Stated contrapositively, if f is a link invariant and L and L' are links, then

$$f(L) \neq f(L') \implies L \neq L'.$$

Thus f recognises the *inequivalence* of two links. However, f does not necessarily recognise the *equivalence* of links since it *does not* follow that $f(L) = f(L') \Rightarrow L = L'$.

A link invariant g with the property that

$$g(L) = g(L') \iff L = L'$$

does recognise both *equivalence and inequivalence* of links, and is therefore called a *complete invariant* of links. Although complete invariants are known, it is hard to test if elements in their codomains are equal or not. This is not surprising since recognising even the unknot is known to be in the computational complexity class NP [74], meaning we believe it to be a computationally demanding problem.

To familiarise the reader with the concept, in this section, we shall provide examples of some easily defined knot and link invariants and discuss approaches for constructing them. The simplest link invariant is the *component number*, which is the number of components (copies of $\mathbb{S}^1$) of the link. It is easy to see that this is a link invariant since if two links are equivalent, then the isotopy between them sets up a bijection between the components of the two links. Although the component number is very easy to compute, it is not a very discriminating invariant since, for example, all knots have the same component number (namely, 1).

In the remainder of this section, we outline three approaches for constructing knot invariants.

Approach I: Constancy on Isotopy Classes

One approach for defining knot invariants is to define them to be constant over isotopy classes of knots, since such functions are, by definition, knot invariants. An example of such an invariant is the *unknotting number*.

Definition 2.2 (Unknotting number). The *unknotting number* is the minimum number of crossing changes required to reduce a knot into the unknot.

Note that by Exercise 1.19, every knot has a finite unknotting number.

As an example, the unknot is the only knot with crossing number zero. By considering the diagram for the trefoil given in Fig. 1.1, it is easy to see that it has crossing number at most one. In fact, the trefoil has unknotting number exactly one, but to show this we need to show that it has no zero crossing diagrams, *i.e.* we need to know that it is not the unknot. This exhibits the difficulty of computing the unknotting number: for while, it is easy in general to find an upper bound for the unknotting number of a knot by looking at any of its diagrams, and to determine the exact unknotting number we also need to show that a lower unknotting number cannot be read from any one of its infinitely many equivalent diagrams.

Exercise 2.3. What is the unknotting number of the knot given in Exercise 1.18?

An invariant with a similar definition to the unknotting number is the *crossing number*.

Definition 2.4 (Crossing number). The *crossing number* of a link L is the minimum of the number of crossings in a diagram of L over all of its diagrams.

Exercise 2.5. What is the crossing number of the knot given in Exercise 1.18? How many knots and links have crossing numbers $1, 2, \ldots, 4$? (At this point you may not have the means to show that the links you exhibit are actually different, so this exercise is really asking you to make a conjecture.)

It is worth noting that the unknotting number and crossing number of a knot may be attained by different diagrams (see [25]).

Approach II: Invariance Under Reidemeister Moves

Another approach to defining a link invariant is to define a function on a link diagram and to show that the value of the function is invariant under the Reidemeister moves. It is this approach that we are mainly concerned with in this book. Often, invariants defined in this way are easier to calculate and are often very powerful (*i.e.* they distinguish between many links), but can have the disadvantage that their topological meaning is unknown.

As an example of this type of invariant, we consider one of the most elementary link invariants, namely 3-colourability.

Definition 2.6 (3-colouring). Regard a knot or link diagram as a set of plane curves, which we call *edges*, between crossings. A *3-colouring* of a link diagram D is the assignment of one of the colours $\{1, 2, 3\}$ to each of the edges of D such that at each crossing either the four incident half-edges have the same colour, or the two incident half-edges forming the over-crossing have the colour a and the remaining two half-edges assigned colours b and c, where $a, b, c \in \{1, 2, 3\}$ and are mutually distinct.

Example 2.7. The knot 6_1 and a 3-colouring of it is

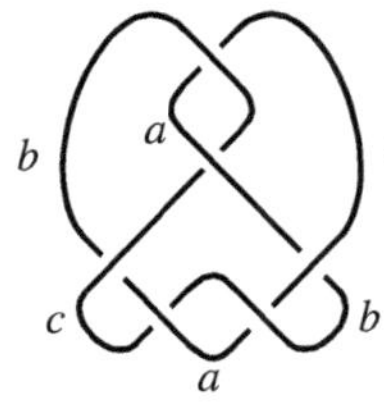

Theorem 2.8. *The number of 3-colourings of a link diagram is an invariant of the link.*

Proof. To prove the theorem, it is sufficient to show that a 3-colouring of a given link diagram uniquely determines a 3-colouring of a link diagram that results from the application of a Reidemeister move. We show that this is the case for the RIII-move.

Suppose that a link diagram D has a specific 3-colouring and that the diagram D' is obtained from D by the application of a single RIII-move. Without loss of generality, let a be the colour assigned to the edges in the arc that crosses over the other arc. Then in the locality of the crossing configuration of D where the RIII-move is applied, the 3-coloured link diagram must be of one of the five forms given on the left-hand side of the moves in Fig. 2.1 (where a, b, c are distinct elements of $\{1, 2, 3\}$). Since outside of this region the 3-colourings of D and D' are identical (the outgoing arcs have the same colour before and after the move), the application of the RIII-move to each of the configurations on the left-hand sides must determine the configurations of the right-hand sides. It is easily checked that this determination is unique.

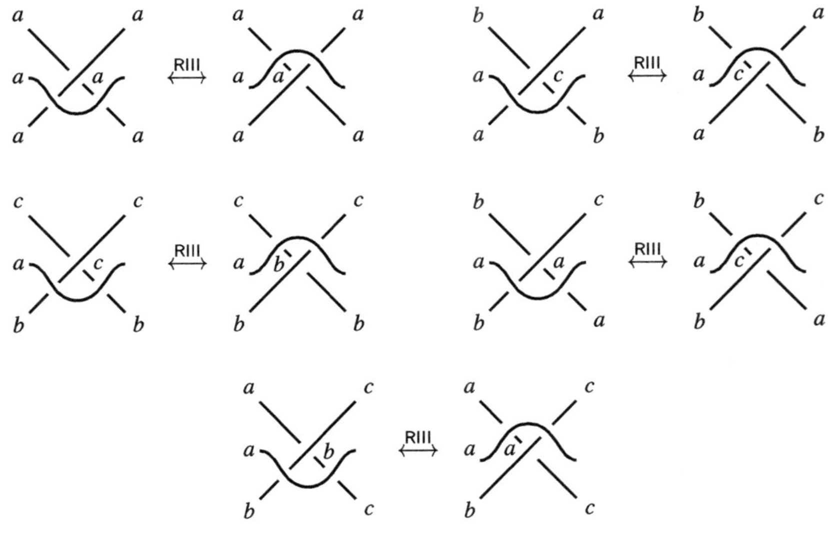

Fig. 2.1 Invariance of 3-colouring under RIII

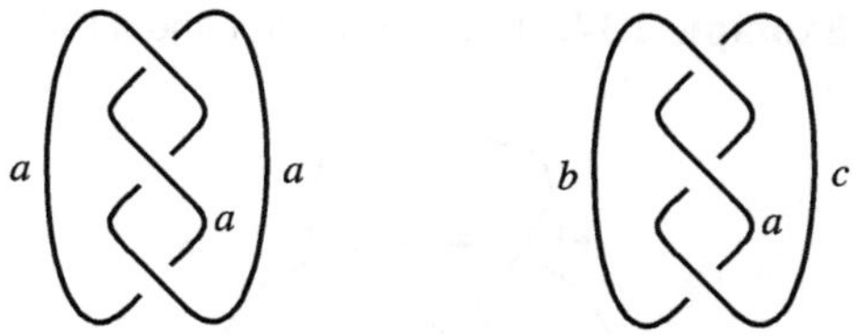

Fig. 2.2 3-colourings of the trefoil

The proof of invariance under the RI-move and the RII-move is similar and is left as an exercise. Completion of this exercise completes the proof. □

Exercise 2.9. Complete the proof of Theorem 2.8 by showing that the number of 3-colourings of a link diagram is invariant under the RI-move and the RII-move.

Example 2.10. There are three 3-colourings of the trefoil with colour $a \in \{1, 2, 3\}$ of the type given in the left of Fig. 2.2, and 3! of the type given on the right of Fig. 2.2, where a, b, c are distinct elements of $\{1, 2, 3\}$. Note that, from Example 1.23, these are indeed diagrams of the trefoil.

Thus the number of 3-colourings of the trefoil is 9. It is easily seen that the number of 3-colourings of the unknot is 3, and hence the trefoil and the unknot are different knots.

Exercise 2.11. Show that the number of 3-colourings of the figure-of-eight knot is 3, and conclude that the figure-of-eight knot and the trefoil are different knots. Note that the number of 3-colourings does not distinguish the figure-of-eight knot from the unknot.

Exercise 2.12. The notion of a 3-colouring can be extended to an n-colouring. An *n-colouring* of a link diagram D is the assignment of an element of $\mathbb{Z}_n$, called a colour, to each of the edges of D such that at each crossing the sum of the colours of the under-crossings equals twice the colour of the over-crossing. Verify that the number of n-colourings is a link invariant.

In 3-colourability, we have just seen an elementary invariant of unoriented links. We shall now discuss the linking number, which is an invariant of oriented links with two or more components.

Definition 2.13 (Sign of a crossing). Let L be an oriented link with components $L_1, \ldots, L_n$ and D be an oriented diagram for L with components $D_1, \ldots, D_n$. We can associate a number $+1$ or -1 to each crossing in the diagram D according to the scheme in Fig. 2.3. This number is called the *sign* of a crossing.

Fig. 2.3 The sign of a crossing

Example 2.14. Below is a knot and a link with the signs of each of their crossings.

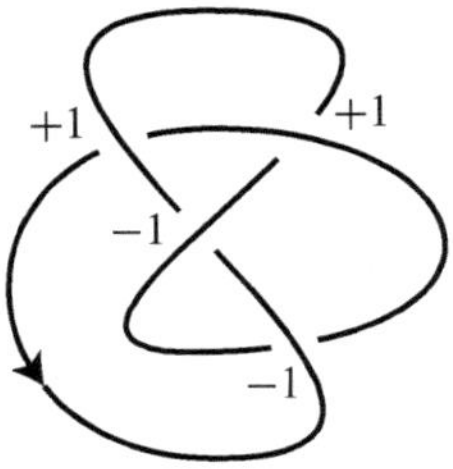

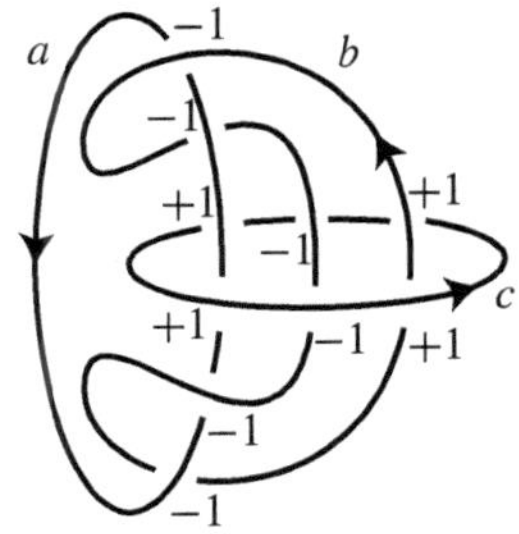

The sign of a crossing can be remembered as follows. If we rotate the overcrossing arc in the positive (counter-clockwise) direction by 90°, then if the directions of the arcs align the crossing is positive, if they do not align it is negative.

Definition 2.15 (Linking number). Let L be an oriented link with components $L_1, \ldots, L_n$ and D be an oriented diagram for L with components $D_1, \ldots, D_n$. The *linking number* $\mathrm{lk}(i, j)$ of L, where $i, j \in \{1, \ldots, n\}$ and $i \neq j$, is the sum of the signs of all crossings between the components D_i and D_j of the diagram D of L.

Example 2.16. The link shown on the right in Example 2.14 has three components labelled a, b and c. We have $\mathrm{lk}(a, b) = -4$, $\mathrm{lk}(a, c) = +2$ and $\mathrm{lk}(b, c) = 0$.

Theorem 2.17. *Linking numbers are independent of the choice of link diagram. Hence the set (and multiset) of linking numbers is an invariant of oriented links.*

Proof. Let L be a link with components $L_1, \ldots, L_n$ and D and D' be two diagrams for L. By Theorem 1.35, we know that D and D' are related through a finite sequence of oriented Reidemeister moves. It is enough to show that when D and D' are related by a single oriented Reidemeister move, the linking numbers $\mathrm{lk}(i, j)$ read from the two diagrams are unchanged for any $i \neq j$.

Suppose that D and D' are related by a single $\overrightarrow{\mathsf{RII}}$-move which eliminates two crossings. Consider $\mathrm{lk}(D; i, j)$, which is the linking number $\mathrm{lk}(i, j)$ of the components D_i and D_j of D; and $\mathrm{lk}(D'; i, j)$, which is the linking number $\mathrm{lk}(i, j)$ of the components D'_i and D'_j of D'. If the $\overrightarrow{\mathsf{RII}}$-move does not change the number of crossings between the D_i and D_j components, then clearly $\mathrm{lk}(D; i, j) = \mathrm{lk}(D'; i, j)$. Suppose now that the $\overrightarrow{\mathsf{RII}}$-move does change the number of crossings between the D_i and D_j components. Then D has two more crossings than D' and one of these crossings will necessarily have a $+1$ sign and the other will have a -1 sign. Therefore, $\mathrm{lk}(D; i, j) = \mathrm{lk}(D'; i, j) + 1 + (-1) = \mathrm{lk}(D'; i, j)$.

Next suppose that D and D' are related by a single oriented $\overrightarrow{\mathsf{RIII}}$-move. Then every crossing of D between two arcs corresponds to a crossing between two arcs of D'. Moreover the corresponding crossings involve the same components of the link diagram and have the same sign. It follows that linking numbers are unchanged by the $\overrightarrow{\mathsf{RIII}}$-move.

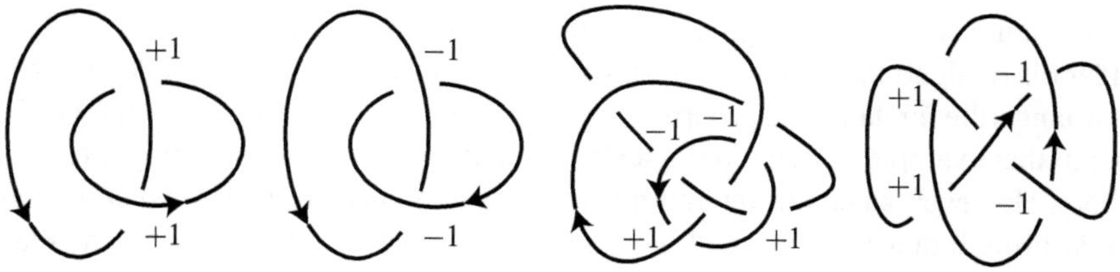

(a) lk = +2 (b) lk = −2 (c) lk = 0 (d) lk = 0

Fig. 2.4 Four links and their linking numbers

Finally, an $\overrightarrow{\mathsf{RI}}$-move adds a crossing to a single component, and since the linking numbers count crossings between distinct components, the linking numbers are unchanged by an $\overrightarrow{\mathsf{RI}}$-move. □

The linking numbers $\mathrm{lk}(i, i)$ are not defined, and any attempt to define these "self-linking numbers" quickly reveals that they are not invariant under the $\overrightarrow{\mathsf{RI}}$-move. We will discuss this point further in Sect. 3.2.

Example 2.18. As an example of the use of linking numbers, consider the links shown in Fig. 2.4 where the components are labelled L_1 and L_2 arbitrarily. It is readily seen that the link in (a) has linking number $\mathrm{lk}(1, 2) = 2$, the link in (b) has linking number $\mathrm{lk}(1, 2) = -2$ and the links in (c) and (d) each have have linking number $\mathrm{lk}(1, 2) = 0$. From this, it follows that the oriented links in (a), (b) and (c) are distinct, and that (a), (b) and (d) are distinct.

However, the linking number does not distinguish between the links (c) and (d). To distinguish between the links (c) and (d), we observe that (c) has a component whose removal will result in a trefoil, but the deletion of either component of (d) will result in an unknot. Since, by 3-colourability, we know that the trefoil and the unknot are distinct knots (see Example 2.10), we can conclude that the links (c) and (d) are distinct. Therefore, the oriented links in Fig. 2.4 are mutually distinct.

As oriented links, (a) and (b) are distinct, but regarded as unoriented links (by forgetting the orientations) they are equivalent (the resulting link is the Hopf link). By the argument above, we also know that (c) and (d) are distinct as unoriented links. To see that, as unoriented links, (a) is distinct from both (c) and (d), observe that any orientation of the components of the links (c) and (d) results in a link with $\mathrm{lk}(1, 2) = 0$, but any orientation of the components of the link in (a) results in a link with non-zero linking number. Therefore, as unoriented links, (a) and (b) are equivalent, and (a), (c) and (d) are distinct.

Approach III: The Topology of the Complement of the Link

Above, we have focussed on knot invariants defined on diagrams in some combinatorial way. Another approach is to consider the complement of a knot or link L in its ambient space, $\mathbb{R}^3 - L$, as a topological space. We can then use the techniques and tools of algebraic topology to construct invariants. A classic example of such an invariant is the fundamental group of the complement, $\pi_1(\mathbb{R}^3 - L)$. (See Remark 2.33 for another example.) Such topologically defined invariants have a deeply interesting theory. However, since our intention is to focus upon invariants defined combinatorially through diagrams, we will not pursue this direction here. We refer the reader to, for example, [31, 120, 158] for good introductions to this aspect of knot theory.

2.2 Skein Relations and Polynomial Invariants

In this section, we describe some very powerful invariants of oriented links. We define these invariants recursively through *skein relations*. These invariants have a different flavour to those we have considered in the previous section. We shall see three knot polynomials here. These are

- the *Jones polynomial*, $J(L) \in \mathbb{Z}[t^{\frac{1}{2}}, t^{-\frac{1}{2}}]$;
- the *HOMFLYPT polynomial*, which is a bivariate generalisation of the other two;
- the *Alexander–Conway polynomial*, $C(L) \in \mathbb{Z}[z]$.

One of the main results in Part II of this book is that these invariants do exist and are instances of a larger class of polynomial invariants of links. For the moment, however, we shall assume that the Jones and HOMFLYPT polynomials are indeed invariants. This will enable us to gain some familiarity with the link polynomials before we see them arise in different settings.

First, some terminology for the definitions of these knot polynomials is required. Let L_+, L_- and L_0 be three link diagrams which are identical except for a small region in which they differ, as shown in Fig. 2.5.

Example 2.19. If L_+ is a trefoil, L_- and L_0 are as follows:

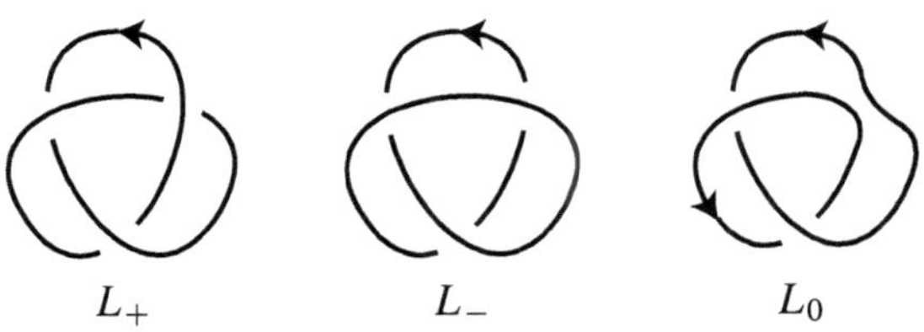

The first polynomial we consider is the Jones polynomial [84, 85], the introduction of which started a revolution in knot theory.

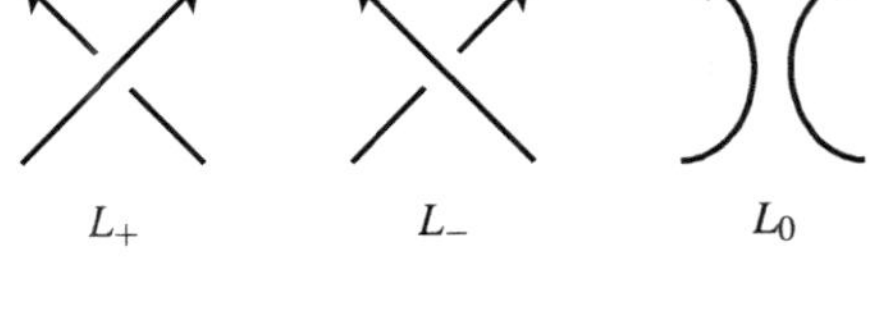

Fig. 2.5 The definitions of L_+, L_- and L_0

Definition 2.20 (Jones polynomial). The *Jones polynomial*, $J(L)$, of a link L is an isotopy invariant of links defined by the *skein relations*

$$\mathbf{SJ1}: \; t^{-1}J(L_+) - tJ(L_-) = (t^{\frac{1}{2}} - t^{-\frac{1}{2}})J(L_0),$$
$$\mathbf{SJ2}: \; J(\mathcal{O}) = 1,$$

where $\mathcal{O}$ is the unknot, and L_+, L_-, and L_0 are links that identical except for in a small region where they differ as indicated in Fig. 2.5 (following Convention 1.21).

We shall see that these skein relations can be used to obtain a recursive procedure for calculating the Jones polynomial, using the unknot as the base case.

Lemma 2.21. *If L is a k component unlink, then*

$$J(L) = \left(-t^{-\frac{1}{2}} - t^{\frac{1}{2}}\right)^{k-1}.$$

Proof. We prove this result for $k = 2$, and leave the proof for arbitrary k to the reader.

$J\Bigg($  $\Bigg) = J\Bigg($ 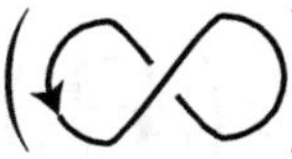 $\Bigg)$ (isotopy invariance)

$= t^2 J\Bigg($ 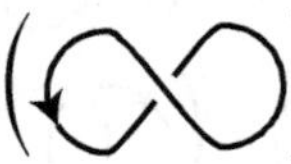 $\Bigg) + t\left(t^{\frac{1}{2}} - t^{-\frac{1}{2}}\right) J\Bigg($  $\Bigg)$ (by **SJ1**)

$= t^2 + t\left(t^{\frac{1}{2}} - t^{-\frac{1}{2}}\right) J\Bigg($ 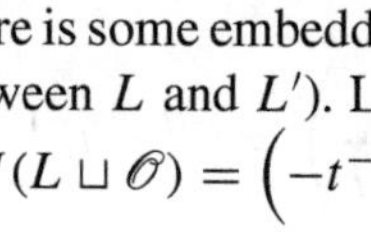 $\Bigg)$ (isotopy invariance, **SJ2**).

But the term on the left is equal to 1, by **SJ2**, and result now follows. □

Exercise 2.22. The *disjoint union* $L \sqcup L'$ of two links L and L' consists of a copy of each of L and L' in $\mathbb{R}^3$ such that there is some embedded ball that contains L but none of L' (so there is no "linking" between L and L'). Let $\mathcal{O}$ be the unknot. Adapt the proof of Lemma 2.21 to prove that $J(L \sqcup \mathcal{O}) = \left(-t^{-\frac{1}{2}} - t^{\frac{1}{2}}\right) J(L)$. Hence complete the proof of Lemma 2.21.

Example 2.23. The Jones polynomial of the Hopf link is

$J\Bigg($  $\Bigg) = -t^{\frac{5}{2}} - t^{\frac{1}{2}}.$

To see this, by the skein relation **SJ1**, we have

$t^{-1} J$ () $= tJ$ () $+ \left(t^{\frac{1}{2}} - t^{-\frac{1}{2}}\right) J$ ()

$= tJ$ (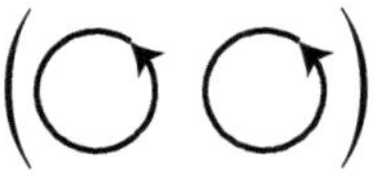) $+ \left(t^{\frac{1}{2}} - t^{-\frac{1}{2}}\right) J$ ()

$$= t\left(-t^{-\frac{1}{2}} - t^{\frac{1}{2}}\right) + \left(t^{\frac{1}{2}} - t^{-\frac{1}{2}}\right),$$

and the result follows.

Example 2.24. The Jones polynomial for the trefoil is

J () $= -t^4 + t^3 + t.$

To see why, by the skein relation **SJ1** we have

J () $= t^2 J$ () $+ t\left(t^{\frac{1}{2}} - t^{-\frac{1}{2}}\right) J$ () (by **SJ1**)

$= t^2 J$ (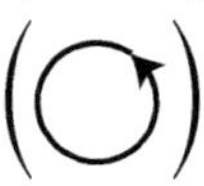) $+ t\left(t^{\frac{1}{2}} - t^{-\frac{1}{2}}\right) J$ ()

$$= t^2 + t\left(t^{\frac{1}{2}} - t^{-\frac{1}{2}}\right)\left(-t^{\frac{5}{2}} - t^{\frac{1}{2}}\right)$$

$$= -t^4 + t^3 + t.$$

Exercise 2.25. Verify that

J () $= t^{-2} - t^{-1} + 1 - t + t^2.$

From the above examples, the reader should see how the skein relations define a knot polynomial. Repeated applications of the skein relation **SJ1** will reduce any link to a sum of Jones polynomials of unlinks. To see why this is so, note that Exercise 1.19 shows that any link can be reduced to an unlink by crossing changes. Let $\mathscr{C}$ denote this set of crossings. To each crossing in $\mathscr{C}$, we apply either the relation

$$J(L_+) = t^2 J(L_-) + t(t^{\frac{1}{2}} - t^{\frac{1}{2}})J(L_0),$$

if the crossing is positive, or the relation

$$J(L_-) = t^{-2} J(L_+) - t^{-1}(t^{\frac{1}{2}} - t^{\frac{-1}{2}})J(L_0),$$

if the crossing is negative. These reduce $J(L)$ to a sum of Jones polynomials of unlinks and link diagrams with fewer crossings. Therefore, the Jones polynomial $J(L)$ of a link L with n crossings is determined by the Jones polynomials of link diagrams of fewer than n crossings. By induction, the construction therefore terminates with a sum of evaluations of the Jones polynomial of unlinks.

This argument certainly produces a polynomial. However, since there is, in general, more than one sequence of skein relations that reduces a link L to the unlink, we do not know at this stage that these sequences produce the same polynomial. Equally seriously, we do not know that, even if it is unique, it is an isotopy invariant of L. We will see later that it is.

Exercise 2.26. Let L be a link and $\overline{L}$ be the link defined as the reflection of L. Prove that $J(\overline{L})$ can be obtained from $J(L)$ by interchanging $t^{\frac{1}{2}}$ and $t^{-\frac{1}{2}}$. Hence prove that the right-hand trefoil and the left-hand trefoil, shown as the knots K_3 and K_4 in Example 1.5, are not equivalent.

The second polynomial on our list of three is defined by a similar skein relation.

Definition 2.27 (HOMFLYPT polynomial). The *HOMFLYPT polynomial*, also known as the *HOMFLY polynomial*, $P(L)$, of a link L, is the isotopy invariant defined by the bivariate skein relations

$$\textbf{SH1}: \quad xP(L_+) - x^{-1}P(L_-) = yP(L_0), \tag{2.2}$$

$$\textbf{SH2}: \quad P(\mathcal{O}) = 1. \tag{2.3}$$

The name HOMFLYPT polynomial comes from the surnames of its discoverers, Freyd, Yetter, Hoste, Lickorish, Millett and Ocneanu [63], who formed four groups of independent discoverers; and Przytycki and Traczyk [154] who did independent work relating to the polynomial.

The derivation of the Jones polynomial for the unlink in Lemma 2.21 can be adjusted to obtain the HOMFLYPT polynomial of the unlink. The result is given in the following lemma.

Lemma 2.28. *Let L be a k component unlink, then*

$$P(L) = \left(\frac{x - x^{-1}}{y}\right)^{k-1}.$$

Similarly, it is straightforward to adjust the computations of the Jones polynomial to show that

$$P\Bigg($$  $$\Bigg) = x^{-1}y + x^{-1}y^{-1} - x^{-3}y^{-1}. \tag{2.4}$$

$$P\left(\right.$$  $$\left.\right) = (2 + y^2)x^{-2} - x^{-4}. \tag{2.5}$$

$$P\left(\right.$$  $$\left.\right) = x^{-2} - 1 + x^2 - y^2. \tag{2.6}$$

Exercise 2.29. Verify (2.4)–(2.6).

Assuming the existence of the HOMFLYPT polynomial, it can be shown that this polynomial is defined by the skein relation by appropriately adjusting the analogous proof for the Jones polynomial.

Note that the Jones polynomial is a special case of the HOMFLYPT polynomial (set $x = t^{-1}$ and $y = t^{\frac{1}{2}} - t^{-\frac{1}{2}}$).

The third polynomial on our list is the Alexander–Conway polynomial, which is another specialisation of the HOMFLYPT polynomial.

Definition 2.30 (Alexander–Conway polynomial). The *Alexander–Conway polynomial* $C(L)$, of a link L, is the isotopy invariant defined by skein relations

$$\textbf{SAC1}: \; C(L_+) - C(L_-) = z\,C(L_0), \tag{2.7}$$

$$\textbf{SAC2}: \; C(\mathscr{O}) = 1. \tag{2.8}$$

Of course, $C(L)$ can be obtained as an evaluation of the HOMFLYPT polynomial $P(L)$ by setting $x = 1$ and $y = z$.

Exercise 2.31. Compute $C\left(\right.$  $\left.\right)$ and $C\left(\right.$  $\left.\right)$ by a direct computation using Definition 2.30 and also as an evaluation of the HOMFLYPT polynomial.

Exercise 2.32. Prove that if L is the disjoint union of two links then $C(L) = 0$.

Remark 2.33. The *Alexander polynomial*, $\Delta(L)$, is a knot polynomial that may be defined as the evaluation of the Alexander–Conway polynomial

$$\Delta(L; t) := C(L; t^{-\frac{1}{2}} - t^{\frac{1}{2}}).$$

Defining the Alexander polynomial as an evaluation of the Alexander–Conway polynomial as we have done here, however, is a little misleading since it was first defined by Alexander in the 1920s [4] *via* algebraic topology. The skein relation of Definition 2.30 is from the late 1960s and is due to Conway [44]. However, it was later noticed that Alexander had given a similar skein relation in his original paper.

The Alexander polynomial is significant since, unlike the Jones and HOMFLYPT polynomials, it has a known and well-understood topological derivation. This topological derivation is outside the scope of this book, and the reader can find details in, for example [46, 120, 158]. However, for the reader familiar with algebraic topology, the construction of the Alexander polynomial is roughly as follows. Let $L \subset \mathbb{R}^3 \cup \{\infty\} = \mathbb{S}^3$. Let $X = \mathbb{S} \setminus L$ be its complement, and X_∞ be an infinite cyclic cover of X with covering transformation t. Then $H_1(X_\infty, \mathbb{Z})$ is a $\mathbb{Z}[t, t^{-1}]$-module. This module is uniquely determined by L, and finitely presentable. It can be shown that its first elementary ideal is principal. The Alexander polynomial is a generator of this ideal.

There are infinitely many pairs of distinct knots with the same HOMFLYPT polynomial. Thus the HOMFLYPT polynomial is not a complete knot invariant. The Jones polynomial and Alexander–Conway polynomials are specializations of the HOMFLYPT polynomial and so are not complete invariants. A long standing conjecture is that a knot K has trivial Jones polynomial $J(K) = 1$ if and only if $K = \mathcal{O}$, the unknot. This result does not hold for links. It is known that there are infinitely many non-trivial links whose Jones polynomial is equal to the Jones polynomial of an unlink [60].

Chapter 3
Framed Links

3.1 Framed Links and Their Diagrams

In studying knots and links, it is often necessary to study other types of "knotted objects". In this chapter, we see the first of these, namely *framed links*. Intuitively, a framed knot or link may be thought of as a knot or link formed from a length of ribbon rather than a length of string. Here we shall insist that the ribbon contains an integral number of twists (so that each component is topologically an annulus, rather than a Möbius band). The significance of framed links for us here is that much of the theory we develop later in the book (specifically Reshetikhin–Turaev invariants, and Vassiliev invariants) occurs most naturally in the setting of framed knots and links. Although we do not discuss the application here, we note that framed links play an important role in 3-manifold topology since closed connected orientable 3-manifolds can be represented by framed links. We refer the interested reader to, for example, [120, 152, 158] for details of the connection with 3-manifold topology.

By reading Chaps. 1 and 2, the reader will have developed some intuition about knots, so we shall be a little briefer and more informal in our treatment of framed links compared to our treatment of links.

Definition 3.1 (Framed link). A *framed link* is a disjoint embedding of n disjoint annuli in $\mathbb{R}^3$. Each annulus is a *component* of the framed link. A framed link with exactly one component is a *framed knot*.

As observed above, a framed link may be regarded informally as a knot or link made with a ribbon, provided the ribbon does not form a Möbius band. Figure 3.1 shows drawings of some framed links.

Framed links are considered up to *ambient isotopy*, where the ambient isotopy for framed links is defined as for links (Definition 1.3) but with the word "links" replaced with "framed links".

The number of "twists in the ribbon" is significant. For example, the framed links in Fig. 3.2 are inequivalent.

D. M. Jackson and I. Moffatt, *An Introduction to Quantum and Vassiliev Knot Invariants*, CMS Books in Mathematics, https://doi.org/10.1007/978-3-030-05213-3_3

Fig. 3.1 Drawings of framed links

Fig. 3.2 Examples of inequivalent framed links

Recall that knots and links that we study here are tame, so they are ambient isotopic to a polygonal, or piecewise-linear, knot made of a finite number of straight line segments. We also assume a tameness condition on framed links by assuming that they are ambient isotopic to one made of finitely many triangles, 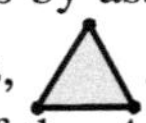. Using this piecewise-linear structure, it is possible to obtain an analogue of the Δ-moves and Theorem 1.10 for framed knots, although we shall exclude the details here, and instead move forward to the diagrammatics of framed knots.

As with knots and links, we can consider projections of framed links and also, by decorating projections with a crossing structure, diagrams of framed link. By using small ambient isotopies on the framed link, we can assume that a projection of a framed link is regular in the sense of (the obvious extension of) Definition 1.12. Then, given a regular projection, we can indicate a crossing structure using line breaks to obtain a diagram of a framed link. Examples of such diagrams are given in Fig. 3.1. In particular, we can assume that every crossing in a diagram is one of the following types:

(a) (b)

The type (a) crossing involving two different segments of the framed link (but possibly segments of the same component); the type (b) crossing involves a single segment and records some "twisting" of the framed link. In fact, we can simplify the drawing by using isotopy to eliminate all of the crossings of type (b).

Since a framed link consists of embedded annuli and no Möbius bands, there can be only an even number of crossings of type (b). Moreover, by isotopy of the framed link, we may assume that the twists occur in adjacent pairs. Then we observe that all of the twists may be eliminated by isotoping the framed link as indicated in the following figure:

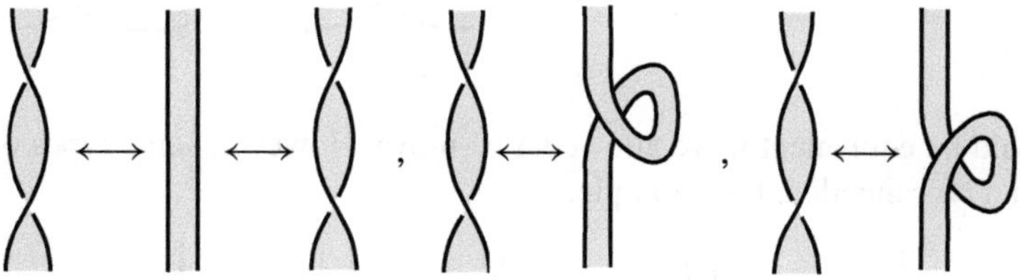

(3.1)

For example,

The result of this discussion is that we can assume that our drawing consists of a "thickening" or, more precisely, a regular neighbourhood, of a link diagrams as indicated in the following figure:

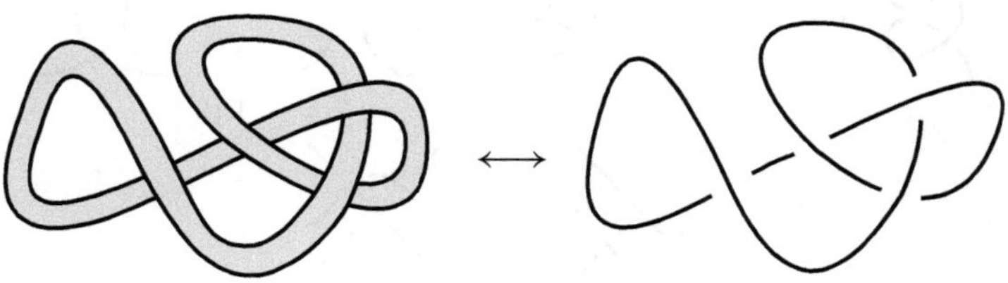

Then every framed link can be represented by a link diagram. Conversely, every link diagram gives rise to a framed link by thickening the drawing through 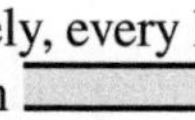 ↔ 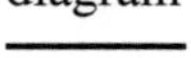.

Definition 3.2 (Framed link diagram). A *framed link diagram* consists of a link diagram. A projection of a framed link is obtained by taking a regular neighbourhood of the diagram (i.e. thickening the lines in the drawing).

As was the case with knots and links, different framed knot diagrams arise from isotopic framed links, and we need to understand how diagrams that represent isotopic framed links relate to one another, i.e. we need an analogue of the Reidemeister moves and Reidemeister's Theorem for framed links. An immediate observation is that the RI move does not hold for framed link diagrams since, for example, the inequivalent framed links shown in Fig. 3.2 are represented by the diagrams

which would be equivalent if we allowed a RI-move. However, some types of pairs of loops can be cancelled. For example,

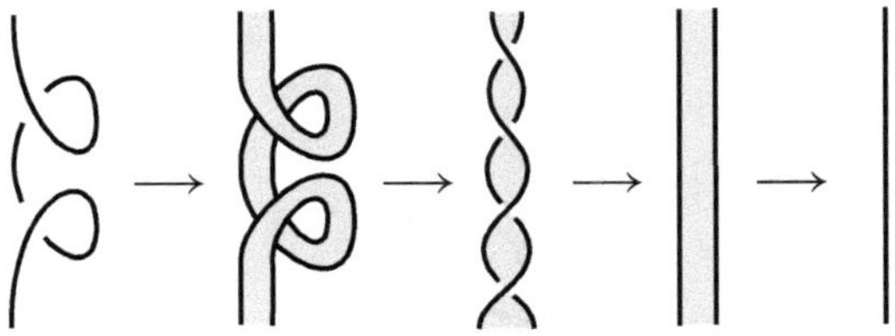

where the second and third moves are by (3.1). In fact, it is this type of cancellation that provides the appropriate modification of the Reidemeister moves.

Definition 3.3 (Framed Reidemeister moves). The *framed Reidemeister moves* consist of planar isotopy and the following three local changes in a diagram.

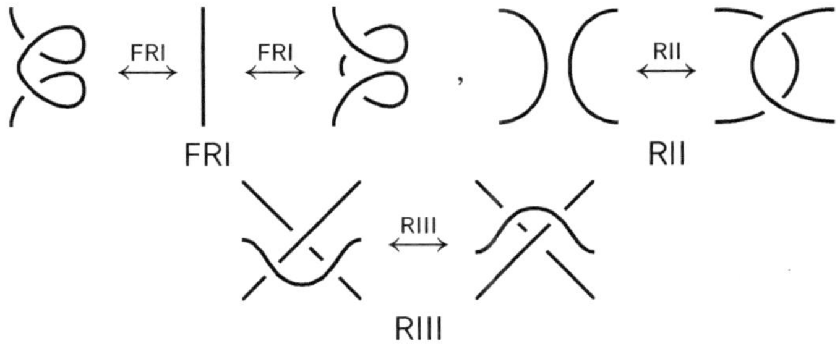

The diagrams are identical outside the local region shown.

Exercise 3.4. Show that the following identity of framed link diagrams is a consequence of the framed Reidemeister moves.

Theorem 3.5 (Framed Reidemeister's Theorem). *Two framed links are equivalent (i.e. ambient isotopic) if and only if their diagrams are related by a finite sequence of framed Reidemeister moves. That is*

$$\frac{\{\textit{framed links}\}}{\textit{isotopy}} \cong \frac{\{\textit{link diagrams}\}}{\textit{framed Reidemeister moves}}.$$

Exercise 3.6. Think about how Theorem 3.5 can be proved. Sketch out a proof of it.

Although we have introduced framed links as embeddings of annuli in $\mathbb{R}^3$, we shall work exclusively with their diagrams. For our purposes, the reader may safely think of framed links as being link diagrams considered up to the framed Reidemeister moves.

As with link diagrams, we can consider framed link diagrams in which each component has been assigned a preferred orientation. A set of framed oriented Reidemeister moves is defined by considering all possible orientations of the arcs involved in the framed Reidemeister moves.

3.2 Framed Link Invariants

As with knots and links, distinguishing framed links is a hard problem. To distinguish between them, we use a *framed link invariant*.

Definition 3.7 (Framed link invariant). A *framed link invariant* (or *framed knot invariant*) f is a function from the set of isotopy classes of framed links: that is,

$$f\colon \frac{\{\text{framed links}\}}{\text{isotopy}} \to S, \tag{3.2}$$

where S is a set.

Also, as with the case of knots and links, by the framed Reidemeister theorem (Theorem 3.5) we can define framed link invariants as functions on framed link diagrams that are invariant under the framed Reidemeister moves. We shall now consider a few examples of framed link invariants, namely:

- linking and self-linking numbers,
- the writhe,
- the Kauffman bracket.

Let L be an oriented link with components $L_1, \ldots, L_n$ and D be an oriented diagram for L with components $D_1, \ldots, D_n$. Recall that, as in Fig. 2.3, we may associate a sign $+1$ or -1 to each crossing of L. Recall also that the linking numbers $\mathrm{lk}(i, j)$ for $i \neq j$ are link invariants. In the case of framed links, we can also consider *self-linking numbers* $\mathrm{lk}(i, i)$.

Definition 3.8 (Linking number). Let L be a framed oriented link with components $L_1, \ldots, L_n$ and D be an oriented framed link diagram for L with components $D_1, \ldots, D_n$. Then for any $i, j \in \{1, \ldots, n\}$, the number $\mathrm{lk}(i, j)$ is the sum of the signs of all crossings between the components D_i and D_j of the diagram D of L.

When $i \neq j$, the number $\mathrm{lk}(i, j)$ is called the *linking number* $\mathrm{lk}(i, j)$ of L. For $i = j$, the number $\mathrm{lk}(i, i)$ is called the *self-linking number* $\mathrm{lk}(i, i)$ of L.

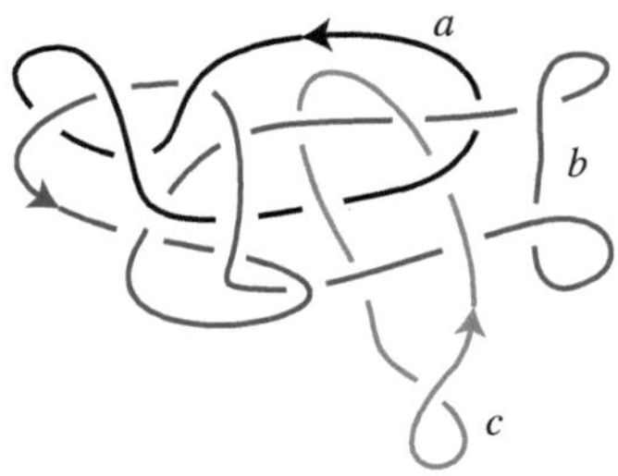

Fig. 3.3 An oriented framed link diagram

For example, the oriented framed link in Fig. 3.3 has linking numbers $\mathrm{lk}(a, b) = \mathrm{lk}(b, a) = -2$, $\mathrm{lk}(a, c) = \mathrm{lk}(c, a) = +2$, $\mathrm{lk}(b, c) = \mathrm{lk}(c, b) = 0$, $\mathrm{lk}(a, a) = -1$, $\mathrm{lk}(b, b) = -2$, and $\mathrm{lk}(c, c) = -1$.

Theorem 3.9. *Linking numbers and self-linking numbers are independent of the choice of framed link diagram and are therefore invariants of oriented framed links.*

Exercise 3.10. Adapt the proof of Theorem 2.17 to show that, for all i, j, $\mathrm{lk}(i, j)$ is invariant under the oriented framed Reidemeister moves, and hence give a proof of Theorem 3.9.

A closely related concept to the linking number is the *writhe* of a link diagram, which is defined as follows.

Definition 3.11 (Writhe). Let D be an oriented link diagram. The *writhe* $\omega(D)$ of D is the sum of the signs of all crossings of D.

As an example, the writhe of the link shown in Fig. 3.3 is -4.

The writhe defines an invariant of framed oriented links and will be important later.

Theorem 3.12. *The writhe is independent of the choice of framed link diagram and is therefore is an invariant of oriented framed links.*

Proof. The proof follows by Theorem 3.9 upon the observation that $\omega(D) = \sum_{i \leq j} \mathrm{lk}(i, j)$. □

Exercise 3.13. Prove that two diagrams of the same oriented link are related by a finite sequence of Reidemeister RII- and RIII-moves if and only if they have the same and linking and self-linking numbers.

The next example of an invariant is the Kauffman bracket, which is a polynomial invariant of (unoriented) framed links.

Definition 3.14 (Kauffman bracket). The *Kauffman bracket* $\langle D\rangle \in \mathbb{Z}[A, A^{-1}]$ of a (framed or unframed) link diagram D is defined by the relations.

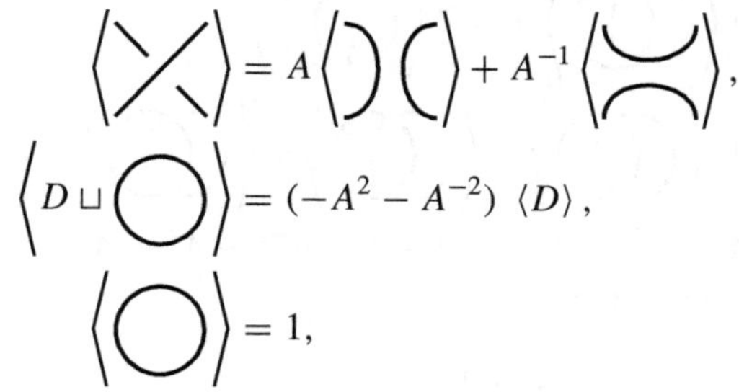

where 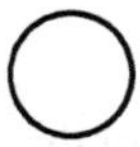 is a diagram with no crossings and $D \sqcup$ 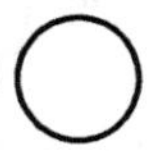 consists of a diagram D together with a (disjoint) diagram 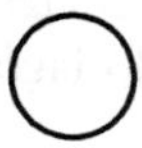.

In the defining relation 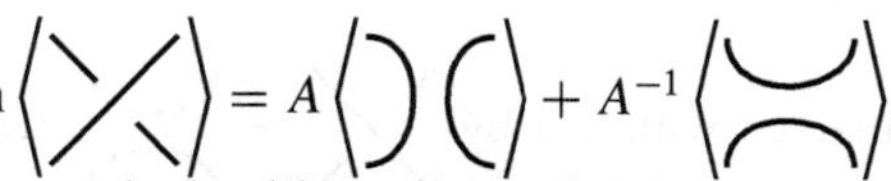 , the term 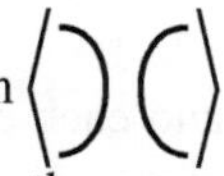 refers to the link diagram obtained by splicing the under-crossing arc to the over-crossing arc with respect to a counterclockwise orientation, while the term 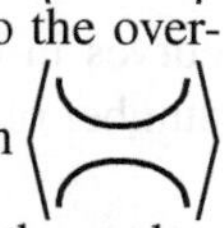 refers to the link diagram obtained by splicing the over-crossing arc to the under-crossing arc with respect to a counterclockwise orientation. All link diagrams are identical outside the region shown.

Example 3.15.

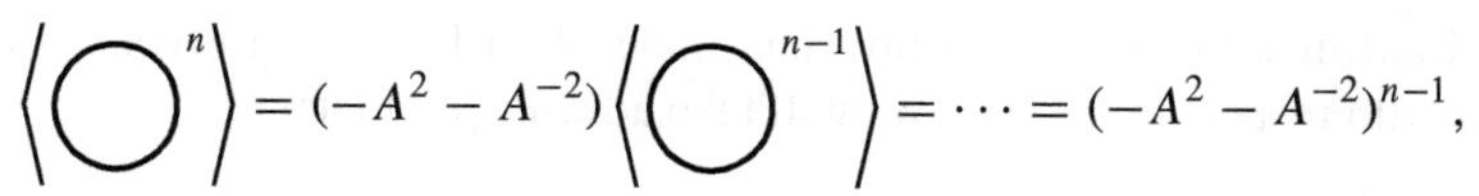

where 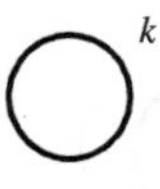 refers to a diagram consisting of k disjoint copies of 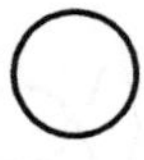.

Example 3.16.

$$= A(-A^2 - A^{-2}) + A^{-1}$$
$$= -A^3.$$

Example 3.17.

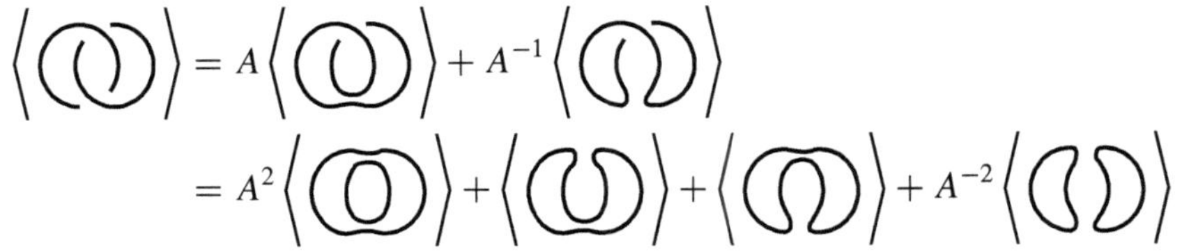

$$= A^2(-A^2 - A^{-2}) + 1 + 1 + A^{-2}(-A^2 - A^{-2})$$
$$= -A^4 - A^{-4}.$$

Exercise 3.18. Verify that  $= -A^5 - A^{-3} + A^{-7}$.

Exercise 3.19. Define a *state*, s, of a link diagram to be an assignment of the symbol A or the symbol B to each crossing. A link diagram with n crossings then has 2^n states. Given a state s of D, changing each crossing marked with an A by 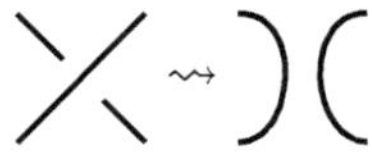 and each crossing marked with a B by , results in a collection of curves in the plane. Let $c(s)$ denote the number of such curves. Let $\alpha(s)$ be the number of A's in a state s, and $\beta(s)$ be the number of B's. Prove that

$$\langle D \rangle = \sum_s A^{\alpha(s)-\beta(s)}(-A^2 - A^{-2})^{c(s)-1},$$

where the sum is over all states s of D. This is called a *state sum* formulation of the Kauffman bracket.

The Kauffman bracket is not invariant under the RI-move. However, it is well-behaved with respect to it, a fact we will take advantage of later.

Lemma 3.20. *The Kauffman bracket satisfies the following identities.*

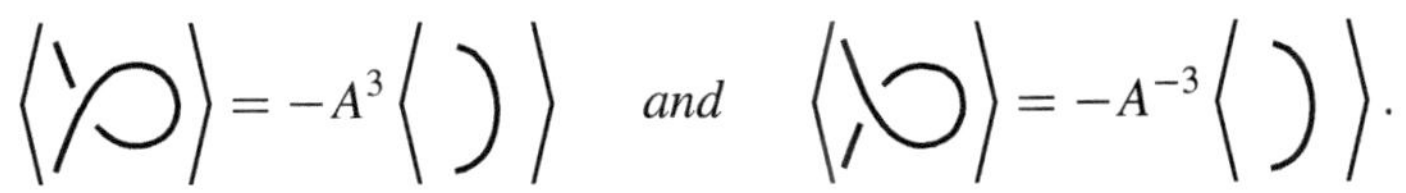

Proof.

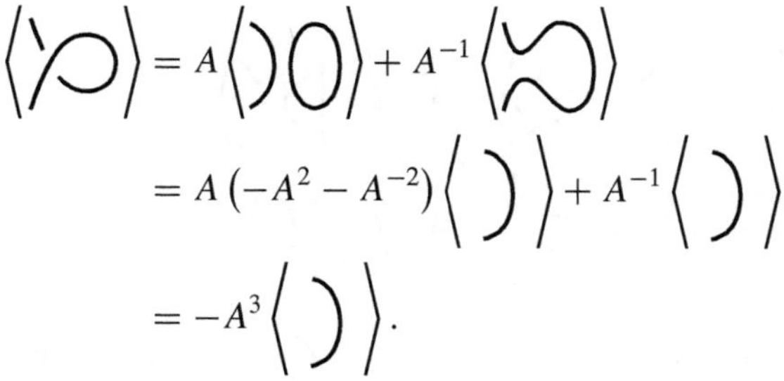

The other case is similar. □

Although the Kauffman bracket is not a link invariant, it is an invariant of framed links.

Theorem 3.21. *$\langle D \rangle$ is invariant under the framed Reidemeister moves, and so defines a framed link invariant.*

Proof. For RI: Invariance of $\langle D \rangle$ under FRI follows by Lemma 3.20.

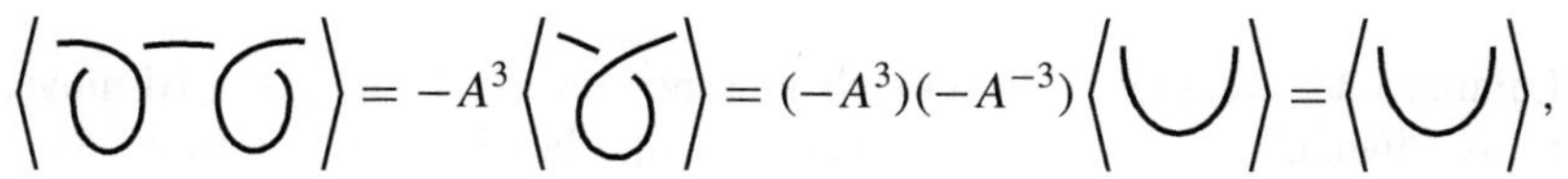

where the second equality is by Lemma 3.20. The other relation in the FRI-move follows similarly.
For RII:

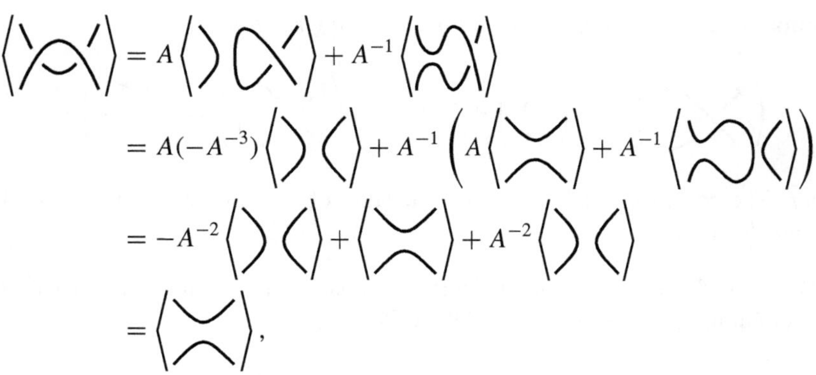

so $\langle D \rangle$ is invariant under RII.

For RIII:

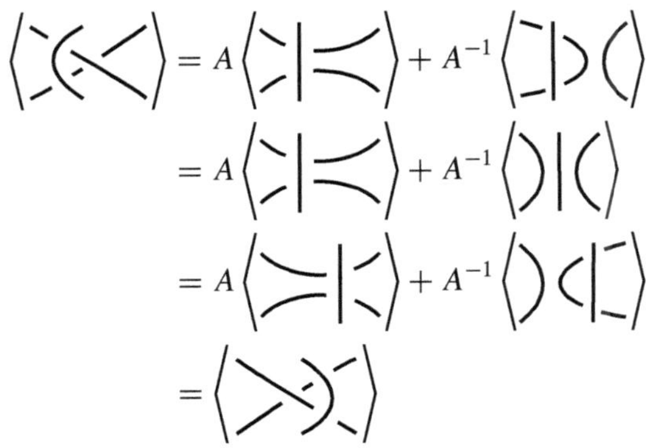

(invariance under RII)

(invariance under RII)

completing the proof. □

3.3 Deframing a Link Invariant

In Lemma 3.20, we saw that although it is not invariant under the RI-move, an RI-move changes the Kaufman bracket by multiplication by a constant. In such cases, there is a standard trick to "deframe" a framed link invariant and thus obtain an invariant of (unframed) oriented links. The trick depends upon the observation that the writhe, although not invariant under the RI-move, does behave predictably under it.

Lemma 3.22. *The writhe ω satisfies the following relations.*

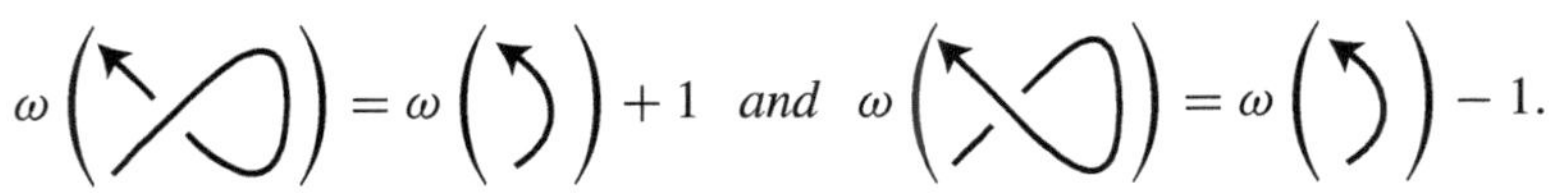

Proof. The proof follows by summing the signs of each of the crossings in each of the link diagrams. □

For the deframing application, suppose we had found a framed link invariant I. If in addition, this invariant satisfied the following relations

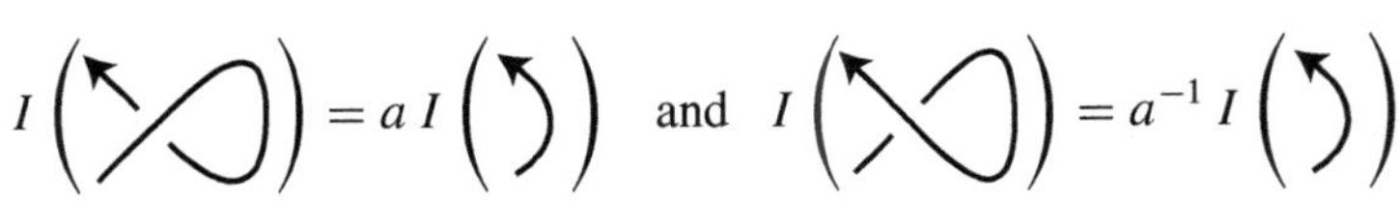

where a is an indeterminate, then it is readily seen that

$$\tilde{I}(D) := a^{-\omega(D)} I(D)$$

is invariant under the RI-move. Since I and ω are both invariant under RII and RIII, (as they are invariants of framed links) so is $\widetilde{I}(D)$. Therefore, $\widetilde{I}(D)$ is an invariant of oriented links. We call this process *deframing*.

To see how deframing works in practice, we return to the Kauffman bracket. The key observation is that if D is oriented, then the sign of the crossing determines the two types of loops in the statement of Lemma 3.20, namely

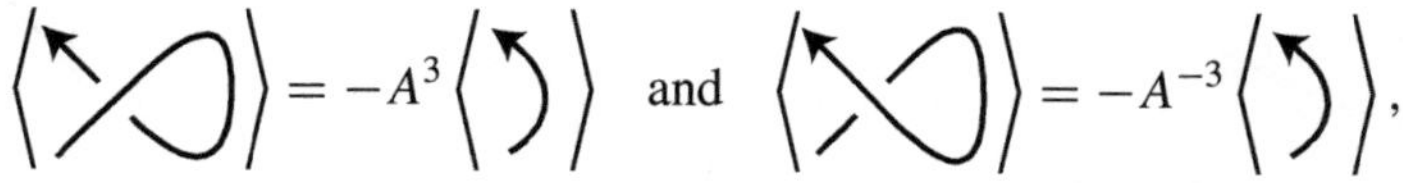

so deframing $\langle\cdot\rangle$ gives an invariant (with $a := -A^3$) of "unframed links".

Theorem 3.23. *Let* $F(D) := (-A)^{-3\omega(D)} \langle D\rangle$. *Then* $F(D)$ *is invariant under the Reidemeister moves, and thus defines a link invariant (of unframed links).*

In fact, we may recognise $F(D)$ as the Jones polynomial (see Definition 2.20). This is the content of the following theorem.

Theorem 3.24. *Let* L *be an oriented link and* D *be a diagram of* L. *Then*

$$J(L) = F(D)|_{t^{-1/2}=A^2}\,.$$

Proof. We start by observing that

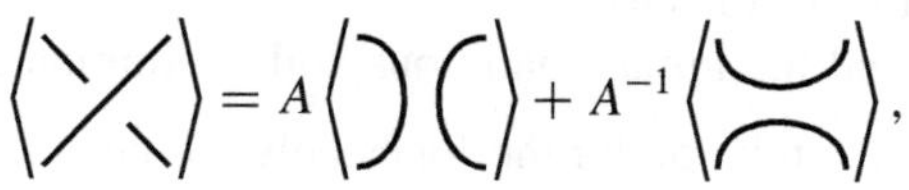

and

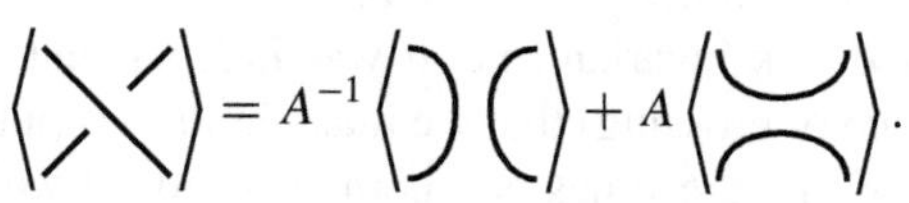

By subtracting these expressions, we obtain

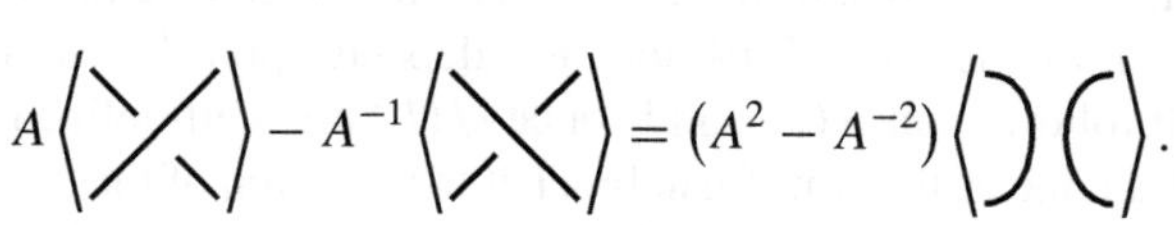

We can apply this expression to an oriented link diagram to obtain

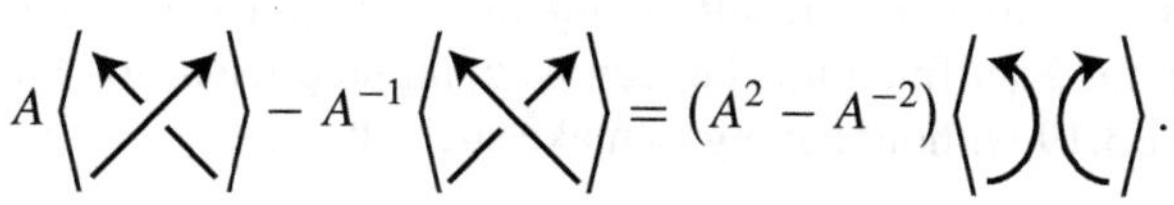

(Note that the Kauffman bracket cannot be applied to oriented link diagrams directly.)

Multiplying by the constant 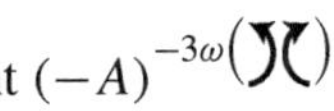 then gives

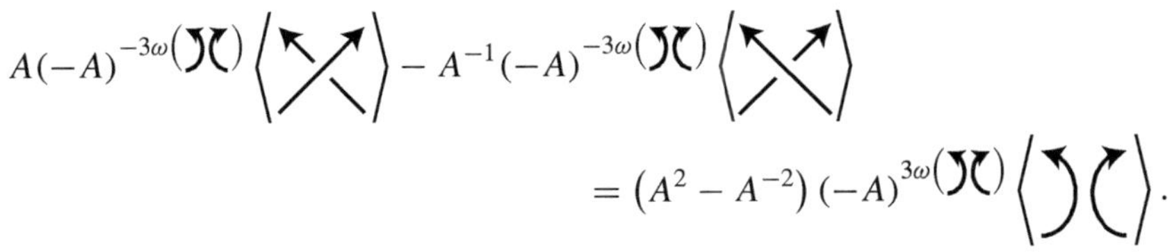

But

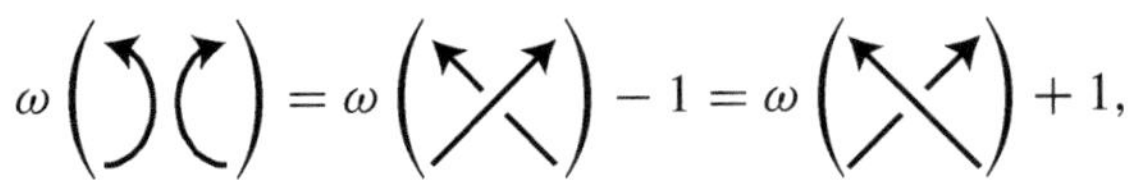

so

which can be rewritten as

$$A^4 F(L_+) - A^{-4} F(L_-) = -\left(A^2 - A^{-2}\right) F(L_0). \tag{3.3}$$

Since F is a link invariant, it takes the same value on every diagram of the unknot. A simple computation, together with invariance under isotopy, then gives $F(\mathscr{O}) = 1$, where $\mathscr{O}$ is any diagram of the unknot.

Comparing (3.3) and the initial conditions with Definition 2.20, we see that we have recovered the skein relation for the Jones polynomial at $t^{-\frac{1}{2}} = A^2$. □

When we introduced the Jones polynomial in Chap. 2, we did not show that it was actually a well-defined link invariant, i.e. it was independent of all of the choices (for example the choice of crossing) that we make in its computation. Theorem 3.23 provides an easy proof that the Jones polynomial is indeed well-defined. Although the computation of the Kauffman bracket $\langle D\rangle$ of a diagram D requires a choice of order of crossings, it is not hard to see that its value is independent of this choice (e.g. the state sum of Exercise 3.19 does not depend upon this choice), and so is well-defined. It follows that $F(D)$, and hence, $J(L)$ are well-defined.

Another advantage of the Kauffman bracket formulation of the Jones polynomial over the skein theoretic one is that when applying the skein relations of Definition 2.20, it is not always clear which crossing to apply the relation to, and choosing a crossing poorly could make a resulting link more complicated than the original. On the other hand, the Kauffman bracket can be computed from, for example, its state sum in Exercise 3.19 without having to make such choices.

Exercise 3.25. Write a proof that the Jones polynomial is a well-defined knot invariant by formalising the above discussion.

Chapter 4
Braids and the Braid Group

It was remarked at the start of Chap. 3 that in studying knots and links, it is often advantageous (or necessary) to consider other types of "knotted object". After seeing both framed and unframed knots and links, we now look at a third type of knotted object, namely a braid. Braids were introduced by Artin in 1925 and are well-studied objects that are encountered in several areas of mathematics. They are, in some sense, simpler objects than knots and links. We shall take advantage of this fact in Part II where we use the structure of a braid to construct knot and link invariants. We go on to use the insights provided by braids to construct other types of link invariant directly from a link diagram.

4.1 Braids and Braid Equivalence

Informally, a braid consists of a box, on a horizontal plane, with n strands (or pieces of string), each with one end anchored at the bottom and the other at the top of the box, and each with the properties that if it is oriented from the bottom of the box to the top, there is no point in the strand where the direction is downwards. In common with the description of many knotted objects, the idea immediately becomes clear with a picture. A picture of a braid on four strands is shown in Fig. 4.1. The lower ends of the braid are at the points $(i, 0, 0)$ on the x-axis, and the upper ends are at the points $(i, 0, 1)$. Each of the four strands is an embedded copy of the interval $[0, 1]$. The intersection of the braid and any plane $\mathbb{R} \times \{i\}$, for $0 \leq i \leq 1$, always consists of the same number of points (in Fig. 4.1, 4) for the braid. This means that no strand of a braid has a local maximum or minimum with respect to the vertical coordinate. The strands of a braid have a natural orientation in the positive Oz-direction.

Definition 4.1 (Braid). An *n-stranded braid* σ is a disjoint union of n copies of the unit interval $I = [0, 1]$ (called *strands*) embedded into $\mathbb{R}^2 \times [0, 1]$ such that the boundary set $\partial(\sigma)$ of the strands is the set $\{1, \ldots, n\} \times \{0\} \times \{0, 1\}$ and such that

D. M. Jackson and I. Moffatt, *An Introduction to Quantum and Vassiliev Knot Invariants*, CMS Books in Mathematics, https://doi.org/10.1007/978-3-030-05213-3_4

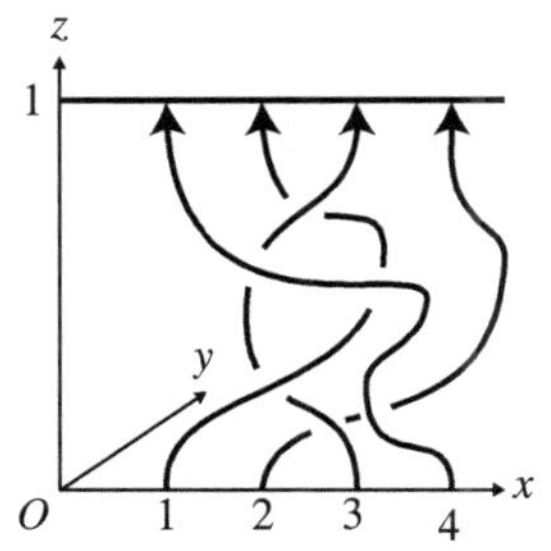

Fig. 4.1 A 4-stranded braid

for each $i \in [0, 1]$ the intersection of $\mathbb{R}^2 \times \{i\}$ and σ consists of exactly n points. All strands of the braid are oriented upwards. (Here all embeddings are piecewise-linear.)

Informally, two braids are equivalent if one may be transformed into the other by deforming the strands of the braids continuously in such a way that no maxima or minima are created and the ends of the braid are fixed (Fig. 4.2).

Equivalence is defined formally as follows.

Definition 4.2 (Braid equivalence). Two braids σ and σ' are said to be *equivalent* if there is a continuous family of (piecewise-linear) homeomorphisms

$$h_t : \mathbb{R}^2 \times [0, 1] \to \mathbb{R}^2 \times [0, 1]$$

such that, for all $t \in [0, 1]$,

1. $h_0 = \mathrm{id}$,
2. $h_1(\sigma) = \sigma'$,
3. $h_t|_{\mathbb{R}^2 \times \{0,1\}} = \mathrm{id}$,
4. $h_t|_{\mathbb{R}^2 \times \{s\}} : \mathbb{R}^2 \times \{s\} \to \mathbb{R}^2 \times \{s\}$, for all $s \in [0, 1]$,
5. $(x, t) \mapsto (h_t(x), t)$ is a homeomorphism of $\mathbb{R}^2 \times [0, 1]^2$ to itself.

Such an isotopy is said to be *level- and boundary-preserving*.

Continuing the analogy with strings, braid equivalence can be thought of as an operation allowing us to push or pull a strand of string in only the *horizontal* direction while the ends of the strands of string remain fixed to the box.

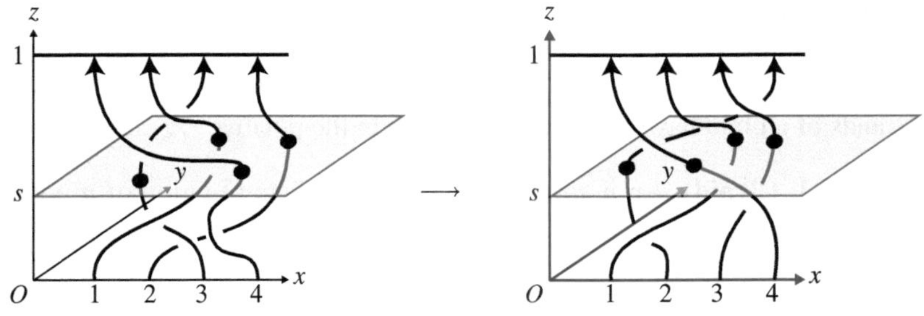

Fig. 4.2 Equivalent braids

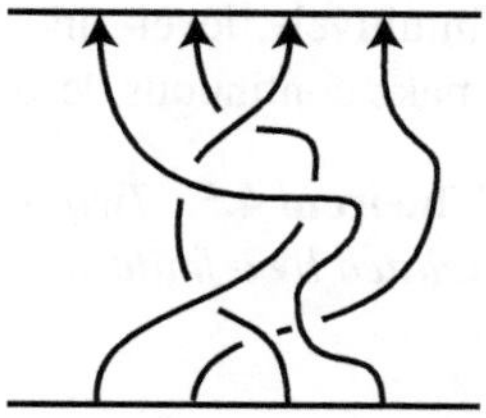

Fig. 4.3 A 4-stranded braid diagram

As with links and framed links, we impose a tameness condition on braids to ensure that the drawings contain only finitely many components. We say a braid is *tame* if it is equivalent to a braid whose components consist of a finite number of straight line segments. We assume that all braids are tame without further comment.

As with knots, braids may be represented by diagrams. An example of a braid diagram for the braid in Fig. 4.1 is given in Fig. 4.3.

Braid diagrams are defined in a way similar to link diagrams. A braid σ can be projected onto the xz-plane by the map that sends (x, y, z) to $(x, 0, z)$. By deforming the braid if necessary, we may assume that the projection is regular in the sense of Definition 1.12. A braid diagram is then obtained by marking an over-/under-crossing structure on the link diagram using line breaks to specify which arc goes over which in the corresponding braid.

Definition 4.3 (Braid diagram). An *n-stranded braid diagram* is the image of a regular projection of a braid on which an under-/over-crossing structure has been assigned to each double point. The crossing structure is indicated by line breaks in the diagram.

Of course, different braid diagrams can represent equivalent braids, so the study braids by diagrams require an analogue of Reidemeister's Theorem.

Definition 4.4 (Braid moves). The *braid moves* consist level- and boundary-preserving isotopy of $\mathbb{R} \times [0, 1]$ (*i.e.* the analogue of conditions (3) and (4) of Definition 4.2 are satisfied in the planar isotopy) and the following three local changes in a diagram.

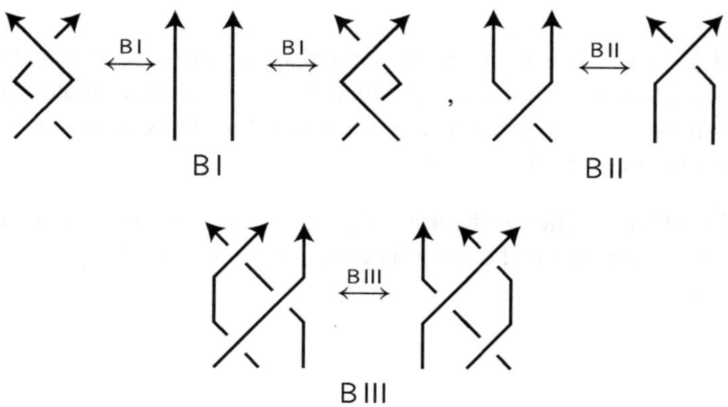

(4.1)

The diagrams are identical outside of the local region shown.

Intuitively, level- and boundary-preserving isotopy of $\mathbb{R} \times [0, 1]$ says that we can make continuous deformations of the braid diagram in the *horizontal* direction.

Theorem 4.5. *Two braid diagrams represent the same braid if and only if they are related by a finite sequence of braid moves. That is*

$$\frac{\{braids\}}{equivalence} \cong \frac{\{braid\ diagrams\}}{braid\ moves}.$$

An example of the use of the braid moves is the following:

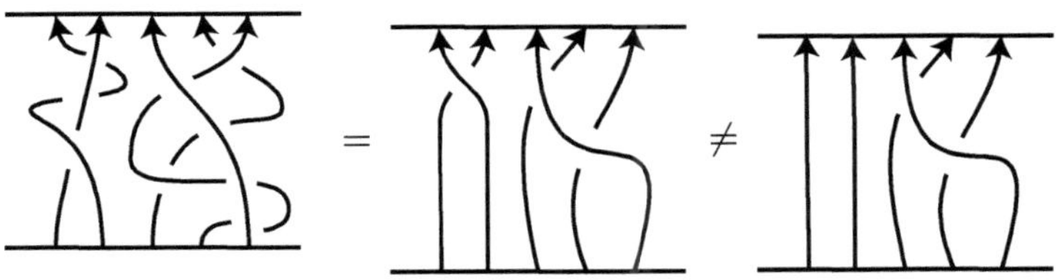

The final inequality is due to the endpoints of a braid being fixed.

The proof of Theorem 4.5 is similar to the proof of Reidemeister's Theorem (Theorem 1.26), and we exclude details. The strategy is first to formulate equivalence in terms of (a restricted type of) Δ-move and then examine the images of these moves under projection, showing that they are all consequences of the braid moves.

Exercise 4.6. Sketch a proof of Theorem 4.5.

4.2 The Braid Group $\mathfrak{B}_n$

Let $\mathscr{B}$ be the set of all braids and let $\mathscr{B}_n$ be the set of all n-stranded braids. Thus,

$$\mathscr{B} = \bigsqcup_{n \geq 0} \mathscr{B}_n.$$

Let σ and σ' be two braids in $\mathscr{B}_n$. In particular, this means that σ and σ' both have n strands. There is a natural *composition* $\sigma \circ \sigma'$, of n-strand braids given by 'stacking' σ on top of σ' and then reparameterising the result by scaling each strand to be a copy of the interval $[0, 1]$ again.

Definition 4.7 (Composition of braids). The *composition* $\sigma \circ \sigma'$ of an n-stranded braid σ with another n-stranded braid σ' is given diagrammatically by

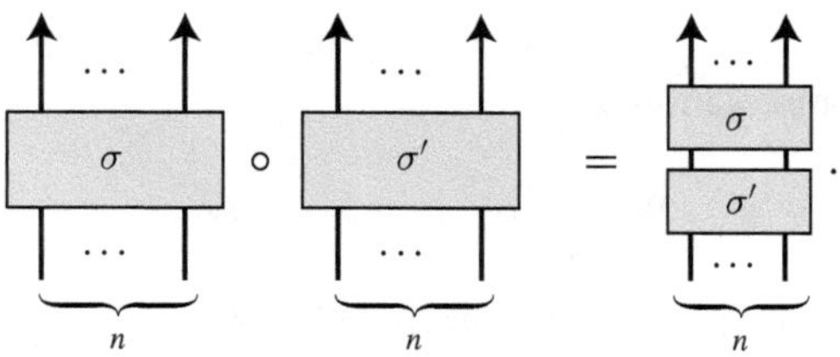

The composition of braids on a different number of strands is undefined.

Sometimes it is convenient to indicate how a braid diagram can be obtained as a composite of braids, which we do by adding dotted horizontal lines to the diagram, as in Example 4.8 below.

The set $\mathscr{B}_n$ of n-stranded braids is clearly closed under composition. It is not difficult to show that this composition is associative and that there is an *identity element* 1_n, the n-stranded braid with no crossings. Thus $(\mathscr{B}_n, \circ)$ is a monoid, and we can describe a set of generators for it as follows.

For $i = 1, \ldots, n-1$, let σ_i denote the n-stranded braid with a single crossing, which is *positive*, between the i-th and $(i+1)$-st strands, and let ς_i denote the n-stranded braid with a single crossing which is *negative* between the i-th and $(i+1)$-st strands. These two braids are

$\sigma_i :=$ 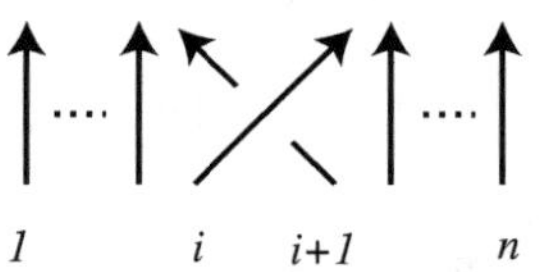

, $\varsigma_i :=$ 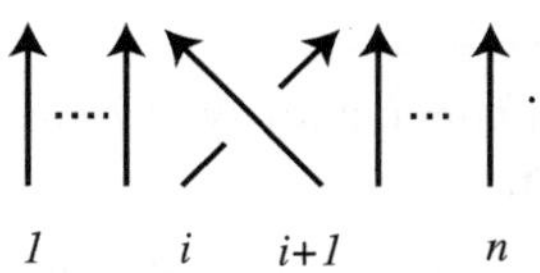

. (4.2)

From the braid move B I given in (4.1), it is clear that $\sigma_i \circ \varsigma_i = 1$, the identity in $\mathscr{B}_n$, so the inverse of σ_i is ς_i, which we therefore denote by σ_i^{-1}. Then $\{\sigma_i, \sigma_i^{-1} : i = 1, \ldots, n-1\}$ is a set of generators for the monoid $\mathscr{B}_n$. A braid $\sigma \in \mathscr{B}_n$ therefore has the form

$$\sigma_{j_1}^{\delta_1} \cdots \sigma_{j_m}^{\delta_m},$$

where $\delta_k = \pm 1$, which therefore has an inverse

$$\sigma = \sigma_{j_m}^{-\delta_m} \cdots \sigma_{j_1}^{-\delta_1}.$$

It follows that $(\mathscr{B}_n, \circ)$ is a group. It is known as the *braid group* and, we shall denote it by $\mathfrak{B}_n$.

Example 4.8. The braid in Fig. 4.3 can be written as:

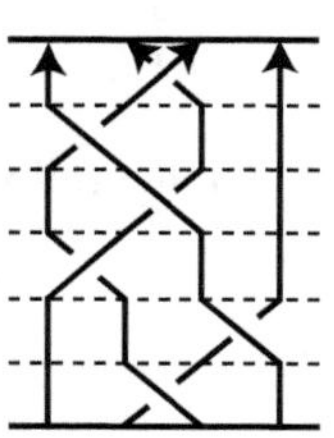

$$= \sigma_2 \circ \sigma_1^{-1} \circ \sigma_2^{-1} \circ \sigma_1 \circ \sigma_3^{-1} \circ \sigma_2^{-1}.$$

The set $\{\sigma_i : i = 1, \ldots, n-1\}$ is a set of generators for $\mathfrak{B}_n$, and Theorem 4.5 provides a group presentation

$$\mathfrak{B}_n = \langle \sigma_1, \ldots, \sigma_{n-1} \mid \mathscr{R} \rangle,$$

where $\mathscr{R}$ is the set of relations, called the *Artin braid relations* or the *braid relations*, given by

$$\mathscr{R} := \begin{cases} \sigma_i \circ \sigma_j = \sigma_j \circ \sigma_i, & \text{if } |i-j| \geq 2, \\ \sigma_i \circ \sigma_{i+1} \circ \sigma_i = \sigma_{i+1} \circ \sigma_i \circ \sigma_{i+1}, & i = 1, \ldots, n-2. \end{cases} \quad (4.3)$$

These arise by writing the braid moves in terms of braid generators.

As an example, in $\mathfrak{B}_4$ is sufficient to show how the relations $\mathscr{R}$ arise. The commutativity of σ_i and σ_j for $i = 1$ and $j = 3$ (the case $|i-j| \geq 2$) (sometimes called *distant commutativity*) follows from

$\sigma_1 \circ \sigma_3 =$ 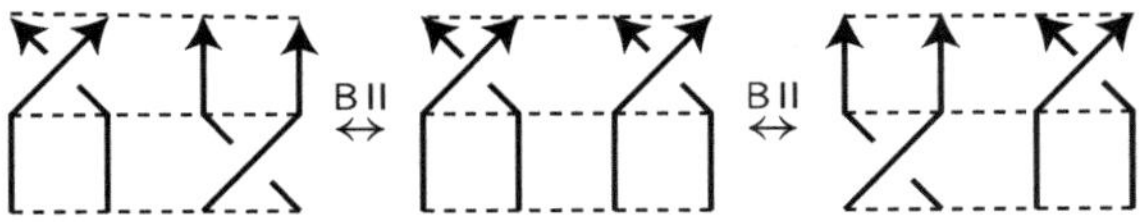 $= \sigma_3 \circ \sigma_1.$

The non-commutativity of σ_i and σ_j for $i = 1$ and $j = 2$ (the case $|i-j| < 2$) follows from

$\sigma_1 \circ \sigma_2 =$ 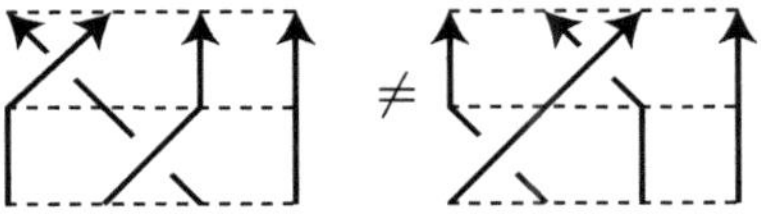 $= \sigma_2 \circ \sigma_1.$

The proof of this is through a simple braid invariant argument. Let $f : \mathscr{B}_4 \to \mathbb{R}$, be the function $f : \sigma \mapsto k$, where the strand of σ starting from the bottom point indexed by 1 ends at the top point indexed by k. This is clearly an invariant of braids since the boundary points are fixed. But $f(\sigma_1 \circ \sigma_2) = 2$ and $f(\sigma_2 \circ \sigma_1) = 3$ so $f(\sigma_1 \circ \sigma_2) \neq f(\sigma_2 \circ \sigma_1)$, whence $\sigma_1 \circ \sigma_2 \neq \sigma_2 \circ \sigma_1$.

In the case of the braid relation B III, we have immediately that

$\sigma_1 \circ \sigma_2 \circ \sigma_1 =$ 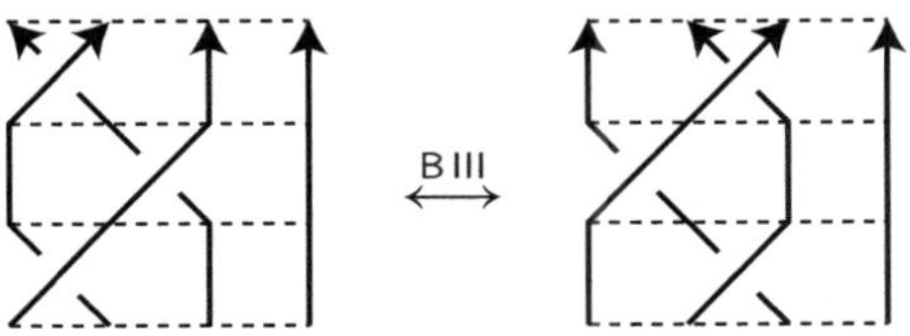 $= \sigma_2 \circ \sigma_1 \circ \sigma_2.$

It is easily seen from the constructions in this section that the following is true.

Theorem 4.9.

$$\frac{\{n\text{-strand braid diagrams}\}}{\text{braid moves}} \cong \mathfrak{B}_n.$$

4.3 The Markov Moves

Braids and links are intimately connected. On one hand, given a braid diagram σ, we can form a link diagram by drawing arcs on the diagram from the top to the bottom of each strand of the braid as in the following definition.

Definition 4.10 (Closure of a braid). The *closure* of the n-stranded braid

$\sigma =$ 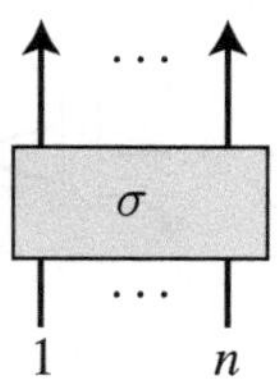

is the link diagram $\widehat{\sigma} =$ 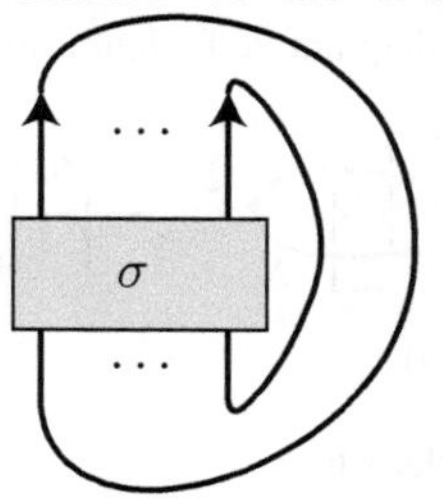

obtained by identifying the upper and lower ends of each strand of σ as shown.

For example,

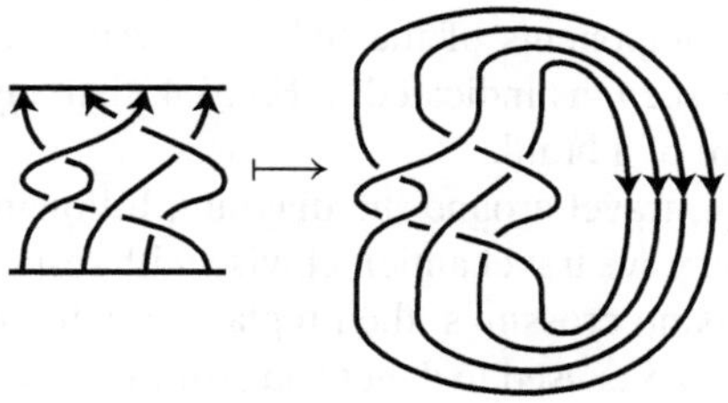

Note that it is not immediately obvious whether closing a braid will result in a knot or a link.

The first result is the following.

Proposition 4.11. *Every braid gives rise to a link.*

Proof. This is done by taking the closure of the braid. □

On the other hand, while Proposition 4.11 tells us how to move from braids to links, it is possible to move in the other direction too, showing that every link can be represented by a diagram $\widehat{\sigma}$ of the closure of some braid σ, and hence can be represented by a braid. Moving from a link diagram to a braid requires a little work, and uses a "combing" process, called *Alexander's trick*, on the link diagram to put it in a required form. This process is described in the proof of Alexander's Theorem below and is illustrated in Fig. 4.6.

Theorem 4.12 (Alexander's Theorem). *Every oriented link arises as the closure of a braid.*

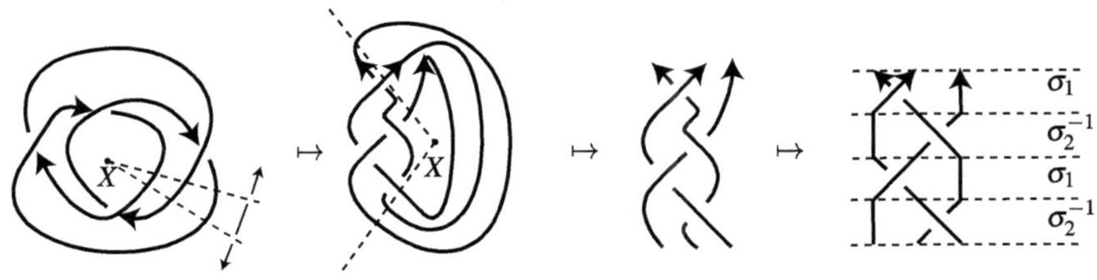

Fig. 4.4 From a diagram of the figure-of-eight knot to its encoding as a braid $(\sigma_1 \circ \sigma_2^{-1})^2$

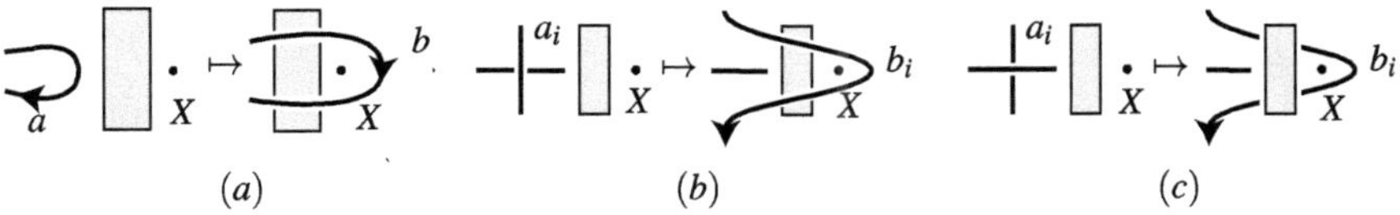

Fig. 4.5 Alexander's trick

Proof. Given an oriented link diagram D, choose a point X in the projection plane but not on a strand. The aim is to construct a link diagram such that, when we follow the orientation of the components of the link we always travel clockwise around the point X. It is easily seen, as indicated in Fig. 4.4, that any such link diagram is equivalent to the closure of a braid.

Now given D and X, travel around the diagram, following the orientation, and mark any arcs of D where we travel anticlockwise with respect to X. If a is such an arc of D and a contains no crossings, then replace a with an arc b as in Fig. 4.5a. The grey box in the Fig. 4.5 is used to denote an arbitrary part of the link diagram. If the arc a does contain crossings, then divide it into subarcs $a_1, a_2, \ldots, a_n$ such that each subarc a_i contains exactly one crossing. Now if a_i passes over some other arc of the link, then replace a_i with the arc b_i shown in Fig. 4.5b, and if a_i passes under some other arc of the link, then replace a_i with the arc b_i shown in Fig. 4.5c.

It is clear that the diagram obtained by this algorithm is equivalent to D and, by following the orientation of the link, we travel clockwise around X, as required. □

Example 4.13. An example of the use of Alexander's trick is given in Fig. 4.6. From it we see that the figure-of-eight knot can be presented by a braid $\sigma_2 \circ \sigma_1 \circ \sigma_1 \circ \sigma_2^{-1} \circ \sigma_2^{-1} \circ \sigma_1^{-1} \in \mathfrak{B}_4$. (Of course it can be presented by other braids.) The first transformation wraps "★" around X, and the second is an application of RI and RII at "★".

We now know that every link gives rise to a braid and every braid gives rise to a link. However, to any given link we can associate many different braids. For example, Figs. 4.4 and 4.6 show that the figure-of-eight knot can be presented as the closure of braids $\sigma_1 \circ \sigma_2^{-1} \circ \sigma_1 \circ \sigma_2^{-1} \in \mathfrak{B}_3$ and $\sigma_2 \circ \sigma_1 \circ \sigma_1 \circ \sigma_2^{-1} \circ \sigma_2^{-1} \circ \sigma_1^{-1} \in \mathfrak{B}_3$. Moreover, a link may be represented by braids with a differing number of strands. For example, for each n, the closure of $\sigma_1 \circ \sigma_2 \circ \cdots \circ \sigma_{n-1} \in \mathfrak{B}_n$ is the unknot. These observations imply that

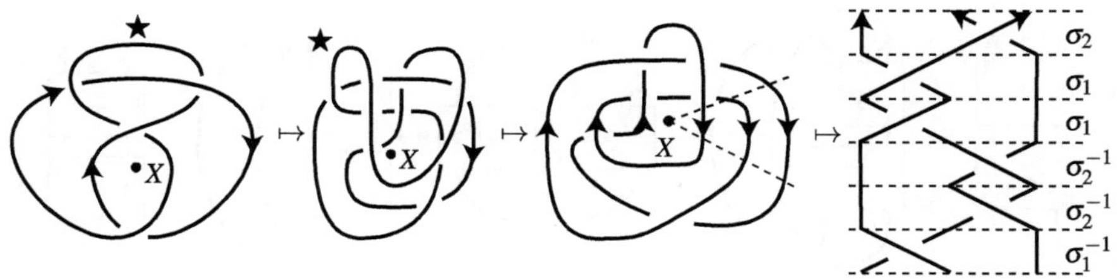

Fig. 4.6 Figure-of-eight knot and a corresponding braid

$$\frac{\{\text{braids}\}}{\text{equivalence}} \not\cong \frac{\{\text{oriented links}\}}{\text{equivalence}}.$$

However, by considering braids up to equivalence and two additional relations, called the Markov moves, we can use braids to obtain a description of the space of oriented links up to their equivalence.

The *Markov moves* act on braids, and there are two of them, which we denote by MI and MII. They are called (unsurprisingly) the *First Markov move* and the *Second Markov move* and are defined diagrammatically as follows.

Definition 4.14 (Markov moves).

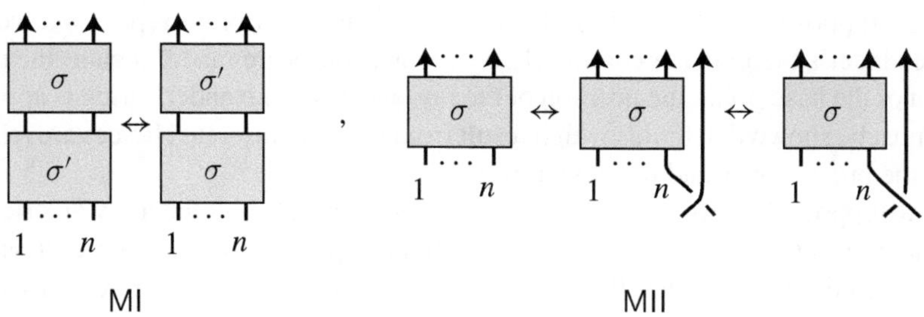

The Markov moves may increase, decrease or leave unchanged the number of strands in a braid.

Theorem 4.15 (Markov's Theorem). *The closure of two braid diagrams σ and σ' represent equivalent links if and only if there is a finite sequence of braid moves and Markov moves taking σ to σ'. That is*

$$\frac{\{\textit{oriented links}\}}{\textit{isotopy}} \cong \frac{\{\textit{braid diagrams}\}}{\textit{braid moves, Markov moves}}. \tag{4.4}$$

In one direction, the theorem is easily seen to be true. If σ and σ' are related by a braid move, then it is readily seen that their closures are related by a Reidemeister move. If σ and σ' are related by an MI-move, then we can use planar isotopy to "slide" σ round the closure component (indicated by the upside-down σ in the middle diagram) as indicated below.

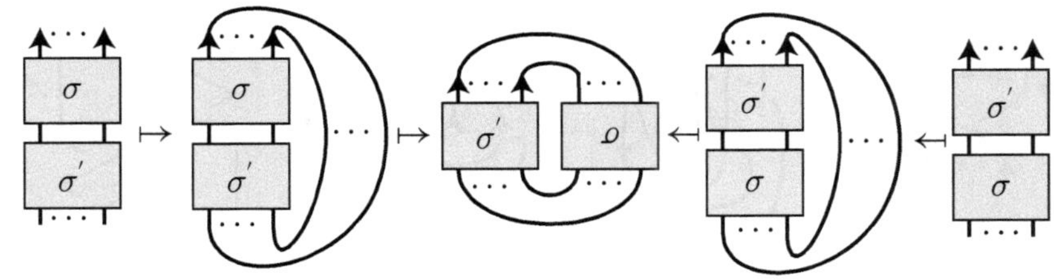

If σ and σ' are related by an MII-move, then their closures are related by an RI-move through

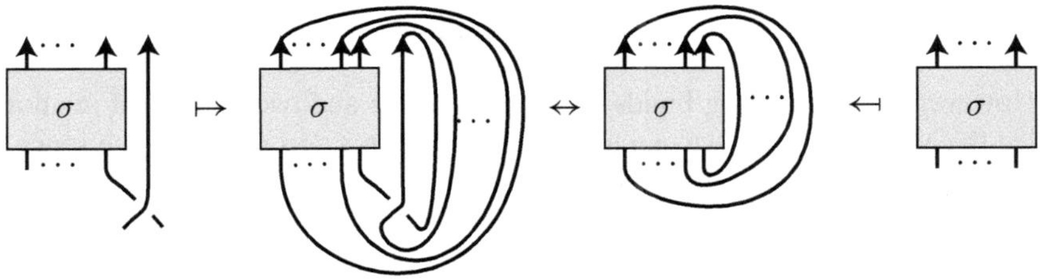

Proving the other direction of the theorem, however, is much more intricate. A proof may be given along the following lines.

- First, suppose that D is a link diagram. In obtaining a braid representative *via* Alexander's theorem (Theorem 4.12), various choices are made, such as the position of the base point, the position of the ray, and how Alexander's trick is applied. It must be shown that braids which result from making different choices are related by the braid moves and Markov moves.
- Next suppose that two oriented link diagrams D and D' are related by an oriented Reidemeister move. Determine how the braid representatives of these diagrams are related and show that this relation is a consequence of the braid moves and Markov moves.

Proving Theorem 4.15 in this way is a case analysis that is entirely diagrammatic. However, it is long and involved, and we therefore omit it. The reader who is interested in working through the details may find them in [22].

The Markov moves can also be described algebraically in terms of the generators of the braid group through a tensor product of braids.

Definition 4.16 (Tensor product of braids). Let σ be an n-stranded braid diagram and σ' be an m-stranded braid diagram. Then the *tensor product* $\sigma \otimes \sigma'$ of σ and σ' is the $(m+n)$-stranded braid obtained by juxtaposing σ' to the right of σ. It is given diagrammatically by

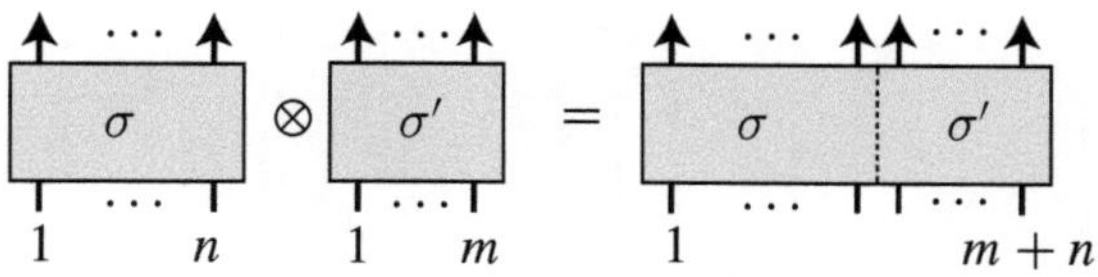

For example,

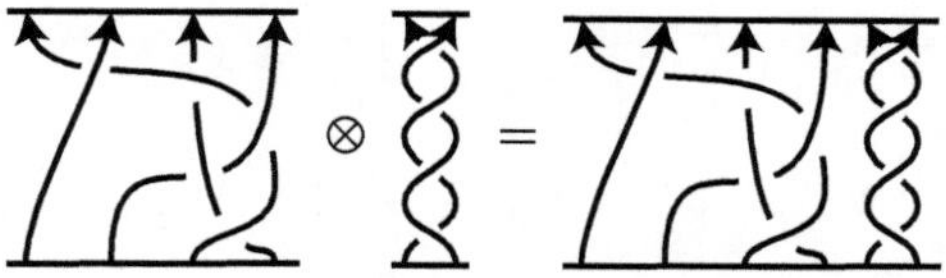

The Markov moves may now be described in terms of braid generators:

$$
\begin{aligned}
&\mathsf{MI}: &\sigma \circ \sigma' = \sigma' \circ \sigma, \qquad &\text{for all } \sigma, \sigma' \in \mathfrak{B}_n,\\
&\mathsf{MII}: &(\sigma \otimes 1) \circ \sigma_n^{\pm 1} = \sigma, \qquad &\text{for all } \sigma \in \mathfrak{B}_n.
\end{aligned}
\tag{4.5}
$$

Part II
Quantum Knot Invariants

Chapter 5
R-Matrix Representations of $\mathfrak{B}_n$

Our ultimate goal in Part II is the construction of a class of link invariants known as Reshetikhin–Turaev invariants, or, more generally for the type of knot invariant described here, *quantum invariants* [156, 157, 180]. Through the theory of quantum groups, the Reshetikhin–Turaev invariants provide a way to construct a link invariant from any semi-simple Lie algebra and any representation of it. We will build up to the Reshetikhin–Turaev invariants slowly (they are defined in Chap. 9), starting with the construction, due to Turaev [180], of some link invariants through representations of the braid group. Our strategy is to construct link invariants by decomposing link diagrams into "elementary" pieces and associating an algebraic object with these pieces. We then develop the conditions required on these mappings to give rise to an invariant of the original diagram.

5.1 A Map on Braid Diagrams

Using the results of Sect. 4.2, we can use braid composition to express an n-stranded braid diagram σ as a product of braid generators:

$\sigma_i :=$ 

, $\sigma_i^{-1} :=$ 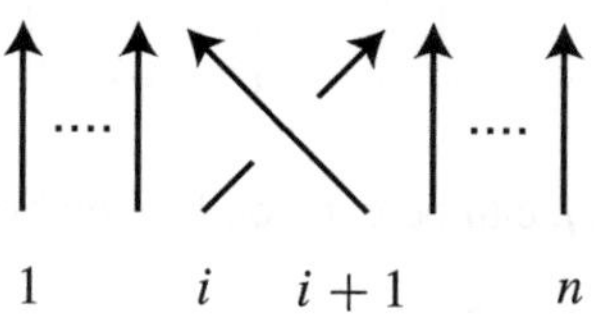

,

and use the tensor product to express each generator in terms of the "*elementary pieces*" 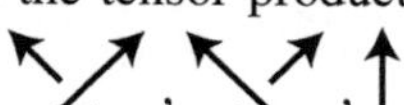 . Thus, we can use the operations of composition and tensor product to express any braid in terms of these three elementary pieces. As an example, where $n = 3$, consider the braid decomposition

D. M. Jackson and I. Moffatt, *An Introduction to Quantum and Vassiliev Knot Invariants*, CMS Books in Mathematics, https://doi.org/10.1007/978-3-030-05213-3_5

$\sigma =$

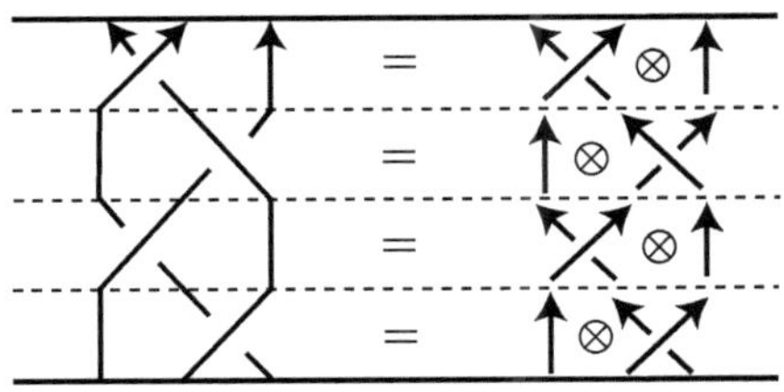

of σ. Then σ may be encoded as

$$\sigma = \sigma_1 \circ \sigma_2^{-1} \circ \sigma_1 \circ \sigma_2$$

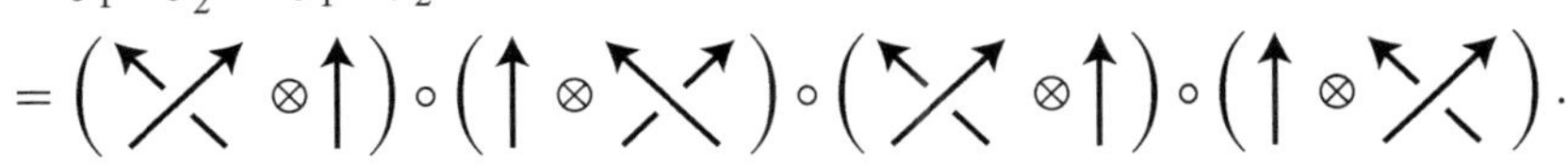

Recall that $\mathrm{Aut}(W)$ denotes the set of invertible linear maps from a space W to itself, and $\mathrm{End}(W)$ denotes the set of all linear maps from W to itself. Given a decomposition of the braid σ, we can now construct a linear map ρ from σ by associating a linear map with each of the elementary pieces as follows:

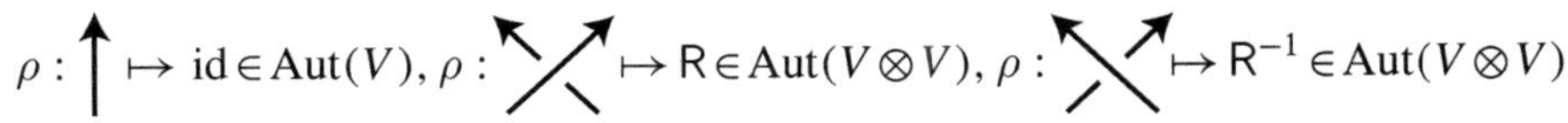

for some $\mathsf{R} \in \mathrm{Aut}(V \otimes V)$. For the moment, we shall take V to be a vector space, but the choice of V will be discussed in more detail shortly. (A reader unfamiliar with the tensor product $\otimes$ many find it helpful to review the material in Appendix A.)

Using the tensor product, we can extend ρ_{R} to the generators of the braid group:

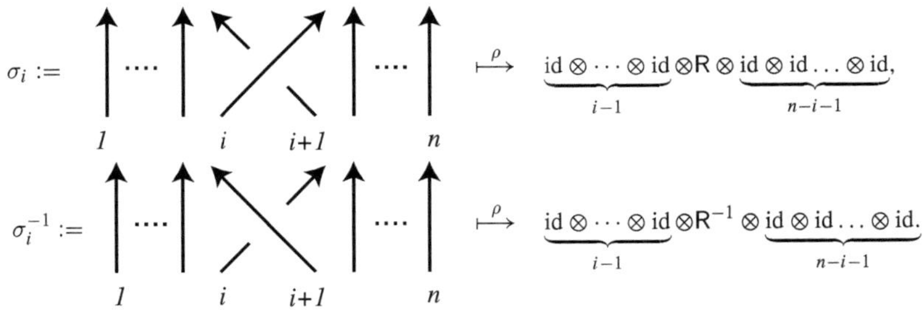

Then ρ can be extended to the braid σ, homomorphically, by setting

$$\rho(\sigma'' \circ \sigma') = \rho(\sigma'') \circ \rho(\sigma'),$$

for any pair of n-stranded braids σ' and σ'':

ρ: 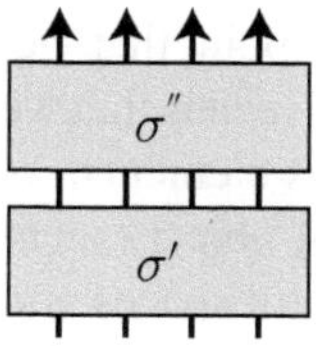 $\longmapsto \rho(\sigma'') \circ \rho(\sigma')$.

Returning to σ in the above example, we see the associations

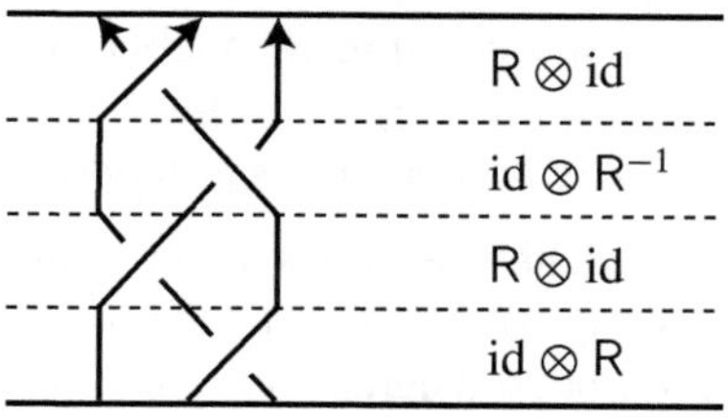

(5.1)

and so, in this case, that the map associated with σ is

$$\begin{aligned}\rho(\sigma) &= \rho(\sigma_1 \circ \sigma_2^{-1} \circ \sigma_1 \circ \sigma_2) = \rho(\sigma_1) \circ \rho(\sigma_2^{-1}) \circ \rho(\sigma_1) \circ \rho(\sigma_2) \\ &= (\mathsf{R} \otimes \mathrm{id}) \circ (\mathrm{id} \otimes \mathsf{R}^{-1}) \circ (\mathsf{R} \otimes \mathrm{id}) \circ (\mathrm{id} \otimes \mathsf{R}) \in \mathrm{Aut}(V \otimes V \otimes V).\end{aligned}$$

Exercise 5.1. Let V be a 2-dimensional vector space over $\mathbb{C}$ with basis $\{e_1, e_2\}$. The vector space $V \otimes V$ is 4-dimensional with basis $\{e_1 \otimes e_1, e_1 \otimes e_2, e_2 \otimes e_1, e_2 \otimes e_2\}$. Suppose R is defined by its action on basis elements *via* $e_1 \otimes e_1 \mapsto e_1 \otimes e_1$, $e_1 \otimes e_2 \mapsto -e_2 \otimes e_1$, $e_2 \otimes e_1 \mapsto -e_1 \otimes e_2$, and $e_2 \otimes e_2 \mapsto e_2 \otimes e_2$. Write down a basis of $V \otimes V \otimes V$, and compute the action of $\rho(\sigma)$ on this basis, where σ is the braid in (5.1). Finally describe the map as a matrix.

Let us now turn our attention to the space V. Since we intend to use linear maps, the obvious choice is to take V to be a vector space so ρ is a vector space automorphism. However, we want to allow scalars to be polynomials or Laurent polynomials so we need to work a little more generally and take V to be a free module of finite rank.

A few words may be helpful for a reader unfamiliar with modules, who might also like to review Appendix A. Here we shall focus upon how to think about modules, rather than upon what they are. Briefly, a module can be thought of as a vector space with scalars in a ring R rather than in a field. Since our module is "free", we can choose a basis $\{e_1, e_2, \ldots, e_n\}$ and represent every element as a linear combination $\lambda_1 e_1 + \lambda_2 e_2 + \ldots + \lambda_n e_n$, where the λ_i are in R. The "finite rank" condition means that every basis has the same (finite) number of elements, which, in contrast to vector spaces, is not always true for modules. Thus, in essence, our choice of V as a free module of finite rank means that we can work with V in terms of bases just as we do with vector spaces, but it will enable us to take scalars in rings such as $\mathbb{C}[t, t^{-1}]$, rather than just fields such as $\mathbb{C}$. In fact, it will do little harm for the reader to think of V as a vector space.

It is perhaps also worth mentioning at this point that we work with vector spaces and modules as freely generated objects. This means that we deal with their elements only as "formal" linear combinations of basis elements, $\lambda_1 e_1 + \lambda_2 e_2 + \ldots + \lambda_n e_n$, without ever realising the e_i as some type of object (such as a vector in $\mathbb{R}^n$). This is a quite different way of thinking about vector spaces and linear algebra from that usually met in undergraduate studies. Readers meeting this way of thinking for the first time might find it comforting to view it as saying that all vector spaces of the same dimension are isomorphic, so it does not matter which particular vector space or which particular basis we choose, and so we might as well represent a dimension n space by choosing the symbols $\{e_1, e_2, \ldots, e_n\}$ to serve as a basis. *It is the maps that act on the spaces that are of paramount importance.*

Returning to ρ, it is immediately seen that when σ and σ' are related by level- and boundary-preserving isotopy of $\mathbb{R} \times [0, 1]$, then $\rho(\sigma) = \rho(\sigma')$. This means that we can extend the map ρ on braid diagrams to a map on diagrams considered up to level- and boundary-preserving isotopy of $\mathbb{R} \times [0, 1]$. Thus, we can make the following definition.

Definition 5.2 (Map ρ_{R}). Let V be a free module of finite rank and $\mathsf{R} \in \operatorname{Aut}(V \otimes V)$. Then ρ_{R} is the map

$$\rho_{\mathsf{R}} : \frac{\{n\text{-stranded braid diagrams}\}}{\text{level and boundary preserving isotopy}} \longrightarrow \operatorname{End}(V^{\otimes n})$$

defined as the multiplicative and tensorial extension (*i.e.* that $\rho(\sigma \circ \sigma') = \rho(\sigma) \circ \rho(\sigma')$ and $\rho(\sigma \otimes \sigma') = \rho(\sigma) \otimes \rho(\sigma')$) of the operations

$\rho :$  $\mapsto \mathrm{id}$, $\rho :$  $\mapsto \mathsf{R}$, $\rho :$  $\mapsto \mathsf{R}^{-1}$.

Example 5.3. We shall work out ρ_{R} in detail when V is a free module of rank 1. Let $\{e_0\}$ be a basis of V. Then $V \otimes V$ is a one-dimensional vector space with basis $\{e_0 \otimes e_0\}$ and so (with respect to this basis) $\mathsf{R} \in \operatorname{Aut}(V \otimes V)$ is represented by a 1×1 matrix $[a]$. Since R is an automorphism, the scalar a must have an inverse and R^{-1} is represented by $[a^{-1}]$.

Given an n-stranded braid σ, we can express it in terms of the generators of the braid group, writing $\sigma = \sigma_{\iota_1}^{\delta_1} \sigma_{\iota_2}^{\delta_2} \cdots \sigma_{\iota_p}^{\delta_p}$, where $\delta_i = \pm 1$. Then

$$\rho(\sigma_i) = \mathrm{id}^{\otimes(i-1)} \otimes \mathsf{R} \otimes \mathrm{id}^{\otimes(n-i-1)} \in \operatorname{End}(V^{\otimes n}).$$

Since V has basis $\{e_0\}$, its n-fold tensor product $V^{\otimes n}$ has basis $\{e_0^{\otimes n}\}$ and so has rank 1. Thus, we can compute the action of $\rho(\sigma_i)$ by

$$\begin{aligned}\mathrm{id}^{\otimes(i-1)} \otimes \mathsf{R} \otimes \mathrm{id}^{\otimes(n-i-1)} : e_0^{\otimes n} \mapsto & \; e_0^{\otimes(i-1)} \otimes \mathsf{R}(e_0 \otimes e_0) \otimes e_0^{\otimes(n-i-1)} \\ &= e_0^{\otimes(i-1)} \otimes a(e_0 \otimes e_0) \otimes e_0^{\otimes(n-i-1)} \\ &= a\, e_0^{\otimes n},\end{aligned}$$

and so $\rho(\sigma_i)$ is represented by the matrix $[a]$. A similar computation shows that $\rho(\sigma_i^{-1})$ is represented by the matrix $[a^{-1}]$. Thus, $\rho(\sigma) = \rho(\sigma_{\iota_1}^{\delta_1}) \circ \rho(\sigma_{\iota_2}^{\delta_2}) \cdots \rho(\sigma_{\iota_p}^{\delta_p})$ is represented by the matrix product $[a^{\delta_1}] \cdot [a^{\delta_2}] \cdots [a^{\delta_p}] = [a^{\delta_1+\delta_2+\cdots+\delta_p}]$. Then, since δ_i is 1 if the corresponding crossing is positive, and -1 if the corresponding crossing is negative, we see that

$$\rho_{\mathsf{R}}(\sigma) = a^{(\text{no. of positive crossings in } \sigma)-(\text{no. of negative crossings in } \sigma)}.$$

The exponent is recognised as the writhe $\omega(\sigma)$ of σ. Therefore, $\rho(\sigma)$ is a braid invariant (since $\omega(\sigma)$ is).

We can then obtain a (very simple) braid polynomial by selecting V to be the module over the ground ring $\mathbb{C}[t, t^{-1}]$ freely generated by $\{e_0\}$, so $V = \{p(t)e_0 : p(t) \in \mathbb{C}[t, t^{-1}]\}$, and $(r(t) \cdot p(t))e_0 + q(t)e_0 = (r(t) \cdot p(t) + q(t))e_0$, and $a = t$. By doing so, we have constructed the polynomial invariant of braids, $t^{\omega(\sigma)}$, *via* linear algebra.

The question is then whether other (more interesting) knot or braid polynomials may be constructed *via* linear algebra in the same way. This part of the book is devoted to showing that, indeed, they can.

5.2 Obtaining Invariants of Braids

We found in Example 5.3 that, in that special case, ρ_{R} defined a braid invariant (*i.e.* was independent of the braid moves of Definition 4.4) for any automorphism R of a free module of rank 1. That we obtained a braid invariant for every R was just a quirk of the fact that V had rank 1. Had we started with a free module of any other rank and an automorphism R of it, ρ_{R} might not have been a braid invariant. We turn to the question of determining under what conditions ρ_{R} will define a braid invariant. That is, we need to understand when the ρ_{R} of Definition 5.2 induces a map

$$\rho_{\mathsf{R}} : \frac{\{n\text{-strand braid diagrams}\}}{\text{braid moves}} \longrightarrow \mathrm{End}(V^{\otimes n}).$$

To do this, we need to find under what conditions $\rho_{\mathsf{R}}(\sigma) = \rho_{\mathsf{R}}(\sigma')$ holds when σ and σ' are related by one of the braid moves of Definition 4.4 (we already know that ρ_{R} is invariant under level- and boundary-preserving isotopy of $\mathbb{R} \times [0, 1]$).

It will be a little easier to work in the language of the braid group. By Theorem 4.9, we know that

$$\frac{\{n\text{-strand braid diagrams}\}}{\text{braid moves}} \cong \mathfrak{B}_n,$$

thus, we can instead determine under what conditions ρ_{R} gives rise to a map

$$\rho_{\mathsf{R}} : \mathfrak{B}_n \to \mathrm{End}(V^{\otimes n}).$$

By construction, we know that

$$\rho_{\mathsf{R}}(\sigma_i \circ \sigma_j) = \rho_{\mathsf{R}}(\sigma_i) \circ \rho_{\mathsf{R}}(\sigma_j) \quad \text{and} \quad \rho_{\mathsf{R}}(\sigma_i^{-1}) = (\rho_{\mathsf{R}}(\sigma_i))^{-1}.$$

This can be rephrased succinctly as saying that ρ_{R} is a *group representation* of the free group generated by $\sigma_1, \ldots, \sigma_{n-1}$.

It remains to determine when ρ_{R} is invariant under the two braid relations $\sigma_i \circ \sigma_j = \sigma_j \circ \sigma_i$ when $|i - j| \geq 2$, and $\sigma_i \circ \sigma_{i+1} \circ \sigma_i = \sigma_{i+1} \circ \sigma_i \circ \sigma_{i+1}$ for $i = 1, \ldots, n-2$.

If $n = 1$ or $n = 2$, then the set of braid relations is empty and so ρ_{R} vacuously satisfies them. Suppose then that $n \geq 3$. We look at the two relations one at a time. For the identity $\sigma_i \circ \sigma_j = \sigma_j \circ \sigma_i$, we assume without loss of generality (because the relation is symmetric in i and j) that $i < j$. Then

$$\begin{aligned}
\rho_{\mathsf{R}}(\sigma_i \circ \sigma_j) &= \rho_{\mathsf{R}}(\sigma_i) \circ \rho_{\mathsf{R}}(\sigma_j) \\
&= (\mathrm{id}^{\otimes(i-1)} \otimes \mathsf{R} \otimes \mathrm{id}^{\otimes(n-i-1)}) \circ (\mathrm{id}^{\otimes(j-1)} \otimes \mathsf{R} \otimes \mathrm{id}^{\otimes(n-j-1)}) \\
&= (\mathrm{id}^{\otimes(i-1)} \otimes \mathsf{R} \otimes \mathrm{id}^{\otimes(j-i-2)} \otimes \mathsf{R} \otimes \mathrm{id}^{\otimes(n-j-1)}) \qquad \text{since } |j-i| \geq 2 \\
&= (\mathrm{id}^{\otimes(j-1)} \otimes \mathsf{R} \otimes \mathrm{id}^{\otimes(n-j-1)}) \circ (\mathrm{id}^{\otimes(i-1)} \otimes \mathsf{R} \otimes \mathrm{id}^{\otimes(n-i-1)}) \\
&= \rho_{\mathsf{R}}(\sigma_j) \circ \rho_{\mathsf{R}}(\sigma_i) = \rho_{\mathsf{R}}(\sigma_j \circ \sigma_i).
\end{aligned}$$

Thus, $\rho_{\mathsf{R}}(\sigma_i \circ \sigma_j) = \rho_{\mathsf{R}}(\sigma_j \circ \sigma_i)$ and so ρ_{R} is invariant under the braid relation $\sigma_i \circ \sigma_j = \sigma_j \circ \sigma_i$ with no additional conditions.

Next we consider the braid relation $\sigma_i \circ \sigma_{i+1} \circ \sigma_i = \sigma_{i+1} \circ \sigma_i \circ \sigma_{i+1}$ for $i = 1, \ldots, n-2$. We need to determine under what conditions $\rho_{\mathsf{R}}(\sigma_i \circ \sigma_{i+1} \circ \sigma_i) = \rho_{\mathsf{R}}(\sigma_{i+1} \circ \sigma_i \circ \sigma_{i+1})$.

$$\begin{aligned}
\rho_{\mathsf{R}}(\sigma_i \circ \sigma_{i+1} \circ \sigma_i) &= \left(\mathrm{id}^{\otimes(i-1)} \otimes \mathsf{R} \otimes \mathrm{id}^{\otimes(n-i-1)}\right) \circ \left(\mathrm{id}^{\otimes i} \otimes \mathsf{R} \otimes \mathrm{id}^{\otimes(n-i-2)}\right) \\
&\qquad \circ \left(\mathrm{id}^{\otimes(i-1)} \otimes \mathsf{R} \otimes \mathrm{id}^{\otimes(n-i-1)}\right). \\
&= \mathrm{id}^{\otimes(i-1)} \otimes ((\mathsf{R} \otimes \mathrm{id}) \circ (\mathrm{id} \otimes \mathsf{R}) \circ (\mathsf{R} \otimes \mathrm{id})) \otimes \mathrm{id}^{\otimes(n-i-2)}.
\end{aligned}$$

Similarly,

$$\rho_{\mathsf{R}}(\sigma_{i+1} \circ \sigma_i \circ \sigma_{i+1}) = \mathrm{id}^{\otimes(i-1)} \otimes ((\mathrm{id} \otimes \mathsf{R}) \circ (\mathsf{R} \otimes \mathrm{id}) \circ (\mathrm{id} \otimes \mathsf{R})) \otimes \mathrm{id}^{\otimes(n-i-2)}.$$

Then the second condition is satisfied if and only if R satisfies the equation

$$(\mathrm{id} \otimes \mathsf{R}) \circ (\mathsf{R} \otimes \mathrm{id}) \circ (\mathrm{id} \otimes \mathsf{R}) = (\mathsf{R} \otimes \mathrm{id}) \circ (\mathrm{id} \otimes \mathsf{R}) \circ (\mathsf{R} \otimes \mathrm{id})$$

for R, which motivates the following definition.

Definition 5.4 (Yang–Baxter equation, and R-matrix). Let V be a free module of finite rank. The *Yang–Baxter equation* in $\mathrm{End}(V^{\otimes 3})$ is

$$(\mathrm{id} \otimes \mathsf{R}) \circ (\mathsf{R} \otimes \mathrm{id}) \circ (\mathrm{id} \otimes \mathsf{R}) = (\mathsf{R} \otimes \mathrm{id}) \circ (\mathrm{id} \otimes \mathsf{R}) \circ (\mathsf{R} \otimes \mathrm{id}).$$

An invertible solution, $\mathsf{R} \in \mathrm{Aut}(V \otimes V)$, of this equation is called an *R-matrix*.

A reader meeting it for the first time should not get "hung-up" on the use of the word "matrix", in the name R-matrix, when it is defined to be a module morphism, since matrices and the module morphisms are equivalent (see Sect. A.3). The term R-matrix is standard terminology for this particular module morphism throughout this area.

The Yang–Baxter equation was introduced independently by the physicists Yang [192] and Baxter [21]. It is important in the theory of exactly solved models in statistical mechanics and in the theory of completely integrable quantum systems. Finding solutions to the Yang–Baxter equation is a difficult problem, and the general form of a solution is unknown, although large families of solutions are known. For example, it can be shown that every complex semi-simple Lie algebra gives rise to an R-matrix. This is a surprising and substantial result and one that we discuss further in Chap. 8.

Exercise 5.5. Let $\mathsf{R} \in \mathrm{End}(V \otimes V)$ be an R-matrix and λ be an invertible scalar. The *twist map* is the map $\tau : a \otimes b \mapsto b \otimes a$. Show that $\lambda\mathsf{R}$, R^{-1} and $\tau \circ \mathsf{R} \circ \tau$ are also R-matrices.

So what we have just shown above is that if $\mathsf{R} \in \mathrm{End}(V \otimes V)$ is an R-matrix, then ρ_R satisfies all of the braid relations, and hence ρ_R defines a representation of the braid group. This is captured in the following theorem.

Theorem 5.6 (Braid group representation, ρ_{R}). *Let V be a free module of finite rank and let $\mathsf{R} \in \mathrm{End}(V \otimes V)$ be an R-matrix. Then the map*

$$\rho_{\mathsf{R}} : \mathfrak{B}_n \to \mathrm{End}(V^{\otimes n}) \colon \sigma_i \mapsto \mathrm{id}^{\otimes(i-1)} \otimes \mathsf{R} \otimes \mathrm{id}^{\otimes(n-i-1)}, \tag{5.2}$$

for $i = 1, \ldots, n-1$ (and extended multiplicatively) affords a representation of the braid group $\mathfrak{B}_n$.

Since the braid group describes braids considered up to isotopy (see Theorem 4.9) we have the following

Corollary 5.7. *Let V be a free module of finite rank and let R be an R-matrix. Then the map*

$$\rho_{\mathsf{R}} : \frac{\{n\text{-strand braid diagrams}\}}{\text{braid moves}} \to \mathrm{End}(V^{\otimes n})$$

defined as the multiplicative and tensorial extension (i.e. that $\rho(\sigma \circ \sigma') = \rho(\sigma) \circ \rho(\sigma')$ and $\rho(\sigma \otimes \sigma') = \rho(\sigma) \otimes \rho(\sigma')$) of the operations

$\rho :$  $\mapsto \mathrm{id} \in \mathrm{End}(V)$, $\rho :$  $\mapsto \mathsf{R} \in \mathrm{End}(V^{\otimes 2})$, $\rho :$  $\mapsto \mathsf{R}^{-1} \in \mathrm{End}(V^{\otimes 2})$

is well-defined and therefore a braid invariant.

5.3 Specific Representations of the Braid Group

The machinery is now in place for constructing some braid invariants. We do this by brute force in the case where the rank of V is 2. This computation will give an indication of the difficulty of obtaining solutions to the Yang–Baxter equation. Moreover , the solutions we obtain will be important later and give rise to non-trivial braid invariants.

Let V be a free module of rank 2 over the ring $\mathbb{C}[t, t^{-1}]$, and $\{e_0, e_1\}$ be a basis for V. The tensor product $V \otimes V$ is four dimensional with canonical ordered basis

$$\{e_0 \otimes e_0,\ e_0 \otimes e_1,\ e_1 \otimes e_0,\ e_1 \otimes e_1\}.$$

Our intention is to construct R-matrices of size 4×4. The ordering of the rows and columns of the matrix is the same as the ordering of the canonically ordered basis of $V \otimes V$. We can describe $\mathsf{R} : V \otimes V \to V \otimes V$ by its action $\mathsf{R} : e_i \otimes e_j \mapsto \sum_{k,l} R_{i,j}^{k,l} e_k \otimes e_l$ on the basis elements of V, which in matrix form is

$$\mathbf{R} := [R_{i,j}^{k,l}] = \begin{bmatrix} R_{00}^{00} & R_{01}^{00} & R_{10}^{00} & R_{11}^{00} \\ R_{00}^{01} & R_{01}^{01} & R_{10}^{01} & R_{11}^{01} \\ R_{00}^{10} & R_{01}^{10} & R_{10}^{10} & R_{11}^{10} \\ R_{00}^{11} & R_{01}^{11} & R_{10}^{11} & R_{11}^{11} \end{bmatrix}.$$

The lower suffices i, j index columns, while the upper suffices k, l index rows. Note that in this book, our convention is to omit or include the comma separating indices depending on which choice in a particular instance adds clarification.

We shall simplify matters by imposing the *charge conservation* condition on the R-matrix. It causes the matrix to have a direct sum decomposition, that restricts it, but makes it more convenient to work with.

Definition 5.8 (Charge conservation). The *charge conservation* condition on an R-matrix is $R_{i,j}^{k,l} = 0 \quad \text{if } i + j \neq k + l.$

After imposing the charge conservation condition, the 4 × 4 R-matrix has the form

$$\mathbf{R} := [R_{i,j}^{k,l}] = \begin{bmatrix} R_{00}^{00} & 0 & 0 & 0 \\ 0 & R_{01}^{01} & R_{10}^{01} & 0 \\ 0 & R_{01}^{10} & R_{10}^{10} & 0 \\ 0 & 0 & 0 & R_{11}^{11} \end{bmatrix} = \begin{bmatrix} a & 0 & 0 & 0 \\ 0 & b & c & 0 \\ 0 & d & e & 0 \\ 0 & 0 & 0 & f \end{bmatrix},$$

where the last equality is simply a change to a more convenient notation.

It remains to find the conditions on the entries of **R** which will make this an R-matrix; that is, a solution of the Yang–Baxter equation. We begin by observing that $V^{\otimes 3}$ has rank eight since V has rank two. Then

$$V^{\otimes 3} = \langle e_p \otimes e_q \otimes e_r \ : \ p, q, r \in \{0, 1\} \rangle = \bigoplus_{i=0}^{3} V_i$$

where, in terms of canonically ordered bases,

$$\begin{aligned} V_0 &= \langle e_0 \otimes e_0 \otimes e_0 \rangle \,, \\ V_1 &= \langle e_0 \otimes e_0 \otimes e_1, e_0 \otimes e_1 \otimes e_0, e_1 \otimes e_0 \otimes e_0 \rangle \,, \\ V_2 &= \langle e_0 \otimes e_1 \otimes e_1, e_1 \otimes e_0 \otimes e_1, e_1 \otimes e_1 \otimes e_0 \rangle \,, \\ V_3 &= \langle e_1 \otimes e_1 \otimes e_1 \rangle \,. \end{aligned}$$

The sum of the suffices of the constituent e_i's in each tensor in the indicated basis of V_k is k, and it is this that provides a grading for the direct sum decomposition of $V^{\otimes 3}$.

The maps $\mathsf{R} \otimes \mathrm{id}$ and $\mathrm{id} \otimes \mathsf{R}$ act on $V^{\otimes 3}$. Since we have imposed the condition that the R-matrix conserves charge, then both $\mathsf{R} \otimes \mathrm{id}$ and $\mathrm{id} \otimes \mathsf{R}$ respect the grading of $V^{\otimes 3}$. To see this, let $p + q + r = k$ so $e_p \otimes e_q \otimes e_r$ is a basis element of V_k. Then

$$\begin{aligned} (\mathsf{R} \otimes \mathrm{id})\,(e_p \otimes e_q \otimes e_r) = \mathsf{R}(e_p \otimes e_q) \otimes \mathrm{id}(e_r) &= \left(\sum_{i,j} \mathsf{R}_{p,q}^{i,j}\, e_i \otimes e_j \right) \otimes e_r \\ &= \left(\sum_{i+j=p+q} \mathsf{R}_{p,q}^{i,j}\, e_i \otimes e_j \right) \otimes e_r \quad \text{(charge conservation)} \\ &= \sum_{i+j+r=k} \mathsf{R}_{p,q}^{i,j}\, e_i \otimes e_j \otimes e_r \in V_k, \end{aligned} \tag{5.3}$$

so V_k is an $(\mathsf{R} \otimes \mathrm{id})_k$-invariant submodule of $V^{\otimes 3}$, where $(\mathsf{R} \otimes \mathrm{id})_k$ denotes the restriction $(\mathsf{R} \otimes \mathrm{id})|_{V_k}$ of $\mathsf{R} \otimes \mathrm{id}$ to V_k. Similarly, V_k is an $(\mathrm{id} \otimes \mathsf{R})_k$-invariant submodule of $V^{\otimes 3}$ since

$$(\mathrm{id} \otimes \mathsf{R})(e_p \otimes e_q \otimes e_r) = \sum_{i+j+p=k} R^{i,j}_{q,r}\, e_p \otimes e_i \otimes e_j \in V_k. \tag{5.4}$$

Next we determine the matrix of $(\mathsf{R} \otimes \mathrm{id})_k$ for $k = 0, \dots, 3$ by means of (5.3). This is a straightforward calculation, but it is instructive to see the details in this case.

For $(\mathsf{R} \otimes \mathrm{id})_1$

$$\begin{aligned}
(\mathsf{R} \otimes \mathrm{id})(e_0 \otimes e_0 \otimes e_1) &= \sum_{i+j=0} R^{i,j}_{0,0}\, e_i \otimes e_j \otimes e_1 = R^{0,0}_{0,0}\, e_0 \otimes e_0 \otimes e_1 = a(e_0 \otimes e_0 \otimes e_1),\\
(\mathsf{R} \otimes \mathrm{id})(e_0 \otimes e_1 \otimes e_0) &= \sum_{i+j=1} R^{i,j}_{0,1}\, e_i \otimes e_j \otimes e_0 = R^{0,1}_{0,1}\, e_0 \otimes e_1 \otimes e_0 + R^{1,0}_{0,1}\, e_1 \otimes e_0 \otimes e_0\\
&= b(e_0 \otimes e_1 \otimes e_0) + d(e_1 \otimes e_0 \otimes e_0),\\
(\mathsf{R} \otimes \mathrm{id})(e_1 \otimes e_0 \otimes e_0) &= \sum_{i+j=1} R^{i,j}_{1,0}\, e_i \otimes e_j \otimes e_0 = R^{0,1}_{1,0}\, e_0 \otimes e_1 \otimes e_0 + R^{1,0}_{1,0}\, e_1 \otimes e_0 \otimes e_0\\
&= c(e_0 \otimes e_1 \otimes e_0) + e(e_1 \otimes e_0 \otimes e_0),
\end{aligned}$$

so

$$(\mathsf{R} \otimes \mathrm{id})_1 = \begin{bmatrix} a & 0 & 0 \\ 0 & b & c \\ 0 & d & e \end{bmatrix}.$$

Exercise 5.9. Complete the case analysis by looking at the $(\mathsf{R} \otimes \mathrm{id})_0$, $(\mathsf{R} \otimes \mathrm{id})_2$ and $(\mathsf{R} \otimes \mathrm{id})_3$ cases. Verify that

$$(\mathsf{R} \otimes \mathrm{id})_0 = [a], \quad (\mathrm{id} \otimes \mathsf{R})_0 = [a], \quad (\mathsf{R} \otimes \mathrm{id})_3 = [f], \quad (\mathrm{id} \otimes \mathsf{R})_3 = [f],$$

$$(\mathrm{id} \otimes \mathsf{R})_1 = \begin{bmatrix} b & c & 0 \\ d & e & 0 \\ 0 & 0 & a \end{bmatrix}, \quad (\mathsf{R} \otimes \mathrm{id})_2 = \begin{bmatrix} b & c & 0 \\ d & e & 0 \\ 0 & 0 & f \end{bmatrix}, \quad \text{and} \quad (\mathrm{id} \otimes \mathsf{R})_2 = \begin{bmatrix} f & 0 & 0 \\ 0 & b & c \\ 0 & d & e \end{bmatrix}.$$

From the Yang–Baxter equation,

$$(\mathrm{id} \otimes \mathsf{R})_k \circ (\mathsf{R} \otimes \mathrm{id})_k \circ (\mathrm{id} \otimes \mathsf{R})_k = (\mathsf{R} \otimes \mathrm{id})_k \circ (\mathrm{id} \otimes \mathsf{R})_k \circ (\mathsf{R} \otimes \mathrm{id})_k$$

for $k = 0, \dots, 3$. Rewriting these as four matrix equations, we see that the cases $k = 0$ and $k = 3$ are clearly trivial, while the cases $k = 1$ and $k = 2$ yield, respectively,

$$\begin{bmatrix} ab^2 + bcd & abc + bce & ac^2 \\ abd + bde & acd + be^2 & ace \\ ad^2 & ade & a^2e \end{bmatrix} = \begin{bmatrix} a^2b & abc & ac^2 \\ abd & b^2e + acd & bce + ace \\ ad^2 & bde + ade & cde + ae^2 \end{bmatrix}$$

and

$$\begin{bmatrix} bf^2 & bcf & c^2 f \\ bdf & b^2 e + cdf & bce + cef \\ d^2 f & bde + def & cde + e^2 f \end{bmatrix} = \begin{bmatrix} b^2 f + bcd & bcf + bce & c^2 f \\ bdf + bde & cdf + be^2 & cef \\ d^2 f & def & ef^2 \end{bmatrix}.$$

Thus, **R** is an R-matrix precisely when its elements satisfy the following system of simultaneous equations:

$$\begin{aligned} b(ab + cd - a^2) &= 0, \\ b(bf + cd - f^2) &= 0, \\ e(cd + ae - a^2) &= 0, \\ e(cd + ef - f^2) &= 0, \end{aligned} \qquad \begin{aligned} bce &= 0, \\ bde &= 0, \\ be(e - b) &= 0. \end{aligned}$$

These equations are invariant under the interchange $b \leftrightarrow e$. By restricting these to the case when $e = 0$ and $b \neq 0$, the above system is equivalent to:

$$\begin{aligned} ab + cd - a^2 &= 0, \\ bf + cd - f^2 &= 0 \end{aligned}$$

whence, by eliminating b between these, we have the quadratic equation

$$(a - f)\left(f + \frac{cd}{a}\right) = 0$$

for f, giving the two solutions $f = a$ and $f = -cd/a$ corresponding, respectively, to the R-matrices

$$\mathbf{R}_1 = \begin{bmatrix} a & 0 & 0 & 0 \\ 0 & a - \frac{cd}{a} & c & 0 \\ 0 & d & 0 & 0 \\ 0 & 0 & 0 & a \end{bmatrix}, \quad \mathbf{R}_2 = \begin{bmatrix} a & 0 & 0 & 0 \\ 0 & a - \frac{cd}{a} & c & 0 \\ 0 & d & 0 & 0 \\ 0 & 0 & 0 & -\frac{cd}{a} \end{bmatrix}. \tag{5.5}$$

Of these two matrices, we shall confine attention to $\mathbf{R}_1$ and shall see that it is related to the Jones polynomial. (We note in passing that the matrix $\mathbf{R}_2$ is related to the Alexander polynomial through a different, but related, approach from that taken here for obtaining knot invariants from R-matrices. An interested reader can find details in [144].)

Chapter 6
Knot Invariants Through R-Matrix Representations of $\mathfrak{B}_n$

We have just constructed in Sect. 5.2 a braid invariant ρ_{R} by using an R-matrix to obtain a representation of the braid group, giving

$$\rho_{\mathsf{R}} : \frac{\{n\text{-stranded braid diagrams}\}}{\text{braid moves}} \longrightarrow \operatorname{End}(V^{\otimes n}).$$

Our aim is to use ρ_{R} to obtain a link invariant. For this we recall from Theorem 4.15 that

$$\frac{\{\text{oriented links}\}}{\text{isotopy}} \cong \frac{\{\text{braids}\}}{\text{braid moves, Markov moves}}$$

so we need to modify ρ_{R} in a way that makes it invariant under the two Markov moves, which we now do. The constructions and results that appear in this chapter are due to Turaev [180].

Throughout this chapter, we shall let V denote a free module of rank n over a commutative unital ring $\mathbb{K}$ with basis $\{e_1, e_2, \ldots, e_n\}$, and let R denote an R-matrix. We recall (see Sect. A.5) that the dual $V^\star$ of V is the free module of rank n with basis $\{e^1, e^2, \ldots, e^n\}$ where each $e^i : V \to \mathbb{K}$ is defined as the linear extension of $e^i(e_j) = \delta_{i,j}$, where $\delta_{i,j}$ is the Kronecker delta function.

6.1 The Diagrammatics of $\rho_{\mathsf{R}}(\sigma)$ and Braid Closure

We start with a diagrammatic presentation of $\rho_{\mathsf{R}}(\sigma)$. For this, let σ be a braid in $\mathfrak{B}_n$. Then $\rho_{\mathsf{R}}(\sigma) \in \operatorname{End}(V^{\otimes n})$. For us, one of the most fundamental and important properties of the tensor product is that, for free modules U and V of finite rank,

$$\operatorname{Hom}(U, V) \cong U^\star \otimes V. \tag{6.1}$$

If the action of $f \in \operatorname{Hom}(U, V)$ in terms of bases is $f : u_i \mapsto \sum_j f_i^j v_j$, then under this isomorphism f is identified with the tensor $\sum_{i,j} f_i^j\, u^i \otimes v_j$ in $U^\star \otimes V$.

D. M. Jackson and I. Moffatt, *An Introduction to Quantum and Vassiliev Knot Invariants*,
CMS Books in Mathematics, https://doi.org/10.1007/978-3-030-05213-3_6

Observe that when f is given as an element of $U^\star \otimes V$, its action on a basis element u_k can be recovered by contraction: $f : u_k \mapsto \sum_{i,j} f_i^j\, u^i(u_k) \otimes v_j = \sum_{i,j} f_i^j\, \delta_k^i \otimes v_j = \sum_j f_k^j\, 1_{\mathbb{K}} \otimes v_j = \sum_j f_k^j\, v_j$, where the last equality uses contraction (see Theorem A.26) to identify $\mathbb{K} \otimes V$ with V.

Exercise 6.1. Prove the isomorphism given in (6.1). (See Lemma A.36 for additional details.)

From (6.1), we see that $\rho_{\mathrm{R}}(\sigma)$ can be viewed as a tensor in $(V^\star)^{\otimes n} \otimes V^{\otimes n}$. We can represent this tensor as a box with n strands entering it at the bottom and n strands leaving at the top. The bottom ends of the strands are each labelled with $V^\star$, and the top ends with V. The diagrammatic representation is therefore

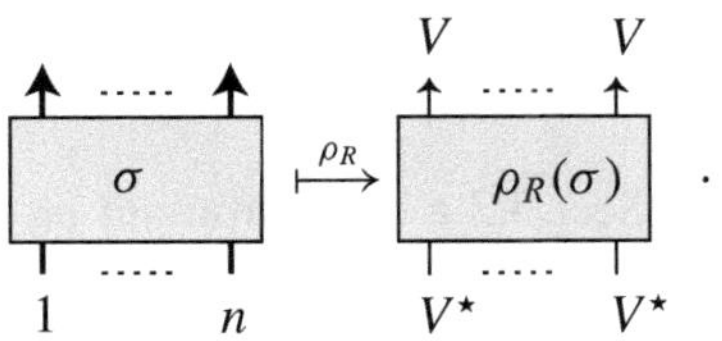

Then, to move from a braid diagram to a link diagram, its closure is taken:

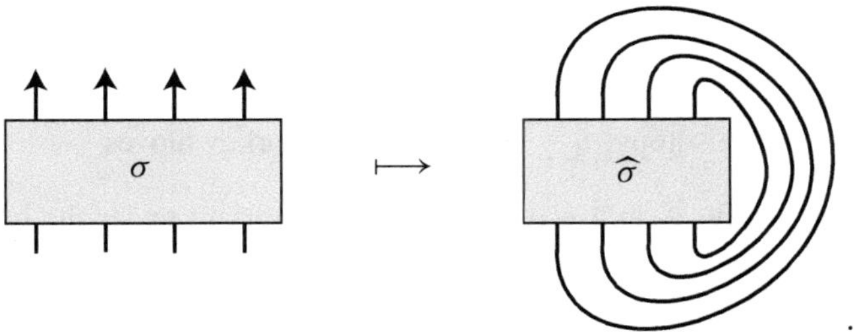

We therefore need an algebraic operation acting on $\rho_{\mathrm{R}}(\sigma) \in (V^\star)^{\otimes n} \otimes V^{\otimes n}$ that mimics the closure operation on the diagrammatic level (we often write ρ for ρ_{R} in diagrams):

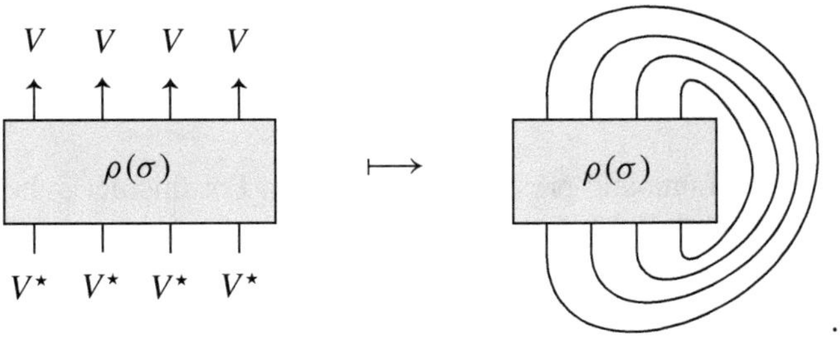

In terms of the diagrammatics, this closure arises from identifying an end marked V and an end marked $V^\star$. Such an identification of ends for an individual strand is achieved using the contraction map κ defined on its basis elements through

$$\kappa : V^\star \otimes V \longrightarrow \mathbb{K} : e^i \otimes e_j \longmapsto e^i(e_j), \tag{6.2}$$

so the algebraic version of closure of n strands is given by the map

$$\kappa : (V^\star)^{\otimes n} \otimes V^{\otimes n} \to \mathbb{K} : (e^{i_1} \otimes \cdots \otimes e^{i_n}) \otimes (e_{j_1} \otimes \cdots \otimes e_{j_n}) \mapsto e^{i_1}(e_{j_1}) \cdots e^{i_n}(e_{j_n}). \tag{6.3}$$

To identify this operation, suppose that W is a free module of rank m with basis $\{w_1, \dots, w_m\}$, dual basis $\{w^1, \dots, w^m\}$ where $w^i(w_j) = \delta_{i,j}$, and that $h \in \mathrm{End}(W)$ is given by $h(w_i) = \sum_{j=1}^m h_i^j w_j$. Then, recalling the action of the isomorphism from (6.1), as an element of $W^* \otimes W$, h is $\sum_{i,j=1}^m h_i^j w^i \otimes w_j$. Applying contraction κ to this element gives

$$\kappa\left(\sum_{i,j=1}^{i} h_i^j \, w^i \otimes w_j\right) = \sum_{i,j=1}^{m} h_i^j w^i(w_j) = \sum_{i=1}^{m} h_i^i.$$

On the other hand, as a matrix h is represented by $\mathbf{H}$ with (i, j)-entry h^i_j, and we see that $\sum_{i=1}^m h_i^i$ is precisely the trace of $\mathbf{H}$. (Recall that the *trace* of a matrix is the sum of its diagonal entries.) The result of this computation is that the trace of a linear map is exactly the composite map

$$\mathrm{Tr}\colon \mathrm{End}(W) \xrightarrow{\cong} W^\star \otimes W \xrightarrow{\kappa} \mathbb{K}. \tag{6.4}$$

Hence, the algebraic analogue of the closure of a braid is the trace.

6.2 Invariance Under the Markov Moves

By considering braid closure, we have just seen that in order to move to knots from braids we should consider $\mathrm{Tr}\,(\rho_{\mathsf{R}}(\sigma))$. We next need to determine under what circumstances $\mathrm{Tr} \circ \rho_{\mathsf{R}}$ is invariant under the Markov moves:

$$\begin{aligned} \mathsf{MI} &: \sigma \circ \sigma' = \sigma' \circ \sigma, && \text{for all } \sigma, \sigma' \in \mathfrak{B}_n, \\ \mathsf{MII} &: (\sigma \otimes 1) \circ \sigma_n^{\pm 1} = \sigma, && \text{for all } \sigma \in \mathfrak{B}_n. \end{aligned} \tag{6.5}$$

To satisfy the first Markov move, MI, we need to show that $\mathrm{Tr}(\rho_{\mathsf{R}}(\xi \circ \sigma)) = \mathrm{Tr}(\rho_{\mathsf{R}}(\sigma \circ \xi))$ for all $\sigma, \xi \in \mathfrak{B}_n$. This can be represented diagrammatically as

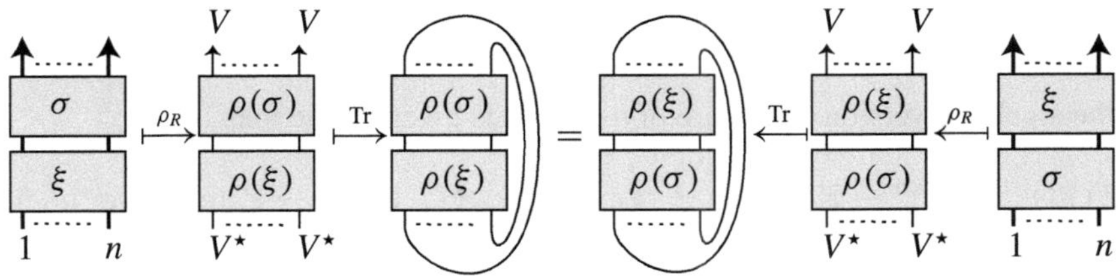

But this holds since, for linear maps, $f, g \in \mathrm{End}(V)$ we have $\mathrm{Tr}(f \circ g) = \mathrm{Tr}(g \circ f)$.

Exercise 6.2. Let $f \in \mathrm{Hom}(U, V)$ and $g \in \mathrm{Hom}(V, W)$ where f and g have actions on basis elements given by $f(u_i) = \sum_{j=1}^{m} f_i^j v_j$ and $g(v_i) = \sum_{j=1}^{l} g_i^j w_j$, where $\{u_1, \dots, u_n\}$, $\{v_1, \dots, v_m\}$ and $\{w_1, \dots, w_l\}$ are bases for U, V and W, respectively. The composite map $g \circ f$ is defined by

$$g \circ f(u_i) = \sum_{j=1}^{m} \sum_{k=1}^{l} f_i^j g_j^k w_k.$$

Write f as an element of $U^* \otimes V$, g as an element of $V^* \otimes W$, and $g \circ f$ as an element of $U^* \otimes W$. Verify that $g \circ f$ can be obtained from f and g by contracting of V and V^*, and hence verify that composition of functions is represented diagrammatically by stacking:

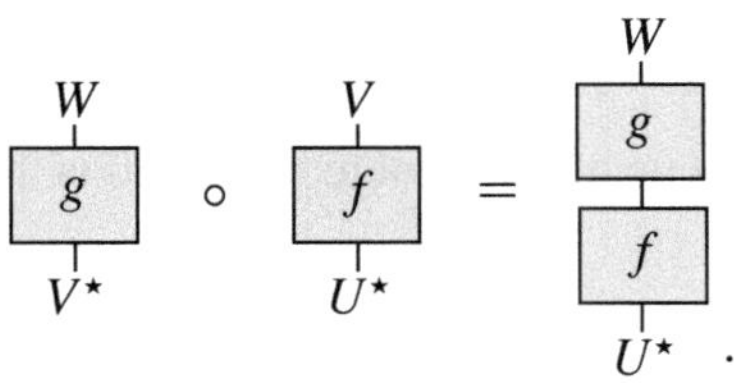

Exercise 6.3. Let $f \in \mathrm{Hom}(U, W)$ and $g \in \mathrm{Hom}(W, U)$. Verify that $\mathrm{Tr}(f \circ g) = \mathrm{Tr}(g \circ f)$ using the functional definition of trace from Eq. (6.4) as well at the matrix definition of trace.

From (6.5), the MII-move expressed in terms of braid group generators is

$$(\sigma \otimes 1) \circ (\sigma_n^{\pm 1}) = \sigma.$$

The representations of the left- and right-hand sides are $\rho_{\mathsf{R}}((\sigma \otimes 1) \circ (\sigma_n^{\pm 1}))$ and $\rho_{\mathsf{R}}(\sigma)$, respectively. Notice that one of these is in $\mathrm{End}(V^{\otimes(n+1)})$ and the other is in $\mathrm{End}(V^{\otimes n})$. Thus we need an operation that will enable us to eliminate the final tensor factor of $V^{\otimes(n+1)}$ in order to relate the two representations. Fortunately, we have just seen the operation of contraction, κ, that is able to do this. Thus we may start with $\rho_{\mathsf{R}}((\sigma \otimes 1) \circ (\sigma_n^{\pm 1})) \in \mathrm{End}(V^{\otimes(n+1)})$, rewrite it as an element of $(V^\star)^{\otimes(n+1)} \otimes V^{\otimes(n+1)}$, use contraction κ to eliminate the relevant tensor factors and then obtain an element of $(V^\star)^{\otimes n} \otimes V^{\otimes n} \cong \mathrm{End}(V^{\otimes n})$. Invariance under the MII-move requires that the map resulting from this process equals $\rho_{\mathsf{R}}(\sigma)$.

This procedure is represented diagrammatically by

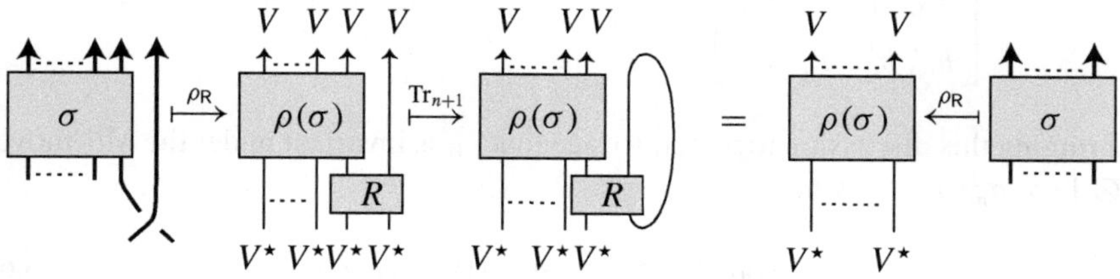

and by a similar diagram with the crossing reversed. Notice that contraction, illustrated in the third figure, closes the ends of the diagram in the $n+1$-st position.

Let us examine the closure operation used to move from the second to the third diagram. We have already seen that closure is the diagrammatic representation of the contraction map κ of (6.3), so we need to apply contraction to the final copies of V and $V^\star$:

$$(e^{i_1} \otimes \cdots \otimes e^{i_n}) \otimes e^{i_{n+1}} \otimes (e_{j_1} \otimes \cdots \otimes e_{j_n}) \otimes e_{j_{n+1}} \\ \mapsto e^{i_{n+1}}(e_{j_{n+1}}) \cdot (e^{i_1} \otimes \cdots \otimes e^{i_n}) \otimes (e_{j_1} \otimes \cdots \otimes e_{j_n}),$$

where in the right-hand side, we have used the fact that $e^{i_{n+1}}(e_{j_{n+1}})$ is a scalar to bring it to the front of the expression. What we have been led to is the *operator trace*

$$\mathrm{Tr}_{n+1} : \mathrm{End}(V^{\otimes(n+1)}) \to \mathrm{End}(V^{\otimes n}).$$

It can be alternatively defined by the following composition, which makes repeated use of (6.1):

$$\begin{aligned}\mathrm{Tr}_{n+1} : \mathrm{End}(V^{\otimes(n+1)}) &\cong \mathrm{End}(V^{\otimes n} \otimes V) \cong (V^{\otimes n} \otimes V)^\star \otimes (V^{\otimes n} \otimes V) \\ &\cong (V^{\otimes n})^\star \otimes V^\star \otimes (V^{\otimes n}) \otimes V \cong (V^{\otimes n})^\star \otimes V^\star \otimes V \otimes (V^{\otimes n}) \\ &\xrightarrow{\mathrm{id}^{\otimes n} \otimes \kappa \otimes \mathrm{id}^{\otimes n}} (V^{\otimes n})^\star \otimes \mathbb{K} \otimes (V^{\otimes n}) \cong (V^{\otimes n})^\star \otimes (V^{\otimes n}) \\ &\cong \mathrm{End}(V^{\otimes n}),\end{aligned}$$

where κ is as in (6.2).

Exercise 6.4. Let $h \in \mathrm{End}(V \otimes V)$ where $h(e_i \otimes e_j) = \sum_{k,l} h_{ij}^{kl} e_k \otimes e_l$. Show that the element-wise action of $\mathrm{Tr}_2(h)$ is $(\mathrm{Tr}_2(h))\,(e_i) = \sum_{j,k} h_{ij}^{kj} e_k$. Then verify that when V has rank 2, in terms of matrices

$$\mathbf{h} = \begin{bmatrix} h_{00}^{00} & h_{01}^{00} & h_{10}^{00} & h_{11}^{00} \\ h_{00}^{01} & h_{01}^{01} & h_{10}^{01} & h_{11}^{01} \\ h_{00}^{10} & h_{01}^{10} & h_{10}^{10} & h_{11}^{10} \\ h_{00}^{11} & h_{01}^{11} & h_{10}^{11} & h_{11}^{11} \end{bmatrix} \quad \text{and} \quad \mathrm{Tr}_2(\mathbf{h}) = \begin{bmatrix} h_{00}^{00} + h_{01}^{01} & h_{10}^{00} + h_{11}^{01} \\ h_{00}^{10} + h_{01}^{11} & h_{10}^{10} + h_{11}^{11} \end{bmatrix}.$$

Bringing this discussion together, we see that ρ_{R} is invariant under the MII-move, $(\sigma \otimes 1) \circ (\sigma_n^{\pm 1}) = \sigma$, when

$$\mathrm{Tr}_{n+1}\left(\rho_{\mathsf{R}}\left((\sigma \otimes 1) \circ (\sigma_n^{\pm 1})\right)\right) = \rho_{\mathsf{R}}(\sigma). \tag{6.6}$$

This is an awkward condition to work with, but the following exercises show that satisfying the much simpler identity

$$\mathrm{Tr}_2(\mathsf{R}^{\pm 1}) = \mathrm{id}_V \tag{6.7}$$

is sufficient to ensure invariance under the MII-move.

Exercise 6.5. Let $f \in \mathrm{End}(V^{\otimes n})$. Prove that

$$\mathrm{Tr}(\mathrm{Tr}_n(f)) = \mathrm{Tr}(f).$$

Exercise 6.6. Let $f \in \mathrm{End}(V^{\otimes(n+1)})$, $g \in \mathrm{End}(V^{\otimes n})$ and $h \in \mathrm{End}(V \otimes V)$. Show that the following identities hold.

1. $\mathrm{Tr}_{n+1}(f \circ (g \otimes \mathrm{id}_V)) = \mathrm{Tr}_{n+1}(f) \circ g$;
2. $\mathrm{Tr}_{n+1}((g \otimes \mathrm{id}_V) \circ f) = g \circ \mathrm{Tr}_{n+1}(f)$;
3. $\mathrm{Tr}_{n+1}(\mathrm{id}_V^{\otimes(n-1)} \otimes h) = \mathrm{id}_V^{\otimes(n-1)} \otimes \mathrm{Tr}_2(h)$.

Hence deduce that $\mathrm{Tr}_{n+1}\left(\rho_{\mathsf{R}}\left((\sigma \otimes 1) \circ (\sigma_n^{\pm 1})\right)\right) = \rho_{\mathsf{R}}(\sigma) \circ (\mathrm{id}^{\otimes(n-1)} \otimes \mathrm{Tr}_2(\mathsf{R}^{\pm 1}))$. Conclude that if $\mathsf{R}^{\pm 1}$ satisfies Eq. (6.7), then ρ_{R} is invariant under the MII-move.

What we have just shown in this section is that if $\mathsf{R} \in \mathrm{End}(V \otimes V)$ is an R-matrix such that $\mathrm{Tr}_2(\mathsf{R}^{\pm 1}) = \mathrm{id}_V$, then $\mathrm{Tr} \circ \rho_{\mathsf{R}}$ is invariant under the braid moves and the Markov moves, and hence defines a knot invariant. However, there is an objection to these invariants in that forcing the identity $\mathrm{Tr}_2(\mathsf{R}^{\pm 1}) = \mathrm{id}_V$ is too strong a condition as it does not allow for important classes of R-matrices (those coming from the theory of quantum groups) to be incorporated into the theory. This objection is resolved in the next section.

Exercise 6.7. Let R be the R-matrix $\mathbf{R}_1$ from (5.5). What form will R have if we insist that $\mathrm{Tr}_2(\mathsf{R}^{\pm 1}) = \mathrm{id}_V$?

6.3 The Link Invariant $\rho_{\mathsf{R},\mu}$

While ρ_R was easily seen to be invariant under the MI-move, condition (6.7) forcing the invariance of ρ_{R} under the MII-move imposes too strong a restriction on the form of the R-matrix. The problem is that in the algebra we have largely ignored the appearance of the closure elements in the diagrams, and, in particular, the maxima and minima these create. To rectify this, we therefore propose to record each maximum algebraically in the braid representation ρ_{R} by associating with each maximum–minimum pair an element $\mu \in \mathrm{End}(V)$, that is to be determined. This is achieved quite simply by replacing ρ_{R} with a new morphism $\rho_{\mathsf{R},\mu}$ defined in terms of ρ_{R} by

$$\rho_{\mathsf{R},\mu}(\sigma) := \mathrm{Tr}(\rho_{\mathsf{R}}(\sigma) \circ \mu^{\otimes n}), \tag{6.8}$$

for $\sigma \in \mathfrak{B}_n$. This may be represented diagrammatically by

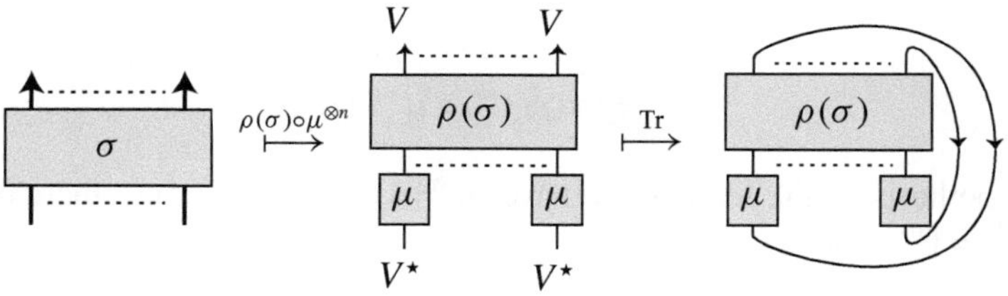

Although $\rho_{\mathsf{R},\mu}$ is not a representation of the braid group $\mathfrak{B}_n$, it is sufficient to our purpose that ρ_{R} is. We now determine the properties that μ must possess to ensure the invariance of $\rho_{\mathsf{R},\mu}$ under the two Markov moves, MI and MII. We shall see that, with this adjustment, this case leads to a powerful theorem (Theorem 6.10) for constructing knot invariants, including, by specialisation, the Jones polynomial and the HOMFLYPT polynomial.

For the MI-move, we need the identity $\rho_{\mathsf{R},\mu}(\sigma \circ \xi) = \rho_{\mathsf{R},\mu}(\xi \circ \sigma)$ to hold. Diagrammatically, this is

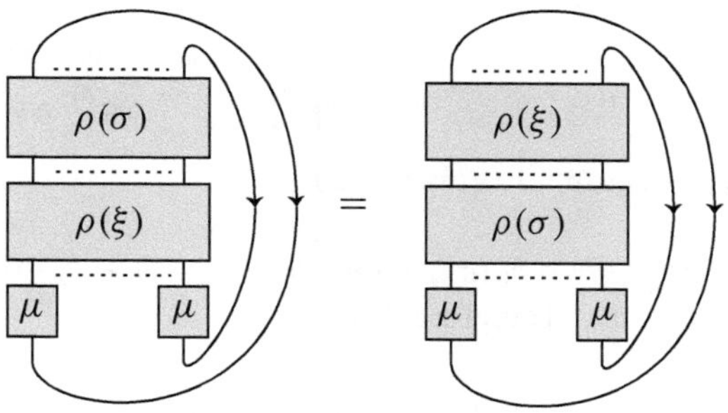

From the diagrammatics, we see that what we need to be able to do is to "slide" $\rho_{\mathsf{R}}(\sigma)$ around the diagram and through $\mu \otimes \cdots \otimes \mu$. Since $\rho_{\mathsf{R}}(\sigma)$ is a composite of expressions of the form $\mathrm{id}^{\otimes(i-1)} \otimes \mathsf{R}^{\pm 1} \otimes \mathrm{id}^{\otimes(n-i-1)}$, we should expect to be able to do this provided this expression commutes with $\mu \otimes \cdots \otimes \mu$, which we should be able to do provided $R^{\pm 1}$ commutes with $\mu \otimes \mu$. The following lemma shows that the intuition provided by the diagrammatics is indeed correct and provides a further simplification.

Lemma 6.8. *Let V be a free module of finite rank, $\mathsf{R} \in \mathrm{End}(V^{\otimes 2})$ be an R-matrix, and $\mu \in \mathrm{End}(V)$ such that $\mu^{\otimes 2} \circ \mathsf{R} = \mathsf{R} \circ \mu^{\otimes 2}$. Then*

$$\rho_{\mathsf{R},\mu}(\sigma \circ \xi) = \rho_{\mathsf{R},\mu}(\xi \circ \sigma),$$

for all $\sigma, \xi \in \mathfrak{B}_n$.

Proof. Using first the fact that ρ_{R} is a representation and then that the trace is invariant under cyclic permutations, we have

$$\begin{aligned} \rho_{\mathsf{R},\mu}(\sigma \circ \xi) &= \mathrm{Tr}(\rho_{\mathsf{R}}(\sigma \circ \xi) \circ \mu^{\otimes n}) && (6.9)\\ &= \mathrm{Tr}(\rho_{\mathsf{R}}(\sigma) \circ \rho_{\mathsf{R}}(\xi) \circ \mu^{\otimes n}) \\ &= \mathrm{Tr}(\rho_{\mathsf{R}}(\xi) \circ \mu^{\otimes n} \circ \rho_{\mathsf{R}}(\sigma)). && (6.10) \end{aligned}$$

By the hypothesis of the lemma, we have, $\mu^{\otimes 2} \circ \mathsf{R} = \mathsf{R} \circ \mu^{\otimes 2}$ and therefore also $\mu^{\otimes 2} \circ \mathsf{R}^{-1} = \mathsf{R}^{-1} \circ \mu^{\otimes 2}$ so

$$\mu^{\otimes 2} \circ \mathsf{R}^{\pm 1} = \mathsf{R}^{\pm 1} \circ \mu^{\otimes 2}. \tag{6.11}$$

For a braid generator $\sigma_i^{\pm 1} \in \mathfrak{B}_{n+1}$, we have

$$\begin{aligned} \rho_{\mathsf{R}}(\sigma_i^{\pm 1}) \circ \mu^{\otimes n} &= \left(\mathrm{id}^{\otimes(i-1)} \otimes \mathsf{R}^{\pm 1} \otimes \mathrm{id}^{\otimes(n-i-1)}\right) \circ \mu^{\otimes n} \qquad \text{(from (5.2))}\\ &= \left(\mathrm{id}^{\otimes(i-1)} \otimes \mathsf{R}^{\pm 1} \otimes \mathrm{id}^{\otimes(n-i-1)}\right) \circ \left(\mu^{\otimes(i-1)} \otimes \mu^{\otimes 2} \otimes \mu^{\otimes(n-i-1)}\right)\\ &= \left(\mathrm{id}^{\otimes(i-1)} \circ \mu^{\otimes(i-1)}\right) \otimes \left(\mathsf{R}^{\pm 1} \circ \mu^{\otimes 2}\right) \otimes \left(\mathrm{id}^{\otimes(n-i-1)} \circ \mu^{\otimes(n-i-1)}\right)\\ &= \left(\mu^{\otimes(i-1)} \circ \mathrm{id}^{\otimes(i-1)}\right) \otimes \left(\mu^{\otimes 2} \circ \mathsf{R}^{\pm 1}\right) \otimes \left(\mu^{\otimes(n-i-1)} \circ \mathrm{id}^{\otimes(n-i-1)}\right)\\ &\qquad \text{(commutation of id and } \mu, \text{ and (6.11))}\\ &= \left(\mu^{\otimes(i-1)} \otimes \mu^{\otimes 2} \otimes \mu^{\otimes(n-i-1)}\right) \circ \left(\mathrm{id}^{\otimes(i-1)} \otimes \mathsf{R}^{\pm 1} \otimes \mathrm{id}^{\otimes(n-i-1)}\right)\\ &= \mu^{\otimes n} \circ \rho_{\mathsf{R}}(\sigma_i^{\pm 1}) \qquad \text{(from (5.2))}. \end{aligned}$$

Thus $\rho_{\mathsf{R}}(\sigma_i^{\pm 1}) \circ \mu^{\otimes n} = \mu^{\otimes n} \circ \rho_{\mathsf{R}}(\sigma_i^{\pm 1})$ for $i = 1, \ldots, n-1$, so $\rho_{\mathsf{R}}(\sigma) \circ \mu^{\otimes n} = \mu^{\otimes n} \circ \rho_{\mathsf{R}}(\sigma)$ for all $\sigma \in \mathfrak{B}_n$. Therefore, from (6.9),

$$\begin{aligned} \rho_{\mathsf{R},\mu}(\sigma \circ \xi) &= \mathrm{Tr}(\rho_{\mathsf{R}}(\xi) \circ \mu^{\otimes n} \circ \rho_{\mathsf{R}}(\sigma)) = \mathrm{Tr}(\rho_{\mathsf{R}}(\xi) \circ \rho_{\mathsf{R}}(\sigma) \circ \mu^{\otimes n})\\ &= \mathrm{Tr}(\rho_{\mathsf{R}}(\xi \circ \sigma) \circ \mu^{\otimes n}) = \rho_{\mathsf{R},\mu}(\xi \circ \sigma), \end{aligned}$$

where the last equality is by (6.8), completing the proof. □

For the MII-move, we can proceed as before. From the diagrammatics

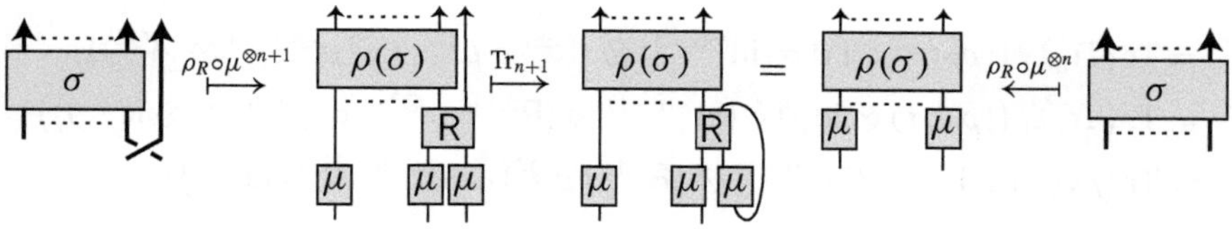

It is clear that we need to prove the identity

$$\mathrm{Tr}_{n+1}\left(\rho_{\mathsf{R}}\left((\sigma\otimes 1)\circ(\sigma_n^{\pm 1})\right)\otimes\mu^{\otimes(n+1)}\right)=\rho_{\mathsf{R}}(\sigma)\otimes\mu^{\otimes n}.$$

But we notice by comparing the third and fourth diagrams that this holds if 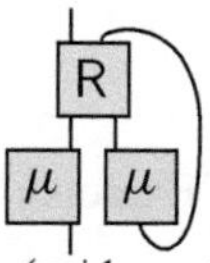 = , which is the diagrammatic version of the identity $\mathrm{Tr}_2\left(\mathsf{R}^{\pm 1}\circ\mu^{\otimes 2}\right)=\mu$. The following result again tells us that the intuition provided by the diagrammatics is correct.

Lemma 6.9. *Let* $\mathsf{R}\in\mathrm{End}(V^{\otimes 2})$ *be an R-matrix. If* $\mu\in\mathrm{End}(V)$ *is such that*

$$\mathrm{Tr}_2\left(\mathsf{R}^{\pm 1}\circ\mu^{\otimes 2}\right)=\mu, \tag{6.12}$$

then $\rho_{\mathsf{R},\mu}\left((\sigma\otimes 1)\circ\sigma_n^{\pm 1}\right)=\rho_{\mathsf{R},\mu}(\sigma)$, *for all* $\sigma\in\mathfrak{B}_n$.

Proof. It is convenient to denote $\rho_{\mathsf{R},\mu}\left((\sigma\otimes 1)\circ\sigma_n^{\pm 1}\right)$ temporarily by A.

$$\begin{aligned}
A &:= \mathrm{Tr}\left(\rho_{\mathsf{R}}(\sigma\otimes 1)\circ\rho_{\mathsf{R}}(\sigma_n^{\pm 1})\circ\mu^{\otimes(n+1)}\right)\\
&= \mathrm{Tr}\left((\rho_{\mathsf{R}}(\sigma\otimes 1))\circ(\mathrm{id}^{\otimes(n-1)}\otimes\mathsf{R}^{\pm 1})\circ\mu^{\otimes(n+1)}\right) && \text{(by (5.2))}\\
&= \mathrm{Tr}\left(\mathrm{Tr}_{n+1}\left((\rho_{\mathsf{R}}(\sigma\otimes 1))\circ(\mathrm{id}^{\otimes(n-1)}\otimes\mathsf{R}^{\pm 1})\circ\mu^{\otimes(n+1)}\right)\right) && \text{(Ex. 6.5).}
\end{aligned}$$

But, rearranging $\mu^{\otimes(n+1)}$ for compatibility with composition of tensors,

$$\begin{aligned}
\mu^{\otimes(n+1)} &= \mu^{\otimes(n-1)}\otimes\mu^{\otimes 2}\\
&= (\mathrm{id}^{\otimes(n-1)}\circ\mu^{\otimes(n-1)})\otimes(\mu^{\otimes 2}\circ\mathrm{id}^{\otimes 2})\\
&= (\mathrm{id}^{\otimes(n-1)}\otimes\mu^{\otimes 2})\circ(\mu^{\otimes(n-1)}\otimes\mathrm{id}^{\otimes 2}).
\end{aligned}$$

Continuing with the simplification of A,

$$
\begin{aligned}
A &= \mathrm{Tr}\left(\mathrm{Tr}_{n+1}\left((\rho_{\mathsf{R}}(\sigma \otimes 1)) \circ (\mathrm{id}^{\otimes(n-1)} \otimes \mathsf{R}^{\pm 1}) \circ (\mathrm{id}^{\otimes(n-1)} \otimes \mu^{\otimes 2})\right.\right.\\
&\qquad\qquad\qquad\qquad\qquad\qquad\qquad\qquad\qquad \left.\left.\circ(\mu^{\otimes(n-1)} \otimes \mathrm{id}^{\otimes 2})\right)\right)\\
&= \mathrm{Tr}\left(\mathrm{Tr}_{n+1}\left((\rho_{\mathsf{R}}(\sigma \otimes 1)) \circ (\mathrm{id}^{\otimes(n-1)} \otimes (\mathsf{R}^{\pm 1} \circ \mu^{\otimes 2})) \circ (\mu^{\otimes(n-1)} \otimes \mathrm{id}^{\otimes 2})\right)\right)\\
&= \mathrm{Tr}\left(\mathrm{Tr}_{n+1}\left((\rho_{\mathsf{R}}(\sigma) \otimes \mathrm{id}_V) \circ (\mathrm{id}^{\otimes(n-1)} \otimes (\mathsf{R}^{\pm 1} \circ \mu^{\otimes 2})) \circ (\mu^{\otimes(n-1)} \otimes \mathrm{id}^{\otimes 2})\right)\right)\\
&= \mathrm{Tr}\left(\rho_{\mathsf{R}}(\sigma) \circ \mathrm{Tr}_{n+1}\left((\mathrm{id}^{\otimes(n-1)} \otimes (\mathsf{R}^{\pm 1} \circ \mu^{\otimes 2})) \circ (\mu^{\otimes(n-1)} \otimes \mathrm{id}^{\otimes 2})\right)\right)\\
&= \mathrm{Tr}\left(\rho_{\mathsf{R}}(\sigma) \circ (\mathrm{Tr}_{n+1}(\mathrm{id}^{\otimes(n-1)} \otimes (\mathsf{R}^{\pm 1} \circ \mu^{\otimes 2}))) \circ (\mu^{\otimes(n-1)} \otimes \mathrm{id})\right)\\
&= \mathrm{Tr}\left(\rho_{\mathsf{R}}(\sigma) \circ (\mathrm{id}^{\otimes(n-1)} \otimes \mathrm{Tr}_2(\mathsf{R}^{\pm 1} \circ \mu^{\otimes 2})) \circ (\mu^{\otimes(n-1)} \otimes \mathrm{id})\right)\\
&= \mathrm{Tr}\left(\rho_{\mathsf{R}}(\sigma) \circ (\mathrm{id}^{\otimes(n-1)} \otimes \mu) \circ (\mu^{\otimes(n-1)} \otimes \mathrm{id})\right)\\
&= \mathrm{Tr}\left(\rho_{\mathsf{R}}(\sigma) \circ (\mu^{\otimes(n-1)} \otimes \mu)\right)\\
&= \mathrm{Tr}\left(\rho_{\mathsf{R}}(\sigma) \circ \mu^{\otimes n}\right)\\
&= \rho_{\mathsf{R},\mu}(\sigma),
\end{aligned}
$$

where the fourth, fifth and sixth equalities are by items (2), (1) and (3) of Exercise 6.6, respectively. □

It follows that if $\mathrm{Tr}_2(\mathsf{R}^{\pm 1} \circ \mu^{\otimes 2}) = \mu$ then $\rho_{\mathsf{R},\mu}$ is invariant under the MII-move. We summarise the content of the previous two lemmas in the following result.

Theorem 6.10. *If an R-matrix* $\mathsf{R} \in \mathrm{End}(V \otimes V)$ *and* $\mu \in \mathrm{End}(V)$ *are such that*

$$(i)\quad \mu^{\otimes 2} \circ \mathsf{R} = \mathsf{R} \circ \mu^{\otimes 2}, \quad \text{and} \quad (ii)\quad \mathrm{Tr}_2(\mathsf{R}^{\pm 1} \circ \mu^{\otimes 2}) = \mu,$$

then $\rho_{\mathsf{R},\mu}(\sigma) := \mathrm{Tr}(\rho_{\mathsf{R}}(\sigma) \circ \mu^{\otimes n})$ *is invariant under both Markov moves.*

Proof. Lemmas 6.8 and 6.9 show that, under the above conditions, $\rho_{\mathsf{R},\mu}$ is invariant under the Markov moves. □

Exercise 6.11. Let R be an R-matrix, and let $\mathsf{R} \in \mathrm{End}(V \otimes V)$ and $\mu \in \mathrm{End}(V)$ be such that $(\mu \otimes \mu) \circ \mathsf{R} = \mathsf{R} \circ (\mu \otimes \mu)$ and $\mathrm{Tr}_2\left(\mathsf{R}^{\pm 1} \circ (\mu \otimes \mu)\right) = \alpha^{\pm 1} \cdot \beta \cdot \mu$ for some $\alpha, \beta \in \mathbb{K}$ where $\alpha, \beta \neq 0$. Show that

$$\rho_{(\mathsf{R},\,\mu,\,\alpha,\,\beta)}(\sigma) := \alpha^{-\omega(\sigma)} \beta^{-n}\, \mathrm{Tr}\left(\rho_{\mathsf{R}}(\sigma) \circ \mu^{\otimes n}\right),$$

where $\sigma \in \mathfrak{B}_n$ and ω is the writhe, is invariant under the Markov moves. Deduce that it defines a link invariant. Show that $\rho_{(\mathsf{R},\,\mu,\,\alpha,\,\beta)} = \rho_{(\alpha^{-1}\mathsf{R},\,\beta^{-1}\mu)}$.

In the light of Theorem 6.10, we can make the following definition.

Definition 6.12 (Invariant $\rho_{\mathsf{R},\mu}$). Let L be an oriented link or an oriented link diagram. Let V be a free module of finite rank, and R and μ be as in Theorem 6.10. Then $\rho_{\mathsf{R},\mu}$ is the link invariant defined by

$$\rho_{\mathsf{R},\mu}(L) := \rho_{\mathsf{R},\mu}(\sigma),$$

where σ is any braid whose closure is equivalent to (a diagram of) L.

Theorem 6.10 (together with Definition 6.12) is the culmination of the first part of this chapter, and it will be used shortly to prove the existence of the Jones polynomial. To make it easier to use, we shall now restate it in a more self-contained form. In fact, the restatement of this theorem will serve a second purpose, namely that of providing at least a partial summary of the route we have taken to arrive at this knot invariant.

Theorem 6.13. *Let L be an oriented link, and let σ an n-stranded braid whose closure is a link equivalent to L. Let V be a free module of finite rank with basis $\{e_1, \ldots, e_n\}$. For $f \in \mathrm{End}(V \otimes V)$, let $f(e_i \otimes e_j) = \sum_{k,l} f_{i,j}^{k,l} e_k \otimes e_l$, and let the operator trace $\mathrm{Tr}_2 \colon \mathrm{End}(V \otimes V) \to \mathrm{End}(V)$ be*

$$\mathrm{Tr}_2(f) \colon e_i \mapsto \sum_{j,k} f_{i,j}^{k,j} e_k.$$

Let $\mathsf{R} \in \mathrm{End}(V \otimes V)$ be an invertible solution of the Yang–Baxter equation

$$(\mathrm{id} \otimes \mathsf{R}) \circ (\mathsf{R} \otimes \mathrm{id}) \circ (\mathrm{id} \otimes \mathsf{R}) = (\mathsf{R} \otimes \mathrm{id}) \circ (\mathrm{id} \otimes \mathsf{R}) \circ (\mathsf{R} \otimes \mathrm{id}),$$

and let $\mu \in \mathrm{End}(V)$ be such that the conditions

$$\mu^{\otimes 2} \circ \mathsf{R} = \mathsf{R} \circ \mu^{\otimes 2} \quad \text{and} \quad \mathrm{Tr}_2(\mathsf{R}^{\pm 1} \circ \mu^{\otimes 2}) = \mu$$

hold.

Let $\rho_{\mathsf{R}} \colon \mathfrak{B}_n \to \mathrm{End}(V^{\otimes n})$ be the representation of $\mathfrak{B}_n$ defined on the braid generators of $\mathfrak{B}_n$ by $\rho_{\mathsf{R}} \colon \sigma_i \mapsto \mathrm{id}^{\otimes(i-1)} \otimes \mathsf{R} \otimes \mathrm{id}^{\otimes(n-i-1)}$ and let

$$\rho_{\mathsf{R},\mu}(\sigma) := \mathrm{Tr}\left(\rho_{\mathsf{R}}(\sigma) \circ \mu^{\otimes n}\right).$$

Then $\rho_{\mathsf{R},\mu}$ is an invariant of the link L.

6.4 Two Polynomial Invariants

Having developed the invariant $\rho_{\mathsf{R},\mu}$, we need to illustrate that it provides a worthwhile theory. That is, we need to show that it does indeed produce interesting invariants. We do this by showing that the Jones polynomial and HOMFLYPT polynomial arise from this framework. This is a significant result since we have not yet shown that the HOMFLYPT polynomial actually exists!

6.4.1 The Jones Polynomial

Let V be a free module of rank 2 over $\mathbb{C}[t^{\frac{1}{2}}, t^{-\frac{1}{2}}]$. Let $\mathsf{R} \in \mathrm{Aut}(V \otimes V)$ and $\mu \in \mathrm{End}(V)$ be the maps given by the matrices

$$\mathbf{R} = \begin{bmatrix} -t^{\frac{1}{2}} & 0 & 0 & 0 \\ 0 & -t^{\frac{1}{2}} + t^{\frac{3}{2}} & -t & 0 \\ 0 & -t & 0 & 0 \\ 0 & 0 & 0 & -t^{\frac{1}{2}} \end{bmatrix} \text{ and } \mu = \begin{bmatrix} -t^{\frac{1}{2}} & 0 \\ 0 & -t^{-\frac{1}{2}} \end{bmatrix}. \tag{6.13}$$

By (5.5), we know that R is an R-matrix.

Exercise 6.14. Verify that the identities $\mu^{\otimes 2} \circ \mathsf{R} = \mathsf{R} \circ \mu^{\otimes 2}$ and $\mathrm{Tr}_2(\mathsf{R}^{\pm 1} \circ \mu^{\otimes 2}) = \mu$ hold.

By Exercise 6.14, R and μ satisfy the conditions required by Theorem 6.10, and so $\rho_{\mathsf{R},\mu}$ is a knot invariant.

Exercise 6.15. Let L be the (positive) Hopf link. Choose a braid representative σ of L (for example, $\sigma_1 \circ \sigma_1 \in \mathfrak{B}_2$). Verify that $\rho_{\mathsf{R},\mu}(L) = (-t^{\frac{1}{2}} - t^{-\frac{1}{2}})(-t^{\frac{5}{2}} - t^{\frac{1}{2}})$.

The reader may have noticed that Exercise 6.15 gives a relation between the $\rho_{\mathsf{R},\mu}$ and the Jones polynomial of the Hopf link from Example 2.23. As we shall now see, this is no coincidence.

Lemma 6.16. *Let* R *and* μ *be as in* (6.13). *Then the invariant* $\rho_{\mathsf{R},\mu}$ *satisfies the skein relations*

$$t^{-1}\rho_{\mathsf{R},\mu}(\sigma_+) - t\rho_{\mathsf{R},\mu}(\sigma_-) = (t^{\frac{1}{2}} - t^{-\frac{1}{2}})\rho_{\mathsf{R},\mu}(\sigma_0)$$

where the braids σ_+, σ_-, *and* σ_0 *are identical except in one region where they differ as shown in Fig. 6.1.*

Fig. 6.1 Local differences in braids

Proof. Because of their differing only at the single crossing, the braids σ_+, σ_-, and σ_0 may be presented in a common form by $\sigma' \circ \sigma_p \circ \sigma$, and $\sigma' \circ \sigma_p^{-1} \circ \sigma$, and $\sigma' \circ 1^{\otimes n} \circ \sigma$, where $p < n$, respectively. Since $\rho_{(\mathsf{R},\mu)} = \mathrm{Tr}(\rho_{\mathsf{R}} \circ \mu^{\otimes n})$ and ρ_{R} is a representation of the braid group $\mathfrak{B}_n$, to prove the lemma it is sufficient to show that

$$t^{-1}\rho_{\mathsf{R}}(\sigma_p) - t\rho_{\mathsf{R}}(\sigma_p^{-1}) = (t^{\frac{1}{2}} - t^{-\frac{1}{2}})\rho_{\mathsf{R}}(1^{\otimes n}).$$

But this is equivalent to showing that $t^{-1}\rho_{\mathsf{R}}(\tau) - t\rho_{\mathsf{R}}(\tau^{-1}) = (t^{\frac{1}{2}} - t^{-\frac{1}{2}})\rho_{\mathsf{R}}(1 \otimes 1)$, where τ is the generator of the braid group $\mathfrak{B}_2$ or, equivalently, that $t^{-1}\mathbf{R} - t\mathbf{R}^{-1} = (t^{\frac{1}{2}} - t^{-\frac{1}{2}})\mathbf{I}_4$, which is readily verified. □

Note that applications of the skein relation in Lemma 6.16 may not necessarily reduce a braid to a linear combination of trivial braids. To see this consider, for example, its application to the element σ_1 of $\mathfrak{B}_2$.

Theorem 6.17. *There exists an invariant of oriented links, denoted by $J(L)$, that is defined by the skein relations*

$$\mathbf{SJ1}: \; t^{-1}J(L_+) - tJ(L_-) = (t^{\frac{1}{2}} - t^{-\frac{1}{2}})J(L_0),$$
$$\mathbf{SJ2}: \; J(\mathscr{O}) = 1$$

where $\mathscr{O}$ is the unknot. Moreover,

$$\rho_{\mathsf{R},\mu}(\sigma) = \left(-t^{\frac{1}{2}} - t^{-\frac{1}{2}}\right) \cdot J(L).$$

where R *and* μ *are as in* (6.13). *The invariant* $J(L)$ *is the* Jones polynomial.

Proof. Define a function T on oriented link diagrams by $T(D) = \rho_{\mathsf{R},\mu}(\sigma)$ where σ is a braid with the property that its closure is equivalent to D under the oriented Reidemeister moves, and R and μ are as in (6.13). Such a braid σ exists by Theorem 4.12. Moreover, $T(D)$ is independent of the choice of σ and is a link invariant by Theorem 4.15.

Suppose that D_+, D_- and D_0 are three oriented link diagrams that are identical except in a small region where they differ as in Fig. 2.5. Let c denote the crossing of D_+ on whose neighbourhood D_- and D_0 differ. By planar isotopy (*e.g.*, by rotating the diagram), we may assume that both strands of D at the crossing c are oriented upwards. We shall apply Theorem 4.15 to D to obtain a braid representative σ_+. For this, choose a base point that is just to the right of the crossing c. Then the combing algorithm described in the proof of Theorem 4.15 will produce a braid in which c is still a positive crossing with both strands directed upwards. Applying exactly the same steps to the diagrams D_- and D_0, which differ from D_+ only in a small neighbourhood of c, will produce braid representatives σ_- and σ_0, respectively, for them. Furthermore, σ_+, σ_- and σ_0 are identical except in one region where they differ as shown in Fig. 6.1. It follows from Lemma 6.16 that

$$t^{-1}\rho_{\mathsf{R},\mu}(\sigma_+) - t\rho_{\mathsf{R},\mu}(\sigma_-) = (t^{\frac{1}{2}} - t^{-\frac{1}{2}})\rho_{\mathsf{R},\mu}(\sigma_0)$$

and hence

$$t^{-1}T(D_+) - tT(D_-) = (t^{\frac{1}{2}} - t^{-\frac{1}{2}})T(D_0).$$

Next, since $\rho_{\mathsf{R},\mu}(1) = -t^{\frac{1}{2}} - t^{-\frac{1}{2}}$ and $T(D)$ is an isotopy invariant, we have

$$T(\mathscr{O}) = -t^{\frac{1}{2}} - t^{-\frac{1}{2}},$$

where $\mathscr{O}$ is any diagram of the unknot.

From this, we see that the normalisation of $T(D)$ defined as

$$J(D) := \left(-t^{\frac{1}{2}} - t^{-\frac{1}{2}}\right)^{-1} \cdot T(D)$$

satisfies the skein relations of the theorem's statement.

We have just shown that there is an invariant of oriented links that satisfies the skein relations **SJ1** and **SJ2** from Definition 2.20. The theorem then follows since, as in Sect. 2.2, the invariant can be computed directly from these skein relations. □

Exercise 6.18. Let L be a link, let L' be the link obtained from L by reversing all of the crossings (so L' is the mirror image of L). Show that $J(L)(t) = J(L')(t^{-1})$ in two ways: (i) by using the definition $\rho_{\mathsf{R},\mu}$, and (ii) by using the skein relation.

Theorem 6.17 demonstrates that the Jones polynomial exists. We had already shown in Sect. 3.3 that the Jones polynomial exists *via* the Kauffman bracket, and so the significance of the proof of existence of the Jones polynomial may not have been apparent.

We shall now consider the HOMFLYPT polynomial and shall show that it can also be obtained as an invariant $\rho_{\mathsf{R},\mu}$, and hence exists. It is important for the reader to recall that we have not yet shown this.

6.4.2 The HOMFLYPT Polynomial

Let V be a free module of rank m over the ring $\mathbb{C}[q, q^{-1}]$. Consider the mapping $\mathsf{R} : V \otimes V \to V \otimes V$ given by $\mathsf{R} : e_i \otimes e_j \mapsto \sum_{k,l} R^{k,l}_{i,j} e_k \otimes e_l$, where

$$R^{k,l}_{i,j} = \begin{cases} q^{-m+1} & \text{if } i = j = k = l, \\ -q^{-m} & \text{if } i = l \neq k = j, \\ q^{-m}(q - q^{-1}) & \text{if } i = k < l = j, \\ 0 & otherwise. \end{cases}$$

Next let $\mu : V \to V$ be defined by $\mu : e_i \mapsto \sum_j \mu^i_j e_j$ where $\mu^i_i = q^{2i-m-1}$.

It can be verified that R is invertible and is a solution to the Yang–Baxter equation, and hence is an R-matrix. We shall omit this verification, but a sufficiently fortitudinous reader should be able to do this by brute force.

Exercise 6.19. Verify that $\rho_{\mathsf{R},\mu}(1) = (q^m - q^{-m})/(q - q^{-1})$, where $1 \in \mathfrak{B}_1$.

Exercise 6.20. Verify that $\mu^{\otimes 2} \circ R = R \circ \mu^{\otimes 2}$ and that $\mathrm{Tr}_2(R^{\pm 1} \circ \mu^{\otimes 2}) = \mu$.

For these values of R and μ, we therefore have by Theorem 6.10 that $\rho_{R,\mu}$ defines a link invariant.

Lemma 6.21. *Let* R *and* μ *be as above. Then invariant* $\rho_{\mathsf{R},\mu}$ *satisfies the skein relation*

$$q^m \rho_{\mathsf{R},\mu}(\sigma_+) - q^{-m} \rho_{\mathsf{R},\mu}(\sigma_-) = (q - q^{-1})\rho_{\mathsf{R},\mu}(\sigma_0),$$

where the braids σ_+, σ_- *and* σ_0 *are identical except in one region where they differ as shown in Fig. 6.1.*

Exercise 6.22. Prove Lemma 6.21 by adapting the proof of Lemma 6.16.

Theorem 6.23. *The invariant of oriented links* $\rho_{\mathsf{R},\mu}(L)$ *is uniquely defined by the skein relations*

$$q^m \rho_{\mathsf{R},\mu}(L_+) - q^{-m} \rho_{\mathsf{R},\mu}(L_-) = (q - q^{-1})\rho_{\mathsf{R},\mu}(L_0), \tag{6.15}$$

where L_+, L_- *and* L_0 *are identical except in the neighbourhood of a single crossing where they differ as in Fig. 2.5, and*

$$\rho_{\mathsf{R},\mu}(\mathcal{O}) = (q^m - q^{-m})/(q - q^{-1}). \tag{6.16}$$

Exercise 6.24. Prove Theorem 6.23 by adapting the proof of Theorem 6.17.

The next theorem establishes the existence of the HOMFLYPT polynomial.

Theorem 6.25. *There exists a knot invariant* P *defined by the skein relation*

$$xP(L_+) - x^{-1}P(L_-) = yP(L_0),$$

with initial condition $P(\mathcal{O}) = 1$.

Proof. Let L be the diagram of a n component link with u crossings. As mentioned in Exercise 1.19, every diagram can be turned into the diagram of an unlink by crossing changes. Therefore, by repeated applications of the skein relation (6.15), we see that the polynomial $\rho_{\mathsf{R},\mu}$ from Theorem 6.23 is the sum of polynomials of the form

$$\pm q^{me}(q - q^{-1})^f \rho_{\mathsf{R},\mu}(\mathcal{O}^d),$$

where $\mathcal{O}^d$ is the d component unlink, $e, f \in \mathbb{Z}$, $0 \le f \le u$, $0 \le d \le u + n$ and $|e| \le u$. By evaluating $\rho_{\mathsf{R},\mu}(\mathcal{O}^d)$, we see this is equal to

$$\pm q^{me}(q^m - q^{-m})^d (q - q^{-1})^{f-d}.$$

Now we would like the exponent of $(q - q^{-1})$ to be positive. To ensure that this happens, we multiply this polynomial by $(q - q^{-1})^{u+n}$ to get the expression

$$\pm q^{me}(q^m - q^{-m})^d(q - q^{-1})^{f+u+n-d}.$$

This implies that $(q - q^{-1})^{u+n}\rho_{\mathsf{R},\mu}$ is the sum of polynomials of this form.

Now noting that $f + u + n - d \leq f + u + n \leq 2u + n$ we see that if we choose

$$m > (2u + n) + (u + n) + u = 4u + 2n$$

this polynomial, and therefore $(q - q^{-1})^{u+n}\rho_{\mathsf{R},\mu}(L)$, can be uniquely written (since $2u + n < m/2$) in the form

$$\sum_{a,b\in\mathbb{Z}} r_{a,b}q^{a+mb},$$

where

- $r_{a,b} \in \mathbb{Z}$;
- $r_{a,b} = 0$ for $|a| > 2u + n$;
- $r_{a,b}$ is independent of the choice of $m > 4u + 2n$.

Now we may set

$$N(L) := (q - q^{-1})^{-u-n} \cdot \sum_{a,b\in\mathbb{Z}} r_{a,b}q^a t^b,$$

where the $r_{a,b,}$ are as above. Then, since $\rho_{\mathsf{R},\mu}$ is an isotopy invariant by Theorem 6.23, we know that the two variable polynomial $N(L)$ must also be an isotopy invariant. Also since $\rho_{\mathsf{R},\mu}$ satisfies the skein relations (6.15) we have that

$$tN(L_+) - t^{-1}N(L_-) = (q - q^{-1})N(L_0),$$

and

$$N(\mathcal{O}) = (t - t^{-1})/(q - q^{-1}).$$

Finally substituting $t = x$, $y = (q - q^{-1})$ in $N(L)$ and multiplying the resulting polynomial by $y/(x - x^{-1})$ recovers the HOMFLYPT polynomial of Definition 2.27 required by the theorem. □

Chapter 7
Operator Invariants

We constructed in Chap. 6 an isotopy invariant $\rho_{(\mathsf{R},\mu)}$ of links. There, Alexander's Theorem 4.12 was used to represent a link L as a closed braid and the value of $\rho_{(R,\mu)}$ was then computed from the representative braid. However, the requirement that we compute the link invariant by first finding a braid representation of the link is limiting (e.g. the theory cannot be applied to framed links). In this chapter, we extend the idea of decomposing a braid diagram into elementary pieces that we used to construct $\rho_{(R,\mu)}$ from braid diagrams, to link diagrams and, in doing so, to give a class of link invariants called *operator invariants*. The invariants and results in this chapter are due to Freyd and Yetter [65], and Turaev [181].

Since operator invariants will be seen to arise very naturally as a generalisation of the construction of $\rho_{(\mathsf{R},\mu)}$, we begin by reviewing the main steps that we used to obtain $\rho_{(\mathsf{R},\mu)}$ in the first place. We then consider what is required to extend this construction so that it may be applied directly to a link diagram.

Step 1: Find a braid $\sigma \in \mathfrak{B}_n$ representing L so that its closure $\widehat{\sigma}$ is a diagram for L.

Step 2: Write the braid in terms of the elementary pieces  using composition and tensor products.

Step 3: Associate an endomorphism to each elementary piece according to the rules

$\rho\Big($  $\Big) = \mathsf{R}$, $\rho\Big($  $\Big) = \mathsf{R}^{-1}$, $\rho\Big($  $\Big) = \mathrm{id}$.

Step 4: Associate an endomorphism $\rho(\sigma) \in \mathrm{End}(V^{\otimes n})$ with σ using the rules

$$\rho(s \otimes t) = \rho(s) \otimes \rho(t) \quad \text{and} \quad \rho(s \circ t) = \rho(s) \circ \rho(t).$$

Step 5: For invariance under the braid moves and Markov moves, normalise $\rho(\sigma)$, take the trace,

$$\rho_{(\mathsf{R},\mu)} = \mathrm{Tr}\left(\rho(\sigma) \circ \mu^{\otimes n}\right),$$

D. M. Jackson and I. Moffatt, *An Introduction to Quantum and Vassiliev Knot Invariants*, CMS Books in Mathematics, https://doi.org/10.1007/978-3-030-05213-3_7

and insist that R is an R-matrix and that $\mu^{\otimes 2} \circ \mathsf{R} = \mathsf{R} \circ \mu^{\otimes 2}$ and $\mathrm{Tr}_2(\mathsf{R}^{\pm 1} \circ \mu^{\otimes 2}) = \mu$.

Of these five steps, the salient one is the construction of the endomorphism $\rho(\sigma)$ in Steps 3 and 4 by defining ρ on elementary pieces. This is the idea that we carry to link diagrams. Thus, our programme involves:

- cutting up a link diagram into "elementary pieces";
- assigning endomorphisms to elementary pieces;
- obtaining a map from link diagrams by performing algebraic operations that are the counterparts of the diagrammatic operations for assembling a knotted diagram;
- obtaining an invariant by imposing conditions on the maps determined by equivalence of the knotted objects.

The first matter to address is what we mean by an "elementary piece" of a link diagram. Notice that if we start "cutting up" a link diagram, for example, by slicing it across the horizontal (*i.e.* "parallel" to the boundary), then what results is no longer a link diagram. Thus to decompose link diagrams, we need to consider a new type of knotted object, called a *tangle*.

7.1 Tangles and Tangle Diagrams

7.1.1 The Definition of a Tangle

Tangles simultaneously generalise the concepts of a link and a braid. Informally, a tangle consists of a box, strands and circles—the ends of strands are attached to the top or bottom of the box, and the strand and circles may be knotted. Below is a tangle diagram.

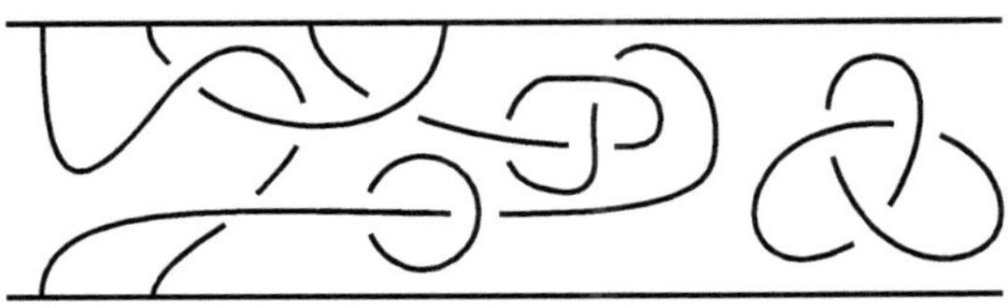

(7.1)

This idea is formalised by the following, where ∂ is used to denote the boundary of an object.

Definition 7.1 (Tangle). An (m, n)*-tangle* T consists of a disjoint union of copies of the unit interval $I = [0, 1]$ and unit circle $\mathbb{S}^1$, together called the *components* of the tangle, embedded into $\mathbb{R}^2 \times I$ such that

1. $\partial(T) = \{(i, 0, 0) : i = 1, \ldots, m\} \sqcup \{(i, 0, 1) : i = 1, \ldots, n\}$.
2. $T \cap \partial(\mathbb{R}^2 \times I) = \partial(T)$.

An (m, n)-tangle is said to be *oriented* if every component is endowed with an orientation.

Condition (2) prevents any intersections between the strands and the boundary of the ambient space $\mathbb{R}^2 \times I$ (*i.e.* the "box") except for the endpoints of the copies of intervals I. Note that knots and links are examples of $(0, 0)$-tangles. The object in (7.1) is a (2,4)-tangle with five components.

The development of the basic theory of tangles mirrors that of knots and links given in Chap. 1. Knots are considered up to isotopy of the ambient space $\mathbb{R}^3$. Tangles, however, are considered up to isotopy of the ambient space $\mathbb{R}^2 \times I$ that fixes the boundary. This means that the ends of any interval components of a tangle are in a fixed position and, informally, two tangles are equivalent if we can continuously deform one into the other by "pushing" the tangle about inside the ambient space $\mathbb{R}^2 \times I$. For example,

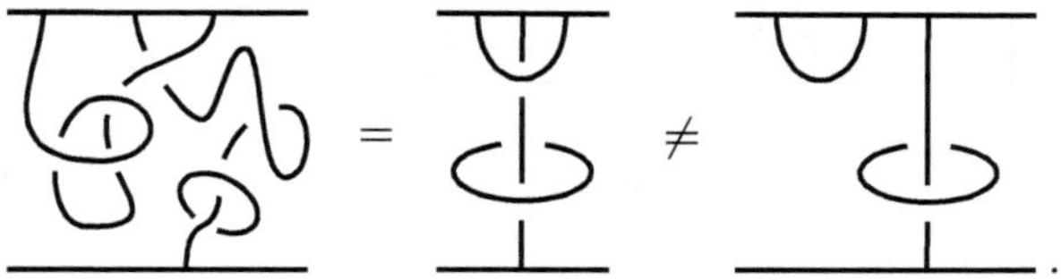

Formally, we have the following.

Definition 7.2 (Equivalence of tangles). Two (oriented) tangles T and T' are said to be *equivalent* if there is a boundary-preserving and orientation-preserving isotopy of $\mathbb{R} \times I$ taking T to T'. That is, there exists a family of homeomorphisms

$$h_t \colon \mathbb{R}^2 \times I \to \mathbb{R}^2 \times I, \qquad t \in [0, 1],$$

such that

1. $h_t|_{\mathbb{R}\times\{0,1\}} = \text{id}$,
2. $h_0 = \text{id}$,
3. $h_1(T) = T'$,
4. $(x, t) \mapsto (h_t(x), t)$ defines a homeomorphism from $\mathbb{R}^2 \times I^2$ to itself.

The first condition of this definition fixes the endpoints of the tangle. It is worth noting that equivalence of tangles, unlike that of braids, allows the creation and annihilation of maxima and minima.

As with braids and links, we consider only *tame* tangles, namely tangles that are isotopic to a polygonal tangle. The tameness condition allows us to reformulate the definition of equivalence in terms of Δ-moves as in Sect. 1.1, with the consequence that two tangles are equivalent if and only if there is a finite sequence of Δ- and Δ^{-1}-moves taking (a polygonal representation of) one to (a polygonal representation of) the other.

As with knots, links and braids, we work with tangles *via* their diagrams. Given an (m, n)-tangle T, we can consider its image under the projection $p : \mathbb{R}^2 \times I \to \mathbb{R} \times I$. By using some very small Δ-moves if necessary, we may assume that the projection is regular as in Definition 1.12. An (m, n)-*tangle diagram* then results by marking an over-/under-crossing structure on the crossing points (represented using line breaks) to record which strand of the tangle passes over which.

The proof of Reidemeister's Theorem (Theorem 1.26) involved examining how a Δ-move on an arc of a link changes its diagram. Since that argument is entirely local to arcs of the link (and is not dependent upon the fact that those arcs are part of a circle $\mathbb{S}^1$ rather than in interval I), the proof of Reidemeister's Theorem extends to the setting of tangles. This gives the following.

Theorem 7.3. *Two tangles are equivalent if and only if their two diagrams are related by a finite sequence of Reidemeister moves*

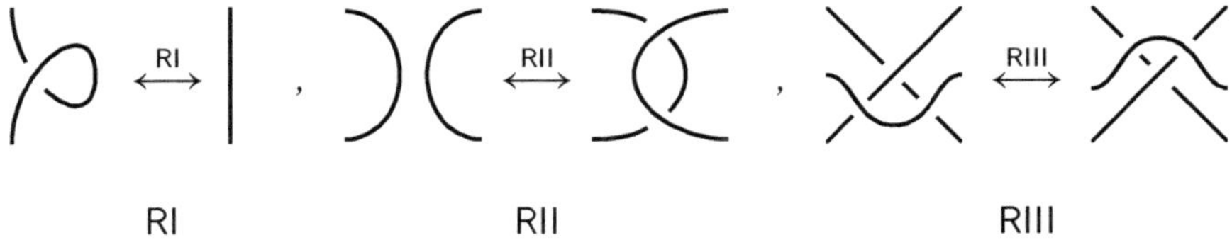

and isotopy of $\mathbb{R} \times I$. *That is,*

$$\frac{\{tangles\}}{equivalence} \cong \frac{\{tangle\ diagrams\}}{Reidemeister\ moves},$$

where isotopy of $\mathbb{R} \times I$ *is included as a Reidemeister move.*

Similarly, if the tangles are oriented, we have the following.

Theorem 7.4. *Two oriented tangles are equivalent if and only if their two diagrams are related by a finite sequence of oriented Reidemeister moves*

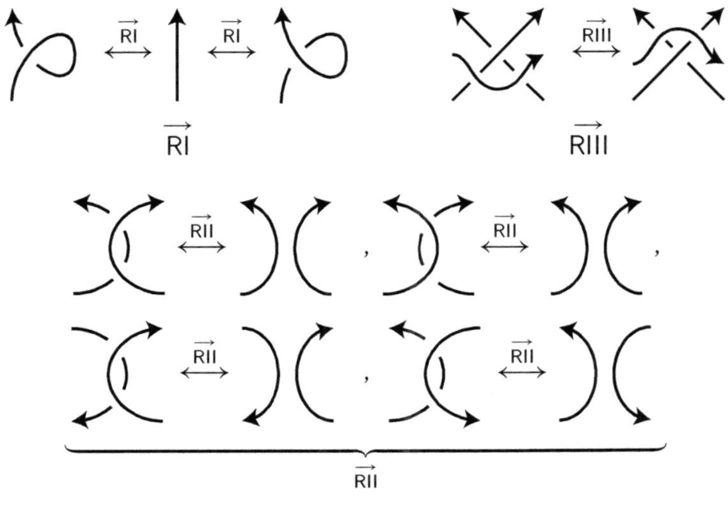

and isotopy of $\mathbb{R} \times I$. *That is*

$$\frac{\{\textit{oriented tangles}\}}{\textit{equivalence}} \cong \frac{\{\textit{oriented tangle diagrams}\}}{\textit{oriented Reidemeister moves}},$$

where isotopy of $\mathbb{R} \times I$ *is included as an oriented Reidemeister move.*

7.1.2 Composition and Tensor Product of Diagrams

Just as with braids, the basic algebraic operations of composition and tensor product can be defined on tangles. Informally, composition involves "stacking" one tangle on top of the other, and the tensor product involves placing tangles next to each other (juxtaposition).

Definition 7.5 (Composition). For an (m, n)-tangle T and a (n, p)-tangle T', the *composition* $T' \circ T$ is the (m, p)-tangle defined diagrammatically by

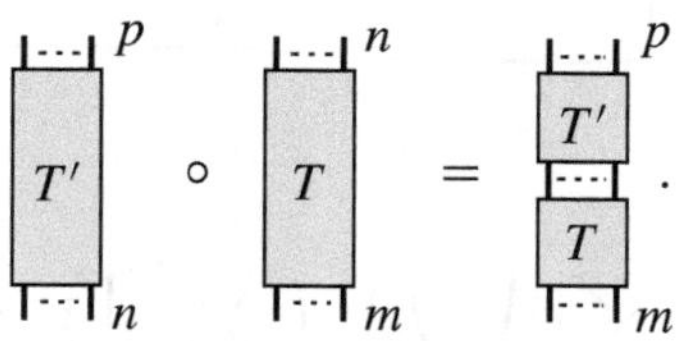

It is implicit in this definition that the composition of the two tangles in Definition 7.5 is to be rescaled to fit $\mathbb{R}^2 \times I$ and that any orientations of T and T' must be compatible for the composition to be defined. For example, consider

In this example, although $T' \circ T$ is compatible, the composite $T \circ T'$ is not and hence is not defined.

Definition 7.6 (Tensor product). For an (m, n)-tangle T and an (p, q)-tangle T', the *tensor* $T' \otimes T$ is the $(m + p, n + q)$-tangle defined diagrammatically by

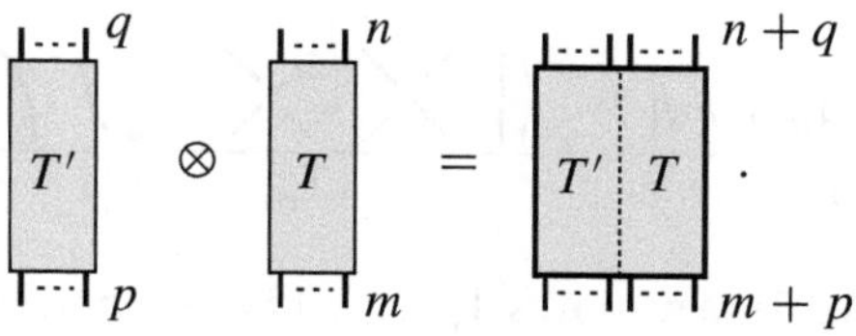

For example,

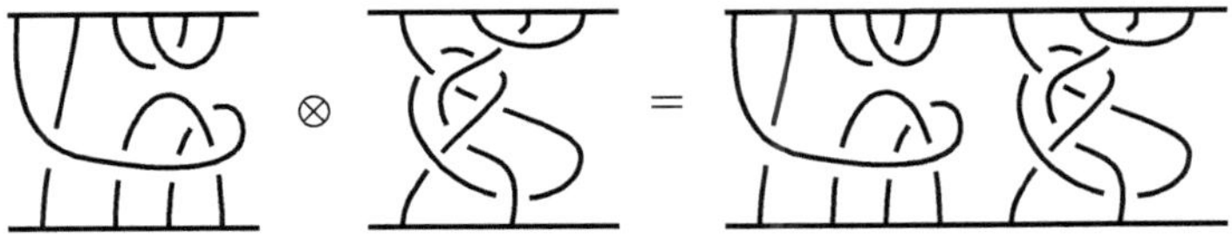

The operations of composition and tensor product enable us to decompose a tangle diagram into the *elementary tangles*  . For example,

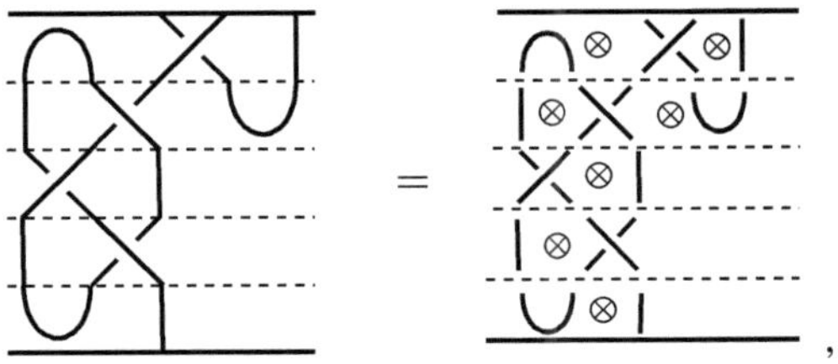

which equals

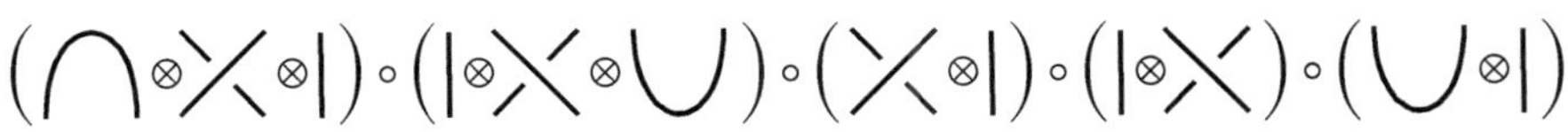

Every link diagram or tangle diagram clearly admits such a decomposition into elementary tangles (possibly after planar isotopy). This observation prompts the following definition.

Definition 7.7 (Tangle generators). A tangle is said to be *generated* by a set S of tangles if it may be obtained by the operations of composition and tensor product of elements in S, and the set of tangles generated by S is denoted by $\langle S\rangle$.

We shall consider tangle diagrams as being the objects generated by 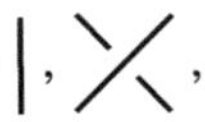  . Doing this does put some restrictions on the diagrams. For example, crossings always look like  or  and so never involve horizontal arcs. This means that a modified set of Reidemeister moves is needed to generate tangle equivalence, as in the following theorem.

Theorem 7.8.

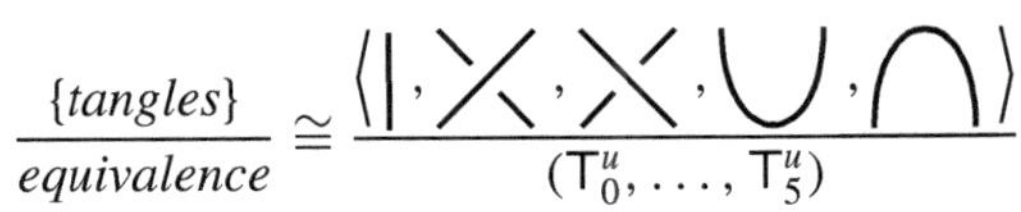

where the unoriented Turaev moves $\mathsf{T}_0^u, \ldots, \mathsf{T}_5^u$ *are defined by level- and boundary-preserving planar isotopy together with*

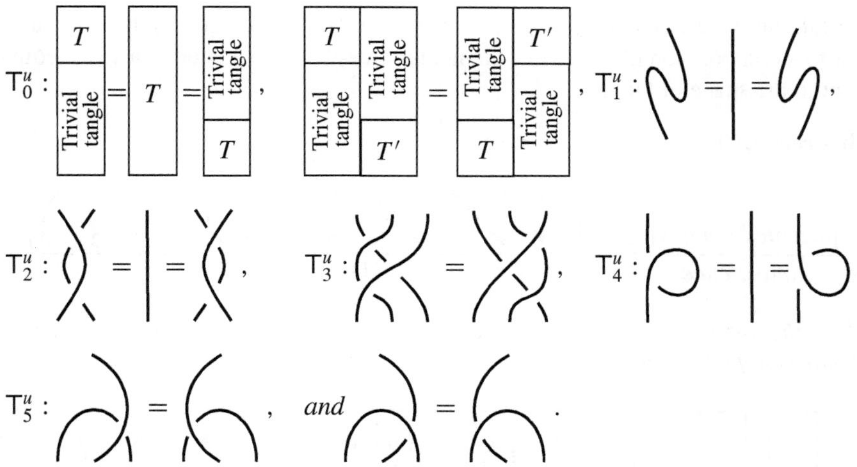

The theorem can be derived from Theorem 7.3 by observing that T_5^u is required for planar isotopy due to the restrictions on how the crossing is drawn in the diagram. A proof can be found in [181].

Definition 7.9 (Trivial tangle). A tangle diagram is said to be *trivial* if it is of the form 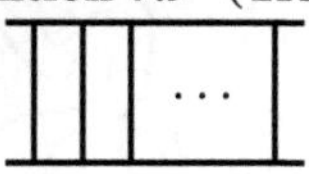 . A tangle is *trivial* if it has a trivial tangle diagram.

7.1.3 Oriented Tangles and the Turaev Moves

Here, we are mainly interested in oriented tangles (the theories developed in the rest of the book are for oriented rather than unoriented tangles). For the construction of tangle invariants, we impose a condition that both of the strands involved in any crossing in a tangle diagram are oriented upwards: 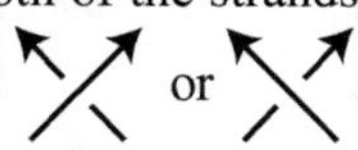 . This condition does not restrict the class of tangles or tangle diagrams we are considering since planar isotopy may be used to rotate a crossing into the required form. For example,

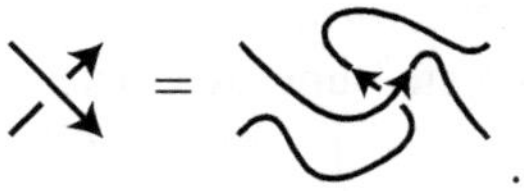

This restriction on the crossings reduces the size of the generating set of oriented tangles to the eight elements

We call these the *elementary tangles*. Since every oriented tangle diagram is equivalent to a composition of tensors of elementary tangles, they generate a space equivalent to the space of tangles.

Theorem 7.10.

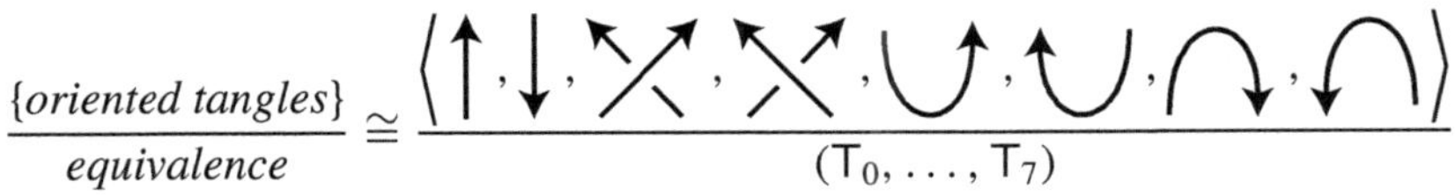

where the Turaev moves $\mathsf{T}_0, \dots, \mathsf{T}_7$ *are defined by level- and boundary-preserving planar isotopy together with*

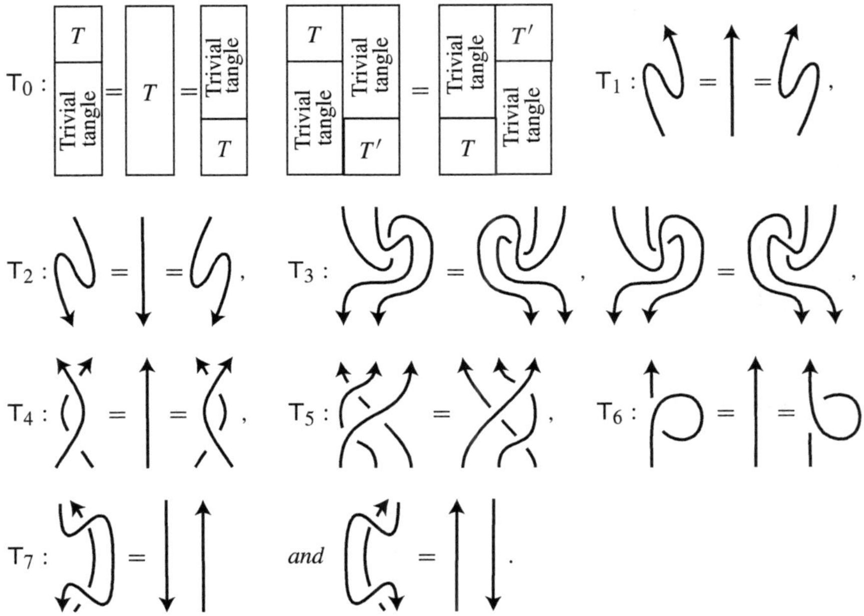

Proof (Sketch). Theorem 7.4 describes oriented tangles in terms of oriented tangle diagrams and the oriented Reidemeister moves. Given an oriented tangle diagram, we can use planar isotopy to rotate each of the crossings so that the two strands in the crossing are directed upwards. The resulting diagram is clearly in

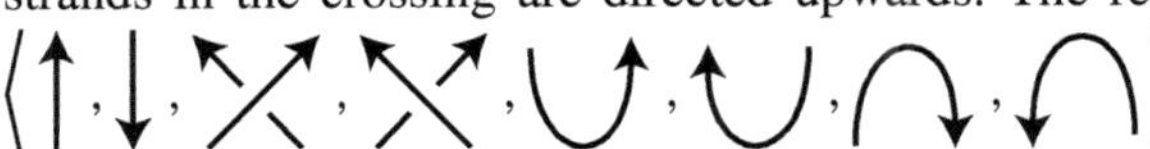

. The Turaev moves arise by modifying the Reidemeister moves so that they lie in this space. Level- and boundary-preserving isotopy and $\mathsf{T}_0, \dots, \mathsf{T}_3$ describe the changes of a tangle diagram that can arise from planar isotopy. The moves $\mathsf{T}_4, \dots, \mathsf{T}_7$ are the oriented Reidemeister moves written in the current setting. (See [181] for a full proof.) □

Convention 7.11. Henceforth, by the term *oriented tangle diagram* we mean an element of the set

and equivalence of oriented tangle diagrams is generated by the Turaev moves $\mathsf{T}_0 - \mathsf{T}_7$.

7.2 A First Definition of an Operator Invariant

7.2.1 *Morphisms Associated with Elementary Pieces*

In the case of ρ_{R}, a morphism was constructed by (i) associating a free module of finite rank, V, with each upward-oriented strand; (ii) associating a homomorphism with each elementary tangle. This association was through, from Definition 5.2,

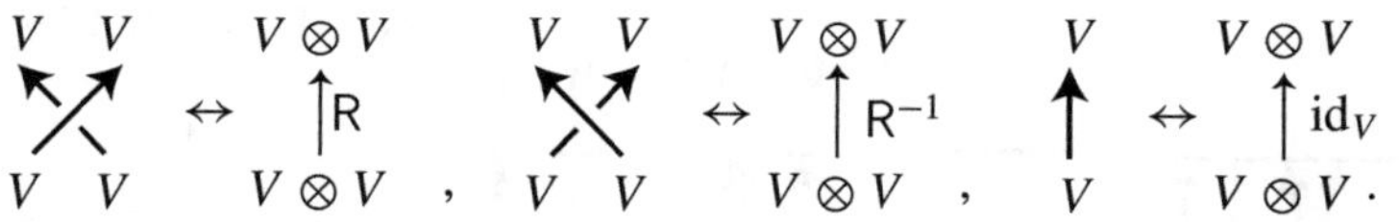

(7.2)

These ideas may be extended to tangles, and we shall introduce a map Q which will evolve into a tangle invariant.

We begin by extending (7.2) to tangles by considering the five remaining generators  and marking:

- The downward pointing ends by another free module W of finite rank.
- The upward pointing ends by a copy of the free module V of finite rank, as before.
- The absence of boundary points by the ground ring $\mathbb{K}$.

This gives the correspondences

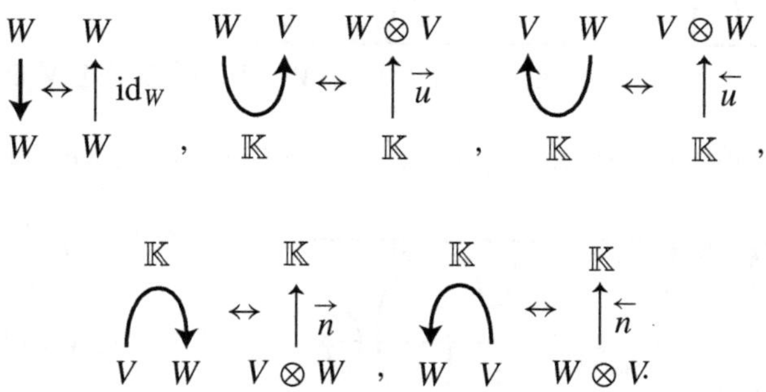

(7.3)

The sets (7.2) and (7.3) of correspondences comprise the complete set of correspondences between the elementary pieces and morphisms. Note that the letters u and n reflect the shape of the cups and caps, and the arrows reflect their orientation.

Given the free modules V and W of finite rank, an R-matrix R, and morphisms $\overrightarrow{u}$, $\overleftarrow{u}$, $\overrightarrow{n}$ and $\overleftarrow{n}$, the correspondences provided in (7.2) and (7.3) give the element-wise action of a map

$$Q \equiv Q_{(V,W,\mathsf{R},\overrightarrow{u},\overleftarrow{u},\overrightarrow{n},\overleftarrow{n})}$$

from elementary pieces to morphisms. This map may be extended to the set of tangles through

$$Q(T \otimes T') := Q(T) \otimes Q(T'), \tag{7.4}$$

$$Q(T \circ T') := Q(T) \circ Q(T'). \tag{7.5}$$

where T and T' are tangles.

Example 7.12. As an illustration, the values of Q on elementary tangles together with $Q(T \otimes T') = Q(T) \otimes Q(T')$ give

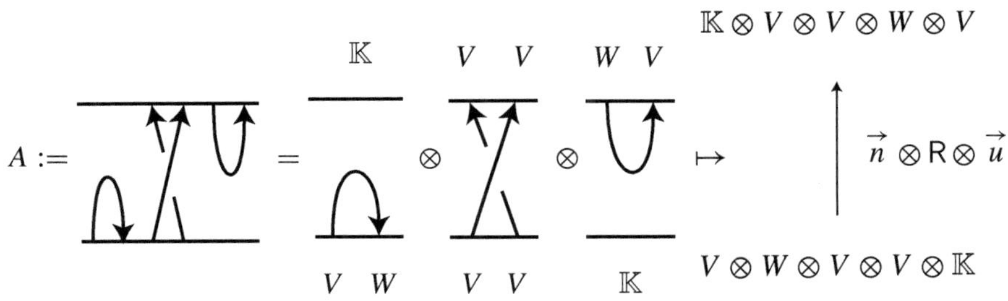

Similarly,

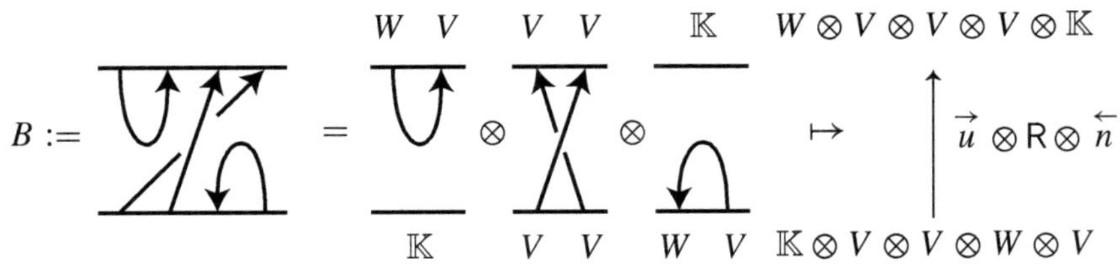

Then, using the property that $Q(T \circ T') = Q(T) \circ Q(T')$, we can compute

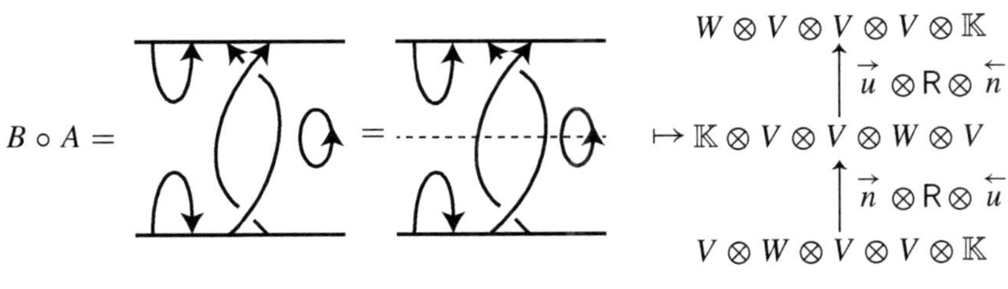

Therefore

$$Q(B \circ A) = (\overrightarrow{u} \otimes \mathsf{R} \otimes \overleftarrow{n}) \circ (\overrightarrow{n} \otimes \mathsf{R} \otimes \overleftarrow{u}).$$

We record this discussion in the following definition.

Definition 7.13 (Map $Q_{(V,W,\mathsf{R},\vec{u},\overleftarrow{u},\vec{n},\overleftarrow{n})}$). Let V and W be free modules of finite rank and R, $\vec{u}$, $\overleftarrow{u}$, $\vec{n}$, $\overleftarrow{n}$ be morphisms as below. Then $Q \equiv Q_{(V,W,\mathsf{R},\vec{u},\overleftarrow{u},\vec{n},\overleftarrow{n})}$ is the map on tangle diagrams defined through the actions below.

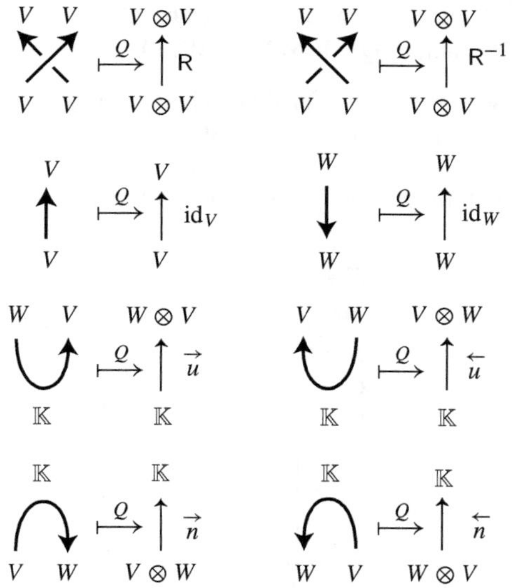

$$T \otimes T' \stackrel{Q}{\longmapsto} Q(T) \otimes Q(T'),$$

$$T \circ T' \stackrel{Q}{\longmapsto} Q(T) \circ Q(T').$$

Before we can continue there is an important computational observation about the domains and codomains of Q. Specifically, there is a subtlety about how we deal with copies of $\mathbb{K}$ in them. The difficulty is illustrated by the following example. Consider the two tangles.

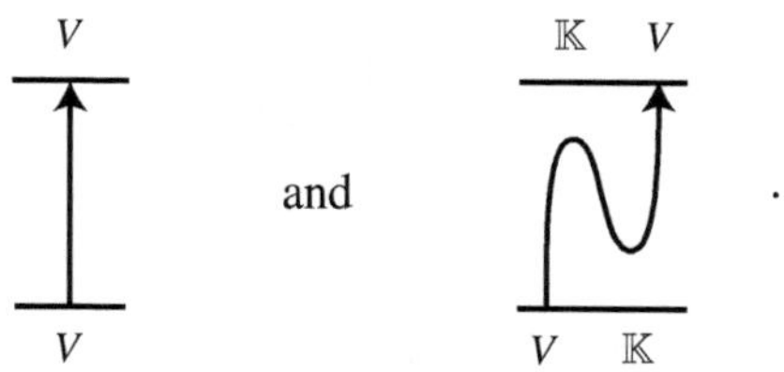

These two tangles are equivalent, so we need Q to take the same value on both tangles (since ultimately we want Q to be a tangle invariant). Yet one tangle gives rise to a map $V \to V$ while the other to $V \otimes \mathbb{K} \to \mathbb{K} \otimes V$, and so the maps are formally different. Fortunately, we have $V \otimes \mathbb{K} \cong V \cong \mathbb{K} \otimes V$ *via* the contraction map κ. Thus, we can resolve this difficulty by using the following convention.

Convention 7.14. In any computation of Q, we adjust the domains and codomains to eliminate, *via* contraction κ, copies of $\mathbb{K}$ (unless the domain or codomain is $\mathbb{K}$) and add them as needed using $\iota := \kappa^{-1}$.

As an illustration of this convention, we regard $Q = (\overrightarrow{u} \otimes \mathsf{R} \otimes \overrightarrow{n}) \circ (\overrightarrow{n} \otimes \mathsf{R} \otimes \overleftarrow{u})$ from Example 7.12 as a map $Q : V \otimes W \otimes V \otimes V \to W \otimes V \otimes V \otimes V$.

The use of Convention 7.14 is critical when considering composition $Q(T) \circ Q(T')$, as illustrated in the following example.

Example 7.15. A computation of the map Q for a tangle T is shown below.

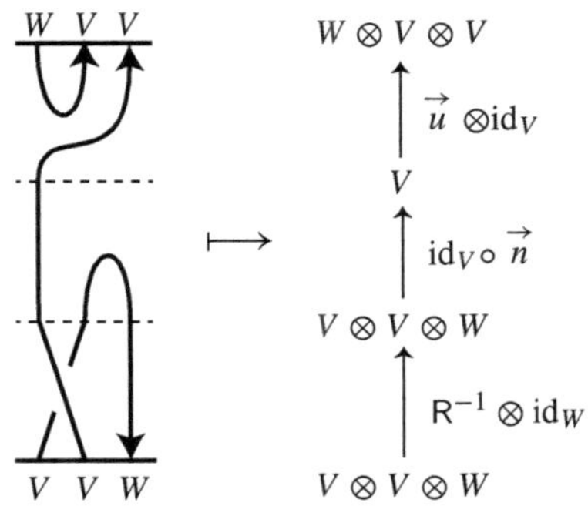

(7.6)

This gives

$$Q(T) = \left(\overrightarrow{u} \otimes \mathrm{id}_V\right) \circ \left(\mathrm{id}_V \otimes \overrightarrow{n}\right) \circ \left(\mathsf{R}^{-1} \otimes \mathrm{id}_W\right). \tag{7.7}$$

To see the use of Convention 7.14, let T_1, T_2 and T_3 be the tangles shown below, so that $T = T_3 \circ T_2 \circ T_1$, and we want to compute $Q(T) = Q(T_3 \circ T_2 \circ T_1) = Q(T_3) \circ Q(T_2) \circ Q(T_1)$.

$T_1 =$  , $T_2 =$ 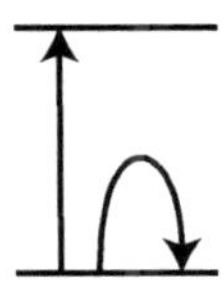 , $T_3 =$ 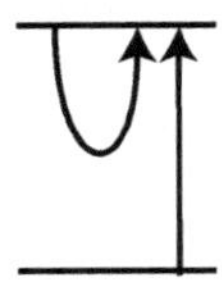

Then, from Definition 7.13,

$$Q(T_1) : V \otimes V \otimes W \to V \otimes V \otimes W, \quad Q(T_2) : V \otimes V \otimes W \to V \otimes \mathbb{K},$$

$$\text{and} \quad Q(T_3) : \mathbb{K} \otimes V \to W \otimes V \otimes V.$$

We cannot immediately form $Q(T_3) \circ Q(T_2) \circ Q(T_1)$ since some of the domains and codomains do not match up. However, our contraction convention gives $Q(T_1)$:

$V \otimes V \otimes W \to V \otimes V \otimes W$, $Q(T_2) : V \otimes V \otimes W \to V$ and $Q(T_3) : V \to W \otimes V \otimes V$ and so we can compose.

7.2.2 *Morphisms Corresponding to the Turaev Moves*

The goal is to consider link invariants, or tangle invariants, rather than simply maps from the set of links or tangles, which so far is all that Definition 7.13 gives. To obtain an invariant, it is necessary that the map Q respects equivalence classes of tangles. This means that, for equivalent tangles T and T', we require that $Q(T) = Q(T')$.

In view of Theorem 7.10, if Q is invariant under the Turaev moves, then it defines a tangle invariant. Our next step is to determine the conditions R, $\overrightarrow{u}$, $\overleftarrow{u}$, $\overrightarrow{n}$, $\overleftarrow{n}$ must satisfy for Q to be invariant under the Turaev moves.

Each Turaev move consists of two tangles, T and T', which are declared to be equal. The map Q associates homomorphisms $Q(T)$ and $Q(T')$ with the tangles T and T', respectively. Obviously, the map Q is invariant under the given Turaev move if and only if $Q(T) = Q(T')$. Since we can calculate the value of $Q(T)$ and $Q(T')$ by slicing and then using juxtaposition (tensor product) and stacking (composition), we may derive a set of conditions on the maps $\overrightarrow{u}$, $\overleftarrow{u}$, $\overrightarrow{n}$, $\overleftarrow{n}$, R, R^{-1} which hold if and only if the corresponding Turaev move holds. Moreover, Q is an invariant of tangles if and only if such an identity holds for every Turaev move. The result of these calculations is a set of algebraic conditions on $\overrightarrow{u}$, $\overleftarrow{u}$, $\overrightarrow{n}$, $\overleftarrow{n}$, R, R^{-1} which hold if and only if Q is invariant under the Turaev moves. We now determine this set of conditions.

<u>T_0 – *Move*:</u> This is trivially satisfied since $\mathrm{id}^{\otimes n} \circ Q(T) = Q(T) = Q(T) \circ \mathrm{id}^{\otimes n}$ and $(Q(T) \otimes \mathrm{id}^{\otimes m}) \circ (\mathrm{id}^{\otimes n} \otimes Q(T')) = Q(T) \otimes Q(T') = (\mathrm{id}^{\otimes n} \otimes Q(T')) \circ (Q(T) \otimes \mathrm{id}^{\otimes m})$.

<u>T_1 – *Move*:</u>

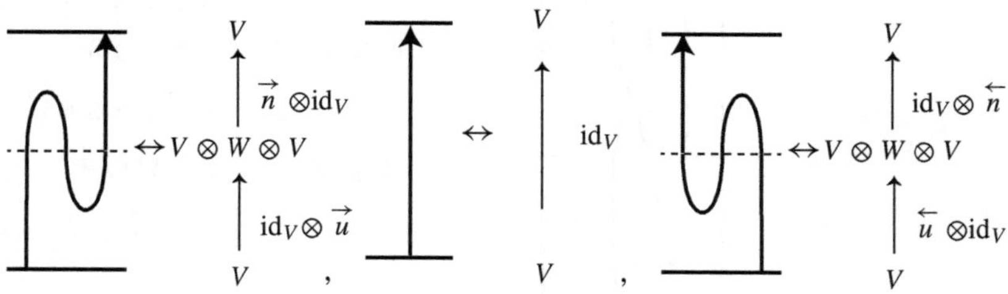

so

Condition from T_1-Move:

$$(\overrightarrow{n} \otimes \mathrm{id}_V) \circ (\mathrm{id}_V \otimes \overrightarrow{u}) = \mathrm{id}_V = (\mathrm{id}_V \otimes \overleftarrow{n}) \circ (\overleftarrow{u} \otimes \mathrm{id}_V). \tag{7.8}$$

<u>$\mathsf{T}_2 - Move$:</u>

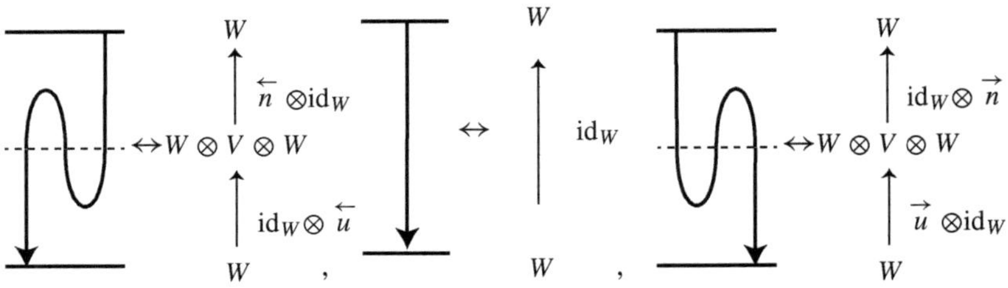

so

Condition from T_2-Move:

$$(\overleftarrow{n} \otimes \mathrm{id}_W) \circ (\mathrm{id}_W \otimes \overleftarrow{u}) = \mathrm{id}_W = (\mathrm{id}_W \otimes \overrightarrow{n}) \circ (\overrightarrow{u} \otimes \mathrm{id}_W). \tag{7.9}$$

<u>$\mathsf{T}_3 - Move$:</u> We show the details of the calculation for the version of the T_2-move with the positive crossing. The result for the version with the negative crossing is obtained by replacing R^{-1} for R in the following calculation.

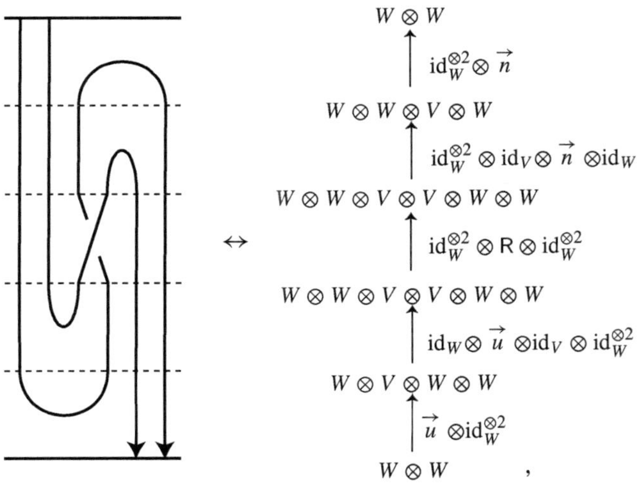

and

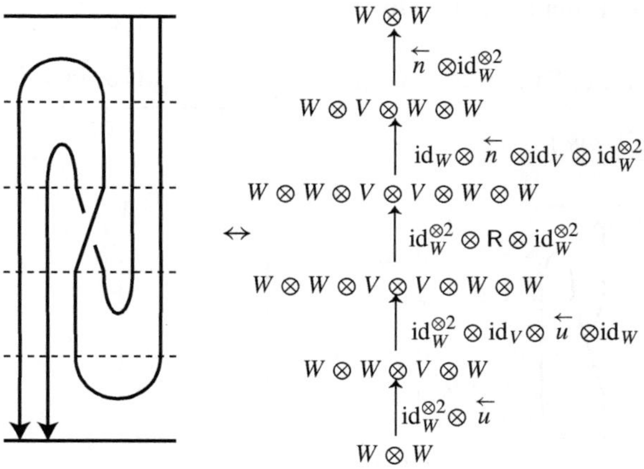

so

Condition from T_3-Move:

$$\left(\mathrm{id}_W^{\otimes 2}\otimes \overrightarrow{n}\right)\circ\left(\mathrm{id}_W^{\otimes 2}\otimes \mathrm{id}_V\otimes \overrightarrow{n}\otimes\mathrm{id}_W\right)\circ\left(\mathrm{id}_W^{\otimes 2}\otimes \mathsf{R}^{\pm 1}\otimes \mathrm{id}_W^{\otimes 2}\right)$$
$$\circ\left(\mathrm{id}_W\otimes \overrightarrow{u}\otimes\mathrm{id}_V\otimes \mathrm{id}_W^{\otimes 2}\right)\circ\left(\overrightarrow{u}\otimes\mathrm{id}_W^{\otimes 2}\right) \qquad (7.10)$$
$$=\left(\overleftarrow{n}\otimes\mathrm{id}_W^{\otimes 2}\right)\circ\left(\mathrm{id}_W\otimes \overleftarrow{n}\otimes\mathrm{id}_V\otimes \mathrm{id}_W^{\otimes 2}\right)\circ\left(\mathrm{id}_W^{\otimes 2}\otimes \mathsf{R}^{\pm 1}\otimes \mathrm{id}_W^{\otimes 2}\right)$$
$$\circ\left(\mathrm{id}_W^{\otimes 2}\otimes \mathrm{id}_V\otimes \overleftarrow{u}\otimes\mathrm{id}_W\right)\circ\left(\mathrm{id}_W^{\otimes 2}\otimes \overleftarrow{u}\right).$$

$\mathsf{T}_4 - Move$:

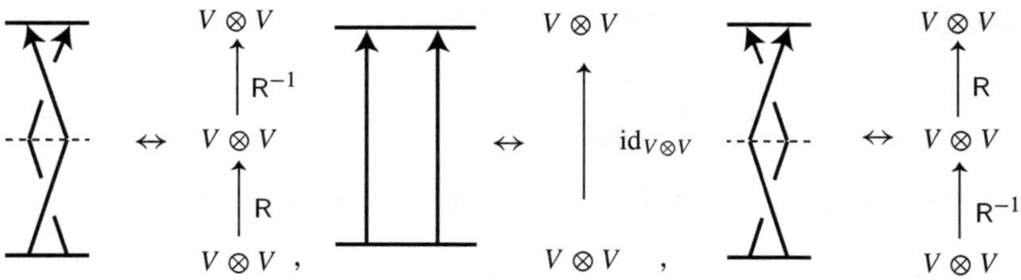

so

Condition from T_4-Move:

$$\mathsf{R}^{-1}\circ\mathsf{R}=\mathrm{id}_{V\otimes V}=\mathsf{R}\circ\mathsf{R}^{-1}. \qquad (7.11)$$

This identity asserts that R is invertible.
$\mathsf{T}_5 - Move$: This is the RIII-move. It has already been shown in Sect. 5.2 that the corresponding condition is the Yang–Baxter Equation (see Def. 5.4), namely,

Condition from T_5-Move:

$$(\mathrm{id} \otimes \mathsf{R}) \circ (\mathsf{R} \otimes \mathrm{id}) \circ (\mathrm{id} \otimes \mathsf{R}) = (\mathsf{R} \otimes \mathrm{id}) \circ (\mathrm{id} \otimes \mathsf{R}) \circ (\mathsf{R} \otimes \mathrm{id}). \quad (7.12)$$

$\mathsf{T}_6 - Move$:

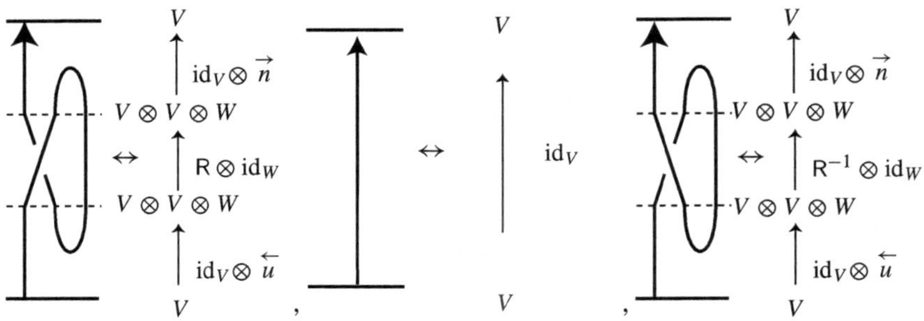

so

Condition from T_6-Move:

$$\left(\mathrm{id}_V \otimes \overrightarrow{n}\right) \circ \left(\mathsf{R}^{\pm 1} \otimes \mathrm{id}_W\right) \circ \left(\mathrm{id}_V \otimes \overleftarrow{u}\right) = \mathrm{id}_V. \quad (7.13)$$

$\mathsf{T}_7 - Move$: Let A and B be the tangles given by

$A :=$ 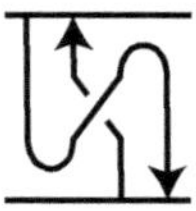 and $B :=$ 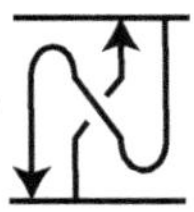 .

Then, the two Turaev moves can be written as

$A \circ B =$  and $B \circ A =$  .

Thus, the conditions for Q to be invariant under the T_7-move are

$$Q(A) \circ Q(B) = \mathrm{id}_W \otimes \mathrm{id}_V \quad \text{and} \quad Q(B) \circ Q(A) = \mathrm{id}_V \otimes \mathrm{id}_W.$$

A computation for $Q(A)$ is

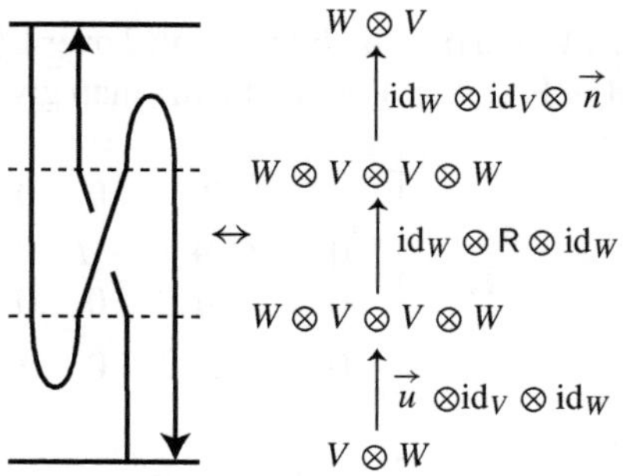

A computation for $Q(B)$ is

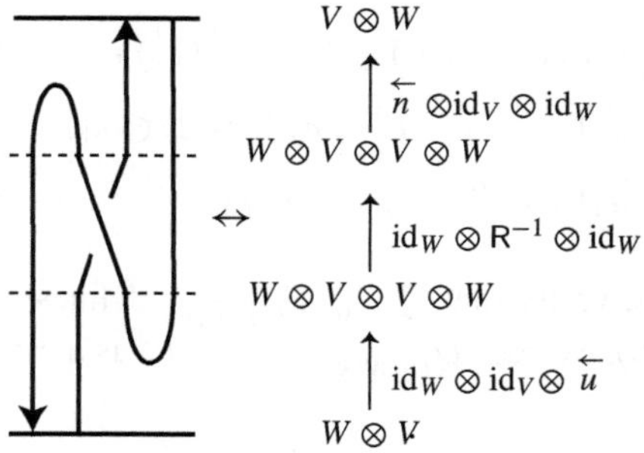

This gives the following.

Condition from T_7-Move:

$$\mathsf{Y} \circ \mathsf{T} = \mathrm{id}_W \otimes \mathrm{id}_V \quad \text{and} \quad \mathsf{T} \circ \mathsf{Y} = \mathrm{id}_V \otimes \mathrm{id}_W, \tag{7.14}$$

where

$$\mathsf{Y} = \left(\mathrm{id}_W \otimes \mathrm{id}_V \otimes \overrightarrow{n}\right) \circ (\mathrm{id}_W \otimes \mathsf{R} \otimes \mathrm{id}_W) \circ \left(\overrightarrow{u} \otimes \mathrm{id}_V \otimes \mathrm{id}_W\right),$$

and

$$\mathsf{T} = \left(\overleftarrow{n} \otimes \mathrm{id}_V \otimes \mathrm{id}_W\right) \circ \left(\mathrm{id}_W \otimes \mathsf{R}^{-1} \otimes \mathrm{id}_W\right) \circ \left(\mathrm{id}_W \otimes \mathrm{id}_V \otimes \overleftarrow{u}\right).$$

We are now able to define a tangle invariant Q.

Definition 7.16 (Operator invariant). The map $Q_{(V,W,\mathsf{R},\overrightarrow{n},\overleftarrow{n},\overrightarrow{u},\overleftarrow{u})}$ constructed in Definition 7.13 is said to be an *operator invariant* if R, $\overrightarrow{n}$, $\overleftarrow{n}$, $\overrightarrow{u}$ and $\overleftarrow{u}$ satisfy Conditions (7.8)–(7.14).

The following result is immediate from the definition.

Theorem 7.17. *An operator invariant is an invariant of oriented tangles.*

Exercise 7.18. Let V be a free module of rank 2 over $\mathbb{C}[t^{\frac{1}{2}}, t^{-\frac{1}{2}}]$ with a basis $\{v_1, v_2\}$, and let $W = V^*$. Let $\mathsf{R} \in \mathrm{Aut}(V \otimes V)$ be the map given by the matrix

$$\mathbf{R} = \begin{bmatrix} -t^{\frac{1}{2}} & 0 & 0 & 0 \\ 0 & -t^{\frac{1}{2}} + t^{\frac{3}{2}} & -t & 0 \\ 0 & -t & 0 & 0 \\ 0 & 0 & 0 & -t^{\frac{1}{2}} \end{bmatrix},$$

and let the maps $\overrightarrow{n}$, $\overleftarrow{n}$, $\overrightarrow{u}$ and $\overleftarrow{u}$ be defined by

$$\begin{aligned} &\overrightarrow{n}:(a\, e_0 + b\, e_1) \otimes (c\, e^0 + d\, e^1) \mapsto -t^{\frac{1}{2}}ac - t^{-\frac{1}{2}}bd, \\ &\overleftarrow{n}:(a\, e_0 + b\, e_1) \otimes (c\, e^0 + d\, e^1) \mapsto ac + bd, \\ &\overrightarrow{u}:1 \mapsto -t^{-\frac{1}{2}}\, e^0 \otimes e_0 - t^{\frac{1}{2}}\, e^1 \otimes e_1, \\ &\overleftarrow{u}:1 \mapsto e^0 \otimes e_0 + e^1 \otimes e_1. \end{aligned}$$

For these maps, verify that $Q_{(V,W,\mathsf{R},\overrightarrow{n},\overleftarrow{n},\overrightarrow{u},\overleftarrow{u})}$ defines an operator invariant. For the tangle T in (7.6), express $Q_{(V,W,\mathsf{R},\overrightarrow{n},\overleftarrow{n},\overrightarrow{u},\overleftarrow{u})}(T)$ as a matrix.

7.3 Simplifying the Definition of Operator Invariants

Definition 7.16 provides a tangle invariant $Q_{(V,W,\mathsf{R},\overrightarrow{n},\overleftarrow{n},\overrightarrow{u},\overleftarrow{u})}$. Defining the invariant involves obtaining five maps R, $\overrightarrow{u}$, $\overleftarrow{u}$, $\overrightarrow{n}$ and $\overleftarrow{n}$ that together satisfy Conditions (7.8)–(7.14). In fact, the definition of Q can be simplified by replacing the four maps $\overrightarrow{u}$, $\overleftarrow{u}$, $\overrightarrow{n}$ and $\overleftarrow{n}$ with just two maps α and β. This simplification requires a fair amount of work (in the form of quite lengthy and intricate computations). To avoid disrupting the narrative, we have put these computations, which constitute the proof of Theorem 7.19, into Appendix B. Here, we shall only provide an overview of the results that are required.

The simplified definition of operator invariants results from the following theorem. For the moment, the operations $(\cdot)^{\cup}$, $(\cdot)^{\cap}$ and $(\cdot)^{t_1}$ are to be regarded simply as ways of constructing new maps from old. Their precise definition will be given a little later in Definitions 7.21 and 7.23.

Theorem 7.19. *Let V and W be free modules of the same finite rank. Then*

$$Q_{(V,W,\mathsf{R},\overrightarrow{n},\overleftarrow{n},\overrightarrow{u},\overleftarrow{u})}$$

is an operator invariant if and only if

1. *there exist isomorphisms $\alpha\colon W^\star \to V$ and $\beta\colon V^\star \to W$ such that*

$$\vec{u} = \alpha^{\cup}, \qquad \vec{n} = (\alpha^{-1})^{\cap}, \qquad \overleftarrow{u} = \beta^{\cup}, \qquad \overleftarrow{n} = (\beta^{-1})^{\cap}.$$

2. R *is an R-matrix.*
3. *If* $\mu := \beta^{\star} \circ \alpha^{-1} \colon V \to V$, *then*

   a. $\left(\tau \circ \mathsf{R}^{-1}\right)^{t_1} \circ (\mathrm{id}_{V^{\star}} \otimes \mu) \circ (\mathsf{R} \circ \tau)^{t_1} \circ (\mathrm{id}_{V^{\star}} \otimes \mu)^{-1} = \mathrm{id}_{V^{\star} \otimes V}$;
   b. $\mathrm{Tr}_2\left(\mathsf{R}^{\pm 1} \circ (\mathrm{id} \otimes \mu)\right) = \mathrm{id}_V$;
   c. $\mathsf{R} \circ (\mu \otimes \mu) = (\mu \otimes \mu) \circ \mathsf{R}$.

In the remainder of this section, we describe the main ideas behind the reformulation given in Theorem 7.19. Its proof can be found in Appendix B.

In the light of Theorem 7.19, the first step is to simplify the notation of operator invariants by introducing the following definition.

Definition 7.20 (Operator invariant $Q_{(\mathsf{R},\alpha,\beta)}$). Let V and W be free modules of the same finite rank. Let $\mathsf{R} : V \otimes V \to V \otimes V$, $\alpha \colon W^{\star} \to V$ and $\beta \colon V^{\star} \to W$ be maps satisfying the conditions in Theorem 7.19. Then

$$Q_{(\mathsf{R},\alpha,\beta)}$$

denotes the operator invariant $Q_{(V,W,\mathsf{R},\vec{n},\overleftarrow{n},\vec{u},\overleftarrow{u})}$ where $\vec{u} = \alpha^{\cup}$, $\vec{n} = (\alpha^{-1})^{\cap}$, $\overleftarrow{u} = \beta^{\cup}$, $\overleftarrow{n} = (\beta^{-1})^{\cap}$.

In fact, we shall see in Sect. 7.4 that if T is a link, then the value of $Q_{(\mathsf{R},\alpha,\beta)}(T)$ depends only upon R and the composite $\beta^{*} \circ \alpha^{-1}$, and not α and β individually.

The idea behind the reformulation runs along the following lines.

- $\vec{u} \colon \mathbb{K} \to W \otimes V$, but $\mathrm{Hom}(\mathbb{K}, W \otimes V) \cong \mathrm{Hom}(W^{*}, V)$, so $\vec{u}$ is equivalent to a map from W^{*} to V. Similar comments hold for $\vec{n}$, $\overleftarrow{u}$ and $\overleftarrow{n}$.
- When V and W are of the same rank, invariance under the Turaev moves T_1 and T_2 forces relationships between these four maps, specifying how they can be obtained from two isomorphisms α and β.
- Invariance under the moves $\mathsf{T}_3, \ldots, \mathsf{T}_6$ forces R to be an R-matrix and the maps R and $\mu := \beta^{*} \circ \alpha^{-1}$ to satisfy the identities

$$\mathrm{Tr}_2\left(\mathsf{R}^{\pm 1} \circ (\mathrm{id} \otimes \mu)\right) = \mathrm{id}_V \quad \text{and} \quad \mathsf{R} \circ (\mu \otimes \mu) = (\mu \otimes \mu) \circ \mathsf{R}.$$

- Invariance under T_7 forces

$$\left(\tau \circ \mathsf{R}^{-1}\right)^{t_1} \circ (\mathrm{id}_{V^{\star}} \otimes \mu) \circ (\mathsf{R} \circ \tau)^{t_1} \circ (\mathrm{id}_{V^{\star}} \otimes \mu)^{-1} = \mathrm{id}_{V^{\star} \otimes V},$$

  where the operation $(\cdot)^{t_1}$ is a natural isomorphism that enables the R-matrix to act on $V^{*} \otimes V$.

The three maps $(\cdot)^{\cap}$, $(\cdot)^{\cup}$ and $(\cdot)^{t_1}$ arise by using the natural isomorphism $\mathrm{Hom}(X, Y) \cong X^{*} \otimes Y$ to change the domains and codomains of linear maps.

For the ∩-map, we observe the following:

$$
\begin{aligned}
\mathrm{Hom}(X, Y) &\cong X^\star \otimes Y && \text{(Lem. (A.36))} \\
&\cong (X \otimes Y^\star)^\star && \text{(Prop. (A.30), Thm. (A.17))} \\
&\cong (X \otimes Y^\star)^\star \otimes \mathbb{K} && \text{(Thm. (A.26))} \\
&\cong \mathrm{Hom}(X \otimes Y^\star, \mathbb{K}) && \text{(Lem. (A.36)).}
\end{aligned}
$$

Definition 7.21 (Map ∩). The ∩-map is the isomorphism

$$
(\cdot)^\cap : \mathrm{Hom}(X, Y) \xrightarrow{\cong} \mathrm{Hom}(X \otimes Y^\star, \mathbb{K}) : f \mapsto (x \otimes g \mapsto g(f(x))),
$$

where $\{x_i\}$ is a basis of X.

Exercise 7.22. By tracking through the defining sequence of isomorphisms, verify that the element-wise action of $(\cdot)^\cap$ is given by $f \mapsto (x \otimes g \mapsto g(f(x)))$.

The ∪-map follows from a similar sequence of isomorphisms:

$$
\begin{aligned}
\mathrm{Hom}(X, Y) &\cong X^\star \otimes Y && \text{(Lem. (A.36))} \\
&\cong \mathbb{K} \otimes (X^\star \otimes Y) && \text{(Thm. (A.26)(3))} \\
&\cong \mathbb{K}^\star \otimes (X^\star \otimes Y) && \text{(Thm. (A.17))} \\
&\cong \mathrm{Hom}(\mathbb{K}, X^\star \otimes Y) && \text{(Lem. (A.36)).}
\end{aligned}
$$

Definition 7.23 (Map ∪). The ∪-map is the isomorphism

$$
(\cdot)^\cup : \mathrm{Hom}(X, Y) \xrightarrow{\cong} \mathrm{Hom}(\mathbb{K}, X^\star \otimes Y) : f \mapsto \left(1_{\mathbb{K}} \mapsto \sum_i x^i \otimes f(x_i)\right),
$$

where $\{x_i\}$ is a basis of X.

Exercise 7.24. By tracking through the defining sequence of isomorphisms, verify that the element-wise action of $(\cdot)^\cup$ is given by $f \mapsto \left(1_{\mathbb{K}} \mapsto \sum_i x^i \otimes f(x_i)\right)$.

Definition 7.25 (Twist map τ). The map τ is the natural isomorphism $\tau : X \otimes Y \to Y \otimes X : x \otimes y \mapsto y \otimes x$. It is called the *twist*.

The $(\cdot)^{t_1}$-map arises from the sequence of isomorphisms

$$
\begin{aligned}
\mathrm{Hom}(V \otimes W, X \otimes Y) &\cong (V \otimes W)^\star \otimes X \otimes Y && \text{(Lem. (A.36))} \\
&\cong V^\star \otimes W^\star \otimes X \otimes Y \\
&\cong (X \otimes W^\star) \otimes (V^\star \otimes Y) \\
&\cong (X^\star \otimes W)^\star \otimes (V^\star \otimes Y) \\
&\cong \mathrm{Hom}(X^\star \otimes W,\ V^\star \otimes Y) && \text{(Lem. (A.36))}
\end{aligned}
$$

Definition 7.26 (Isomorphism $(\cdot)^{t_1}$). The $(\cdot)^{t_1}$-map is the isomorphism

$$\begin{array}{rccc} (\cdot)^{t_1}: & \mathrm{Hom}(V\otimes W, X\otimes Y) & \xrightarrow{\cong} & \mathrm{Hom}(X^{\star}\otimes W, V^{\star}\otimes Y) \\ : & \left(v_i\otimes w_j \mapsto \sum_{k,l} f_{ij}^{kl}\, x_k\otimes y_l\right) & \mapsto & \left(x^k\otimes w_j \mapsto \sum_{i,l} f_{ij}^{kl}\, v^i\otimes y_l\right), \end{array}$$

where $\{v_i\}$, $\{w_i\}$, $\{x_i\}$, $\{y_i\}$ are bases for V, W, X and Y, respectively.

Exercise 7.27. By tracking through the defining sequence of isomorphisms, verify that the element-wise action of $(\cdot)^{t_1}$ is given by $\left(v_i\otimes w_j \mapsto \sum_{k,l} f_{ij}^{kl}\, x_k\otimes y_l\right) \mapsto \left(x^k\otimes w_j \mapsto \sum_{i,l} f_{ij}^{kl}\, v^i\otimes y_l\right)$.

Exercise 7.28. Let V be a free module of rank 2 over $\mathbb{C}[t^{\frac{1}{2}}, t^{-\frac{1}{2}}]$, and let $W = V^*$. Let $\mathsf{R}\in \mathrm{Aut}(V\otimes V)$ be the map given by the matrix

$$\mathbf{R} = \begin{bmatrix} -t^{\frac{1}{2}} & 0 & 0 & 0 \\ 0 & -t^{\frac{1}{2}}+t^{\frac{3}{2}} & -t & 0 \\ 0 & -t & 0 & 0 \\ 0 & 0 & 0 & -t^{\frac{1}{2}} \end{bmatrix}, \quad \alpha = \begin{bmatrix} -t^{-\frac{1}{2}} & 0 \\ 0 & -t^{\frac{1}{2}} \end{bmatrix}, \quad \beta = \begin{bmatrix} 1 & 0 \\ 0 & 1 \end{bmatrix}.$$

Compute $\overrightarrow{u} = \alpha^{\cup}$, $\overrightarrow{n} = (\alpha^{-1})^{\cap}$, $\overleftarrow{u} = \beta^{\cup}$, $\overleftarrow{n} = (\beta^{-1})^{\cap}$, and show that $Q_{(\mathsf{R},\alpha,\beta)}$ coincides with the operator invariant of Exercise 7.18.

7.4 The Relationship Between Operator Invariants and $\rho_{(\mathsf{R},\mu)}$

Up to this point in Part II, we have constructed two types of link invariant:

- First, in Chap. 6 we obtained, as in Theorem 6.13, an invariant

$$\rho_{(\mathsf{R},\mu)}(\sigma) := \mathrm{Tr}\left(\rho_{\mathsf{R}}(\sigma)\circ \mu^{\otimes n}\right);$$

  given an oriented link diagram D, the computation of this invariant required finding a braid σ whose closure is equivalent to D.
- Second, in the present chapter, we have constructed an invariant

$$Q_{(\mathsf{R},\alpha,\beta)},$$

  where the notation follows that of Definition 7.20.

These two invariants were constructed in a similar way, by "slicing" a diagram into elementary pieces. Furthermore, there is a notable similarity in the conditions on the maps denoted by μ in Theorems 6.13 and 7.19. There is a small difference in the Tr_2 condition between the two theorems, but the following lemma indicates that this difference is cosmetic.

Lemma 7.29. *If* $\mathsf{R}^{\pm 1}\colon V\otimes V\to V\otimes V$ *and* $\mu\colon V\to V$, *where* μ *is invertible, then*

$$\mathrm{Tr}_2(\mathsf{R}^{\pm 1}\circ(\mu\otimes\mu))=\mu \quad\Longleftrightarrow\quad \mathrm{Tr}_2(\mathsf{R}^{\pm 1}\circ(\mathrm{id}\otimes\mu))=\mathrm{id}_V.$$

Proof. By Part 1 of Exercise 6.6, we have

$$\mathrm{Tr}_2(\mathsf{R}\circ(\mu\otimes\mu))=\mathrm{Tr}_2((\mathsf{R}\circ(\mathrm{id}\otimes\mu))\circ(\mu\otimes\mathrm{id}))=(\mathrm{Tr}_2(\mathsf{R}\circ(\mathrm{id}\otimes\mu)))\circ\mu$$

so

$$\begin{aligned}\mathrm{Tr}_2(\mathsf{R}\circ(\mu\otimes\mu))=\mu &\Longleftrightarrow (\mathrm{Tr}_2(\mathsf{R}\circ(\mathrm{id}\otimes\mu)))\circ\mu=\mu\\ &\Longleftrightarrow \mathrm{Tr}_2(\mathsf{R}\circ(\mathrm{id}\otimes\mu))=\mathrm{id}_V\end{aligned}$$

where the latter bi-implication is by composition with μ^{-1}. The result for R^{-1} holds since R was arbitrary. □

If V is a free module of finite rank, $W=V^*$, and $\mathsf{R}:V\otimes V\to V\otimes V,\alpha:V\to V$, and $\beta:V^*\to V^*$ are maps satisfying the conditions stated in Theorem 7.19 for $\mathcal{Q}_{(\mathsf{R},\alpha,\beta)}$ to be an operator invariant, then for $\mu=\beta^\star\circ\alpha^{-1}$ the pair $\mathsf{R}:V\otimes V\to V\otimes V$ and $\mu:V\to V$ satisfy the conditions of Theorem 6.13, and hence $\rho_{(\mathsf{R},\mu)}$ is a link invariant. This immediately gives the following.

Theorem 7.30. *Let* V *be a free module of finite rank, and* $W=V^*$. *If* $\mathcal{Q}_{(\mathsf{R},\alpha,\beta)}$ *is an operator invariant, then* $\rho_{(\mathsf{R},\beta^\star\circ\alpha^{-1})}$ *is a link invariant.*

In the light of Theorem 7.30, it is natural to ask how the values of the invariants $\mathcal{Q}_{(\mathsf{R},\alpha,\beta)}$ and $\rho_{(\mathsf{R},\beta^\star\circ\alpha^{-1})}$ compare. The following theorem provides the answer.

Theorem 7.31. *Let* D *be a link diagram. Let* $\mathcal{Q}_{(\mathsf{R},\alpha,\beta)}$ *be an operator invariant. Then,* $\rho_{(\mathsf{R},\mu)}$ *is defined through* $\mu=\beta^\star\circ\alpha^{-1}$ *and*

$$\mathcal{Q}_{(\mathsf{R},\alpha,\beta)}(D)(1_{\mathbb{K}})=\rho_{(\mathsf{R},\mu)}(D). \tag{7.15}$$

Moreover, if D *is a link diagram and* $\mathcal{Q}_{(\mathsf{R},\alpha,\beta)}$ *and* $\mathcal{Q}_{(\mathsf{R},\alpha',\beta')}$ *are operator invariants such that*

$$\beta^\star\circ\alpha^{-1}=\mu=(\beta')^\star\circ(\alpha')^{-1}$$

then

$$\mathcal{Q}_{(\mathsf{R},\alpha,\beta)}(D)=\mathcal{Q}_{(\mathsf{R},\alpha',\beta')}(D). \tag{7.16}$$

We prove this theorem after a preliminary discussion.

There are two points to be noted. First, the codomain of $\rho_{(\mathsf{R},\beta^\star\circ\alpha^{-1})}$ is $\mathbb{K}$, while the codomain of $\mathcal{Q}_{(\mathsf{R},\alpha,\beta)}(D)$ is $\mathrm{End}(\mathbb{K})$, so, to make the codomains agree, we evaluate $\mathcal{Q}_{(\mathsf{R},\alpha,\beta)}(D)$ at a specific scalar, namely $1_{\mathbb{K}}$. The second point concerns the construction of $\rho_{(\mathsf{R},\beta^\star\circ\alpha^{-1})}$. To calculate $\rho_{(\mathsf{R},\beta^\star\circ\alpha^{-1})}(D)$, it is first necessary to express the link diagram D as a closed braid $\widehat{\sigma}$, where $\sigma\in\mathfrak{B}_n$, so that

$$\rho_{(\mathsf{R},\beta^\star\circ\alpha^{-1})}(D) = \rho_{(\mathsf{R},\beta^\star\circ\alpha^{-1})}(\widehat{\sigma}) = \operatorname{Tr}\left(\rho_{\mathsf{R}}(\sigma)\circ(\beta^\star\circ\alpha^{-1})^{\otimes n}\right).$$

By isotopy invariance, we also have

$$Q_{(\mathsf{R},\alpha,\beta)}(D) = Q_{(\mathsf{R},\alpha,\beta)}(\widehat{\sigma}).$$

The form of $\widehat{\sigma}$ is particularly convenient for calculating $Q_{(\mathsf{R},\alpha,\beta)}(\widehat{\sigma})$, and we shall prove the theorem by calculating and comparing $Q_{(\mathsf{R},\alpha,\beta)}(\widehat{\sigma})$ and $\rho_{(\mathsf{R},\beta^\star\circ\alpha^{-1})}(\widehat{\sigma})$.

Let $\widehat{\sigma}$ be the closure of the n-stranded braid σ, so its slicing is

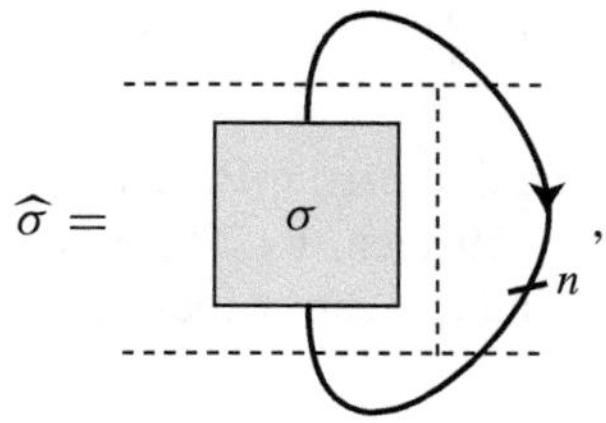

with the convention that the *striation* (*i.e.* the diagonal mark) on a strand that is marked by n indicates a collection of n parallel such strands. The diagram has been sliced into four parts: the braid σ, n strands forming n maxima, n strands forming n minima, and n parallel strands joining the maxima and minima. Denoting $Q_{(\mathsf{R},\alpha,\beta)}$ by Q, it follows from Definition 7.13 that

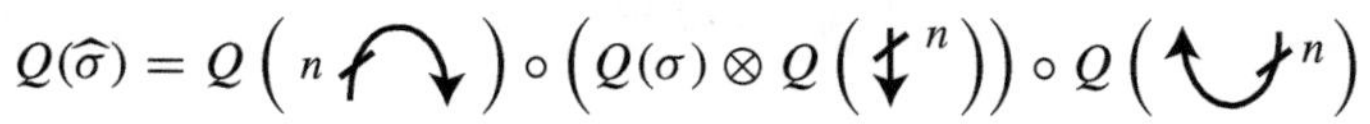 (7.17)

where Q  corresponds to the upper slice of the diagram, Q 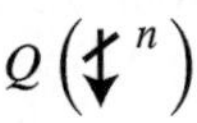 to the right-middle slice, and Q  to the lower slice. But both $\rho_{(\mathsf{R},\mu)}$ and Q are constructed from their values on the elementary pieces specified by

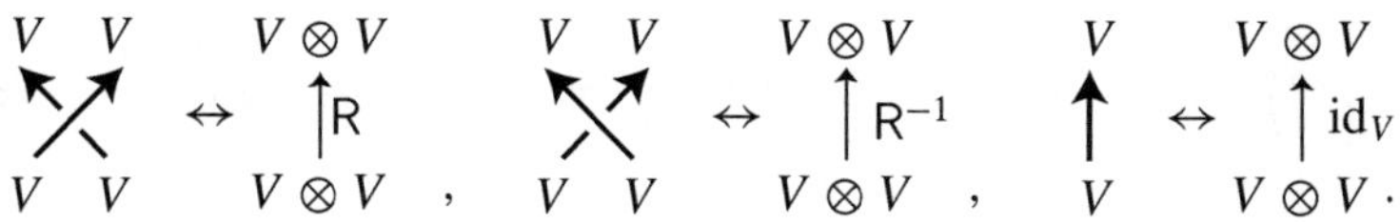

so, clearly,

$$Q(\sigma) = \rho_{\mathsf{R}}(\sigma), \qquad \text{and} \qquad Q\left(\ldots\right) = \mathrm{id}_W^{\otimes n}. \tag{7.18}$$

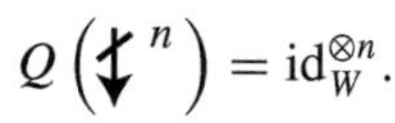

Next, we can write

and so

$$Q\left(\right.$$ 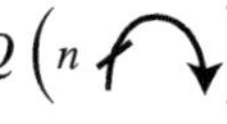 $$\left.\right) = Q\left(\right.$$  $$\left.\right) \circ \cdots \circ Q\left(\right.$$  $$\left.\right),$$

giving

$$Q\left(\right.$$  $$\left.\right) = (\vec{n}) \circ \cdots \circ \left(\mathrm{id}_V^{\otimes(n-1)} \otimes \vec{n} \otimes \mathrm{id}_W^{\otimes(n-1)}\right). \tag{7.19}$$

A similar argument shows that

$$Q\left(\right.$$  $$\left.\right) = \left(\mathrm{id}_V^{\otimes(n-1)} \otimes \overleftarrow{u} \otimes \mathrm{id}_W^{\otimes(n-1)}\right) \circ \cdots \circ (\overleftarrow{u}). \tag{7.20}$$

Proof (**Theorem** 7.31)**.** For the first part, if R, μ satisfy the conditions for operator invariants given in Theorem 7.19 then, by Theorem 7.30, $\rho_{(\mathsf{R},\mu)}$ is defined. The element-wise actions of $(\alpha^{-1})^{\cap}$ and $\beta^{\cup}$ can be found in Lemma B.4. Now, from (7.17),

$$Q\left(\right.$$ 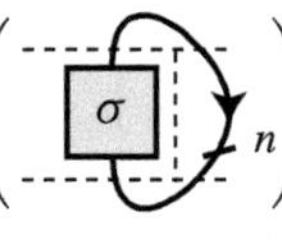 $$\left.\right) = Q\left(\right.$$  $$\left.\right) \circ \left(Q(\sigma) \otimes Q\left(\right.$$ 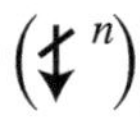 $$\left.\right)\right) \circ Q\left(\right.$$  $$\left.\right)$$

where, as before, $Q_{(\mathsf{R},\alpha,\beta)}$ is denoted by Q. Moreover, from (7.18),

$$Q\left(\right.$$ 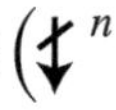 $$\left.\right) = \mathrm{id}_W^{\otimes n}$$

and, from (7.20), and that $\overleftarrow{u} = \beta^{\cup}$,

$$Q\left(\right.$$  $$\left.\right) = \left(\mathrm{id}_V^{\otimes(n-1)} \otimes \beta^{\cup} \otimes \mathrm{id}_W^{\otimes(n-1)}\right) \circ \cdots \circ (\beta^{\cup})$$

so,

$$Q\left(\right.$$  $$\left.\right) : 1_{\mathbb{K}} \mapsto \sum_{i_1,\ldots,i_{2n}} \beta^{i_1 i_{2n}} \beta^{i_2 i_{2n-1}} \ldots \beta^{i_n i_{n+1}} v_{i_1} \otimes \cdots \otimes v_{i_n} \otimes w_{i_{n+1}} \otimes \cdots \otimes w_{i_{2n}},$$

where $\beta \colon v^i \mapsto \sum_j \beta^{ij} w_j$, so $\beta^{\cup} \colon 1_{\mathbb{K}} \mapsto \sum_{i,j} \beta^{ij} v_i \otimes w_j$.

Similarly,

$$Q\left(\right.$$  $$\left.\right) = \left(\alpha^{-1}\right)^{\cap} \circ \cdots \circ \left(\mathrm{id}_V^{\otimes(n-1)} \otimes \left(\alpha^{-1}\right)^{\cap} \otimes \mathrm{id}_W^{\otimes(n-1)}\right),$$

so

$$Q\left(\right.$$  $$\left.\right) : v_{i_1} \otimes \cdots \otimes v_{i_n} \otimes w_{i_{n+1}} \otimes \cdots \otimes w_{i_{2n}} \mapsto \overline{\alpha}_{i_n\, i_{n+1}} \overline{\alpha}_{i_{n-1}\, i_{n+2}} \cdots \overline{\alpha}_{i_1\, i_{2n}},$$

where $\alpha^{-1} \colon v_i \mapsto \sum_j \overline{\alpha}_{ij} w^j$ so $(\alpha^{-1})^{\cap} \colon v_i \otimes w_j \mapsto \overline{\alpha}_{ij}$.

For the element-wise action of Q 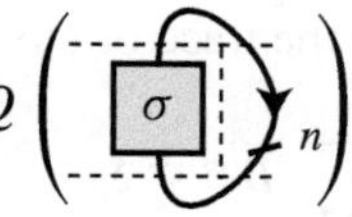, we have

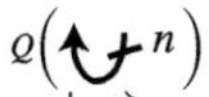

$$1_{\mathbb{K}} \overset{Q(\cdot)}{\longmapsto} \sum_{i_1,\dots,i_{2n}} \beta^{i_1\,i_{2n}}\,\beta^{i_2\,i_{2n-1}}\dots\beta^{i_n\,i_{n+1}} v_{i_1}\otimes\dots\otimes v_{i_n}\otimes w_{i_{n+1}}\cdots\otimes w_{i_{2n}} \tag{7.21}$$

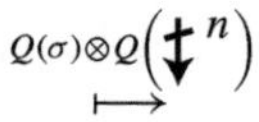

$$\overset{Q(\sigma)\otimes Q(\cdot)}{\longmapsto} \sum_{\substack{i_1,\dots,i_{2n},\\ j_1,\dots,j_n}} \rho^{j_1\dots j_n}_{i_1\dots i_n}\,\beta^{i_1\,i_{2n}}\,\beta^{i_2\,i_{2n-1}}\dots\beta^{i_n\,i_{n+1}}\, v_{j_1}\otimes\dots\otimes v_{j_n}\otimes w_{i_{n+1}}\otimes\cdots w_{i_{2n}}$$

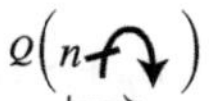

$$\overset{Q(\cdot)}{\longmapsto} \sum_{\substack{i_1,\dots,i_{2n},\\ j_1,\dots,j_n}} \rho^{j_1\dots j_n}_{i_1\dots i_n}\,\beta^{i_1\,i_{2n}}\,\beta^{i_2\,i_{2n-1}}\dots\beta^{i_n\,i_{n+1}}\,\overline{\alpha}_{j_n\,i_{n+1}}\,\overline{\alpha}_{j_{n-1}\,i_{n+2}}\cdots\overline{\alpha}_{j_1\,i_{2n}},$$

where

$$\rho(\sigma)\colon v_{i_1}\otimes\dots\otimes v_{i_n}\mapsto \sum_{j_1,\dots,j_n} \rho^{j_1\dots j_n}_{i_1\dots i_n}\, v_{j_1}\otimes\dots\otimes v_{j_n}.$$

On the other hand, the element-wise action of $\rho_{\mathsf{R}}(\sigma)\circ\mu^{\otimes n}$ is obtained as follows. Note that

$$\mu\colon v_i\mapsto \sum_{j,k}\overline{\alpha}_{ij}\,\beta^{kj}\,v_k.$$

Then

$$v_{i_1}\otimes\dots\otimes v_{i_n} \overset{\mu^{\otimes n}}{\longmapsto} \sum_{j_1,\dots,j_n} \overline{\alpha}_{i_1\,j_1}\,\beta^{k_1\,j_1}\dots\overline{\alpha}_{i_n\,j_n}\,\beta^{k_n\,j_n}\,v_{k_1}\otimes\dots\otimes v_{k_n}$$

$$\overset{\rho_{\mathsf{R}}(\sigma)}{\longmapsto} \sum_{l_1,\dots,l_n} B^{l_1\dots l_n}_{i_1\dots i_n}\, v_{l_1}\otimes\dots\otimes v_{l_n} \tag{7.22}$$

where

$$B^{l_1\dots l_n}_{i_1\dots i_n} := \sum_{\substack{j_1,\dots,j_n,\\ k_1,\dots,k_n}} \overline{\alpha}_{i_1\,j_1}\,\beta^{k_1\,j_1}\dots\overline{\alpha}_{i_n\,j_n}\,\beta^{k_n\,j_n}\,\rho^{l_1\dots l_n}_{k_1\dots k_n}.$$

As a tensor, the map in (7.22) is

$$\sum_{\substack{i_1,\dots,i_n,\\ l_1,\dots,l_n}} B^{l_1\dots l_n}_{i_1\dots i_n}\,(v_{i_1}\otimes\dots\otimes v_{i_n})^{\star}\otimes(v_{l_1}\otimes\dots\otimes v_{l_n}).$$

Then, contracting this to obtain the trace of this expression, we have

$$\sum_{\substack{i_1,\ldots,i_n,\\ l_1,\ldots,l_n}} B^{l_1\ldots l_n}_{i_1\ldots i_n}\,\delta_{i_1,l_1}\cdots\delta_{i_n,l_n},$$

and therefore

$$\mathrm{Tr}\left(\rho_{\mathsf{R}}(\sigma)\circ\mu^{\otimes n}\right)=\sum_{\substack{i_1,\ldots,i_n,\\ j_1,\ldots,j_n,\\ k_1,\ldots,k_n}}\overline{\alpha}_{i_1\,j_1}\,\beta^{k_1\,j_1}\cdots\overline{\alpha}_{i_n\,j_n}\,\beta^{k_n\,j_n}\,\rho^{i_1\ldots i_n}_{k_1\ldots k_n}. \tag{7.23}$$

Reindexing this summation, for $p=1,\ldots,n$, through $j_p\mapsto i_{2n-p+1}$, $k_p\mapsto i_p$, and $i_p\mapsto j_p$ gives (7.21). Thus, the element-wise actions are the same so the first part of the theorem follows.

For the second part, given D, there is a presentation of it as a closed braid $\widehat{\sigma}$: this may always be done by Theorem 4.12. Then

$$\begin{aligned}
Q_{(\mathsf{R},\alpha,\beta)}(D)(1_{\mathbb{K}}) &= Q_{(\mathsf{R},\alpha,\beta)}(\widehat{\sigma})(1_{\mathbb{K}}) && \text{(isotopy invariance)}\\
&= Q_{(\mathsf{R},\alpha',\beta')}(\widehat{\sigma})(1_{\mathbb{K}}) && \text{(Part 1 of the Theorem)}\\
&= Q_{(\mathsf{R},\alpha',\beta')}(D)(1_{\mathbb{K}}) && \text{(isotopy invariance).}
\end{aligned}$$

Since a linear map from $\mathbb{K}$ to $\mathbb{K}$ is determined by its action on $1_{\mathbb{K}}$, the result follows. □

Since for links, as opposed to tangles, $Q_{(\mathsf{R},\alpha,\beta)}$ is dependent only upon $\beta^\star\circ\alpha^{-1}=\mu$, we may denote $Q_{(\mathsf{R},\alpha,\beta)}$ by $Q_{(\mathsf{R},\mu)}$. In general, however, the operator invariant of a tangle will depend upon the choice of α and β. For example, for the $(2,0)$-tangle , we have $Q_{(\mathsf{R},\alpha,\beta)}\Big($  $\Big)=(\alpha^{-1})^{\cap}$, which clearly depends on α.

We have shown that, for every $\mathsf{R}\in\mathrm{Aut}(V\otimes V)$ and $\mu\in\mathrm{Aut}(V)$ that satisfy

- The Yang–Baxter Equation,
- $\mathsf{R}\circ(\mu\otimes\mu)=(\mu\otimes\mu)\otimes\mathsf{R}$,
- $\mathrm{Tr}_2\left(R^{\pm1}\circ(\mathrm{id}_V\otimes\mu)\right)=\mathrm{id}_V$,
- $(\tau\circ\mathsf{R}^{-1})^{t_1}\otimes(\mathrm{id}_{V^\star}\otimes\mu)\circ(\mathsf{R}\circ\tau)^{t_1}\circ(\mathrm{id}_{V^\star}\otimes\mu)^{-1}=\mathrm{id}_{V^\star\otimes V}$.

there exists a *link invariant* $Q_{(\mathsf{R},\mu)}$. Moreover, this invariant is equivalent to $\rho_{(\mathsf{R},\mu)}$.

In summary, this means:

- We have extended the construction of representations of the braid group using R-matrices to oriented link diagrams, at the expense of additional conditions on the R-matrix.
- We have also extended the construction to tangle invariants $Q_{(\mathsf{R},\alpha,\beta)}$.

7.5 Framed Tangles

In our exposition of operator invariants to this point, we have considered oriented knots and links. Recall from Chap. 3 that framed links form another class of knotted objects. In that chapter, we saw two ways of thinking about framed links: the first as "knotted ribbons" in $\mathbb{R}^3$; the second as link diagrams modulo the framed Reidemeister moves of Definition 3.3. The concept of a framed link may be extended to the concept of a *framed tangle*. We shall see in Chap. 9 that much of the theory of operator invariants resides naturally in the setting of *framed oriented tangles*.

Although we could approach framed tangles through the geometric approach of "forming tangles in $\mathbb{R}^3$ out of ribbons", we shall instead give a purely diagrammatic approach which essentially involves replacing the RI-move by the FRI-move in the work on tangle diagrams in Sect. 7.1.

A *framed* (m, n)*-tangle diagram* consists of an (m, n)-tangle diagram. As with tangle diagrams, framed tangle diagram can be oriented or unoriented, with the orientation indicated by an arrow on each component of the tangle. Framed tangle diagrams are considered up to a different equivalence relation from that used for tangle diagrams.

Definition 7.32 (Equivalence). Two framed tangle diagrams are *equivalent* if and only if their two diagrams are related by a finite sequence of framed Reidemeister moves. These consist of isotopy of $\mathbb{R} \times I$ and the following three local changes in a diagram:

FRI RII

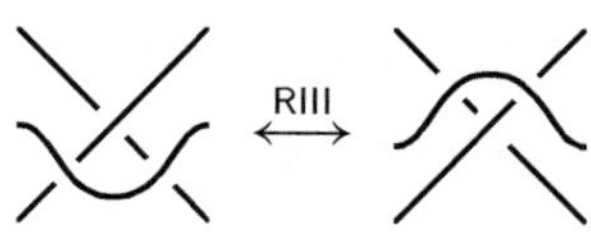

RIII

For oriented tangles, the moves preserve orientation of the strands shown.

(The above definition should be compared with Theorem 7.3.)

Composition, $\circ$, and the tensor product, $\otimes$, of framed tangle diagrams are defined just as for tangles in Definitions 7.5 and 7.6. Since an oriented framed tangle diagram is just an oriented tangle diagram, we can follow the constructions of Sect. 7.1, consider a generating set for oriented tangle diagrams and obtain an analogue of Theorem 7.10 that expresses oriented framed tangles in terms of a generating set and equivalence relations.

Theorem 7.33.

$$\frac{\{\textit{oriented framed tangle diagrams}\}}{\textit{framed Reidemeister moves}} \cong$$

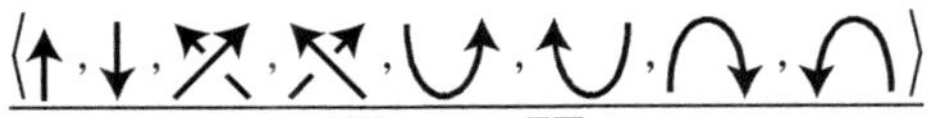

$$(\mathsf{FT}_0, \ldots, \mathsf{FT}_7)$$

where the framed Turaev moves *are*

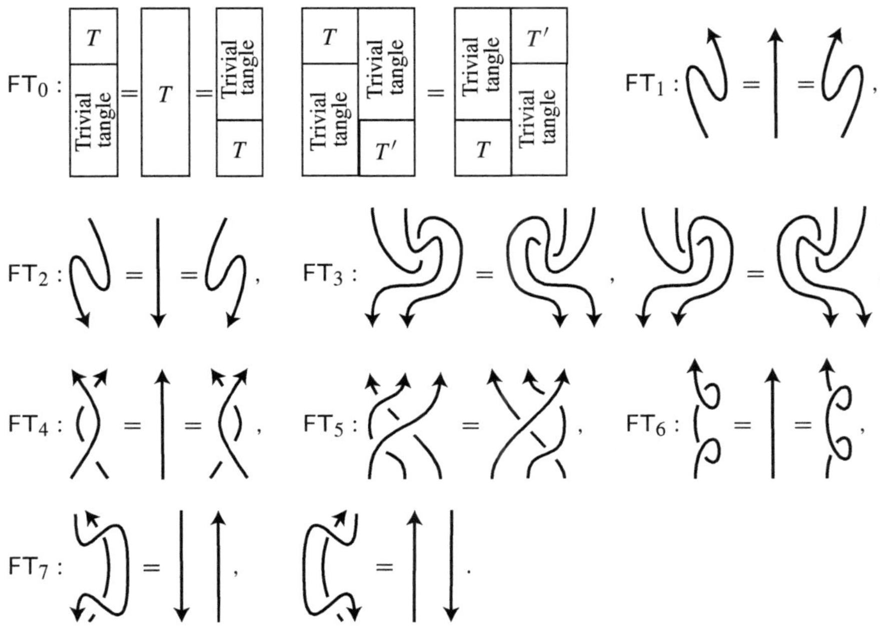

As in Sect. 7.2, we may construct a map

$$Q^f \equiv Q^f_{(V,W,\mathsf{R},\vec{u},\overleftarrow{u},\vec{n},\overleftarrow{n})}$$

from framed tangles by associating morphism to elementary tangles.

Definition 7.34 (Map $Q^f_{(V,W,\mathsf{R},\vec{u},\overleftarrow{u},\vec{n},\overleftarrow{n})}$). Let V and W be free modules of finite rank and morphisms R, $\vec{u}$, $\overleftarrow{u}$, $\vec{n}$, $\overleftarrow{n}$ as below, $Q^f \equiv Q^f_{(V,W,\mathsf{R},\vec{u},\overleftarrow{u},\vec{n},\overleftarrow{n})}$ is the map on tangle diagrams defined through the actions below.

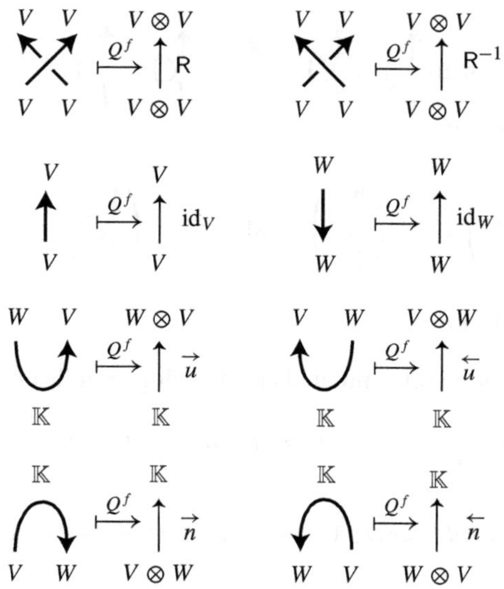

(7.23)

$$T \otimes T' \overset{Q}{\longmapsto} Q^f(T) \otimes Q^f(T'),$$

$$T \circ T' \overset{Q}{\longmapsto} Q^f(T) \circ Q^f(T').$$

Again, as in Sect. 7.2, we can find a set of conditions on R, $\vec{n}$, $\overleftarrow{n}$, $\vec{u}$, $\overleftarrow{u}$ that will make Q^f into an invariant of oriented framed tangles.

Theorem 7.35. *If the maps* R, $\vec{n}$, $\overleftarrow{n}$, $\vec{u}$, $\overleftarrow{u}$ *satisfy the conditions (7.8)–(7.12), (7.14) for the* $\mathsf{T}_1 - \mathsf{T}_5$, T_7*-moves, then* Q^f *is invariant under* FT_6 *if and only if*

$$\Big((\mathrm{id}_V \otimes \vec{n}) \circ (\mathsf{R}^{\pm 1} \otimes \mathrm{id}_W) \circ (\mathrm{id}_V \otimes \overleftarrow{u})\Big) \circ \Big((\mathrm{id}_V \otimes \vec{n}) \circ (\mathsf{R}^{\mp 1} \otimes \mathrm{id}_W) \circ (\mathrm{id}_V \otimes \overleftarrow{u})\Big) = \mathrm{id}_V. \tag{7.25}$$

Proof. As in Sect. 7.2, if the maps satisfy the stated conditions, then Q^f is invariant under $\mathsf{FT}_0, \ldots, \mathsf{FT}_5$ and FT_7. It remains to find a condition that the maps must satisfy in order that Q^f satisfies FT_6. By the calculations for (7.13), the T_6-move, we have

 $\overset{Q^f}{\longmapsto} (\mathrm{id}_V \otimes \overrightarrow{n}) \circ (\mathsf{R} \otimes \mathrm{id}_W) \circ (\mathrm{id}_V \otimes \overleftarrow{u})$ and

 $\overset{Q^f}{\longmapsto} (\mathrm{id}_V \otimes \overrightarrow{n}) \circ (\mathsf{R}^{-1} \otimes \mathrm{id}_W) \circ (\mathrm{id}_V \otimes \overleftarrow{u})$.

Then, FT_6 is equivalent to 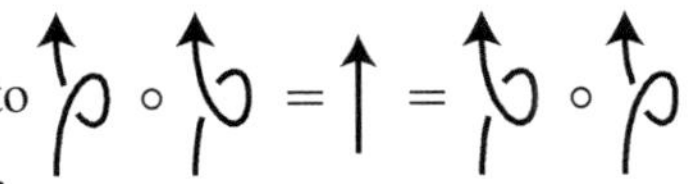. Therefore, Q^f is invariant under FT_6 if and only if

Q^f 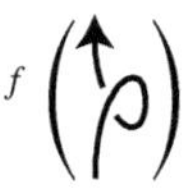 $\circ\, Q^f$  $= \mathrm{id}_V = Q^f$ 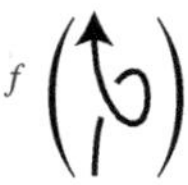 $\circ\, Q^f$ 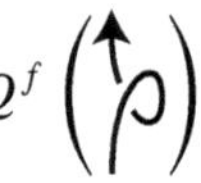,

and the result now follows. □

This theorem may be reformulated as an analogue of Theorem 7.19 as follows.

Theorem 7.36. $Q^f_{(V,W,\mathsf{R},\overrightarrow{n},\overleftarrow{n},\overrightarrow{u},\overleftarrow{u})}$ *is an invariant of framed oriented tangles if and only if*

1. *There exist isomorphisms* $\alpha\colon W^\star \to V$ *and* $\beta\colon V^\star \to W$ *such that*

$$\overrightarrow{u} = \alpha^\cup, \qquad \overrightarrow{n} = (\alpha^{-1})^\cap, \qquad \overleftarrow{u} = \beta^\cup, \qquad \overleftarrow{n} = (\beta^{-1})^\cap.$$

2. R *is an invertible R-matrix.*
3. *If* $\mu := \beta^\star \circ \alpha^{-1}\colon V \to V$, *then*
   a. $\left(\tau \circ \mathsf{R}^{-1}\right)^{t_1} \circ (\mathrm{id}_{V^\star} \otimes \mu) \circ (\mathsf{R} \circ \tau)^{t_1} \circ (\mathrm{id}_{V^\star} \otimes \mu)^{-1} = \mathrm{id}_{V^\star \otimes V}$;
   b. $\mathrm{Tr}_2\,(\mathsf{R} \circ (\mathrm{id}_V \otimes \mu)) = \left(\mathrm{Tr}_2\left(\mathsf{R}^{-1} \circ (\mathrm{id}_V \otimes \mu)\right)\right)^{-1}$;
   c. $\mathsf{R} \circ (\mu \otimes \mu) = (\mu \otimes \mu) \circ \mathsf{R}$.

Proof. By the proof of Theorem 7.19, it is necessary only to show that condition 3(b) of the theorem implies invariance under FT_6. By Lemma B.8, we know that the element-wise action of

$$(\mathrm{id}_V \otimes \overrightarrow{n}) \circ (\mathsf{R}^{\pm 1} \otimes \mathrm{id}_W) \circ (\mathrm{id}_V \otimes \overleftarrow{u})$$

is the same as the element-wise action of

$$\mathrm{Tr}_2\left(\mathsf{R}^{\pm 1} \circ (\mathrm{id}_V \otimes \mu)\right).$$

The result follows since the relevant maps in (7.25) are invertible. This completes the proof. □

Definition 7.37 (Framed operator invariant). Let V and W be free modules of the same finite rank. Let $\mathsf{R} : V \otimes V \to V \otimes V$, $\alpha\colon W^\star \to V$ and $\beta\colon V^\star \to W$ be maps satisfying the condition in Theorem 7.36. Then

$$Q^f_{(\mathsf{R},\alpha,\beta)}$$

denotes the *framed operator invariant* $Q^f_{(V,W,\mathsf{R},\vec{n},\overleftarrow{n},\vec{u},\overleftarrow{u})}$ where $\vec{u} = \alpha^{\cup}$, $\vec{n} = (\alpha^{-1})^{\cap}$, $\overleftarrow{u} = \beta^{\cup}$, $\overleftarrow{n} = (\beta^{-1})^{\cap}$.

Exercise 7.38. Let $Q^f_{(\mathsf{R},\vec{n},\overleftarrow{n},\vec{u},\overleftarrow{u})}$ be an operator invariant of framed tangles, and suppose that for some invertible $c \in \mathbb{K}$,

$$\mathrm{Tr}_2\left(\mathsf{R}^{\pm 1} \circ (\mathrm{id}_V \otimes \mu)\right) = c^{\pm 1}\,\mathrm{id}_V.$$

By using the deframing approach of Sect. 3.3, prove that

$$c^{-\omega(D)}\, Q^f(D)$$

(where $\omega(D)$ is the writhe of D) is an invariant of (unframed) oriented tangles.

Chapter 8
Ribbon Hopf Algebras

The invariants $\rho_{\mathsf{R},\mu}$ (see Theorem 6.13) and $Q_{\mathsf{R},\alpha,\beta}$ (see Theorem 7.31) require us to provide a set of morphisms R, μ, α and β in order to define them. Moreover, these morphisms must satisfy a quite technical set of conditions for us to obtain a knot invariant. So far, very little has been said about how we might set about finding a suitable collection of such morphisms. In this and the following chapter, we shall address this issue by showing how the theory of quantum groups gives rise to knot invariants.

The idea is to obtain the morphisms R, μ, α and β as representations of suitable elements of a suitable algebra. Of course, since R, μ, α and β satisfy a strong set of constraints, we should expect the need for sophisticated algebras. Indeed, this proves to be the case. The type of algebra we need to consider is a ribbon Hopf algebra. These are (Hopf) algebras equipped with special elements, denoted by R and $\mathbf{v}$, whose representations give rise to the morphisms R, μ, α and β. (Note the double use of the symbol R here to denote an element of an algebra and a morphism. In practice, this should not cause any confusion.)

This chapter gives an introduction to ribbon Hopf algebras and the concepts (such as algebras, coalgebras and Hopf algebras) required to understand them. In Chapter 9, we describe how ribbon Hopf algebras and their representations give rise to framed tangle invariants.

At first it might appear that this process is a sleight of hand, since we have just moved the problem of constructing morphisms R, μ, α and β to one of constructing ribbon Hopf algebras. However, as discussed at the end of this chapter, ribbon Hopf algebras may be constructed from semi-simple Lie algebras. Thus, by passing through the framework of ribbon Hopf algebras we are able to construct a knot invariant from any semi-simple Lie algebra. This is a deep and important fact, and one that, in Part IV, will bind the theory of operator invariants we have seen in Part II of the book to the very different theory of Vassiliev invariants that we shall see in Part III.

D. M. Jackson and I. Moffatt, *An Introduction to Quantum and Vassiliev Knot Invariants*,
CMS Books in Mathematics, https://doi.org/10.1007/978-3-030-05213-3_8

8.1 Algebras, Coalgebras and Hopf Algebras

Before considering ribbon Hopf algebras, we must understand what an algebra, a coalgebra and a Hopf algebra are. In this chapter, $\mathbb{K}$ will always be a commutative ring with unity.

8.1.1 Algebras

Throughout this section, we let $\mathfrak{A}$ be a free module of finite rank over a ring with unity $\mathbb{K}$. Then a *product* on $\mathfrak{A}$ is a module morphism

$$m\colon \mathfrak{A} \otimes \mathfrak{A} \to \mathfrak{A}$$

and $\mathfrak{A}$ endowed with m is an *algebra* $(\mathfrak{A}, m)$. Thus, an algebra is simply a vector space or module equipped with a concept of multiplication.

As is usual in algebra, except when we wish to emphasise the multiplication, we denote $(\mathfrak{A}, m)$ by just $\mathfrak{A}$. Similar comments hold for the other types of algebraic objects we consider here.

According to this definition of an algebra, we should write the product of elements x and y of $\mathfrak{A}$ as $m(x \otimes y)$. However, it is customary to write it as $m(x, y)$ or $x \cdot y$ or xy, so we shall follow this custom.

The algebra $(\mathfrak{A}, m)$ is *associative* if

$$m(m(x, y), z) = m(x, m(y, z)) \quad \text{or} \quad (x \cdot y) \cdot z = x \cdot (y \cdot z), \qquad \text{for all } x, y, z \in \mathfrak{A}.$$

This condition is better expressed in terms of maps, in which case $(\mathfrak{A}, m)$ is *associative* if the diagram

$$\begin{array}{ccc} \mathfrak{A} \otimes \mathfrak{A} \otimes \mathfrak{A} & \xrightarrow{\mathrm{id}\otimes m} & \mathfrak{A} \otimes \mathfrak{A} \\ \big\downarrow{\scriptstyle m\otimes\mathrm{id}} & & \big\downarrow{\scriptstyle m} \\ \mathfrak{A} \otimes \mathfrak{A} & \xrightarrow{\quad m \quad} & \mathfrak{A} \end{array} \tag{8.1}$$

is a commutative diagram (which happens if and only if $m \circ (m \otimes \mathrm{id}) = m \circ (\mathrm{id} \otimes m)$).

In this text, we shall always assume that the algebras are associative without further comment.

An element $x \in \mathfrak{A}$ is *central* if $x \cdot y = y \cdot x$ for all $y \in \mathfrak{A}$. The *centre* of $\mathfrak{A}$ is the set of all its central elements.

A *unit* of the algebra $(\mathfrak{A}, m)$ is an element $1_{\mathfrak{A}}$ such that

$$m(x, 1_{\mathfrak{A}}) = x = m(1_{\mathfrak{A}}, x) \quad \text{or} \quad x \cdot 1_{\mathfrak{A}} = x = 1_{\mathfrak{A}} \cdot x \quad \text{for all } x \in \mathfrak{A}. \tag{8.2}$$

If the algebra $(\mathfrak{A}, m)$ has a unit, then the unit is unique, for if u is also a unit then, from (8.2), $1_{\mathfrak{A}} = m(1_{\mathfrak{A}}, u) = u$. If $1_{\mathfrak{A}}$ exists then $(\mathfrak{A}, m)$ is said to be a *unital algebra*.

Exercise 8.1. Let V be a module. Show that $\mathrm{End}(V)$ forms an algebra with unit under composition.

Exercise 8.2. Consider the set of braids on n strands. We can form a vector space on this set by considering all formal finite linear combinations of braids over $\mathbb{C}$. (See Definition 10.1 for a discussion of this vector space in the case of knots.) Show that this vector space of braids forms an algebra with unit, where multiplication is given by braid composition.

Exercise 8.3. Let $(\mathfrak{A}, m)$ be a algebra. Define a map $m^{\mathrm{op}} : \mathfrak{A} \otimes \mathfrak{A} \to \mathfrak{A}$ as $m^{\mathrm{op}} := m \circ \tau$, where τ is the twist map of Definition 7.25. Verify that $(\mathfrak{A}, m^{\mathrm{op}})$ is a algebra. Show that $(\mathfrak{A}, m)$ has a unit if and only if $(\mathfrak{A}, m^{\mathrm{op}})$ does, and that both units are the same element of $\mathfrak{A}$. The algebra $(\mathfrak{A}, m^{\mathrm{op}})$ is called the *opposite algebra* of $(\mathfrak{A}, m)$.

Units can be characterised in terms of maps as follows. A map $\eta \in \mathrm{Hom}(\mathfrak{A}, \mathbb{K})$ is a *unit* if the diagram

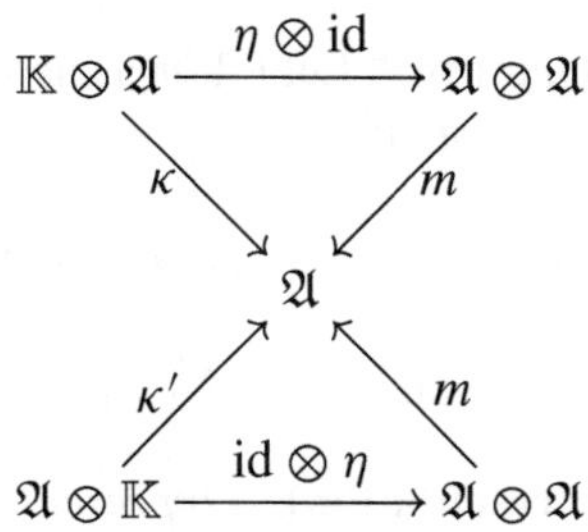

(8.3)

commutes, where

$$\kappa : \mathbb{K} \otimes \mathfrak{A} \xrightarrow{\cong} \mathfrak{A} : k \otimes a \mapsto ka \quad \text{and} \quad \kappa' : \mathfrak{A} \otimes \mathbb{K} \xrightarrow{\cong} \mathfrak{A} : a \otimes k \mapsto ka \tag{8.4}$$

are the contraction maps.

Exercise 8.4. Show that the two definitions of a unit are equivalent, and, in particular, each $\eta(1_{\mathbb{K}}) = 1_{\mathfrak{A}}$ defines either type of unit from the other.

At times, we shall want to define algebras through their generators.

Definition 8.5 (Generators). The *associative unital algebra generated* by elements $E_1, E_2, \ldots, E_n$ over a unital ring $\mathbb{K}$ is the algebra defined as follows. The underlying module is that generated by the set of all words in $E_1, E_2, \ldots, E_n$. Multiplication is defined by concatenation of words $m : w_1 \otimes w_2 \mapsto w_1 w_2$, and the unit is the empty word.

For example, the associative unital algebra over $\mathbb{C}$ generated by elements X and Y will consist of all finite formal linear combinations of elements of

$$\{X^{n_1}Y^{n_2}X^{n_3}\cdots X^{n_{k-1}}Y^{n_k} : k, n_1, n_k \in \mathbb{N}_0, n_2, \ldots, n_{k-1} \in \mathbb{N}\},$$

the set of all distinct words on the alphabet $\{X, Y\}$. As an example of multiplication, $m : Y^2X^3Y^2X \otimes X^2Y^4X^2 \mapsto Y^2X^3Y^2X^3Y^4X^2$.

Definition 8.6 (Graded algebra). An algebra $(\mathfrak{A}, m)$ is said to be *graded* if $\mathfrak{A} = \bigoplus_{k\geq 0} \mathfrak{A}_k$, where each $\mathfrak{A}_k$ is a free module of finite rank and

$$m\colon \mathfrak{A}_p \otimes \mathfrak{A}_q \to \mathfrak{A}_{p+q}.$$

Example 8.7. Let V be a vector space. The *tensor algebra*, $T(V)$, of V is defined to be the algebra on $\bigoplus_{r\geq 0} V^{\otimes r}$, where $V^{\otimes 0} = \mathbb{K}$, with multiplication given by $(x_1 \otimes \ldots \otimes x_p) \otimes (y_1 \otimes \ldots \otimes y_q) \mapsto x_1 \otimes \ldots \otimes x_p \otimes y_1 \otimes \ldots \otimes y_q$. Note that the elements of $T(V)$ consist of finite linear combinations of tensor products of elements of V. The unit of $T(V)$ is 1 and $T(V)$ is a graded algebra.

Definition 8.8 (Algebra morphism). Let $(\mathfrak{A}, m_{\mathfrak{A}})$ and $(\mathfrak{B}, m_{\mathfrak{B}})$ be algebras. Then $f\colon \mathfrak{A} \to \mathfrak{B}$ is an *algebra morphism* if

$$f\left(m_{\mathfrak{A}}(x \otimes y)\right) = m_{\mathfrak{B}}\left(f(x) \otimes f(y)\right).$$

That is, $f \circ m_{\mathfrak{A}} = m_{\mathfrak{B}} \circ (f \otimes f)$, or $f(x \cdot y) = f(x) \cdot f(y)$.

Definition 8.9 (Algebra representation). A *representation* of an algebra $\mathfrak{A}$ consists of an algebra morphism $\rho\colon \mathfrak{A} \to \mathrm{End}(V)$, where V is a free module of finite rank over the same ring as $\mathfrak{A}$.

A related concept we shall use is that of an anti-homomorphism.

Definition 8.10 (Anti-homomorphism). Let $(\mathfrak{A}, m_{\mathfrak{A}})$ and $(\mathfrak{B}, m_{\mathfrak{B}})$ be algebras. Then $f\colon \mathfrak{A} \to \mathfrak{B}$ is an *algebra anti-homomorphism* if

$$f(x \cdot y) = f(y) \cdot f(x).$$

That is, $f \circ m_{\mathfrak{A}} = m_{\mathfrak{B}} \circ (f \otimes f) \circ \tau$, or $f\left(m_{\mathfrak{A}}(x \otimes y)\right) = m_{\mathfrak{B}}\left(f(y) \otimes f(x)\right)$.

Example 8.11. Let V be a free module of finite rank n, and consider the algebra $\mathrm{End}(V)$ from Exercise 8.1 as the algebra of $n \times n$ matrices over $\mathbb{K}$. Then the map that sends a matrix $\mathbf{A} = [a_{i,j}]$ to its *transpose* $\mathbf{A}^T = [a_{j,i}]$ defines an anti-homomorphism since, for matrices, $(\mathbf{AB})^T = \mathbf{B}^T\mathbf{A}^T$.

8.1.2 Coalgebras

The next concept we need is that of a coalgebra. Coalgebras arise by reversing the directions of the arrows in the definition of an algebra. Again, we let $\mathfrak{A}$ be a free module of finite rank over a commutative unital ring $\mathbb{K}$.

A *coproduct* on $\mathfrak{A}$ is a module morphism

$$\Delta\colon \mathfrak{A} \to \mathfrak{A} \otimes \mathfrak{A},$$

and $\mathfrak{A}$ endowed with Δ is a *coalgebra* $(\mathfrak{A}, \Delta)$ that we often denote by $\mathfrak{A}$.

A coalgebra $(\mathfrak{A}, \Delta)$ is *coassociative* if the diagram

$$\begin{array}{ccc} \mathfrak{A} & \xrightarrow{\ \Delta\ } & \mathfrak{A} \otimes \mathfrak{A} \\ \big\downarrow{\scriptstyle \Delta} & & \big\downarrow{\scriptstyle \Delta\otimes\mathrm{id}} \\ \mathfrak{A} \otimes \mathfrak{A} & \xrightarrow{\mathrm{id}\otimes\Delta} & \mathfrak{A} \otimes \mathfrak{A} \otimes \mathfrak{A} \end{array}$$

commutes. Equivalently,

$$(\mathrm{id} \otimes \Delta) \circ \Delta = (\Delta \otimes \mathrm{id}) \circ \Delta. \tag{8.5}$$

In this text, we shall always assume that coalgebras are coassociative without further comment.

Observe that coassociativity is defined by reversing the directions of the arrows in the definition of associativity of the commutative diagram (8.1). The definition of a counit arises similarly by reversing directions of the arrows in the definition of a unit of the commutative diagram (8.3).

A *counit* of $(\mathfrak{A}, \Delta)$ is a module morphism

$$\varepsilon\colon \mathfrak{A} \to \mathbb{K}$$

such that the diagram

$$\begin{array}{ccccc} \mathfrak{A} \otimes \mathfrak{A} & & \xrightarrow{\ \varepsilon \otimes \mathrm{id}\ } & & \mathbb{K} \otimes \mathfrak{A} \\ & \nwarrow{\scriptstyle \Delta} & & \nearrow{\scriptstyle \iota} & \\ & & \mathfrak{A} & & \\ & \swarrow{\scriptstyle \Delta} & & \searrow{\scriptstyle \iota'} & \\ \mathfrak{A} \otimes \mathfrak{A} & & \xrightarrow{\ \mathrm{id} \otimes \varepsilon\ } & & \mathfrak{A} \otimes \mathbb{K} \end{array}$$

commutes, where

$$\iota\colon \mathfrak{A} \xrightarrow{\cong} \mathbb{K} \otimes \mathfrak{A}\colon a \mapsto 1_{\mathbb{K}} \otimes a \quad \text{and} \quad \iota'\colon \mathfrak{A} \xrightarrow{\cong} \mathfrak{A} \otimes \mathbb{K}\colon a \mapsto a \otimes 1_{\mathbb{K}}. \tag{8.6}$$

Equivalently,

$$(\varepsilon \otimes \mathrm{id}) \circ \Delta(a) = 1_{\mathbb{K}} \otimes a \quad \text{and} \quad (\mathrm{id} \otimes \varepsilon) \circ \Delta(a) = a \otimes 1_{\mathbb{K}} \quad \text{for all } a \in \mathfrak{A}. \tag{8.7}$$

We note that $\kappa^{-1} = \iota$.

Definition 8.12 (Graded coalgebra). A coalgebra $(\mathfrak{A}, \Delta)$ is said to be *graded* if $\mathfrak{A} = \bigoplus_{k \geq 0} \mathfrak{A}_k$, where each $\mathfrak{A}_k$ is a free module of finite rank and

$$\Delta \colon \mathfrak{A}_k \to \bigoplus_{p+q=k} \mathfrak{A}_p \otimes \mathfrak{A}_q.$$

Example 8.13. Let V be a vector space and let $T(V)$ be its tensor algebra (see Example 8.7). A coproduct can be defined on $T(V)$ by setting $\Delta(x_1 \otimes \ldots \otimes x_r) = \sum_{i=0}^{r} (x_1 \otimes \ldots \otimes x_i) \otimes (x_{i+1} \otimes \ldots \otimes x_r)$. With this, $T(V)$ forms a graded coassociative coalgebra. It has counit given by $\varepsilon(x) = x$ when $x \in \mathbb{K}$ and $\varepsilon(x) = 0$ otherwise.

Definition 8.14 (Coalgebra morphism). Let $(\mathfrak{A}, \Delta_{\mathfrak{A}})$ and $(\mathfrak{B}, \Delta_{\mathfrak{B}})$ be coalgebras. Then $f \colon \mathfrak{A} \to \mathfrak{B}$ is *coalgebra morphism* if

$$\Delta_{\mathfrak{B}} \circ f = (f \otimes f) \circ \Delta_{\mathfrak{A}}.$$

Exercise 8.15. Let $(\mathfrak{A}, \Delta)$ be a coalgebra. Define a map $\Delta^{\mathrm{op}} : \mathfrak{A} \to \mathfrak{A} \otimes \mathfrak{A}$ as $\Delta^{\mathrm{op}} := \tau \circ \Delta$, where τ is the twist map. Verify that $(\mathfrak{A}, \Delta^{\mathrm{op}})$ is a coalgebra. Show that $(\mathfrak{A}, \Delta)$ has a counit if and only if $(\mathfrak{A}, \Delta^{\mathrm{op}})$ does. The coalgebra $(\mathfrak{A}, \Delta^{\mathrm{op}})$ is called the *opposite coalgebra* of $(\mathfrak{A}, \Delta)$.

Example 8.16. Let $(\mathfrak{A}, m)$ be a graded algebra. Then $(\mathfrak{A}^\star, \beta \circ m^\star)$ is a coalgebra, where β is the natural isomorphism from $(\mathfrak{A} \otimes \mathfrak{A})^\star$ to $\mathfrak{A}^\star \otimes \mathfrak{A}^\star$. (See Proposition A.30 for the isomorphism.) To see why this is observe that $m : \mathfrak{A} \otimes \mathfrak{A} \to \mathfrak{A}$, so $m^\star : \mathfrak{A}^\star \to (\mathfrak{A} \otimes \mathfrak{A})^\star$ and $\beta \circ m^\star : \mathfrak{A}^\star \to \mathfrak{A}^\star \otimes \mathfrak{A}^\star$. It remains to show that $\beta \circ m^\star$ is coassociative. By the associativity of m, we have $(m \circ (m \otimes \mathrm{id}_{\mathfrak{A}}))^\star = (m \circ (\mathrm{id}_{\mathfrak{A}} \otimes m))^\star$. For module morphisms f and g, we have $(f \circ g)^\star = g^\star \circ f^\star$ and $(f \otimes g)^\star = (f^\star \otimes g^\star) \circ \beta$. Making use of this, we see that $(\mathrm{id}_{\mathfrak{A}^\star} \otimes \beta) \circ (\mathrm{id}_{\mathfrak{A}^\star} \otimes m^\star) \circ \beta \circ m^\star = (\mathrm{id}_{\mathfrak{A}^\star} \otimes \beta) \circ (\mathrm{id}_{\mathfrak{A}^\star} \otimes m^\star) \circ \beta \circ m^\star$ are equal as maps from $\mathfrak{A}^\star$ to $\mathfrak{A}^\star \otimes \mathfrak{A}^\star \otimes \mathfrak{A}^\star$. Coassociativity then follows.

Exercise 8.17. Let $(\mathfrak{A}, \Delta)$ be a graded coalgebra. Show that $(\mathfrak{A}^\star, \Delta^\star \circ \alpha)$ is an algebra, where α is the natural isomorphism from $\mathfrak{A}^\star \otimes \mathfrak{A}^\star$ to $(\mathfrak{A} \otimes \mathfrak{A})^\star$. Moreover, if $(\mathfrak{A}, \Delta)$ has a counit ε, show that $\varepsilon^\star \circ \lambda$ is a unit of $(\mathfrak{A}^\star, \Delta^\star \circ \alpha)$ where $\lambda \colon 1_{\mathbb{K}} \mapsto (1_{\mathbb{K}})^\star$.

Exercise 8.18. Let $(\mathfrak{A}, m, \eta)$ be an algebra with unit, and $(\mathfrak{B}, \Delta, \varepsilon)$ be a coalgebra with counit. For $f, g \in \mathrm{Hom}(\mathfrak{B}, \mathfrak{A})$, the *convolution product*, $f * g \in \mathrm{Hom}(\mathfrak{B}, \mathfrak{A})$, of f and g is defined as the composition

$$\mathfrak{B} \xrightarrow{\Delta} \mathfrak{B} \otimes \mathfrak{B} \xrightarrow{f \otimes g} \mathfrak{A} \otimes \mathfrak{A} \xrightarrow{m} \mathfrak{A}.$$

Show that $(\mathrm{Hom}(\mathfrak{B}, \mathfrak{A}), *, \eta \circ \varepsilon)$ is an algebra with unit $\eta \circ \varepsilon$.

8.1.3 Bialgebras and Hopf Algebras

Of course, it is possible that a module or vector space has both an algebra structure and a coalgebra structure. Such situations give rise to bialgebras and Hopf algebras. These consist of a structure which is simultaneously an algebra with unit and a coalgebra with counit, together with some compatibility relations between the algebra and coalgebra structures.

Definition 8.19 (Bialgebra). $(\mathfrak{A}, m, \Delta, \eta, \varepsilon)$ is a *bialgebra* if

1. $(\mathfrak{A}, m)$ is an algebra with unit η;
2. $(\mathfrak{A}, \Delta)$ is a coalgebra with counit ε;

together with the compatibility relations

3. Δ, ε are algebra morphisms;
4. m, η are coalgebra morphisms.

Exercise 8.20. Prove that conditions (3) and (4) in the definition of bialgebra are equivalent, and hence either one of them may be dropped from the definition.

Exercise 8.21. Prove that if $\mathfrak{A} = (\mathfrak{A}, m, \Delta, \eta, \varepsilon)$ is a bialgebra, then so are $\mathfrak{A}^{\mathrm{op}} = (\mathfrak{A}, m^{\mathrm{op}}, \Delta, \eta, \varepsilon)$ and $\mathfrak{A}^{\mathrm{cop}} = (\mathfrak{A}, m, \Delta^{\mathrm{op}}, \eta, \varepsilon)$.

Exercise 8.22. Let $(\mathfrak{A}, m, \Delta, \eta, \varepsilon)$ be a bialgebra. Use Example 8.16 and Exercise 8.17 to show that the dual space of $\mathfrak{A}$ admits a bialgebra structure.

A Hopf algebra consists of a bialgebra equipped with a special automorphism S called the *antipode*.

Definition 8.23 (Hopf algebra). A *Hopf algebra*

$$\mathfrak{A} := (\mathfrak{A}, m, \Delta, \eta, \varepsilon, S)$$

is a bialgebra on $\mathfrak{A}$, with product m, a coproduct Δ, a unit η and counit ε, together with an *anti-homomorphism* $S\colon \mathfrak{A} \to \mathfrak{A}$, so $S(xy) = S(y)S(x)$, called an *antipode* such that the *compatibility relations*

$$\begin{aligned} m \circ (S \otimes \mathrm{id}) \circ \Delta &= \eta \circ \varepsilon, \\ m \circ (\mathrm{id} \otimes S) \circ \Delta &= \eta \circ \varepsilon \end{aligned} \tag{8.8}$$

are satisfied.

Example 8.24. Let G be a group and $\mathbb{K}$ be a field, and let $\mathbb{K}[G]$ be the group algebra. (The group algebra $\mathbb{K}[G]$ can be thought of as acting like the ring of polynomials but with elements of G instead of indeterminates.) The group algebra $\mathbb{K}[G]$ forms a Hopf algebra with coproduct given by taking $\Delta(g) = g \otimes g, \varepsilon = 1$, and $S(g) = g^{-1}$, for all $g \in G$.

Example 8.25. Let V be a vector space and let $T(V)$ be its tensor algebra from Example 8.7. Define a coproduct by setting $\Delta(v) = 1 \otimes v + v \otimes 1$, for all $v \in V$, then extending multiplicatively, $T(V)$ becomes a bialgebra, with a counit given by the zero map. By defining an antipode by $S(v) = -v$, for all $v \in V$ and extending anti-multiplicatively (so the order of the elements is reversed) $T(V)$ becomes a Hopf algebra.

Bialgebras and Hopf algebras are *graded* if they are graded as both as algebras and coalgebras.

Exercise 8.26. Let $\mathfrak{A} = (\mathfrak{A}, m, \Delta, \eta, \varepsilon, S)$ be a Hopf algebra such that S is bijective. Prove that $\mathfrak{A}^{\mathrm{op}} = (\mathfrak{A}, m^{\mathrm{op}}, \Delta, \eta, \varepsilon, S^{-1})$ is a Hopf algebra. Similarly show that $\mathfrak{A}^{\mathrm{cop}} = (\mathfrak{A}, m, \Delta^{\mathrm{op}}, \eta, \varepsilon, S^{-1})$ is.

Exercise 8.27. Let $\mathfrak{A} = (\mathfrak{A}, m, \Delta, \eta, \varepsilon, S)$ be a graded Hopf algebra. Show that the dual bialgebra of Exercise 8.22 forms a Hopf algebra with idempotent $S^{\star}$.

We shall use of the following result later. A consequence of it is that S is a bialgebra morphism from $\mathfrak{A}$ to $\mathfrak{A}^{\mathrm{op\ cop}}$.

Lemma 8.28. *Let* $\mathfrak{A} = (\mathfrak{A}, m, \Delta, \eta, \varepsilon, S)$ *be a Hopf algebra. Then*

$$\Delta \circ S = (S \otimes S) \circ \tau \circ \Delta. \tag{8.9}$$

Proof. Recall the convolution product from Exercise 8.18. Write $f := \Delta \circ S$ and $g := (S \otimes S) \circ \tau \circ \Delta$, so $f, g \in \mathrm{Hom}(\mathfrak{A}, \mathfrak{A} \otimes \mathfrak{A})$. We shall show that $f * \Delta = \Delta * g = (\eta \otimes \eta) \circ \varepsilon$. The result follows from this identity since, by Exercise 8.18, $\mathrm{Hom}(\mathfrak{A}, \mathfrak{A} \otimes \mathfrak{A})$ is an algebra with product $*$ and unit $(\eta \otimes \eta) \circ \varepsilon$, so

$$f = f * ((\eta \otimes \eta) \circ \varepsilon) = f * \Delta * g = ((\eta \otimes \eta) \circ \varepsilon) * g = g.$$

We first show $f * \Delta = (\eta \otimes \eta) \circ \varepsilon$. Let $x \in \mathfrak{A}$ and write $\Delta(x) = \sum x' \otimes x''$. Then

$$\begin{aligned}
(f * \Delta)(x) &= \sum \Delta(S(x')) \cdot \Delta(x'') \\
&= \Delta\left(\sum S(x') \cdot x''\right) && (\Delta \text{ an alg. morphism}) \\
&= \Delta\left(\eta \circ \varepsilon(x)\right) && (\text{by } (8.8)) \\
&= \Delta\left(\varepsilon(x) 1_{\mathfrak{A}}\right) \\
&= \varepsilon(x) \Delta\left(1_{\mathfrak{A}}\right)
\end{aligned}$$

$$= \varepsilon(x)\,(1_{\mathfrak{A}} \otimes 1_{\mathfrak{A}}) \qquad (\Delta \text{ an alg. morphism})$$
$$= ((\eta \otimes \eta) \circ \varepsilon)(x).$$

Next we show $\Delta * g = (\eta \otimes \eta) \circ \varepsilon$. For this, write $\Delta(x) = \sum x' \otimes x''$, $((\Delta \otimes \mathrm{id}) \circ \Delta)(x) = \sum y' \otimes y'' \otimes y'''$, and $((\Delta \otimes \mathrm{id} \otimes \mathrm{id}) \circ (\Delta \otimes \mathrm{id}) \circ \Delta)(x) = \sum z' \otimes z'' \otimes z''' \otimes z''''$.

By cocommutativity, $\sum z' \otimes z'' \otimes z''' \otimes z'''' = \sum y' \otimes \Delta(y'') \otimes y'''$. Then

$$\begin{aligned}
\sum z' \otimes z'' S(z''') \otimes S(z'''') &= \sum y' \otimes (m \circ (\mathrm{id} \otimes S) \circ \Delta)(y'') \otimes S(y''') \\
&= \sum y' \otimes (\eta \circ \varepsilon)(y'') \otimes S(y''') && \text{(by (8.8))} \\
&= \sum y' \otimes \varepsilon(y'') 1_{\mathfrak{A}} \otimes S(y''').
\end{aligned}$$

From this, it follows that

$$\begin{aligned}
(\Delta * g)(x) &= \sum z' S(z'''') \otimes z'' S(z''') \\
&= \sum y' S(y''') \otimes \varepsilon(y'') 1_{\mathfrak{A}} \qquad (8.10) \\
&= \sum y' \varepsilon(y'') S(y''') \otimes 1_{\mathfrak{A}}.
\end{aligned}$$

We also have

$$\begin{aligned}
\sum y' \otimes \varepsilon(y'') \otimes S(y''') &= ((\mathrm{id} \otimes \mathrm{id} \otimes S) \circ (\mathrm{id} \otimes \varepsilon \otimes \mathrm{id}) \circ (\Delta \otimes \mathrm{id}) \circ \Delta)(x) \\
&= \sum ((\mathrm{id} \otimes \varepsilon) \circ \Delta(x')) \otimes S(x'') \\
&= \sum x' \otimes 1_{\mathbb{K}} \otimes S(x'') && \text{(by (8.7))}
\end{aligned}$$

and so

$$\begin{aligned}
(8.10) &= \sum x' S(x'') \otimes 1_{\mathfrak{A}} \\
&= (m \circ (\mathrm{id} \otimes S) \circ \Delta)(x) \otimes 1_{\mathfrak{A}} \\
&= \varepsilon(x) 1_{\mathfrak{A}} \otimes 1_{\mathfrak{A}} \\
&= ((\eta \otimes \eta) \circ \varepsilon)(x).
\end{aligned}$$

Thus $f * \Delta = \Delta * g = (\eta \otimes \eta) \circ \varepsilon$, as required. □

8.2 Quasi-triangular Hopf Algebras and R-Matrices

A quasi-triangular Hopf algebra, introduced by Drinfel'd [56], consists of a Hopf algebra together with a special element R called a *universal R-matrix*. The raison d'être of quasi-triangular Hopf algebra is, as suggested by the term "universal R-matrix", that they give rise to solutions of the Yang–Baxter equation. In this section, we introduce quasi-triangular Hopf algebras and explain their connection with the Yang–Baxter equation.

Definition 8.29 (Quasi-triangular Hopf algebra). A *quasi-triangular Hopf algebra* is a pair $(\mathfrak{A}, \mathsf{R})$ with $\mathsf{R} \in \mathfrak{A} \otimes \mathfrak{A}$, where $\mathfrak{A} = (\mathfrak{A}, m, \Delta, \varepsilon, \eta, S)$ is a Hopf algebra, S is a bijection, and R is an invertible element satisfying the conditions

1. $(\tau \circ \Delta)(a) = \mathsf{R} \cdot \Delta(a) \cdot \mathsf{R}^{-1}$, for all $a \in \mathfrak{A}$,
2. $(\Delta \otimes \mathrm{id}_{\mathfrak{A}})(\mathsf{R}) = \mathsf{R}_{13} \cdot \mathsf{R}_{23}$,
3. $(\mathrm{id}_{\mathfrak{A}} \otimes \Delta)(\mathsf{R}) = \mathsf{R}_{13} \cdot \mathsf{R}_{12}$,

in which R_{12}, R_{23} and R_{13} are defined in terms of R by

4. $\mathsf{R} =: \sum_i \alpha_i \otimes \beta_i$, for some $\alpha_i, \beta_i \in \mathfrak{A}$,
5. $\mathsf{R}_{12} := \mathsf{R} \otimes 1_{\mathfrak{A}} = \sum_i \alpha_i \otimes \beta_i \otimes 1_{\mathfrak{A}}$,
6. $\mathsf{R}_{13} := (\mathrm{id}_{\mathfrak{A}} \otimes \tau)(\mathsf{R} \otimes 1_{\mathfrak{A}}) = \sum_i \alpha_i \otimes 1_{\mathfrak{A}} \otimes \beta_i$,
7. $\mathsf{R}_{23} := 1_{\mathfrak{A}} \otimes \mathsf{R} = \sum_i 1_{\mathfrak{A}} \otimes \alpha_i \otimes \beta_i$.

The element R is called the *universal R-matrix*.

As is standard practice, we shall denote quasi-triangular Hopf algebra by just $\mathfrak{A}$, or $(\mathfrak{A}, \mathsf{R})$, or $(\mathfrak{A}, m, \Delta, \varepsilon, \eta, S, \mathsf{R})$ depending on whether or not we wish to emphasise particular terminology.

The reader should note the distinction between the names "R-matrix" and "universal R-matrix": the former is a morphism (see Definition 5.4), while the latter is an element of an algebra.

Exercise 8.30. Let $(\mathfrak{A}, \mathsf{R})$ be a quasi-triangular Hopf algebra, where $\mathfrak{A}$ is the Hopf algebra $(\mathfrak{A}, m, \Delta, \varepsilon, \eta, S)$. Recall the Hopf algebra $\mathfrak{A}^{\mathrm{op}}$ from Exercise 8.26. Prove that $(\mathfrak{A}^{\mathrm{op}}, \mathsf{R}_{21})$ is a quasi-triangular Hopf algebra where $\mathsf{R}_{21} := \tau(\mathsf{R})$.

As one would expect, the universal R-matrix R interacts nicely with the Hopf algebra structure of a quasi-triangular Hopf algebra. The following proposition provides an illustration of this, as well as providing some identities that we shall make use of in applications to knot theory in Chapter 9.

Proposition 8.31. *Let $\mathfrak{A} = (\mathfrak{A}, m, \Delta, \varepsilon, \eta, S, \mathsf{R})$ be a quasi-triangular Hopf algebra. Then*

1. $((\eta \circ \varepsilon) \otimes \mathrm{id}_{\mathfrak{A}})(\mathsf{R}) = 1_{\mathfrak{A}\otimes\mathfrak{A}} = (\mathrm{id}_{\mathfrak{A}} \otimes (\eta \circ \varepsilon))(\mathsf{R})$,
2. $(S \otimes \mathrm{id}_{\mathfrak{A}})(\mathsf{R}) = \mathsf{R}^{-1} = (\mathrm{id}_{\mathfrak{A}} \otimes S^{-1})(\mathsf{R})$,
3. $(S \otimes S)(\mathsf{R}) = \mathsf{R}$.

Proof. Here we shall prove only Part (1) of the proposition in order to illustrate some computations within a quasi-triangular Hopf algebra. Proofs for the remaining parts can be found in Appendix C.

We have that ε is a counit and $\iota^{-1} = \kappa$, so, from (8.7),

$$\kappa \circ (\varepsilon \otimes \mathrm{id}_{\mathfrak{A}}) \circ \Delta = \mathrm{id}_{\mathfrak{A}} = \kappa \circ (\mathrm{id}_{\mathfrak{A}} \otimes \varepsilon) \circ \Delta.$$

Then

$$\begin{aligned}
\mathsf{R} &= (\mathrm{id}_{\mathfrak{A}} \otimes \mathrm{id}_{\mathfrak{A}})(\mathsf{R}) \\
&= ((\kappa \circ (\varepsilon \otimes \mathrm{id}_{\mathfrak{A}}) \circ \Delta) \otimes \mathrm{id}_{\mathfrak{A}})\,(\mathsf{R}) \\
&= ((\kappa \circ (\varepsilon \otimes \mathrm{id}_{\mathfrak{A}})) \otimes \mathrm{id}_{\mathfrak{A}}) \circ (\Delta \otimes \mathrm{id}_{\mathfrak{A}}))\,(\mathsf{R}) \\
&= (\kappa \circ (\varepsilon \otimes \mathrm{id}_{\mathfrak{A}})) \otimes \mathrm{id}_{\mathfrak{A}})(\mathsf{R}_{13} \cdot \mathsf{R}_{23}) \qquad \text{(Def. 8.29)} \\
&= (\kappa \circ (\varepsilon \otimes \mathrm{id}_{\mathfrak{A}})) \otimes \mathrm{id}_{\mathfrak{A}}) \sum_{i,j} \alpha_i \otimes \alpha_j \otimes (\beta_i \cdot \beta_j) \\
&= \sum_{i,j} \big(\varepsilon(\alpha_i)\alpha_j\big) \otimes (\beta_i \cdot \beta_j) \\
&= \left(\sum_i \varepsilon(\alpha_i) 1_{\mathfrak{A}} \otimes \beta_i\right) \cdot \left(\sum_j \alpha_j \otimes \beta_j\right) \\
&= (((\eta \circ \varepsilon) \otimes \mathrm{id}_{\mathfrak{A}})(\mathsf{R})) \cdot \mathsf{R},
\end{aligned}$$

where the last equality uses that $\eta(1_{\mathbb{K}}) = 1_{\mathfrak{A}}$, as in Exercise 8.4.

But R is invertible, so $((\eta \circ \varepsilon) \otimes \mathrm{id}_{\mathfrak{A}})(\mathsf{R}) = 1_{\mathfrak{A}\otimes\mathfrak{A}}$. The other identity follows similarly, completing the proof of Part (1). □

We now turn our attention to the most important property of quasi-triangular Hopf algebras: that R-matrices can be recovered from them. The key to the recovery of R-matrices is an algebraic relation between the elements R_{12}, R_{13} and R_{23}, given in the next result. Its proof is an exercise in using the definition of a quasi-triangular Hopf algebra.

Lemma 8.32. *The elements* R_{12}, R_{13} *and* R_{23} *of a quasi-triangular Hopf algebra are related by* $\mathsf{R}_{12} \cdot \mathsf{R}_{13} \cdot \mathsf{R}_{23} = \mathsf{R}_{23} \cdot \mathsf{R}_{13} \cdot \mathsf{R}_{12}$.

Proof.

$$\begin{aligned}
\mathsf{R}_{12} \cdot \mathsf{R}_{13} \cdot \mathsf{R}_{23} &= \mathsf{R}_{12} \cdot (\Delta \otimes \mathrm{id}_{\mathfrak{A}})(\mathsf{R}) && \text{(Def. 8.29(2))} \\
&= (\mathsf{R} \otimes 1_{\mathfrak{A}}) \cdot (\Delta \otimes \mathrm{id}_{\mathfrak{A}})(\mathsf{R}) && \text{(Def. 8.29(5))} \\
&= (\mathsf{R} \otimes 1_{\mathfrak{A}}) \cdot \left(\sum_i \Delta(\alpha_i) \otimes \beta_i\right) && \text{(Def. 8.29(4))} \\
&= \sum_i (\mathsf{R} \cdot \Delta(\alpha_i)) \otimes \beta_i \\
&= \sum_i ((\tau \circ \Delta(\alpha_i)) \cdot \mathsf{R}) \otimes \beta_i && \text{(Def. 8.29(1))} \\
&= \sum_i ((\tau \circ \Delta(\alpha_i)) \otimes \beta_i) \cdot (\mathsf{R} \otimes 1_{\mathfrak{A}}) \\
&= \left(((\tau \circ \Delta) \otimes \mathrm{id}_{\mathfrak{A}}) \sum_i \alpha_i \otimes \beta_i\right) \cdot (\mathsf{R} \otimes 1_{\mathfrak{A}})
\end{aligned}$$

$$
\begin{aligned}
&= ((\tau \circ \Delta) \otimes \mathrm{id}_{\mathfrak{A}})(\mathsf{R}) \cdot \mathsf{R}_{12} && \text{(Def. 8.29(4) \& (5))}\\
&= ((\tau \otimes \mathrm{id}_{\mathfrak{A}}) \circ (\Delta \otimes \mathrm{id}_{\mathfrak{A}})(\mathsf{R})) \cdot \mathsf{R}_{12}\\
&= ((\tau \otimes \mathrm{id}_{\mathfrak{A}})(\mathsf{R}_{13} \cdot \mathsf{R}_{23})) \cdot \mathsf{R}_{12} && \text{(Def. 8.29(2))}.
\end{aligned}
$$

But

$$
\begin{aligned}
(\tau \otimes \mathrm{id}_{\mathfrak{A}})(\mathsf{R}_{13} \cdot \mathsf{R}_{23}) &= (\tau \otimes \mathrm{id}_{\mathfrak{A}}) \left(\left(\sum_i \alpha_i \otimes 1_{\mathfrak{A}} \otimes \beta_i\right) \cdot \left(\sum_j 1_{\mathfrak{A}} \otimes \alpha_j \otimes \beta_j\right)\right)\\
& \qquad \text{(by Def. 8.29(6) \& (7))}\\
&= (\tau \otimes \mathrm{id}_{\mathfrak{A}}) \left(\sum_{i,j} \alpha_i \otimes \alpha_j \otimes (\beta_i \cdot \beta_j)\right)\\
&= \sum_{i,j} \alpha_j \otimes \alpha_i \otimes (\beta_i \cdot \beta_j)\\
&= \left(\sum_i 1_{\mathfrak{A}} \otimes \alpha_i \otimes \beta_i\right) \cdot \left(\sum_j \alpha_j \otimes 1_{\mathfrak{A}} \otimes \beta_j\right)\\
& \qquad \text{(by Def. 8.29(6) \& (7))}\\
&= \mathsf{R}_{23} \cdot \mathsf{R}_{13}.
\end{aligned}
$$

The result now follows. □

The identity $\mathsf{R}_{12} \cdot \mathsf{R}_{13} \cdot \mathsf{R}_{23} = \mathsf{R}_{23} \cdot \mathsf{R}_{13} \cdot \mathsf{R}_{12}$ of Lemma 8.32 is an algebraic analogue of the functional Yang–Baxter equation from Definition 5.4. To recover a solution to the functional Yang–Baxter equation, we need to move from the algebra to maps. This is done by means of a representation.

Here a *representation* of a quasi-triangular Hopf algebra $\mathfrak{A}$ consists of an algebra morphism $\rho\colon \mathfrak{A} \to \mathrm{End}(V)$, where V is a free module of finite rank over the same ring as $\mathfrak{A}$. Making use of the twist map, $\tau : V \otimes V \to V \otimes V : x \otimes y \mapsto y \otimes x$, we can recover a (functional) R-matrix by considering representations as follows.

Proposition 8.33. *Let $\mathfrak{A}$ be a quasi-triangular Hopf algebra, let V be a free module of finite rank and let $\rho\colon \mathfrak{A} \to \mathrm{End}(V)$ be a representation of $\mathfrak{A}$. If R is the universal* R*-matrix, then $\tau \circ (\rho \otimes \rho)(\mathsf{R})$ is invertible and is a solution of the (functional) Yang–Baxter equation of Definition 5.4.*

Proof. The element-wise action of $\tau \circ ((\rho \otimes \rho)(\mathsf{R})) \in \mathrm{End}(V \otimes V)$ is

$$
\tau \circ ((\rho \otimes \rho)(\mathsf{R}))\colon v \otimes v' \mapsto \sum_i \rho_{\beta_i}(v') \otimes \rho_{\alpha_i}(v)
$$

where

$$
\rho_{\alpha_i} := \rho(\alpha_i) \in \mathrm{End}(V), \quad \rho_{\beta_i} := \rho(\beta_i) \in \mathrm{End}(V), \quad \text{and} \quad \mathsf{R} = \sum_i \alpha_i \otimes \beta_i. \tag{8.11}
$$

Let

$$F := \tau \circ ((\rho \otimes \rho)(\mathsf{R}))\colon V \otimes V \to V \otimes V$$

and consider the map $(F \otimes \mathrm{id}) \circ (\mathrm{id} \otimes F) \circ (F \otimes \mathrm{id})$. Applying this map in three steps to $x \otimes y \otimes z$ gives

$$\begin{aligned}
x \otimes y \otimes z &\overset{F\otimes\mathrm{id}}{\longmapsto} \sum_i \rho_{\beta_i}(y) \otimes \rho_{\alpha_i}(x) \otimes z \\
&\overset{\mathrm{id}\otimes F}{\longmapsto} \sum_{i,j} \rho_{\beta_i}(y) \otimes \rho_{\beta_j}(z) \otimes \left(\rho_{\alpha_j} \circ \rho_{\alpha_i}(x)\right) \\
&\overset{F\otimes\mathrm{id}}{\longmapsto} \sum_{i,j,k} \left(\rho_{\beta_k} \circ \rho_{\beta_j}(z)\right) \otimes \left(\rho_{\alpha_k} \circ \rho_{\beta_i}(y)\right) \otimes \left(\rho_{\alpha_j} \circ \rho_{\alpha_i}(x)\right) \qquad (8.12)\\
&= \sum_{i,j,k} \left(\rho_{\beta_k\cdot\beta_j}(z) \otimes \rho_{\alpha_k\cdot\beta_i}(y) \otimes \rho_{\alpha_j\cdot\alpha_i}(x)\right) \\
&= \quad (\rho \otimes \rho \otimes \rho) \left(\sum_{i,j,k} (\beta_k \cdot \beta_j) \otimes (\alpha_k \cdot \beta_i) \otimes (\alpha_j \cdot \alpha_i) \right) (z \otimes y \otimes x).
\end{aligned}$$

Similarly, the action of $(\mathrm{id} \otimes F) \circ (F \otimes \mathrm{id}) \circ (\mathrm{id} \otimes F)$ is given by

$$x \otimes y \otimes z \mapsto (\rho \otimes \rho \otimes \rho) \left(\sum_{i,j,k} (\beta_j \cdot \beta_i) \otimes (\beta_k \cdot \alpha_i) \otimes (\alpha_k \cdot \alpha_j) \right) (z \otimes y \otimes x). \tag{8.13}$$

But, from Definition 8.29,

$$\begin{aligned}
\mathsf{R}_{12} \cdot \mathsf{R}_{13} \cdot \mathsf{R}_{23} &= \left(\sum_k \alpha_k \otimes \beta_k \otimes 1_{\mathfrak{A}} \right) \cdot \left(\sum_j \alpha_j \otimes 1_{\mathfrak{A}} \otimes \beta_j \right) \cdot \left(\sum_i 1_{\mathfrak{A}} \otimes \alpha_i \otimes \beta_i \right) \\
&= \sum_{i,j,k} (\alpha_k \cdot \alpha_j) \otimes (\beta_k \cdot \alpha_i) \otimes (\beta_j \cdot \beta_i)
\end{aligned}$$

and

$$\begin{aligned}
\mathsf{R}_{23} \cdot \mathsf{R}_{13} \cdot \mathsf{R}_{12} &= \left(\sum_k 1_{\mathfrak{A}} \otimes \alpha_k \otimes \beta_k \right) \cdot \left(\sum_j \alpha_j \otimes 1_{\mathfrak{A}} \otimes \beta_j \right) \cdot \left(\sum_i \alpha_i \otimes \beta_i \otimes 1_{\mathfrak{A}} \right) \\
&= \sum_{i,j,k} (\alpha_j \cdot \alpha_i) \otimes (\alpha_k \cdot \beta_i) \otimes (\beta_k \cdot \beta_j).
\end{aligned}$$

Then, from Lemma 8.32,

$$\sum_{i,j,k}(\alpha_k \cdot \alpha_j) \otimes (\beta_k \cdot \alpha_i) \otimes (\beta_j \cdot \beta_i) = \sum_{i,j,k}(\alpha_j \cdot \alpha_i) \otimes (\alpha_k \cdot \beta_i) \otimes (\beta_k \cdot \beta_j).$$

Thus, permuting the terms in the tensor products on each side by the same permutation gives

$$\sum_{i,j,k}(\beta_j \cdot \beta_i) \otimes (\beta_k \cdot \alpha_i) \otimes (\alpha_k \cdot \alpha_j) = \sum_{i,j,k}(\beta_k \cdot \beta_j) \otimes (\alpha_k \cdot \beta_i) \otimes (\alpha_j \cdot \alpha_i).$$

By applying $\rho \otimes \rho \otimes \rho$ to both sides, it therefore follows from (8.12) and (8.13) that

$$(F \otimes \mathrm{id}) \circ (\mathrm{id} \otimes F) \circ (F \otimes \mathrm{id}) = (\mathrm{id} \otimes F) \circ (F \otimes \mathrm{id}) \circ (\mathrm{id} \otimes F),$$

so F satisfies the Yang–Baxter equation.

To show that F is invertible, recall from Definition 8.29 that R is invertible and that $\rho \otimes \rho$ is a representation of $\mathfrak{A} \otimes \mathfrak{A}$. Thus, $(\rho \otimes \rho)(\mathsf{R}^{-1})$ is the inverse of $(\rho \otimes \rho)(\mathsf{R})$. Also $\tau^{-1} = \tau$, so

$$(\tau \circ ((\rho \otimes \rho)(\mathsf{R})))^{-1} = \big((\rho \otimes \rho)(\mathsf{R}^{-1})\big) \circ \tau,$$

completing the proof. □

8.3 Ribbon Hopf Algebras

Our ultimate aim is to use a quasi-triangular Hopf algebra $\mathfrak{A}$ and a representation ρ of it to construct an operator invariant $Q_{(V,W,\mathsf{R},\vec{n},\overleftarrow{n},\vec{u},\overleftarrow{u})}$. Of course, to do this we need to construct maps R, $\vec{n}$, $\overleftarrow{n}$, $\vec{u}$ and $\overleftarrow{u}$ from $\mathfrak{A}$ and ρ. Proposition 8.33 showed how an R-matrix can be constructed, so we now turn our attention to $\vec{n}$, $\overleftarrow{n}$, $\vec{u}$ and $\overleftarrow{u}$. These maps arise by considering representations of some special elements denoted by $\mathbf{u}$ and $\mathbf{v}$. The element $\mathbf{u}$ is an element in the quasi-triangular Hopf algebra arising from the universal R-matrix, and $\mathbf{v}$ may be loosely thought of as a "square root" of $\mathbf{u}$. In general, a quasi-triangular Hopf algebra may not have the desired element $\mathbf{v}$ (although it always contains $\mathbf{u}$). As defined at the end of this section, a quasi-triangular Hopf algebra with this element $\mathbf{v}$ is called a *ribbon Hopf algebra*, and is due to Reshetikhin and Turaev [157].

Definition 8.34 (The element u). Let $(\mathfrak{A}, m, \Delta, \varepsilon, \eta, S, \mathsf{R})$ be a quasi-triangular Hopf algebra. The element $\mathbf{u} \in \mathfrak{A}$ is defined by

$$\mathbf{u} := \sum_i S(\beta_i) \cdot \alpha_i$$

where $\sum_i \alpha_i \otimes \beta_i = \mathsf{R}$, the universal R-matrix.

As we shall see, $\mathbf{u}$ has many nice algebraic properties. A fundamental one is the following.

Lemma 8.35 (Almost-centrality). *Let $\mathfrak{A}$ be a quasi-triangular Hopf algebra, and let $\mathbf{u}$ be as given in Definition 8.34. Then $\mathbf{u}$ is* almost central. *That is,*

$$\mathbf{u} \cdot x = S^2(x) \cdot \mathbf{u}, \qquad \text{for all } x \in \mathfrak{A}.$$

A proof of Lemma 8.35 can be found in Appendix C (see Lemma C.2).

A consequence of the almost-centrality of $\mathbf{u}$ is that it is invertible.

Lemma 8.36 (Invertibility, quasi-idempotency). *Let $\mathfrak{A}$ be a quasi-triangular Hopf algebra, and let $\mathbf{u}$ be as given in Definition 8.34. Then*

1. *The element $\mathbf{u}$ is invertible, with inverse*

$$\mathbf{u}^{-1} = \sum_i S^{-1}(\overline{\beta}_i) \cdot \overline{\alpha}_i$$

where $\overline{\alpha}_j$ and $\overline{\beta}_j$ are defined by

$$\mathsf{R}^{-1} = \sum_i \overline{\alpha}_i \otimes \overline{\beta}_i; \tag{8.14}$$

2. *S is* quasi-idempotent. *That is, $S^2(a) = \mathbf{u}\, a\, \mathbf{u}^{-1}$ for all $a \in \mathfrak{A}$.*

Proof. For the first part, consider

$$1_{\mathfrak{A}} \otimes 1_{\mathfrak{A}} \overset{\tau}{\longmapsto} 1_{\mathfrak{A}} \otimes 1_{\mathfrak{A}} \overset{S\otimes \mathrm{id}_{\mathfrak{A}}}{\longmapsto} 1_{\mathfrak{A}} \otimes 1_{\mathfrak{A}} \overset{m}{\longmapsto} 1_{\mathfrak{A}},$$

recalling that S is an anti-homomorphism so $S(1_{\mathfrak{A}}) = 1_{\mathfrak{A}}$. Then

$$\begin{aligned} 1_{\mathfrak{A}} &= (m \circ (S \otimes \mathrm{id}_{\mathfrak{A}}) \circ \tau)(1_{\mathfrak{A}} \otimes 1_{\mathfrak{A}}) = (m \circ (S \otimes \mathrm{id}_{\mathfrak{A}}) \circ \tau)(\mathsf{R} \cdot \mathsf{R}^{-1}) \\ &= \sum_{i,j} S(\beta_j \overline{\beta}_i)\, \alpha_j\, \overline{\alpha}_i = \sum_{i,j} S(\overline{\beta}_i)\, S(\beta_j)\, \alpha_j\, \overline{\alpha}_i = \sum_i S(\overline{\beta}_i) \cdot \mathbf{u} \cdot \overline{\alpha}_i \end{aligned}$$

where the final equality is from Definition 8.34. To obtain the inverse of $\mathbf{u}$, we use the almost-centrality property of $\mathbf{u}$ that is stated in Lemma 8.35 to move $\mathbf{u}$ to the left in the above expression. Then, resuming, we have

$$\begin{aligned} 1_{\mathfrak{A}} = \sum_i S(\overline{\beta}_i) \cdot \mathbf{u} \cdot \overline{\alpha}_i &= \sum_i S^2\left(S^{-1}(\overline{\beta}_i)\right) \cdot \mathbf{u} \cdot \overline{\alpha}_i \\ &= \sum_i \left(\mathbf{u} \cdot \left(S^{-1}(\overline{\beta}_i)\right)\right) \cdot \overline{\alpha}_i = \mathbf{u} \cdot \sum_i \cdot \left(S^{-1}(\overline{\beta}_i)\right) \cdot \overline{\alpha}_i \end{aligned}$$

from Definition 8.35. Thus $\mathbf{u}$ has a right inverse. By almost-centrality, it follows that it has a left inverse, and therefore an inverse. Part (1) now follows.

The second part follows immediately from Lemma 8.35 since, from Part (1), $\mathbf{u}$ is invertible. □

As should be expected given the definition of $\mathbf{u}$, there are various relationships between $\mathbf{u}$, S and R. We record some of these here for use later, but postpone their proofs until Appendix C.

Proposition 8.37. *Let $\mathfrak{A}$ be a quasi-triangular Hopf algebra, and let $\mathbf{u}$ be as given in Definition 8.34. Then*

1. $S^2(\mathbf{u}) = \mathbf{u}$,
2. $\mathbf{u} \cdot S(\mathbf{u}) = S(\mathbf{u}) \cdot \mathbf{u}$,
3. $\mathbf{u}\, S(\mathbf{u})$ *is in the centre of* $\mathfrak{A}$,
4. $\Delta(\mathbf{u}) = (\tau(\mathsf{R}) \cdot \mathsf{R})^{-1} \cdot (\mathbf{u} \otimes \mathbf{u}) = (\mathbf{u} \otimes \mathbf{u}) \cdot (\tau(\mathsf{R}) \cdot \mathsf{R})^{-1}$,
5. $\mathbf{u}^{-1} = \sum_i \beta_i\, S^2(\alpha_i)$,
6. $\Delta(S(\mathbf{u})) = (\tau(\mathsf{R}) \cdot \mathsf{R})^{-1} \cdot (S(\mathbf{u}) \otimes S(\mathbf{u})) = (S(\mathbf{u}) \otimes S(\mathbf{u})) \cdot (\tau(\mathsf{R}) \cdot \mathsf{R})^{-1}$,
7. $\varepsilon(\mathbf{u}) = 1_{\mathbb{K}}$.

As noted in the introduction to this section, our knot theoretic applications will require us to be able to take the "square root" of the element $\mathbf{u}$. We need to consider quasi-triangular Hopf algebras that contain a special element $\mathbf{v}$ that satisfies certain axioms. Such a structure is called a ribbon Hopf algebra, and it is this algebraic structure that admits the knot theoretic applications.

Definition 8.38 (Ribbon Hopf algebra). A *ribbon Hopf algebra* is a pair $(\mathfrak{A}, \mathbf{v})$ where $\mathfrak{A}$ is a quasi-triangular Hopf algebra, and $\mathbf{v} \in \mathfrak{A}$ is an element that satisfies the following conditions:

1. $\mathbf{v}$ is central,
2. $\mathbf{v}^2 = S(\mathbf{u}) \cdot \mathbf{u}$,
3. $S(\mathbf{v}) = \mathbf{v}$,
4. $\varepsilon(\mathbf{v}) = 1_{\mathbb{K}}$,
5. $\Delta(\mathbf{v}) = (\tau(\mathsf{R}) \cdot \mathsf{R})^{-1} \cdot (\mathbf{v} \otimes \mathbf{v})$.

Before proceeding, we record one result about $\mathbf{v}$.

Lemma 8.39. *Let $(\mathfrak{A}, \mathbf{v})$ be a ribbon Hopf algebra. Then $\mathbf{v}$ is invertible.*

Proof. From Lemma 8.36, $\mathbf{u}$ is invertible and, since $\mathfrak{A}$ is a quasi-triangular Hopf algebra, S is invertible from Definition 8.29. Thus $\mathbf{v}^2$ is invertible by Definition 8.38(2). Let $\mathbf{w} := (\mathbf{v}^2)^{-1}$. Then $\mathbf{v}(\mathbf{v}\mathbf{w}) = 1_{\mathfrak{A}}$. But $\mathbf{v}$ is central by Part (1), so $(\mathbf{v}\mathbf{w})\mathbf{v} = 1_{\mathfrak{A}}$. Thus $\mathbf{v}$ is invertible. □

8.4 Examples of Ribbon Hopf Algebras

So far, we have defined and studied quasi-triangular Hopf algebras and ribbon Hopf algebras, and saw their importance in the present context, namely that they give rise to R-matrices. However, there is one critical point that is yet to be addressed. This is whether such algebras actually do exist.

In fact, they do exist. Remarkable independent work of Drinfel'd and Jimbo in the late 1980s showed that every semi-simple Lie algebra over $\mathbb{C}$ gives rise to such an algebra. Here we shall describe the ribbon Hopf algebra that arises from the Lie algebra $\mathfrak{sl}_2$. The theory of Lie algebras and quantum groups is fascinating and involved, and a detailed discussion of it is outside the scope of this text. Accordingly, we shall briefly describe only those results that are pertinent to our present purpose. In particular, we shall consider only the Lie algebra $\mathfrak{sl}_2$ in detail. The results given here, however, do in fact extend to a larger class of Lie algebras, called semi-simple Lie algebras. For further background, the reader may consult, for example, [61, 71, 78, 80] for additional Lie algebra background or [33, 90, 127] for quantum groups. The algebras $U_h(\mathfrak{g})$ considered here are due to Drinfel'd [54, 55], Kulish and Reshetikhin for the special case of $U_h(\mathfrak{sl}_2)$ [109], and Jimbo [83] for $U_q(\mathfrak{g})$.

8.4.1 *Lie Algebras*

A Lie algebra is a particular type of non-associative algebra over a field. We assume the field is of characteristic zero here, and do so without further comment.

Definition 8.40 (Lie algebra). A *Lie algebra* $\mathfrak{g}$ consists of a vector space $\mathfrak{g}$ equipped with a product $[\cdot,\cdot] : \mathfrak{g} \otimes \mathfrak{g} \to \mathfrak{g}$ that satisfies:

1. the *Jacobi identity*

$$[x,[y,z]] + [y,[z,x]] + [z,[x,y]] = 0, \quad \text{for all } x, y, z \in \mathfrak{g},$$

2. the *anti-symmetry relation*

$$[x,y] = -[y,x], \quad \text{for all } x, y \in \mathfrak{g}.$$

The product in a Lie algebra is almost always denoted using square brackets, as above, and is called a *Lie bracket*.

Exercise 8.41. Show that the axiom $[x,y] = -[y,x]$ for all $x, y \in \mathfrak{g}$ is equivalent to the condition $[x,x] = 0$ for all $x \in \mathfrak{g}$. (Provided the field is not of characteristic two.)

Example 8.42. The following are Lie algebras.

1. Any vector space V with Lie bracket given by $[x,y] = 0$, for all $x, y \in V$.

2. Any associative algebra $(\mathfrak{A}, m)$ with Lie bracket given by the commutator $[x, y] = x \cdot y - y \cdot x$.
3. If V is an n-dimensional vector space over a field $\mathbb{K}$, then the vector space $\mathrm{End}(V)$ with Lie bracket given by $[f, g] = f \circ g - g \circ f$ forms a Lie algebra. This Lie algebra is denoted by $\mathfrak{gl}_n(\mathbb{K})$.
4. $\mathfrak{gl}_n(\mathbb{K})$ can be defined in terms of matrices as the vector space of $n \times n$ matrices over $\mathbb{K}$ with Lie bracket given by $[A, B] = A \cdot B - B \cdot A$.
5. $\mathfrak{sl}_n(\mathbb{K})$ is the Lie algebra consisting of the vector space of all $n \times n$ matrices over $\mathbb{K}$ that have trace zero with Lie bracket given by $[A, B] = A \cdot B - B \cdot A$.

Exercise 8.43. Verify that the above examples of Lie algebras are indeed Lie algebras.

Although we shall not make use of the fact here, we note in passing that every semisimple Lie algebra over $\mathbb{C}$ is isomorphic to one of the three *classical Lie algebras*, denoted by $\mathfrak{sl}_n(\mathbb{C})$, $\mathfrak{so}_n(\mathbb{C})$, $\mathfrak{sp}_{2n}(\mathbb{C})$, or one of five *exceptional Lie algebras* known as $\mathfrak{e}_6$, $\mathfrak{e}_7$, $\mathfrak{e}_8$, $\mathfrak{f}_4$ and $\mathfrak{g}_2$.

In this text, we shall be concerned primarily with the Lie algebra $\mathfrak{sl}_2(\mathbb{C})$. Our algebraic frameworks are set up for working with algebraic objects in terms of bases and generators so we shall define versions of $\mathfrak{sl}_2(\mathbb{C})$ and related Lie algebras in terms of their generators. The following definition specifies the Lie algebras $\mathfrak{sl}_2$, $\mathfrak{gl}_n$ and $\mathfrak{sl}_n$ in terms of their bases. The definition of the action of the Lie brackets of $\mathfrak{gl}_n$ and $\mathfrak{sl}_n$ as it is presented in Definition 8.44 is rather unenlightening. However, Exercise 8.45 will bring meaning to them.

Definition 8.44 (The Lie algebras $\mathfrak{sl}_2$, $\mathfrak{gl}_n$ and $\mathfrak{sl}_n$).

1. $\mathfrak{sl}_2$ is the Lie algebra that consists of the vector space over $\mathbb{C}$ consisting of formal linear combinations of the symbols x, y and h and with Lie bracket given by the linear extension of

$$[h, x] = 2x, \qquad [h, y] = -2y, \qquad [x, y] = h. \tag{8.15}$$

2. $\mathfrak{gl}_n$ is the Lie algebra that consists of the vector space over $\mathbb{C}$ consisting of formal linear combinations of elements of the set $\{e_{i,j} : 1 \leq i, j \leq n\}$ and with Lie bracket given by the linear extension of

$$\left[e_{i,j}, e_{k,l}\right] = \begin{cases} e_{i,l} & \text{if } j = k \text{ and } i \neq l, \\ -e_{k,j} & \text{if } i = l \text{ and } j \neq k, \\ e_{i,i} - e_{j,j} & \text{if } j = k \text{ and } i = l, \\ 0 & \text{otherwise.} \end{cases} \tag{8.16}$$

3. $\mathfrak{sl}_n$ is the Lie algebra that consists of the vector space over $\mathbb{C}$ consisting of formal linear combinations of elements of the set

$$\{e_{i,j} : 1 \leq i, j \leq n, i \neq j\} \cup \{h_i : 1 \leq i \leq n-1\}$$

and with Lie bracket given as follows.
The values $\left[e_{i,j}, e_{k,l}\right]$ are as in (8.16), $\left[h_i, h_j\right] = 0$ for all i, j, and

$$\left[h_i, e_{j,k}\right] = \begin{cases} e_{i,k} & \text{if } i = j \neq k \text{ and } i+1 \neq k, \\ -e_{i+1,k} & \text{if } i+1 = j \neq k \text{ and } i \neq k, \\ -e_{j,i} & \text{if } i = k \neq j \neq i+1, \\ e_{j,i+1} & \text{if } i+1 = k \neq j \neq i, \\ 2e_{i,i+1} & \text{if } i = j \text{ and } i+1 = k, \\ -2e_{i+1,i} & \text{if } i = k \text{ and } i+1 = j, \\ 0 & \text{otherwise.} \end{cases} \tag{8.17}$$

Exercise 8.45. Let $\mathfrak{g}_1$ and $\mathfrak{g}_2$ be Lie algebras over the same field. A map $\varphi : \mathfrak{g}_1 \to \mathfrak{g}_2$ is a *Lie algebra morphism* if it is a linear map and $\varphi([x, y]) = [\varphi(x), \varphi(y)]$, for all $x, y \in \mathfrak{g}_1$. If in addition φ is bijective, then it is an *isomorphism*. Show that each of the pairs of Lie algebras $\mathfrak{sl}_2$ and $\mathfrak{sl}_2(\mathbb{C})$, $\mathfrak{sl}_n$ and $\mathfrak{sl}_n(\mathbb{C})$, and $\mathfrak{gl}_n$ and $\mathfrak{gl}_n(\mathbb{C})$ are isomorphic. (For $\mathfrak{sl}_n$ and $\mathfrak{gl}_n$, let $\mathbf{E}_{i,j}$ be the $n \times n$ matrix with (i, j)-entry equal to 1, and all other entries zero. Define isomorphisms that send $e_{i,j} \mapsto \mathbf{E}_{i,j}$ and $h_i \mapsto \mathbf{E}_{i,i} - \mathbf{E}_{i+1,i+1}$.)

When a Lie algebra has a basis $\{e_1, e_2, \dots, e_n\}$, its Lie bracket may be described by its action on basis elements:

$$[e_i, e_j] = \sum_{k=1}^{n} \gamma_{i,j}{}^k e_k. \tag{8.18}$$

The scalars $\gamma_{i,j}{}^k$ are called *structure constants*. (Note that here we are using offset indices to emphasise the order of i, j and k. This will be helpful later.)

Definition 8.46 (Universal enveloping algebra $U(\mathfrak{g})$). Let $\mathfrak{g}$ be a Lie algebra with basis $\{e_1, e_2, \dots, e_n\}$ and structure constants $\gamma_{i,j}{}^k$. The *universal enveloping algebra* of $\mathfrak{g}$, denoted by $U(\mathfrak{g})$, is the associative unital algebra generated by elements $E_1, E_2, \dots, E_n$ and subject to the relations

$$E_i E_j - E_j E_i = \sum_{k=1}^{n} \gamma_{i,j}{}^k E_k, \quad \text{for } 1 \leq i, j \leq n.$$

Definition 8.47 (Universal enveloping algebras $U(\mathfrak{sl}_n)$, $U(\mathfrak{gl}_n)$).

1. $U(\mathfrak{sl}_2)$ is the associative algebra over $\mathbb{C}$ generated by X, Y and H, with relations

$$HX - XH = 2X, \qquad HY - YH = -2Y, \qquad XY - YX = H. \tag{8.19}$$

2. $U(\mathfrak{gl}_n)$ is the associative algebra over $\mathbb{C}$ generated by the elements of $\{E_{i,j} : 1 \leq i, j \leq n\}$ and subject to the relations given by (8.16) where lower case e's are

replaced by upper case E's, where $[\cdot,\cdot]$ denotes the *commutator*, so $[a,b] := ab - ba$.
3. $U(\mathfrak{sl}_n)$ is the associative algebra over $\mathbb{C}$ generated by elements of $\{E_{i,j} : 1 \leq i, j \leq n, i \neq j\} \cup \{H_i : 1 \leq i \leq n-1\}$ and subject to the relations given by (8.16) and (8.17), where lower case e's and h's are replaced by upper case E's and H's, and where $[\cdot,\cdot]$ denotes the commutator.

Exercise 8.48. Show that $U(\mathfrak{sl}_2)$ forms a Hopf algebra with comultiplication, antipode and unit given by

$$\Delta(A) = A \otimes 1 + 1 \otimes A, \quad S(A) = -A, \quad \text{and} \quad \varepsilon(A) = 0,$$

for all $A \in U(\mathfrak{sl}_2)$.

Specialising the definition of a representation of an algebra to the present setting gives the following. A *representation* of $U(\mathfrak{g})$ on a vector space V is an algebra morphism

$$\rho : U(\mathfrak{g}) \to \mathrm{End}(V),$$

where $\mathrm{End}(V)$ is the associative algebra of linear maps on V.

We can obtain standard representations of $U(\mathfrak{sl}_n)$ and $U(\mathfrak{gl}_n)$ by what is in essence sending the generators to the matrices from which the Lie algebras first arose.

Definition 8.49 (Standard representation of $U(\mathfrak{sl}_n)$ and $U(\mathfrak{gl}_n)$).

1. The *standard representation of* $U(\mathfrak{sl}_2)$ is defined by setting

$$\rho(X) = \begin{bmatrix} 0 & 1 \\ 0 & 0 \end{bmatrix}, \qquad \rho(Y) = \begin{bmatrix} 0 & 0 \\ 1 & 0 \end{bmatrix}, \qquad \rho(H) = \begin{bmatrix} 1 & 0 \\ 0 & -1 \end{bmatrix}. \tag{8.20}$$

2. The *standard representation of* $U(\mathfrak{gl}_n)$ is defined by setting $\rho(E_{i,j}) = \mathbf{E}_{i,j}$, where $\mathbf{E}_{i,j}$ is the $n \times n$ matrix with (i,j)-entry equal to 1, and all other entries zero.
3. The *standard representation of* $U(\mathfrak{sl}_n)$ is defined by the is defined by setting $\rho(E_{i,j}) = \mathbf{E}_{i,j}$ and $\rho(H_i) = \mathbf{E}_{i,i} - \mathbf{E}_{i+1,i+1}$, where $\mathbf{E}_{i,j}$ is the $n \times n$ matrix with (i,j)-entry equal to 1, and all other entries zero.

Of course, these Lie algebras have representations other than the standard representations. For example, the *n-dimensional irreducible representation of* $\mathfrak{sl}_2$ is

$$[\rho(X)]_{r,s} := (n-r)\,\delta_{r,s-1}, \quad [\rho(Y)]_{r,s} := s\,\delta_{r-1,s}, \quad [\rho(H)]_{r,s} := (n-2r+1)\,\delta_{r,s} \tag{8.21}$$

Written in full, the matrices $\rho(X)$, $\rho(Y)$ $\rho(H)$ are

$$\rho(X) = \begin{bmatrix} 0 & n-1 & 0 & \cdots & 0 & 0 \\ 0 & 0 & n-2 & \cdots & 0 & 0 \\ \vdots & \vdots & & \ddots & \vdots & \vdots \\ 0 & 0 & 0 & \cdots & 2 & 0 \\ 0 & 0 & 0 & \cdots & 0 & 1 \\ 0 & 0 & 0 & \cdots & 0 & 0 \end{bmatrix}_{n\times n}, \quad \rho(Y) = \begin{bmatrix} 0 & 0 & 0 & \cdots & 0 & 0 \\ 1 & 0 & 0 & \cdots & 0 & 0 \\ 0 & 2 & 0 & \cdots & 0 & 0 \\ \vdots & \vdots & \vdots & \ddots & \vdots & \vdots \\ 0 & 0 & 0 & \cdots & 0 & 0 \\ 0 & 0 & 0 & \cdots & n-1 & 0 \end{bmatrix}_{n\times n},$$

and $\rho(H) = \mathrm{diag}(n-1, n-3, \cdots, -(n-1))$.

Exercise 8.50. Verify that the matrices given in (8.21) constitute a representation of $U(\mathfrak{sl}_2)$.

8.4.2 *The Quantized Universal Enveloping Algebra $U_h(\mathfrak{sl}_2)$*

The *quantized universal enveloping algebra* $U_h(\mathfrak{sl}_2)$ is obtained by "deforming" the universal enveloping algebra $U(\mathfrak{sl}_2)$ of $\mathfrak{sl}_2$.

We define $U_h(\mathfrak{sl}_2)$ to be the algebra over $\mathbb{C}[[h]]$ generated by X, Y and H and subject to the relations

$$[H, X] = 2X, \qquad [H, Y] = -2Y, \qquad [X, Y] = \frac{\sinh\left(\frac{1}{2}hH\right)}{\sinh\left(\frac{1}{2}h\right)}. \tag{8.22}$$

Here we are using $[\cdot,\cdot]$ to denote the commutator, so, for example, $[H, Y] := HY - YX$. The commutator $[X, Y]$, as a formal series in $\mathbb{C}[H][[h]]$, is

$$[X, Y] = \frac{e^{\frac{1}{2}hH} - e^{-\frac{1}{2}hH}}{e^{\frac{1}{2}h} - e^{-\frac{1}{2}h}} = H + \left(\frac{H^3 - H}{24}\right) h^2 + \cdots$$

In computations and results about $U_h(\mathfrak{sl}_2)$, some expressions will appear frequently, and so we set up the following notation for them.

$$q := e^{\frac{1}{2}h}, \quad K := e^{\frac{1}{4}hH}, \quad \overline{q} := q^{-1}, \quad \overline{K} := K^{-1}. \tag{8.23}$$

In addition, we will use the *quantum integer* $[n]$ and the *quantum factorial* $[n]!$ which are defined as follows, for an integer natural number n:

$$[n] := \frac{q^n - \overline{q}^n}{q - \overline{q}}, \quad [n]! := [1]\cdot[2]\cdots[n]. \tag{8.24}$$

Exercise 8.51. Verify that the following identities hold:

$$\begin{aligned}[-x] &= -[x],\\ [x]\,[y] - [x-a]\,[y+a] &= [a]\,[y-x+a],\\ [x]\,[y] + [a]\,[x+y+a] &= [x+a]\,[y+a].\end{aligned}$$

The universal enveloping algebra $U_h(\mathfrak{sl}_2)$ may be endowed with Hopf structure.

Theorem 8.52. *The quantized universal enveloping algebra $U_h(\mathfrak{sl}_2)$ is a Hopf algebra with comultiplication, antipode, counit and unit defined by*

1. Comultiplication:

$$\begin{aligned}\Delta(X) &:= X \otimes K + \overline{K} \otimes X,\\ \Delta(Y) &:= Y \otimes K + \overline{K} \otimes Y,\\ \Delta(H) &:= H \otimes 1 + 1 \otimes H,\end{aligned}$$

2. Antipode: $S(X) := -qX, \quad S(Y) := -\overline{q}\,Y, \quad S(H) := -H,$
3. Counit: $\varepsilon(X) := \varepsilon(Y) := \varepsilon(H) := 0,$
4. Unit: $\eta(1) := 1,$

where Δ and ε are extended as algebra morphisms, and m and η are extended as coalgebra morphisms.

Exercise 8.53. Show that in $U_h(\mathfrak{sl}_2)$, $\Delta(K) = K \otimes K$, $S(K) = \overline{K}$, and $\varepsilon(K) = 1$.

Exercise 8.54. Prove Theorem 8.52 by verifying that each of the generators of $U_h(\mathfrak{sl}_2)$ satisfies the defining properties of a Hopf algebra.

The Hopf algebra $U_h(\mathfrak{sl}_2)$ can be turned into a ribbon Hopf algebra by adjoining a universal R-matrix and element $\mathbf{v}$.

Theorem 8.55. $(U_h(\mathfrak{sl}_2), \mathsf{R}, \mathbf{v})$ *forms a ribbon Hopf algebra with universal R-matrix*

$$\mathsf{R} := e^{\frac{h}{4}\,H\otimes H} \cdot \sum_{n\geq 0} \frac{(q-\overline{q}\,)^n}{[n]!} q^{n(n-3)/2}\,(KX)^n \otimes (\overline{K}\,Y)^n,$$

and

$$\mathbf{v} := e^{-\frac{h}{4}H^2} \cdot \sum_{n\geq 0} \frac{(q-\overline{q}\,)^n}{[n]!} \overline{q}^{\,n(n+3)/2}\,\overline{K}^{\,2(n+1)}\,Y^n X^n.$$

The proof of this result can be found in, for example, [90], or see [34] for a combinatorial approach.

8.4.3 The Quantized Universal Enveloping Algebra $U_q(\mathfrak{sl}_2)$

For the knot theory applications, we need to work within the "subalgebra" of $U_h(\mathfrak{sl}_2)$ generated by the elements X, Y, K and $\overline{K}$, over the ring of convergent power series in h, where K is to be regarded as an entirely new symbol. That is, in this algebra K is *not* determined by H, for the latter is not in this algebra. An inverse for K will be needed, so it has to be added as a generator, $\overline{K}$.

Definition 8.56 (Algebra $U_q(\mathfrak{sl}_2)$). The *q-analogue algebra* of $U(\mathfrak{sl}_2)$, denoted by $U_q(\mathfrak{sl}_2)$, is the algebra generated by the symbols X, Y, K and $\overline{K}$ and subject to the relations:

$$K\overline{K} = 1 = \overline{K}\,K, \quad KX = qXK, \quad KY = \overline{q}\,YK, \quad [X,Y] = \frac{K^2 - \overline{K}^2}{q - \overline{q}}. \tag{8.25}$$

Theorem 8.57. *The q-analogue $U_q(\mathfrak{sl}_2)$ of the algebra $U(\mathfrak{sl}_2)$ is a Hopf algebra with comultiplication, antipode, counit and unit defined by*

1. Comultiplication:

$$\begin{aligned} \Delta(X) &:= X \otimes K + \overline{K} \otimes X, & \Delta(Y) &:= Y \otimes K + \overline{K} \otimes Y, \\ \Delta(K) &:= K \otimes K, & \Delta(\overline{K}) &:= \overline{K} \otimes \overline{K}, \end{aligned}$$

2. Antipode: $S(X) := -qX$, $\quad S(Y) := -\overline{q}\,Y$, $\quad S(K) := \overline{K}$, $\quad S(\overline{K}) := K$,
3. Counit: $\varepsilon(X) := \varepsilon(Y) := 0$, $\quad \varepsilon(K) := \varepsilon(\overline{K}) := 1$,
4. Unit: $\eta(1) := 1$,

where Δ and ε are extended as algebra morphisms, and m and η are extended as coalgebra morphisms, and S as an anti-homomorphism.

Exercise 8.58. Prove Theorem 8.57.

The previous theorem asserts that $U_q(\mathfrak{sl}_2)$ is a Hopf algebra. To this, a universal R-matrix may be adjoined so that $(U_q(\mathfrak{sl}_2), \mathsf{R})$ is a ribbon Hopf algebra.

Theorem 8.59. $(U_q(\mathfrak{sl}_2), \mathsf{R}, \mathbf{v})$ *is a Ribbon Hopf algebra where*

$$\mathsf{R} := e^{\frac{h}{4}\,H\otimes H} \cdot \sum_{n\geq 0} \frac{(q-\overline{q}\,)^n}{[n]!} q^{n(n-3)/2}\,(KX)^n \otimes (\overline{K}\,Y)^n,$$

and

$$\mathbf{v} = e^{-\frac{h}{4}H^2} \cdot \sum_{n\geq 0} \frac{(\overline{q}-q)^n}{[n]!}\,\overline{q}^{\,n(n+3)/2}\overline{K}^{\,2(n+1)}\,Y^n X^n.$$

See, for example, [90] for a proof of this result.

The reader probably will have noticed that the statement of Theorem 8.59 is not correct. The problem with the statement is that R and $\mathbf{v}$ do not actually belong to $U_q(\mathfrak{sl}_2)$ since the elements $e^{\frac{h}{4}\,H\otimes H}$ and $e^{-\frac{h}{4}H^2}$ do not. With some care and by working in a suitable completion of $U_q(\mathfrak{sl}_2)$, it is possible to obtain a mathematically correct statement of the theorem (see, e.g., [90] for details). A consequence of the latter is that we may use, with impunity, the "not quite correct statement" of Theorem 8.59.

8.4.4 Representations

We can obtain representations of $U_h(\mathfrak{sl}_2)$ and $U_q(\mathfrak{sl}_2)$ by deforming representations of $U(\mathfrak{sl}_2)$. The deformation applied to the matrices of the standard representation of $\mathfrak{sl}_2$ given in (8.21) affects only $\rho(X)$ and $\rho(Y)$, but not $\rho(H)$. For example, the standard three-dimensional representation for $U_h(\mathfrak{sl}_2)$ is

$$\rho(X) = \begin{bmatrix} 0 & [2] & 0 \\ 0 & 0 & [1] \\ 0 & 0 & 0 \end{bmatrix}, \quad \rho(Y) = \begin{bmatrix} 0 & 0 & 0 \\ [1] & 0 & 0 \\ 0 & [2] & 0 \end{bmatrix}, \quad \rho(H) = \begin{bmatrix} 2 & 0 & 0 \\ 0 & 0 & 0 \\ 0 & 0 & -2 \end{bmatrix}.$$

The general statement is contained in the following result.

Theorem 8.60. *The matrices*

$$[\rho(X)]_{r,s} := [n-r]\,\delta_{r,s-1}, \quad [\rho(Y)]_{r,s} := [s]\,\delta_{r-1,s}, \quad [\rho(H)]_{r,s} := (n-2r+1)\,\delta_{r,s}$$

for $1 \le r, s \le n$ *afford a representation of* $U_h(\mathfrak{sl}_2)$.

Proof. Recalling (8.22), it is necessary to show that the three conditions

$$\rho([H,X]) = 2\rho(X), \quad \rho([H,Y]) = -2\rho(Y), \quad \rho([X,Y]) = \frac{\sinh\left(\frac{1}{2}h\rho(H)\right)}{\sinh\left(\frac{1}{2}h\right)}$$

hold. For the first condition, we have

$$\begin{aligned} [\rho([H,X])]_{r,s} &= [\rho(H)\rho(X) - \rho(X)\rho(H)]_{r,s} \\ &= \sum_{k=1}^{n} \left((n-2r+1)\,[n-k]\,\delta_{r,k}\delta_{k,s-1} - [n-r]\,(n-2k+1)\,\delta_{r,k-1}\delta_{k,s}\right) \\ &= (n-2r+1)\,[n-r]\,\delta_{r,s-1} - [n-r]\,(n-2r-1)\,\delta_{r,s-1} \\ &= 2\,[n-r]\delta_{r,s-1} = 2\,[\rho(X)]_{r,s}\,, \end{aligned}$$

so the first condition holds. Similarly, the second condition holds. For the third condition,

$$\begin{aligned} [\rho([X,Y])]_{r,s} &= [\rho(X)\rho(Y) - \rho(Y)\rho(X)]_{r,s} \\ &= \sum_{k=0}^{n} \left([n-r]\,[s]\,\delta_{r,k-1}\,\delta_{k-1,s} - [k]\,[n-k]\,\delta_{r-1,k}\,\delta_{k,s-1}\right) \\ &= ([n-r]\,[r] - [r-1]\,[n-r+1])\;\delta_{r,s} = [n-2r+1]\,\delta_{r,s} \end{aligned}$$

by Exercise 8.51. On the other hand

$$\left[\frac{\sinh\left(\frac{1}{2}h\rho(H)\right)}{\sinh\left(\frac{1}{2}h\right)}\right]_{r,s} = \left[\frac{q^{\rho(H)} - \overline{q}^{\,\rho(H)}}{q-\overline{q}}\right]_{r,s}$$
$$= \frac{q^{n-2r+1} - \overline{q}^{\,n-2r+1}}{q-\overline{q}}\,\delta_{r,s} = [n-2r+1]\,\delta_{r,s}.$$

Thus, this condition also holds, which completes the proof. □

A representation of the q-analogue algebra $U_q(\mathfrak{sl}_2)$ follows by a similar deformation of representation of $U(\mathfrak{sl}_2)$. We consider the map ρ defined by

$$\rho(X) = \begin{bmatrix} 0 & [n-1] & 0 & \cdots & 0 & 0 \\ 0 & 0 & [n-2] & \cdots & 0 & 0 \\ \vdots & \vdots & \vdots & \ddots & \vdots & \vdots \\ 0 & 0 & 0 & \cdots & [2] & 0 \\ 0 & 0 & 0 & \cdots & 0 & [1] \\ 0 & 0 & 0 & \cdots & 0 & 0 \end{bmatrix}, \quad \rho(Y) = \begin{bmatrix} 0 & 0 & 0 & \cdots & 0 & 0 & 0 \\ [1] & 0 & 0 & \cdots & 0 & 0 & 0 \\ 0 & [2] & 0 & \cdots & 0 & 0 & 0 \\ \vdots & \vdots & & \ddots & \vdots & \vdots & \vdots \\ 0 & 0 & 0 & \cdots & [n-2] & 0 & 0 \\ 0 & 0 & 0 & \cdots & 0 & [n-1] & 0 \end{bmatrix}.$$

and $\rho(K) = \operatorname{diag}\left(q^{(n-1)/2}, q^{(n-3)/2}, \cdots, q^{-(n-1)/2}\right)$.

Theorem 8.61. *The matrices with (r,s)-element defined by*

$$[\rho(X)]_{r,s} := [n-r]\,\delta_{r,s-1}, \qquad [\rho(Y)]_{r,s} := [s]\,\delta_{r-1,s},$$
$$[\rho(K)]_{r,s} := q^{(n-2r+1)/2}\,\delta_{r,s}, \qquad \rho(\overline{K}) = (\rho(K))^{-1},$$

for $1 \le r, s \le n$ afford a representation of $U_q(\mathfrak{sl}_2)$.

Proof. From (8.25), we need to show that

1. $\rho(K\overline{K}) = \rho(\overline{K}\,K) = \rho(1)$, 2. $\rho(KX) = \rho(qXK)$,
3. $\rho(KY) = \rho(\overline{q}\,YK)$, 4. $\rho([X,Y]) = \rho\left(\frac{K^2-\overline{K}^2}{q-\overline{q}}\right)$.

The first condition follows immediately from the definition of $\rho(\overline{K})$ in the statement of the theorem and the fact that ρ is a homomorphism. For the second condition, since ρ is a homomorphism,

$$[\rho(KX)]_{r,s} = \sum_{k=0}^{n} [\rho(K)]_{r,k}\,[\rho(X)]_{k,s} = \sum_{k=0}^{n} q^{(n-2r+1)/2}\delta_{r,k}\,[n-k]\,\delta_{k,s-1}$$
$$= q^{(n-2r+1)/2}\,[n-r]\,\delta_{r,s-1},$$

and

$$[\rho(XK)]_{r,s} = \sum_{k=0}^{n} [\rho(X)]_{r,k}\,[\rho(K)]_{k,s} = \sum_{k=0}^{n} [n-r]\,\delta_{r,k-1}\,q^{(n-2k+1)/2}\delta_{k,s}$$
$$= q^{(n-2r-1)/2}\,[n-r]\,\delta_{r,s-1},$$

so the condition holds. The proof of the third condition is similar to the second and is left as an exercise. For the fourth condition,

$$
\begin{aligned}
[\rho([X,Y])]_{r,s} &= \sum_{k\geq 0}\left([\rho(X)]_{r,k}\,[\rho(Y)]_{k,s} - [\rho(Y)]_{r,k}\,[\rho(X)]_{k,s}\right)\\
&= \sum_{k\geq 0}\left([n-r]\,\delta_{r,k-1}\,[k]\,\delta_{k-1,s} - [r]\,\delta_{r-1,k}\,[n-k]\,\delta_{k,s-1}\right)\\
&= ([n-r]\,[r] - [r-1]\,[n-r+1])\;\delta_{r,s} = [n-2r+1]\,\delta_{r,s},
\end{aligned}
$$

where we have used Exercise 8.51. On the other hand

$$
\left[\rho\!\left(\frac{K^2-\overline{K}^2}{q-\overline{q}}\right)\right]_{r,s} = \frac{[\rho(K)]^2_{r,s} - \left[\rho(\overline{K})\right]^2_{r,s}}{q-\overline{q}} = \frac{q^{n-2r+1}-\overline{q}^{\,n-2r+1}}{q-\overline{q}}\delta_{r,s} = [n-2r+1]\,\delta_{r,s}.
$$

□

Exercise 8.62. Complete the proof of Theorem 8.61 by verifying that $\rho(KY) = \rho(\overline{q}\,YK)$.

Chapter 9
Reshetikhin–Turaev Invariants

Previously, we constructed operator invariants by associating morphisms with the elementary pieces. Our aim now is to provide a construction for such morphisms by associating them with representations of elements of a ribbon Hopf algebra. The resulting invariants, from [157], are called Reshetikhin–Turaev invariants.

9.1 Coloured Tangles

We fix a ribbon Hopf algebra $\mathfrak{A} := (\mathfrak{A}, m, \Delta, \eta, \varepsilon, S, \mathsf{R}, \mathbf{v})$ and shall use it to construct tangle invariants by following the approach used to construct operator invariants in Sect. 7.2. To do this, we need to associate morphisms with the elementary tangles

We shall do this by using representations of $\mathfrak{A}$ to construct the required maps from $\mathfrak{A}$.

In fact, we can be a little more general in our construction than we were in the construction of operator invariants, by the simple device of associating different maps with the elementary tangles arising from different components of the tangle. This is done by associating a different representation with each component of the tangle, as in the following definition.

Definition 9.1 (Coloured tangle). A *coloured tangle* (resp., *coloured tangle diagram*) is a tangle (respectively, tangle diagram) such that each component i of the tangle (respectively, tangle diagram) is assigned a representation $\rho^{V_i} : \mathfrak{A} \to \mathrm{End}(V_i)$ of the ribbon Hopf algebra $\mathfrak{A}$, where V_i is a free module of finite rank.

D. M. Jackson and I. Moffatt, *An Introduction to Quantum and Vassiliev Knot Invariants*, CMS Books in Mathematics, https://doi.org/10.1007/978-3-030-05213-3_9

Note that different components of a coloured tangle can be assigned the same representation.

Coloured tangles and coloured tangle diagrams may be oriented, unoriented, framed or unframed. However, here we shall work primarily with framed oriented tangles as these are the correct sort to associate with ribbon Hopf algebras.

Our previous definitions and results on tangles and their diagrams may be extended easily to the setting of coloured diagrams by adding the requirement that the representation of $\mathfrak{A}$ associated with a component is preserved by any operations on them.

Two (framed/unframed/oriented/unoriented) coloured tangles or tangle diagrams are *equivalent* if they are equivalent as (framed/unframed/oriented/unoriented) tangles or tangle diagrams, and the representation associated with each component is preserved under this equivalence.

The operations of *tensor product* and *composition* of coloured tangle diagrams are defined just as in the non-coloured case. However, for composition, we include the additional requirement that only components with the same representation of $\mathfrak{A}$ can be stacked on top of each other.

In this chapter, following the notation in Definition A.14, whenever we index a basis using subscripts, its dual basis is described by replacing the subscripts with superscripts. For example, if $\{e_i\}_i$ is a basis for V, then $\{e^i\}_i$ is the basis for $V^\star$ given by $e^i(e_j) = \delta^i_j$.

9.2 Construction of the Invariants

Following the construction of operator invariants, we now construct a framed, oriented, coloured tangle invariant from elementary pieces in the usual way, by associating a linear map with each of the coloured pieces under the action of $Q^{\mathfrak{A}}$. The resulting map $Q^{\mathfrak{A}}$ is called a *Reshetikhin–Turaev invariant.*

Definition 9.2 (Reshetikhin–Turaev invariants). Let $(\mathfrak{A}, \mathbf{v})$ be a ribbon Hopf algebra, $\mathbf{u}$ be as in Definition 8.34, and $\rho^{V_i} : \mathfrak{A} \to \mathrm{End}(V_i)$ be a representation of $\mathfrak{A}$. For notational convenience, let $\rho^V : \mathfrak{A} \to \mathrm{End}(V)$ and $\rho^W : \mathfrak{A} \to \mathrm{End}(W)$ be two of these representations. Let $\{e_i\}_i$ be a basis of V and $\{e^i\}_i$ be the dual basis of $V^\star$. Then the *Reshetikhin–Turaev invariant* $Q^{\mathfrak{A}}$ is the map on coloured tangle diagrams given by the following actions.

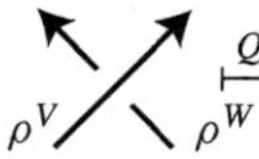

$$\xrightarrow{Q^{\mathfrak{A}}} \tau \circ (\rho^V \otimes \rho^W)(\mathsf{R})\colon V \otimes W \to W \otimes V$$

$$: x \otimes y \mapsto \sum_i \rho^W_{\beta_i}(y) \otimes \rho^V_{\alpha_i}(x),$$

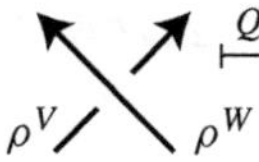

$$\xrightarrow{Q^{\mathfrak{A}}} (\rho^W \otimes \rho^V)(\mathsf{R}^{-1}) \circ \tau\colon V \otimes W \to W \otimes V$$

$$: x \otimes y \mapsto \sum_i \rho^W_{S(\alpha_i)}(y) \otimes \rho^V_{\beta_i}(x),$$

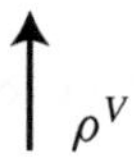

$$\xrightarrow{Q^{\mathfrak{A}}} \mathrm{id}_V,$$

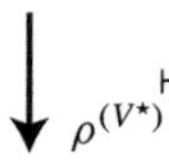

$$\xrightarrow{Q^{\mathfrak{A}}} \mathrm{id}_{V^\star}, \qquad (9.1)$$

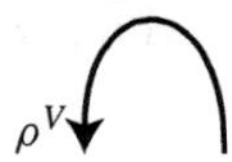

$$\xrightarrow{Q^{\mathfrak{A}}} \big(f \otimes x \mapsto f(x)\colon V^\star \otimes V \to \mathbb{K}\big),$$

$$\xrightarrow{Q^{\mathfrak{A}}} \big(x \otimes f \mapsto (f \circ \rho^V_{\mathbf{v}^{-1}\mathbf{u}}(x))\colon V \otimes V^\star \to \mathbb{K}\big),$$

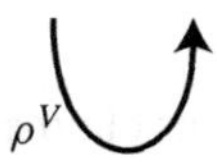

$$\xrightarrow{Q^{\mathfrak{A}}} \left(1_{\mathbb{K}} \mapsto \sum_i e^i \otimes \rho^V_{\mathbf{u}^{-1}\mathbf{v}}(e_i)\colon \mathbb{K} \to V^\star \otimes V\right),$$

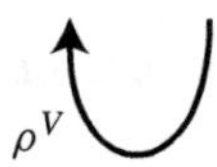

$$\xrightarrow{Q^{\mathfrak{A}}} \left(1_{\mathbb{K}} \mapsto \sum_i e_i \otimes e^i : \mathbb{K} \to V \otimes V^\star\right),$$

$$T \otimes T' \xrightarrow{Q^{\mathfrak{A}}} Q^{\mathfrak{A}}(T) \otimes Q^{\mathfrak{A}}(T'),$$

$$T \circ T' \xrightarrow{Q^{\mathfrak{A}}} Q^{\mathfrak{A}}(T) \circ Q^{\mathfrak{A}}(T').$$

To keep track of the domain and range of $Q^{\mathfrak{A}}$ for a given tangle, following the notation in Definition 7.13, it can be helpful, for each V, to decorate the ends of each ρ^V coloured tangle component with V if it is oriented upwards, and V^* if it is oriented downwards.

Let us compute the explicit element-wise actions, stated in the above list, for the maps associated with the tangles  and 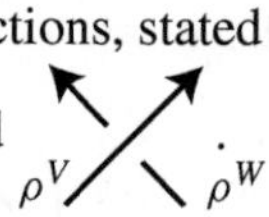.

For 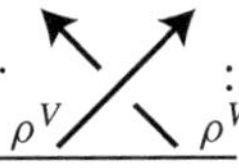: Straightforwardly, and using the notation in (8.11),

$$\left(\tau \circ (\rho^V \otimes \rho^W)(\mathsf{R})\right)(x \otimes y) = \tau \circ \sum_i \left((\rho^V \otimes \rho^W)(\alpha_i \otimes \beta_i)\right)(x \otimes y)$$
$$= \tau \circ \sum_i \rho^V_{\alpha_i}(x) \otimes \rho^W_{\beta_i}(y) = \sum_i \rho^W_{\beta_i}(y) \otimes \rho^V_{\alpha_i}(x),$$

For 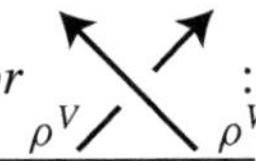: First note that, for $x \otimes y \in V \otimes W$,

$$\left((\rho^W \otimes \rho^V)(\mathsf{R}^{-1}) \circ \tau\right)(x \otimes y) = (\rho^W \otimes \rho^V)(\mathsf{R}^{-1})(y \otimes x).$$

But

$$\begin{aligned}(\rho^W \otimes \rho^V)(\mathsf{R}^{-1}) &= (\rho^W \otimes \rho^V) \circ (S \otimes \mathrm{id}_{\mathfrak{A}})(\mathsf{R}) \qquad \text{(Prop. 8.31)}\\ &= (\rho^W \otimes \rho^V) \circ \sum_i S(\alpha_i) \otimes \beta_i\\ &= \sum_i \rho^W_{S(\alpha_i)} \otimes \rho^V_{\beta_i}.\end{aligned}$$

Thus $(\rho^W \otimes \rho^V)(\mathsf{R}^{-1}) \circ \tau : x \otimes y \mapsto \sum_i \rho^W_{S(\alpha_i)}(y) \otimes \rho^V_{\beta_i}(x)$, as claimed.

Theorem 9.3. *The Reshetikhin–Turaev invariant, $Q^{\mathfrak{A}}$, is an invariant of framed oriented coloured tangles.*

Proof. We must show that $Q^{\mathfrak{A}}$ is invariant under each of the Turaev moves given in Theorem 7.33. For each such move, the appropriate tangle diagram is sliced into elementary pieces.

Here we will show invariance under FT_1 and FT_3 to illustrate the methods involved. Details of the invariance under the remaining Turaev moves can be found in Appendix D.

For FT_1: Since the diagram 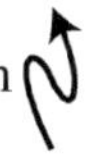 has only one component, there is only one colour ρ^V, say. By slicing, we have

$Q^{\mathfrak{A}}\Big($  $\Big) = Q^{\mathfrak{A}}\Big($  $\Big) \circ Q^{\mathfrak{A}}\Big($  $\Big).$

The colouring of each elementary piece induced through the slicing is necessarily consistent, so the composition is defined. Now, recalling the domain correction conventions from Sect. 7.2,

$e_i \overset{\iota}{\longmapsto} e_i \otimes 1_{\mathbb{K}}$

$\overset{Q^{\mathfrak{A}}}{\longmapsto}$ 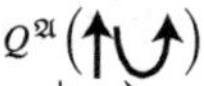 $\sum_j e_i \otimes e^j \otimes \rho^V_{\mathbf{u}^{-1}\mathbf{v}}(e_j)$ (see List (9.1))

$\overset{Q^{\mathfrak{A}}}{\longmapsto}$ 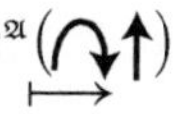 $\sum_j \underbrace{\left(e^j \circ \rho^V_{\mathbf{v}^{-1}\mathbf{u}}\right)(e_i)}_{\in\mathbb{K}} \otimes \rho^V_{\mathbf{u}^{-1}\mathbf{v}}(e_j)$ (see List (9.1))

$\overset{\kappa}{\longmapsto} \sum_j \underbrace{\left(e^j \circ \rho^V_{\mathbf{v}^{-1}\mathbf{u}}\right)(e_i)}_{\in\mathbb{K}} \cdot \rho^V_{\mathbf{u}^{-1}\mathbf{v}}(e_j).$

Let $\rho^V_{\mathbf{v}^{-1}\mathbf{u}}(e_p) = \sum_q n^p_q\, e_q$ and $\rho^V_{\mathbf{u}^{-1}\mathbf{v}}(e_p) = \sum_r m^p_r\, e_r$. Then

$Q^{\mathfrak{A}}$  $(e_i) = \sum_{j,r} e^j \left(\sum_q n^i_q\, e_q\right) \cdot m^j_r\, e_r = \sum_{j,r} n^i_j\, m^j_r\, e_r = \left(\rho^V_{\mathbf{u}^{-1}\mathbf{v}} \circ \rho^V_{\mathbf{v}^{-1}\mathbf{u}}\right)(e_i)$

$= \rho^V_{\mathbf{u}^{-1}\mathbf{v}\mathbf{v}^{-1}\mathbf{u}}(e_i) = \rho^V_{1_{\mathfrak{A}}}(e_i).$

since ρ^V is a representation of $\mathfrak{A}$. Thus $Q^{\mathfrak{A}}$  $= \mathrm{id}_V$.

Similarly, for $Q^{\mathfrak{A}}$ , slicing gives

$Q^{\mathfrak{A}}$  $= Q^{\mathfrak{A}}$  $\circ\, Q^{\mathfrak{A}}$ .

Then

$e_i \overset{\iota}{\longmapsto} 1_{\mathbb{K}} \otimes e_i \overset{Q^{\mathfrak{A}}}{\longmapsto}$  $\sum_j e_j \otimes e^j \otimes e^i \overset{Q^{\mathfrak{A}}}{\longmapsto}$  $\sum_j e_j \otimes e^j(e_i) \overset{\kappa}{\longmapsto} e_i,$

so $Q^{\mathfrak{A}}$  $= \mathrm{id}_V$.

It follows that $Q^{\mathfrak{A}}$  $= Q^{\mathfrak{A}}$ 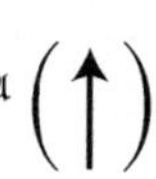 $= Q^{\mathfrak{A}}$ 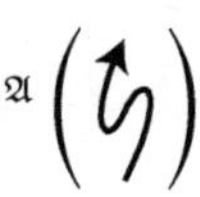 so $Q^{\mathfrak{A}}$ is invariant under FT_1.

<u>*For* FT_3</u>: The tangle  on the left-hand side of the FT_3-move has two components that are coloured by representations as shown. By slicing, we obtain a colouring of the elementary pieces that is consistent with composition, which leads to the relation

$$Q^{\mathfrak{A}}\left(\right.$$  $$\left.\right) = Q^{\mathfrak{A}}(\downarrow\downarrow\curvearrowright) \circ Q^{\mathfrak{A}}(\downarrow\downarrow\uparrow\curvearrowright\downarrow) \circ Q^{\mathfrak{A}}(\downarrow\downarrow\nwarrow\!\nearrow\downarrow\downarrow)$$

$$\circ\, Q^{\mathfrak{A}}(\downarrow\cup\uparrow\downarrow\downarrow) \circ Q^{\mathfrak{A}}(\cup\downarrow\downarrow) : V^\star \otimes W^\star \to W^\star \otimes V^\star.$$

Then

$$\begin{aligned}
e^i \otimes w^j &\overset{\iota}{\longmapsto} 1_{\mathbb{K}} \otimes e^i \otimes w^j \\
&\overset{Q^{\mathfrak{A}}(\cup\downarrow\downarrow)}{\longmapsto} \textstyle\sum_k w^k \otimes \rho^W_{\mathbf{u}^{-1}\mathbf{v}}(w_k) \otimes e^i \otimes w^j \\
&\overset{\iota}{\longmapsto} \textstyle\sum_k w^k \otimes 1_{\mathbb{K}} \otimes \rho^W_{\mathbf{u}^{-1}\mathbf{v}}(w_k) \otimes e^i \otimes w^j \\
&\overset{Q^{\mathfrak{A}}(\downarrow\cup\uparrow\downarrow\downarrow)}{\longmapsto} \textstyle\sum_{k,l} w^k \otimes e^l \otimes \rho^V_{\mathbf{u}^{-1}\mathbf{v}}(e_l) \otimes \rho^W_{\mathbf{u}^{-1}\mathbf{v}}(w_k) \otimes e^i \otimes w^j \\
&\overset{Q^{\mathfrak{A}}(\downarrow\downarrow\nwarrow\nearrow\uparrow\downarrow\downarrow)}{\longmapsto} \textstyle\sum_{k,l,n} w^k \otimes e^l \otimes \rho^W_{\beta_n}\left(\rho^W_{\mathbf{u}^{-1}\mathbf{v}}(w_k)\right) \otimes \rho^V_{\alpha_n}\left(\rho^V_{\mathbf{u}^{-1}\mathbf{v}}(e_l)\right) \otimes e^i \otimes w^j \\
&= \textstyle\sum_{k,l,n} w^k \otimes e^l \otimes \rho^W_{\beta_n\mathbf{u}^{-1}\mathbf{v}}(w_k) \otimes \rho^V_{\alpha_n\mathbf{u}^{-1}\mathbf{v}}(e_l) \otimes e^i \otimes w^j \\
&\overset{Q^{\mathfrak{A}}(\downarrow\downarrow\uparrow\curvearrowright\downarrow)}{\longmapsto} \textstyle\sum_{k,l,n} w^k \otimes e^l \otimes \rho^W_{\beta_n\mathbf{u}^{-1}\mathbf{v}}(w_k) \otimes \underbrace{e^i\left(\rho^V_{\mathbf{v}^{-1}\mathbf{u}\alpha_n\mathbf{u}^{-1}\mathbf{v}}(e_l)\right)}_{\in\mathbb{K}} \otimes w^j \\
&\overset{\kappa}{\longmapsto} \textstyle\sum_{k,l,n} \underbrace{e^i\left(\rho^V_{\mathbf{v}^{-1}\mathbf{u}\alpha_n\mathbf{u}^{-1}\mathbf{v}}(e_l)\right)}_{\in\mathbb{K}} \cdot w^k \otimes e^l \otimes \rho^W_{\beta_n\mathbf{u}^{-1}\mathbf{v}}(w_k) \otimes w^j \\
&\overset{Q^{\mathfrak{A}}(\downarrow\downarrow\curvearrowright)}{\longmapsto} \textstyle\sum_{k,l,n} \underbrace{e^i\left(\rho^V_{\mathbf{v}^{-1}\mathbf{u}\alpha_n\mathbf{u}^{-1}\mathbf{v}}(e_l)\right)}_{\in\mathbb{K}} \cdot w^k \otimes e^l \otimes \underbrace{w^j\left(\rho^W_{\mathbf{v}^{-1}\mathbf{u}\beta_n\mathbf{u}^{-1}\mathbf{v}}(w_k)\right)}_{\in\mathbb{K}} \\
&\overset{\kappa}{\longmapsto} \textstyle\sum_{k,l,n} \underbrace{e^i\left(\rho^V_{\mathbf{v}^{-1}\mathbf{u}\alpha_n\mathbf{u}^{-1}\mathbf{v}}(e_l)\right)}_{\in\mathbb{K}} \cdot \underbrace{w^j\left(\rho^W_{\mathbf{v}^{-1}\mathbf{u}\beta_n\mathbf{u}^{-1}\mathbf{v}}(w_k)\right)}_{\in\mathbb{K}} \cdot w^k \otimes e^l
\end{aligned}$$

since $\mathbf{v}$ is central and, by Lemma 8.35, $\mathbf{u}a\mathbf{u}^{-1} = S^2(a)$ for any $a \in \mathfrak{A}$. Thus

$$e^i \otimes w^j \overset{Q^{\mathfrak{A}}(\ \cdot\ )}{\longmapsto} \sum_{k,l,n} \underbrace{e^i\left(\rho^V_{S^2(\alpha_n)}(e_l)\right)}_{\in\mathbb{K}} \cdot \underbrace{w^j\left(\rho^W_{S^2(\beta_n)}(w_k)\right)}_{\in\mathbb{K}} \cdot (w^k \otimes e^l). \tag{9.2}$$

The tangle on the right-hand side of the FT_3-move is  which has two components that are coloured by representations as shown. By slicing, we obtain a colouring of the elementary pieces that is consistent with composition, which leads to the relation

$$Q^{\mathfrak{A}}\Big($$  $$\Big) = Q^{\mathfrak{A}}\left(\curvearrowleft\downarrow\downarrow\right) \circ Q^{\mathfrak{A}}\left(\downarrow\frown\uparrow\downarrow\downarrow\right) \circ Q^{\mathfrak{A}}\left(\downarrow\downarrow\nwarrow\!\!\!\nearrow\downarrow\downarrow\right)$$
$$\circ\, Q^{\mathfrak{A}}\left(\downarrow\downarrow\uparrow\smile\downarrow\right) \circ Q^{\mathfrak{A}}\left(\downarrow\downarrow\uparrow\smile\right) : V^\star \otimes W^\star \to W^\star \otimes V^\star.$$

Proceeding as before, we have

$$\begin{aligned}
e^i \otimes w^j &\overset{\iota}{\longmapsto} e^i \otimes w^j \otimes 1_{\mathbb{K}} \\
&\overset{Q^{\mathfrak{A}}(\downarrow\downarrow\uparrow\smile)}{\longmapsto} \textstyle\sum_l e^i \otimes w^j \otimes e_l \otimes e^l \\
&\overset{\iota}{\longmapsto} \textstyle\sum_l e^i \otimes w^j \otimes e_l \otimes 1_{\mathbb{K}} \otimes e^l \\
&\overset{Q^{\mathfrak{A}}(\downarrow\downarrow\uparrow\smile\downarrow)}{\longmapsto} \textstyle\sum_{k,l} e^i \otimes w^j \otimes e_l \otimes w_k \otimes w^k \otimes e^l \\
&\overset{Q^{\mathfrak{A}}(\downarrow\downarrow\nwarrow\!\!\!\nearrow\downarrow\downarrow)}{\longmapsto} \textstyle\sum_{k,l,n} e^i \otimes w^j \otimes \rho^W_{\beta_n}(w_k) \otimes \rho^V_{\alpha_n}(e_l) \otimes w^k \otimes e^l \\
&\overset{Q^{\mathfrak{A}}(\downarrow\frown\uparrow\downarrow\downarrow)}{\longmapsto} \textstyle\sum_{k,l,n} e^i \otimes \underbrace{w^j\left(\rho^W_{\beta_n}(w_k)\right)}_{\in\mathbb{K}} \otimes \rho^V_{\alpha_n}(e_l) \otimes w^k \otimes e^l \\
&\overset{\kappa}{\longmapsto} \textstyle\sum_{k,l,n} \underbrace{w^j\left(\rho^W_{\beta_n}(w_k)\right)}_{\in\mathbb{K}} \cdot e^i \otimes \rho^V_{\alpha_n}(e_l) \otimes w^k \otimes e^l \\
&\overset{Q^{\mathfrak{A}}(\curvearrowleft\downarrow\downarrow)}{\longmapsto} \textstyle\sum_{k,l,n} \underbrace{e^i\left(\rho^V_{\alpha_n}(e_l)\right)}_{\in\mathbb{K}} \otimes \underbrace{w^j\left(\rho^W_{\beta_n}(w_k)\right)}_{\in\mathbb{K}} \cdot (w^k \otimes e^l) \\
&\overset{\kappa}{\longmapsto} \textstyle\sum_{k,l,n} \underbrace{e^i\left(\rho^V_{\alpha_n}(e_l)\right)}_{\in\mathbb{K}} \cdot \underbrace{w^j\left(\rho^W_{\beta_n}(w_k)\right)}_{\in\mathbb{K}} \cdot (w^k \otimes e^l)
\end{aligned}$$

so

$$e^i \otimes w^j \overset{Q^{\mathfrak{A}}(\,\cdot\,)}{\longmapsto}$$  $$\sum_{k,l,n} \underbrace{e^i\left(\rho^V_{\alpha_n}(e_l)\right)}_{\in\mathbb{K}} \cdot \underbrace{w^j\left(\rho^W_{\beta_n}(w_k)\right)}_{\in\mathbb{K}} \cdot (w^k \otimes e^l) \tag{9.3}$$

The invariance of $Q^{\mathfrak{A}}$ under this FT_3-move is therefore established by demonstrating that the right-hand sides of (9.2) and (9.3) are equal, which may be carried out as follows. From Proposition 8.31(3), we have $(S \otimes S)(\mathsf{R}) = \mathsf{R}$ so applying $S \otimes S$ again gives $(S^2 \otimes S^2)(\mathsf{R}) = \mathsf{R}$ and therefore

$$\sum_n S^2(\alpha_n) \otimes S^2(\beta_n) = \sum_n \alpha_n \otimes \beta_n.$$

Thus, by applying the homomorphism $\rho^V \otimes \rho^W$ to this relation,

$$\sum_n \rho^V_{S^2(\alpha_n)} \otimes \rho^W_{S^2(\beta_n)} = \sum_n \rho^V_{\alpha_n} \otimes \rho^W_{\beta_n},$$

and by applying this, in turn, to the element $e_l \otimes w_k$ gives

$$\sum_n \rho^V_{S^2(\alpha_n)}(e_l) \otimes \rho^W_{S^2(\beta_n)}(w_k) = \sum_n \rho^V_{\alpha_n}(e_l) \otimes \rho^W_{\beta_n}(w_k).$$

Finally, by applying the linear map $\kappa \circ (e^i \otimes w^j)$ to this relation, we have

$$\sum_n \underbrace{e^i\left(\rho^V_{S^2(\alpha_n)}(e_l)\right)}_{\in\mathbb{K}} \cdot \underbrace{w^j\left(\rho^W_{S^2(\beta_n)}(w_k)\right)}_{\in\mathbb{K}} = \sum_n \underbrace{e^i\left(\rho^V_{\alpha_n}(e_l)\right)}_{\in\mathbb{K}} \cdot \underbrace{w^j\left(\rho^W_{\beta_n}(w_k)\right)}_{\in\mathbb{K}}.$$

But these are the coefficients in (9.2) and (9.3) of the image of $e^i \otimes w^j$ under $Q^{\mathfrak{A}}$ (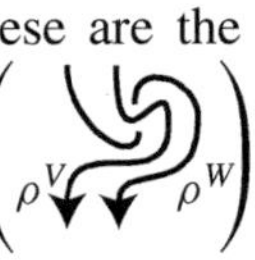) and $Q^{\mathfrak{A}}$ (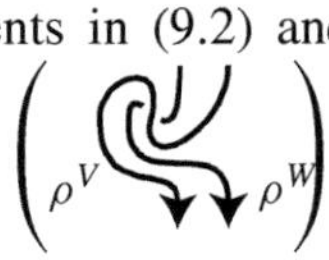) with respect to the basis of $W^\star \otimes V^\star$, so

$Q^{\mathfrak{A}}$ () $(e^i \otimes w^j) = Q^{\mathfrak{A}}$ () $(e^i \otimes w^j)$

for all i, j and so $Q^{\mathfrak{A}}$ () $= Q^{\mathfrak{A}}$ (). Therefore, $Q^{\mathfrak{A}}$ is invariant under one FT_3-move.

For the FT_3-move with the negative crossing, similar calculations give that the action of $Q^{\mathfrak{A}}$ () is given by

$$e^i \otimes w^j \mapsto \sum_{k,l,n} e^i\left(\rho^V_{S^2(\beta_n)}(e_l)\right) \cdot w^i\left(\rho^W_{S^2(S(\alpha_n))}(w_k)\right) \cdot w^k \otimes e^l,$$

and the action of $Q^{\mathfrak{A}}$ () is given by

$$e^i \otimes w^j \mapsto \sum_{k,l,n} e^i\left(\rho^V_{\beta_n}(e_l)\right) \cdot w^i\left(\rho^W_{S(\alpha_n)}(w_k)\right) \cdot w^k \otimes e^l.$$

Proceeding as in the previous case, $S^2 \otimes S^2(\mathsf{R}) = \mathsf{R}$, and applying $S \otimes \mathrm{id}$ to each side gives $S^3 \otimes S^2(\mathsf{R}) = S \otimes \mathrm{id}(\mathsf{R})$. Thus

$$\sum_n S^2(S(\alpha_n)) \otimes S^2(\beta_n) = \sum_n S(\alpha_n) \otimes \beta_n,$$

from which it follows that $Q^{\mathfrak{A}}$ () $= Q^{\mathfrak{A}}$ (). Therefore, $Q^{\mathfrak{A}}$ is invariant under both FT_3-moves.

We have shown invariance under two of the Turaev moves. A proof of the invariance under the remaining Turaev moves can be found in Appendix D. □

9.3 $Q^{\mathfrak{A}}$ is an Operator Invariant

The Reshetikhin–Turaev invariants, $Q^{\mathfrak{A}}$, were constructed by choosing a ribbon Hopf algebra $\mathfrak{A}$, colouring the tangle with representations $\rho^{V_i}: \mathfrak{A} \to \mathrm{End}(V_i)$ of $\mathfrak{A}$, and then associating an elementary linear map with each elementary tangle. There is a clear similarity between the construction of $Q^f_{(\mathsf{R},\alpha,\beta)}$ and $Q^{\mathfrak{A}}$ (compare (7.24) and (9.1)). This is made precise in the next theorem.

Theorem 9.4. *Let T be a tangle diagram, let $(\mathfrak{A}, \mathsf{R}, \mathbf{v})$ be a ribbon Hopf algebra, and let $\rho: \mathfrak{A} \to \mathrm{End}(V)$ be a representation of $\mathfrak{A}$. Then*

$$Q^{\mathfrak{A}} = Q^f_{(\tau\circ(\rho^V\otimes\rho^V)(\mathsf{R}),\, \rho^V_{\mathbf{u}^{-1}\mathbf{v}},\, \mathrm{id}_{V^\star})},$$

where, in the definition of $Q^{\mathfrak{A}}$, each component of T is coloured by ρ^V, and, in the definition of the framed invariant Q^f, we set $W = V^\star$. In particular, $\mu = \rho^V_{\mathbf{v}^{-1}\mathbf{u}}$.

Proof. Let

$\mathsf{M} = Q^{\mathfrak{A}}$ (), $\vec{u} = Q^{\mathfrak{A}}$ (), $\overleftarrow{u} = Q^{\mathfrak{A}}$ (),

$\vec{n} = Q^{\mathfrak{A}}$ (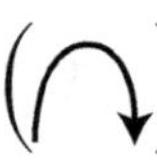), $\overleftarrow{n} = Q^{\mathfrak{A}}$ ().

Then $Q^{\mathfrak{A}}$ and $Q^f_{(V,V^\star,\mathsf{M},\vec{n},\overleftarrow{n},\vec{u},\overleftarrow{u})}$ are tangle invariants constructed in identical ways, and are therefore equal. The maps α and β in the definition of Q^f are defined, according to (B.1), by

$$\alpha: V \to V : e_i \mapsto \sum_j \vec{u}^{ij}\, e_j \quad \text{and} \quad \beta: V^\star \to V^\star : e^i \mapsto \sum_j \overleftarrow{u}^{ij}\, e^j,$$

where

$$\overrightarrow{u}: \mathbb{K} \to V^\star \otimes V: 1_\mathbb{K} \mapsto \sum_{i,j} \overrightarrow{u}^{ij}\ e^i \otimes e_j,$$

$$\overleftarrow{u}: \mathbb{K} \to V \otimes V^\star: 1_\mathbb{K} \mapsto \sum_{i,j} \overleftarrow{u}^{ij}\ e_i \otimes e^j.$$

Since

$$\overleftarrow{u} = Q^\mathfrak{A}\Big($$ 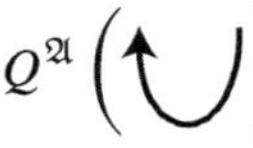 $$\Big): 1_\mathbb{K} \mapsto \sum_i e_i \otimes e^i$$

we see that $\overleftarrow{u}^{ij} = \delta_{ij}$ so $\beta\colon e^i \mapsto e^i$ whence $\beta = \mathrm{id}_{V^\star}$.

Similarly,

$$\overrightarrow{u} = Q^\mathfrak{A}\Big($$  $$\Big): 1_\mathbb{K} \mapsto \sum_i e^i \otimes \rho_{\mathbf{u}^{-1}\mathbf{v}}(e_i).$$

Then $\overrightarrow{u}^{ij} = [\rho_{\mathbf{u}^{-1}\mathbf{v}}]_i^j$ (an element of a matrix) where $\rho_{\mathbf{u}^{-1}\mathbf{v}}(e_i) = \sum_j [\rho_{\mathbf{u}^{-1}\mathbf{v}}]_i^j\ e_j$ so

$$\alpha\colon e_i \mapsto \sum_j [\rho_{\mathbf{u}^{-1}\mathbf{v}}]_i^j\ e_j = \rho_{\mathbf{u}^{-1}\mathbf{v}}(e_i)$$

so $\alpha = \rho_{\mathbf{u}^{-1}\mathbf{v}}$, completing the proof. □

9.4 A Knot Invariant from $U_q(\mathfrak{sl}_2)$

We illustrate the construction of a framed tangle invariant from the ribbon Hopf algebra $U_q(\mathfrak{sl}_2)$ and its standard two-dimensional representation. Recall from Definition 8.56 that $U_q(\mathfrak{sl}_2)$ is the algebra generated by symbols X, Y, K and $\overline{K}$ and subject to the relations:

$$K\overline{K} = 1 = \overline{K}\,K, \quad KX = qXK, \quad KY = \overline{q}\,YK, \quad [X, Y] := \frac{K^2 - \overline{K}^2}{q - \overline{q}}.$$

From Theorem 8.59, these form a ribbon Hopf algebra by taking

$$\mathsf{R} := e^{\frac{h}{4}H\otimes H} \cdot \sum_{n\geq 0} \frac{(q - \overline{q})^n}{[n]!} q^{n(n-3)/2} (KX)^n \otimes (\overline{K}\,Y)^n,$$

and

$$\mathbf{v} = e^{-\frac{h}{4}H^2} \cdot \sum_{n\geq 0} \frac{(\overline{q} - q)^n}{[n]!}\ \overline{q}^{\,n(n+3)/2}\ \overline{K}^{\,2(n+1)} Y^n X^n,$$

where $\overline{q} = q^{-1}$, $q = e^{h/2}$, $\overline{K} = K^{-1}$, and $K = e^{hH/4}$.

We consider the standard two-dimensional representation of $U_q(\mathfrak{sl}_2)$. For this, let V be a free module over $\mathbb{C}[q^{1/2}, q^{-1/2}]$ of rank 2 with basis $\{e_1, e_2\}$. The two-dimensional representation of $U_q(\mathfrak{sl}_2)$, namely ρ^V, is defined on the generators X, Y, K and $\overline{K}$ of $U_q(\mathfrak{sl}_2)$ by

$$\rho^V_X = \begin{bmatrix} 0 & 1 \\ 0 & 0 \end{bmatrix}, \ \rho^V_Y = \begin{bmatrix} 0 & 0 \\ 1 & 0 \end{bmatrix}, \ \rho^V_K = \begin{bmatrix} q^{\frac{1}{2}} & 0 \\ 0 & \overline{q}^{\frac{1}{2}} \end{bmatrix}, \ \rho^V_{\overline{K}} = \begin{bmatrix} \overline{q}^{\frac{1}{2}} & 0 \\ 0 & q^{\frac{1}{2}} \end{bmatrix}.$$

We shall also need to use $\rho^V_H = \begin{bmatrix} 1 & 0 \\ 0 & -1 \end{bmatrix}$. With ρ^V so defined, its action on the element H is such that $\rho^V_K = \rho^V_{e^{hH/4}}$.

It is to be recalled that the less awkward notation $\rho^V_{KX}(x)$, for example, is used instead of $\left(\rho^V(KX)\right)(x)$ to denote the application to $x \in V$ of the operator ρ^V_{KX} corresponding to the element KX of $\mathfrak{sl}_2$ through the representation ρ^V whose carrier space is V.

We colour all strands of the tangles by ρ^V which, henceforth, is denoted by ρ. By Theorems 8.55 and 9.3, we know that $U_q(\mathfrak{sl}_2)$ and ρ give rise to a Reshetikhin–Turaev invariant. We shall denote this invariant of framed oriented tangles by $Q^{\mathfrak{sl}_2,\rho}$. Since each strand is coloured by the same representation, the representation may be omitted from the figures, so $Q^{\mathfrak{sl}_2,\rho}$ is an invariant of tangles, rather than coloured tangles.

Our first step towards understanding $Q^{\mathfrak{sl}_2,\rho}$ is to compute its actions on each elementary tangle as in (9.1). It will be convenient to describe some of these actions by matrices, and for this, we choose the ordered basis

$$\{e_1 \otimes e_1, e_1 \otimes e_2, e_2 \otimes e_1, e_2 \otimes e_2\}$$

of $V \otimes V$ and construct matrices with respect to this ordered basis. We now compute the actions.

$Q^{\mathfrak{sl}_2,\rho}$  $: V \otimes V \to V \otimes V : x \otimes y \mapsto (\tau \circ (\rho \otimes \rho)(\mathrm{R}))\ (x \otimes y)$,

obtained from (9.1). Then

$$(\rho^V \otimes \rho^V)(\mathrm{R}) = e^{\frac{h}{4}\rho_H \otimes \rho_H} \sum_{n \geq 0} \frac{(q - \overline{q})^n}{[n]!} q^{n(n-3)/2} \rho_{(KX)^n} \otimes \rho_{(\overline{K}Y)^n}.$$

It is easily checked that

$$\rho_{KX} = \rho_K \circ \rho_X = \begin{bmatrix} 0 & q^{\frac{1}{2}} \\ 0 & 0 \end{bmatrix} \quad \text{and} \quad \rho_{\overline{K}Y} = \rho_{\overline{K}} \circ \rho_Y = \begin{bmatrix} 0 & 0 \\ q^{\frac{1}{2}} & 0 \end{bmatrix}.$$

It follows that $\rho_{(KX)^n} = (\rho_{(KX)})^n = \mathbf{0}$ and $\rho_{(\overline{K}Y)^n} = (\rho_{(\overline{K}Y)})^n = \mathbf{0}$ when $n > 1$. Thus,

$$(\rho \otimes \rho)(\mathsf{R}) = e^{\frac{h}{4}\rho_H \otimes \rho_H}\left((\rho_1 \otimes \rho_1) + (1 - \overline{q}^{\,2})(\rho_{KX} \otimes \rho_{\overline{K}\,Y})\right),$$

so

$Q^{\mathfrak{sl}_2,\rho}$  $: x \otimes y \mapsto \tau \circ e^{\frac{h}{4}\rho_H \otimes \rho_H}\left((\mathrm{id} \otimes \mathrm{id}) + (1 - \overline{q}^{\,2})(\rho_{KX} \otimes \rho_{\overline{K}\,Y})\right)(x \otimes y).$

As matrices,

$$\tau = \begin{bmatrix} 1&0&0&0 \\ 0&0&1&0 \\ 0&1&0&0 \\ 0&0&0&1 \end{bmatrix}, \quad e^{\frac{h}{4}\rho_{H \otimes H}} = \begin{bmatrix} q^{\frac{1}{2}} & 0 & 0 & 0 \\ 0 & \overline{q}^{\,\frac{1}{2}} & 0 & 0 \\ 0 & 0 & \overline{q}^{\,\frac{1}{2}} & 0 \\ 0 & 0 & 0 & q^{\frac{1}{2}} \end{bmatrix}, \quad \rho_{KX \otimes \overline{K}\,Y} = \begin{bmatrix} 0&0&0&0 \\ 0&0&q&0 \\ 0&0&0&0 \\ 0&0&0&0 \end{bmatrix}.$$

Therefore, the matrix for the operator corresponding to the positive crossing is

$Q^{\mathfrak{sl}_2,\rho}$  $= \begin{bmatrix} q^{\frac{1}{2}} & 0 & 0 & 0 \\ 0 & 0 & \overline{q}^{\,\frac{1}{2}} & 0 \\ 0 & \overline{q}^{\,\frac{1}{2}} & q^{\frac{1}{2}} - \overline{q}^{\,\frac{3}{2}} & 0 \\ 0 & 0 & 0 & q^{\frac{1}{2}} \end{bmatrix}.$ (9.4)

Instead of using the element action given in (9.1) to obtain the matrix for the negative crossing, which of course may be done, it is easier in the present instance simply to take the inverse of the matrix for the positive crossing. Thus, making use of the fact that $Q^{\mathfrak{sl}_2,\rho}$ is invariant under FT_4,

$Q^{\mathfrak{sl}_2,\rho}$  $= \Big(Q^{\mathfrak{sl}_2,\rho}$  $\Big)^{-1} = \begin{bmatrix} \overline{q}^{\,\frac{1}{2}} & 0 & 0 & 0 \\ 0 & \overline{q}^{\,\frac{1}{2}} - q^{\frac{3}{2}} & q^{\frac{1}{2}} & 0 \\ 0 & q^{\frac{1}{2}} & 0 & 0 \\ 0 & 0 & 0 & \overline{q}^{\,\frac{1}{2}} \end{bmatrix}.$ (9.5)

From (9.1), the element-wise action of $Q^{\mathfrak{sl}_2,\rho}$ 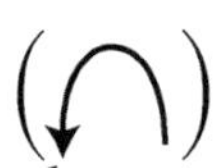 $: V^* \otimes V \to \mathbb{K}$ is given by $f \otimes x \mapsto f(x)$. Therefore, its action on basis elements of $V^* \otimes V$ is

$Q^{\mathfrak{sl}_2,\rho}$  $: e^i \otimes e_j \mapsto \delta_{i,j}.$ (9.6)

Computing the action of $Q^{\mathfrak{sl}_2,\rho}$  is a little more involved. From (9.1), the element-wise action of $Q^{\mathfrak{sl}_2,\rho}$  $: \mathbb{K} \to V^* \otimes V$ is

$$Q^{\mathfrak{sl}_2,\rho}$$

$$: 1 \mapsto \sum_{i=1}^{2} e^i \otimes \rho_{\mathbf{u}^{-1}\mathbf{v}}(e_i).$$

It can be shown (see for example [90]) that the identity $\mathbf{v} = \overline{K}^2\mathbf{u}$ holds in $U_q(\mathfrak{sl}_2)$. Using this, $\rho_{\mathbf{u}^{-1}\mathbf{v}} = \rho_{\overline{K}^2}$, and $\rho_{\overline{K}^2} = \begin{bmatrix} \overline{q} & 0 \\ 0 & q \end{bmatrix}$, so, $\rho_{\mathbf{u}^{-1}\mathbf{v}}(e_1) = \overline{q}\, e_1$ and $\rho_{\mathbf{u}^{-1}\mathbf{v}}(e_2) = q\, e_2$. Then

$$Q^{\mathfrak{sl}_2,\rho}$$

$$: 1 \mapsto \overline{q}\, e^1 \otimes e_1 + q\, e^2 \otimes e_2. \qquad (9.7)$$

Since $Q^{\mathfrak{sl}_2,\rho}$  and $Q^{\mathfrak{sl}_2,\rho}$  are identity maps, we now have the actions of $Q^{\mathfrak{sl}_2,\rho}$ on the elementary tangles. However, we can go further in our understanding of $Q^{\mathfrak{sl}_2,\rho}$ and show that it has a skein theoretic definition.

By using the matrices in (9.4) and (9.5), and proceeding as in Lemma 6.16, it is easily checked that

$$q^{1/2}\begin{bmatrix} q^{\frac{1}{2}} & 0 & 0 & 0 \\ 0 & 0 & \overline{q}^{\frac{1}{2}} & 0 \\ 0 & \overline{q}^{\frac{1}{2}} & q^{\frac{1}{2}} - \overline{q}^{\frac{3}{2}} & 0 \\ 0 & 0 & 0 & q^{\frac{1}{2}} \end{bmatrix} - \overline{q}^{1/2}\begin{bmatrix} \overline{q}^{\frac{1}{2}} & 0 & 0 & 0 \\ 0 & \overline{q}^{\frac{1}{2}} - q^{\frac{3}{2}} & q^{\frac{1}{2}} & 0 \\ 0 & q^{\frac{1}{2}} & 0 & 0 \\ 0 & 0 & 0 & \overline{q}^{\frac{1}{2}} \end{bmatrix} = (q - \overline{q})\mathbf{I}_4.$$

It follows that

$$q^{1/2}Q^{\mathfrak{sl}_2,\rho}$$

$$-\,\overline{q}^{1/2}Q^{\mathfrak{sl}_2,\rho}$$

$$= (q - \overline{q})Q^{\mathfrak{sl}_2,\rho}$$

$$. \qquad (9.8)$$

We can also remove positive twists from our diagrams as follows.

From (D.2), we have that $Q^{\mathfrak{sl}_2,\rho}$ 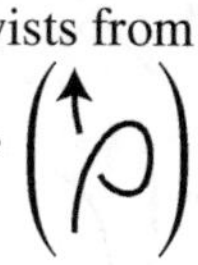 $= \rho_{\mathbf{v}^{-1}}$. We have, using that $(\rho_X)^n = \mathbf{0}$, for $n > 1$,

$$\begin{aligned} \rho_{\mathbf{v}} &= e^{-\frac{h}{4}\rho(H)^2} \cdot \sum_{n\geq 0} \frac{(\overline{q} - q)^n}{[n]!} \overline{q}^{\,n(n+3)/2} \rho(\overline{K})^{2(n+1)} \rho(Y)^n \rho(X)^n \\ &= e^{-\frac{h}{4}\rho(H)^2} \left(\rho(\overline{K})^2 + (\overline{q} - q)\overline{q}^{\,2}\rho(\overline{K})^4\rho(Y)\rho(X)\right) \\ &= \begin{bmatrix} q^{-\frac{1}{2}} & 0 \\ 0 & q^{-\frac{1}{2}} \end{bmatrix}\left(\begin{bmatrix} \overline{q} & 0 \\ 0 & q \end{bmatrix} + (\overline{q}^{\,3} - \overline{q})\begin{bmatrix} 0 & 0 \\ 0 & q^2 \end{bmatrix}\right) \\ &= \begin{bmatrix} q^{-\frac{3}{2}} & 0 \\ 0 & q^{-\frac{3}{2}} \end{bmatrix} = q^{-\frac{3}{2}}\mathbf{I}. \end{aligned}$$

Thus, using the fact that ρ is a representation, $\rho_{\mathbf{v}^{-1}} = q^{3/2}\mathbf{I}$. Then, since $Q^{\mathfrak{sl}_2,\rho}$  $= \mathbf{I}$, we have

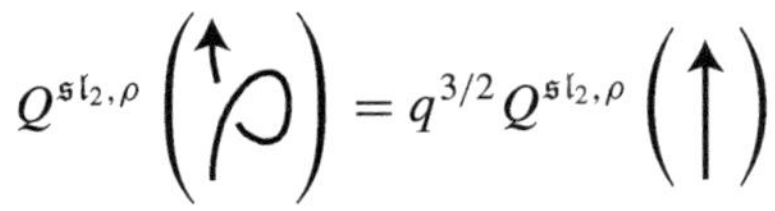

(9.9)

The value of $Q^{\mathfrak{sl}_2,\rho}$ on the unknot is found from (9.6) and (9.7) thus.

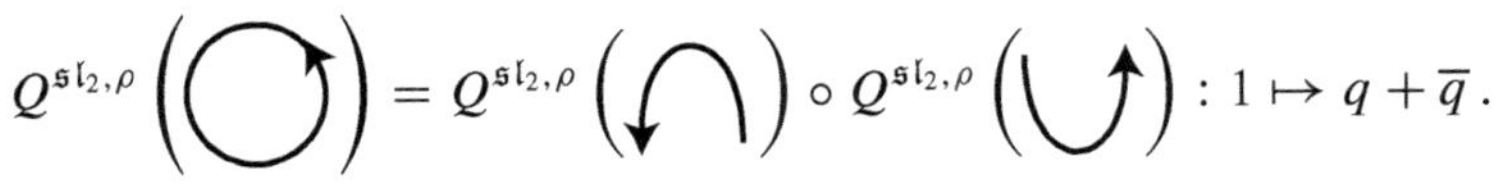

(9.10)

In examining (9.8) and (9.10), it is likely that the skein relations for, say, the Jones polynomial will come to mind, and the reader will notice that we have a skein theoretic formulation for the Reshetikhin–Turaev invariant $Q^{\mathfrak{sl}_2,\rho}$. However, some care needs to be taken here. There are two important points to notice. First, for the relation (9.8) to reduce the computation of $Q^{\mathfrak{sl}_2,\rho}(T)$ to its values on unlinks we need to restrict the domain of $Q^{\mathfrak{sl}_2,\rho}$ to links, rather than tangles in general. Secondly, once we have restricted to links, relation (9.8) will reduce the computation of $Q^{\mathfrak{sl}_2,\rho}(T)$ to its values on unlinks. However, these are *framed* unlinks. Thus, we need both (9.10) and (9.9) to compute the value of $Q^{\mathfrak{sl}_2,\rho}(T)$ on unlinks. By collecting these observations, we obtain the following.

Theorem 9.5. *There is a polynomial invariant $Q^{\mathfrak{sl}_2}$ of framed oriented links defined by the skein relations*

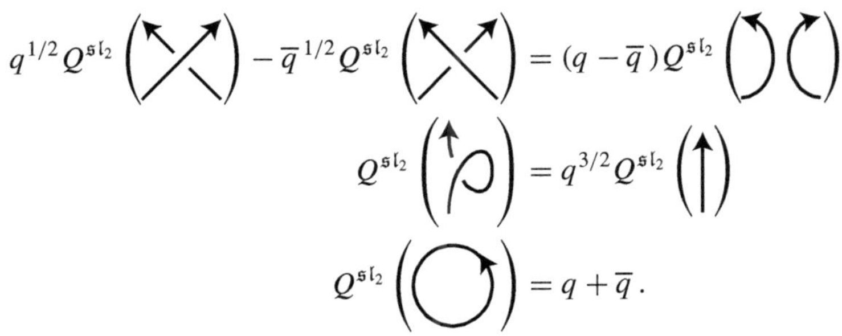

Furthermore, this invariant is equivalent to the Reshetikhin–Turaev invariant $Q^{\mathfrak{sl}_2,\rho}$: $\mathbb{C}[\overline{q}^{\,1/2}, q^{1/2}] \to \mathbb{C}[\overline{q}^{\,1/2}, q^{1/2}]$ via $Q^{\mathfrak{sl}_2,\rho}(L) : 1 \mapsto Q^{\mathfrak{sl}_2}(L)$.

Exercise 9.6. Use the deframing process described in Sect. 3.3 to construct an invariant of unframed oriented links or tangles from $Q^{\mathfrak{sl}_2}$. Obtain a skein relation for this invariant and compare it to that for the Jones polynomial.

Exercise 9.7. Above we studied the Reshetikhin–Turaev invariant arising from the two-dimensional representation of $U_q(\mathfrak{sl}_2)$. The purpose of this exercise is to determine the invariant arising from the one-dimensional representation. For this, let $\{e_1\}$ be a basis of V, and let $\{e^1\}$ be the corresponding dual basis of $V^\star$. The map ρ^V defined on the generators X, Y, K and $\overline{K}$ of $U_q(\mathfrak{sl}_2)$ by $\rho^V(X) = \rho^V(Y) = [0]_{1\times 1}$, $\rho^V(K) = \rho^V(\overline{K}) = [1]_{1\times 1}$, affords a representation of $U_q(\mathfrak{sl}_2)$. Also $\rho^V(H) = 0$. Proceeding as in the two-dimensional case, compute the values of $Q^{\mathfrak{sl}_2,\rho^V}$ on each of the elementary tangles. Find a skein relation that determines the values of $Q^{\mathfrak{sl}_2,\rho^V}$ on links. Hence show that in this case $Q^{\mathfrak{sl}_2,\rho^V}$ is a trivial invariant.

As mentioned previously, although our focus here is on the (framed) Reshetikhin–Turaev invariant $Q^{\mathfrak{sl}_2,\rho}$ coming from $U_q(\mathfrak{sl}_2)$ and its standard two-dimensional representation, the construction here does work for other representations (e.g. the knot invariant arising from the n-dimensional representation of $U_q(\mathfrak{sl}_2)$ is known as the *coloured Jones polynomial*), and for quantum groups coming from other Lie algebras (e.g. the HOMFLYPT polynomial arises from the $U_q(\mathfrak{sl}_n)$ and its standard representation). The key point is that Reshetikhin–Turaev invariants provide a framework that unifies an important class of knot invariants. We could then ask if there is a single invariant that unifies all Reshetikhin–Turaev invariants. It turns out that one does exist. This invariant is known as the Kontsevich invariant and we shall meet it in Part IV, but first we adopt a different perspective and consider Vassiliev knot invariants.

Part III
Vassiliev Invariants

Chapter 10
The Fundamentals of Vassiliev Invariants

This part of the text sees an apparent departure from the invariants of Part II. So far our focus has been on constructing link invariants by considering the combinatorics and algebra of link diagrams. In this part, we consider instead the space of all numerically valued knot invariants, or, dually, of all knots. By considering crossing changes on link diagrams, we get a filtration on these spaces, and these filtrations give rise to the notion of Vassiliev invariants. Not all knot invariants are Vassiliev invariants. However, many important ones are. In fact, they include (in an appropriate sense) the Reshetikhin–Turaev invariants we have seen in Part II.

Vassiliev invariants were introduced independently by Vassiliev [185, 186] and by Goussarov [69, 70]. They are also known in the literature as Vassiliev–Goussarov invariants, although here we use the shorter, more common name. Here we follow the combinatorial approach to Vassiliev invariants of Birman and Lin [24]. We shall see, through an intricate use of standard linear algebra, how Vassiliev invariants are equivalent to combinatorial objects called chord diagrams, and how this translation of the knot invariants into combinatorial objects is essential in developing the subject and in making contact with the Reshetikhin–Turaev invariants.

10.1 Vassiliev Invariants and Singular Knots

The first step in the theory of Vassiliev invariants is the formation of a vector space of knots.

Definition 10.1 (Vector space of knots). Let $\mathscr{K}$ denote the vector space of finite formal linear combinations of isotopy classes of oriented knots over $\mathbb{C}$, *i.e.* $\mathscr{K}$ is the (infinite dimensional) vector space over $\mathbb{C}$ with basis consisting of the isotopy classes of knots. We call $\mathscr{K}$ the *vector space of knots*.

We usually represent an isotopy class of a knot K by its member K (or any other member of the class), or by a diagram of K. Thus, generally, when we specify

D. M. Jackson and I. Moffatt, *An Introduction to Quantum and Vassiliev Knot Invariants*, CMS Books in Mathematics, https://doi.org/10.1007/978-3-030-05213-3_10

a knot K, we mean the "isotopy class of K". An example of vector space operations in $\mathscr{K}$ is given by

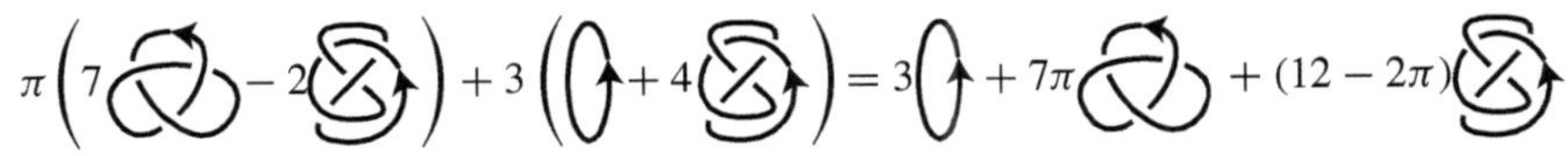

where the scalar multiplication and addition are just formal operations, and no topological meaning should be inferred from the linear combinations.

A *numerical knot invariant* evaluated in $\mathbb{C}$ is a knot invariant that takes its values in $\mathbb{C}$. Any such knot invariant may be extended linearly to give a map from $\mathscr{K}$ to $\mathbb{C}$. A vector space structure over $\mathbb{C}$ may be defined on the set of knot invariants *via* the usual addition and scalar multiplication of functions. The next definition captures a trivial yet important observation. (A review of dual spaces can be found in Sect. A.5.)

Definition 10.2 ($\mathbb{C}$-valued knot invariant). A $\mathbb{C}$*-valued knot invariant* is an element of $\mathscr{K}^\star$, *i.e.* a function from $\mathscr{K}$ to $\mathbb{C}$.

It is convenient when working with Vassiliev invariants to make use of the following generalisation of a knot, called a *singular knot*.

Intuitively, a singular knot is a knot that is allowed to intersect itself a finite number of times, and at each of these intersection points only two strands can intersect, and must do so transversally, thus:  . To make the singular point clear in our drawings, each double point of a singular knot is marked by a "•" in a corresponding singular knot diagram. Thus, singular crossings are drawn as  . An example of a singular knot (diagram) is

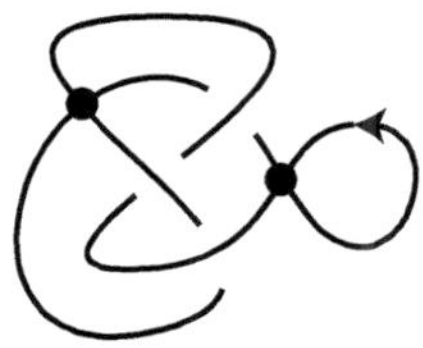

.

A formal definition of a singular knot requires the concept of an immersion. A continuous function $f : \mathbb{S}^1 \to \mathbb{R}^3$ is an *immersion* if every point $x \in \mathbb{S}^1$ has a neighbourhood N_x such that the restriction of f to N_x is a homeomorphism. For convenience, we shall define our singular knots to be tame from the outset.

Definition 10.3 (Singular knot). A *polygonal singular knot* is an immersion of a polygon into $\mathbb{R}^3$ that has finitely many multiple points and is such that edges only intersect in points, no intersection point is a vertex of the polygon, and each intersection only includes two edges.

A *singular knot* is an immersion of $\mathbb{S}^1$ into $\mathbb{R}^3$ that is ambient isotopic to a polygonal singular knot.

As with knots, two singular knots are *equivalent* if they are ambient isotopic, as in Definition 1.3. Again, as with knots, we usually work with diagrams of singular knots, which are considered up to the Reidemeister moves.

Definition 10.4 (Singular knot diagram). A *singular knot diagram* is the image of a regular projection of a singular knot in which the images of intersection points are marked with a • and an under-/over-crossing structure has been assigned to all other double points. The crossing structure is indicated by line breaks in the diagram.

Theorem 10.5. *Two singular knots are equivalent if and only if their diagrams are related by a finite sequence of Reidemeister moves. That is*

$$\frac{\{\textit{singular knots}\}}{\textit{isotopy}} \cong \frac{\{\textit{singular knot diagrams}\}}{\textit{Reidemeister moves}}.$$

The proof of Theorem 10.5 is very similar to the proof of Theorem 1.26 and so is omitted.

It is to be emphasised that the cyclic order of the arcs at a singular point in a singular knot is fixed under isotopy. Intuitively, we may think of two arcs as being welded together at a singular point and so, although we can move the singular point around in space, we cannot change the arcs entering and leaving it.

For the moment, we shall not study singular knots in their own right (although we shall do so in the next chapter). Instead, here:

- singular knots provide a compact notation for certain linear combinations of knots in $\mathscr{K}$.

An oriented singular knot is identified with a unique element of $\mathscr{K}$ by eliminating each of its singular points through repeated application of the *Vassiliev relation*, which is defined as follows.

Definition 10.6 (Vassiliev relation).

As usual, the three diagrams are identical outside of the region shown.

We shall use the term "to desingularise" a singular knot K to mean the exhaustive application of the Vassiliev relation to remove all the singular points of K. The element of $\mathscr{K}$ resulting from desingularising a singular knot is called a *Vassiliev resolution* of K, which we denote by $V_{\text{res}}(K)$. Similarly, we speak of a *desingularisation* of a knot.

As an example of desingularisation:

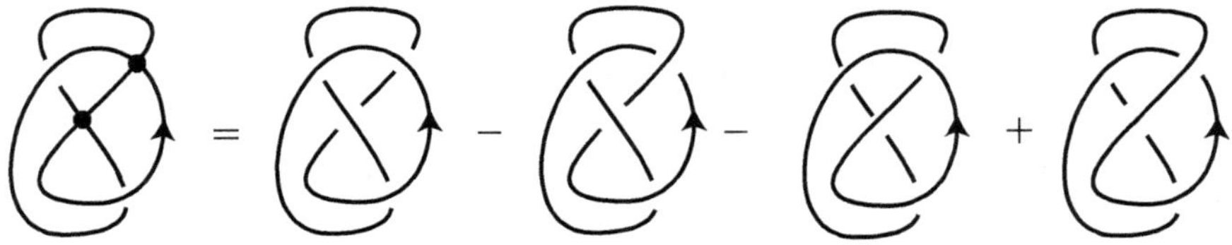

(10.1)

(the right-hand side being a Vassiliev resolution).

The following lemma tells us that K has a unique Vassiliev resolution: that is, the result of desingularising a singular knot is independent of the order in which each of the singular points is resolved using the Vassiliev relation. In addition, it provides a useful formula for the resolution of a singular knot.

Lemma 10.7. *Let S_K be the set of all singular points of a singular knot K. For each $A \subseteq S_K$, let K_A be the knot in which the singular points of K that are in A are resolved negatively and the remaining singular points of K are resolved positively. Then the Vassiliev resolution of K is*

$$V_{\text{res}}(K) = \sum_{A \subseteq S_K} (-1)^{|A|} K_A.$$

Exercise 10.8. Prove Lemma 10.7 by using induction on the number of singular points of a singular knot.

As already noted, we adopt the following convention in this chapter.

Convention 10.9. When working within the vector space $\mathscr{K}$, we shall make the identification

$$K \longleftrightarrow V_{\text{res}}(K) \qquad (10.2)$$

between a singular knot and its Vassiliev resolution. That is to say, a singular knot K is compact notation for a linear combination of knots in this chapter.

In Chap. 11, however, we shall consider singular knots as objects in their own right and therefore shall not make this identification. Thus, observing that we are working in the context of $\mathscr{K}$ is important. We shall see that the use of singular knots has profound consequences.

Keeping in mind the identification of a singular knot in $\mathscr{K}$ with its Vassiliev resolution, we make the following definition.

Definition 10.10 (Vector space $\mathscr{K}_m$). $\mathscr{K}_m$ is the (infinite dimensional) subspace of $\mathscr{K}$ spanned by the set of all oriented singular knots with (at least) m singular points.

It follows from the definition of $\mathscr{K}_m$ that the space $\mathscr{K}$ of all knots admits a filtration

$$\mathscr{K} = \mathscr{K}_0 \geq \mathscr{K}_1 \geq \mathscr{K}_2 \geq \mathscr{K}_3 \geq \cdots. \qquad (10.3)$$

We have defined $\mathscr{K}_m$ to be the subspace of $\mathscr{K}$ spanned by the set of all singular knots with *at least* m singular points. Equivalently, we may in fact define it to be the subspace spanned by set of all singular knots with *exactly* m singular points, as the following result shows.

Lemma 10.11. *$\mathscr{K}_m$ is the vector space over $\mathbb{C}$ spanned by the set of all oriented knots K with exactly m singular points, with the identification given in (10.2) for K.*

Proof. Let $K \in \mathscr{K}_n$ where $n > m$. Then we may apply Vassiliev's relation to obtain $K = \mathfrak{b} - \mathfrak{c}$ where $\mathfrak{b}, \mathfrak{c} \in \mathscr{K}_{n-1}$. The result follows by iterating this a finite number of times to express K as an element of $\mathscr{K}_m$. □

The existence of the filtration (10.3) enables us to form quotient spaces $\mathscr{K}_m / \mathscr{K}_{m+1}$, a construction that will be crucial to later arguments, both knot theoretically and algebraically.

With the concept of the vector spaces $\mathscr{K}_m$, we may now define Vassiliev invariants.

Definition 10.12 (Vassiliev invariant—1 st.).

1. A knot invariant θ is a *Vassiliev invariant of degree m* (or *finite type invariant of degree m*) if it is a linear map $\theta : \mathscr{K} \to \mathbb{C}$ with the property that

$$\theta|_{\mathscr{K}_{m+1}} = 0,$$

   where 0 is the zero map.
2. $\mathscr{V}$ denotes the set of all Vassiliev invariants.
3. $\mathscr{V}_m$ denotes the set of all Vassiliev invariants of degree m.

Exercise 10.13. Verify that a degree m Vassiliev invariant θ is also a degree n Vassiliev invariant for each $n > m$.

A second definition of Vassiliev invariants will given in Definition 11.4.

10.2 Examples of Vassiliev Invariants

With these preliminaries completed, we must now show that

1. Vassiliev invariants exist, and that
2. there are highly non-trivial ones;

and answer a residual basic question

3. are there knot invariants that are not Vassiliev invariants?

The next exercise establishes that Vassiliev invariants exist by exhibiting a particular, albeit uninteresting, Vassiliev invariant.

Exercise 10.14. This exercise classifies the degree 0 Vassiliev invariants.

1. Let $r \in \mathbb{C}$ and $\kappa_r : \mathscr{K} \to \mathbb{C}$ be the map defined on its basis elements by $\kappa_r(K) = r$ for each knot K. Show that κ_r is a degree 0 Vassiliev invariant. κ_r is called a *constant invariant*.
2. Let θ be a degree 0 Vassiliev invariant. Show that θ satisfies the skein relation 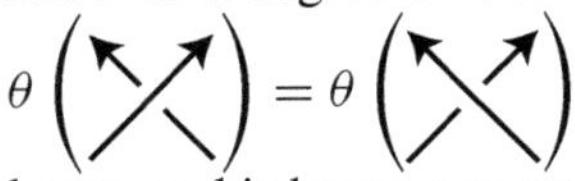

. Hence deduce that θ takes the same value on all knots, and is hence a constant invariant.

With the question of existence resolved, we now turn to establishing that there are highly non-trivial Vassiliev invariants. The following examples of Vassiliev invariants arise from knot polynomials. In the context of Vassiliev invariants, knot polynomials are to be considered as mappings from $\mathscr{K}$ to a polynomial ring, defined as the linear extensions to the vector space $\mathscr{K}$.

<u>*The Alexander–Conway polynomial.*</u> Recall from Definition 2.30 that the Alexander–Conway polynomial $C(K) \in \mathbb{Z}[z]$ may be defined by the skein relation

SAC1 : 

,

SAC2 : $C(\mathscr{O}) = 1.$

Applying **SAC1** to a singular point of a singular knot gives

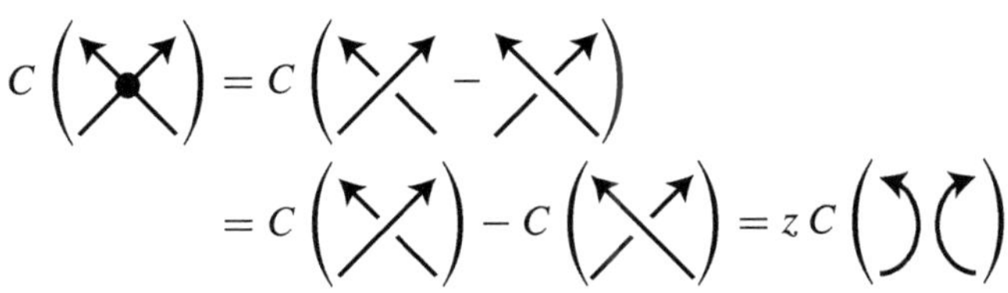

(10.4)

By iterating this, we see that the Alexander–Conway polynomial of a knot with $m + 1$ singular points has the form $C(K) = z^{m+1} f(z)$ where $f(z)$ is some polynomial in z. Thus,

$$[z^m]\, C(K) = 0,$$

where $[z^m]$ denotes the *coefficient operator* which, when applied to $C(K)$, gives the coefficient of z^m in $C(K)$. Therefore, $[z^m]\, C : \mathscr{K} \to \mathbb{C}$ is a degree m Vassiliev invariant.

Theorem 10.15. *The coefficient of the z^m term of the Alexander–Conway polynomial is a degree m Vassiliev invariant.*

The HOMFLYPT polynomial. Next, we consider the HOMFLYPT polynomial. Recall from Definition 2.27 that the HOMFLYPT polynomial is given by the skein relation

SH1 : 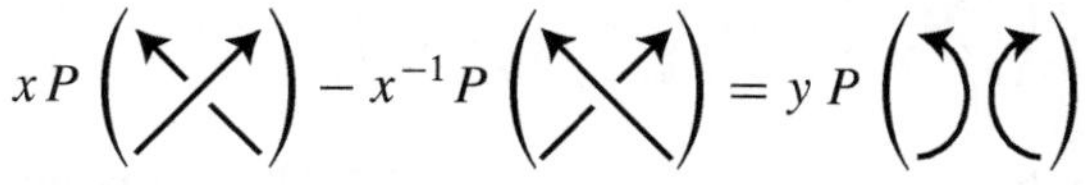

SH2 : $P(\mathcal{O}) = 1.$

Consider its 1-variable specializations (one for each value of n)

 (10.5)

obtained by setting $x = q^n$ and $y = q - q^{-1}$, which the reader will recognise from Theorem 6.23. Changing the indeterminate of this Laurent polynomial from q to h by setting

$$q = e^{h/2},$$

and therefore $q^{-1} = e^{-h/2}$, we obtain a formal power series $Q_{n,h} := P_n(e^{h/2})$ over $\mathbb{C}$ in the indeterminate h.

For example, the 2-variable HOMFLYPT polynomial of the right-hand trefoil is, from (2.5),

$$(2 + y^2)x^{-2} - x^{-4}.$$

The substitution $x = q^n$ and $y = (q - q^{-1})$ gives

$$q^{-2n+2} + q^{-2n-2} - q^{-4n},$$

and setting $q = e^{h/2}$ and denoting the right-hand trefoil by K gives the formal power series

$$Q_{n,h}(K) = 1 + \left(1 - n^2\right) h^2 + \left(n^3 - n\right) h^3 + \left(\frac{1}{2}n^2 + \frac{1}{12} - \frac{7}{12}n^4\right) h^4 + \cdots$$

in the indeterminate h.

Theorem 10.16. *The coefficients of* $Q_{n,h} := P_n(e^{h/2})$, *where* $P_n(q)$ *is the HOMFLYPT polynomial of* (10.5), *define Vassiliev invariants for each integer* n. *In particular,* $[h^m]Q_{n,h} : \mathcal{K} \to \mathbb{C}$, *the coefficient of* h^m *in* $Q_{n,h}$, *gives a degree* m *Vassiliev invariant.*

Proof. We begin by considering a singular knot and applying the Vassiliev relation to one of its singular points. From the Vassiliev relation and the linearity of $Q_{n,h}$

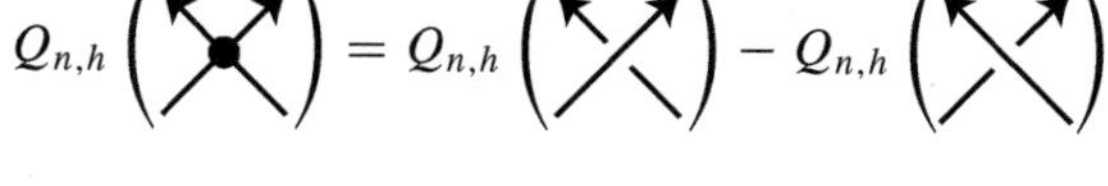

so

$$[h^0]\, Q_{n,h}$$  $$= [h^0]\, Q_{n,h}$$  $$- [h^0]\, Q_{n,h}$$  $$. \qquad (10.6)$$

Next, from the skein relation (10.5) for the HOMFLYPT polynomial, we have

$$e^{nh/2} Q_{n,h}$$  $$- e^{-nh/2} Q_{n,h}$$  $$= \left(e^{h/2} - e^{-h/2}\right) Q_{n,h}$$  $$.$$

Taking the constant term in h on each side, we have,

$$[h^0]\, Q_{n,h}$$  $$- [h^0]\, Q_{n,h}$$  $$= 0. \qquad (10.7)$$

Equations (10.6) and (10.7) together give $[h^0]\, Q_{n,h}$  $= 0$. Thus,

$$Q_{n,h}$$  $$= h \cdot f(n, h)$$

for some formal power series $f(n, h)$ in h and n. Therefore, if $K \in \mathscr{K}_{m+1}$, it follows that

$$Q_{n,h}(K) = h^{m+1} \cdot g(n, h)$$

for some formal power series $g(n, h)$ in h and n. In particular, $[h^m]Q_{n,h}(K) = 0$, so $[h^m]Q_{n,h}(K)$ vanishes on $\mathscr{K}_{m+1}$ and hence is a Vassiliev invariant of degree m. □

Exercise 10.17. Use Theorem 10.16 to deduce how the coefficients of the Jones polynomial give rise to Vassiliev invariants. Adapt the proof of Theorem 10.16 to give an independent proof of your answer.

Having seen that non-trivial Vassiliev invariants exist (Theorem 10.15 is from [12], and Theorem 10.16 is from [24]), it is natural to ask whether or not all numerical knot invariants are Vassiliev invariants. An instance of a numerical knot invariant that is not a Vassiliev invariant is the *crossing number*, which is the minimum number of crossings that *any* diagram of a given knot can attain. Other examples of invariants that are not Vassiliev invariants include the signature, genus, bridge number and unknotting number. The reader who would like to read further on this topic may wish to consult the articles [24, 49, 178].

10.3 The Vector Space of Vassiliev Invariants

The next result produces a vector space from the set $\mathscr{V}_m$ of degree m Vassiliev invariants. This is, of course, the vector space over $\mathbb{C}$ generated by the set $\mathscr{V}_m$, but we set it down explicitly.

Proposition 10.18. *The set $\mathscr{V}_m$ of all degree m Vassiliev invariants equipped with addition and scalar multiplication defined by*

$$(\theta + \theta')(K) := \theta(K) + \theta'(K), \quad (\lambda\,\theta)(K) := \lambda(\theta(K)) \tag{10.8}$$

for all $\theta, \theta' \in \mathscr{V}_m$ and $\lambda \in \mathbb{C}$ and $K \in \mathscr{K}$, is a vector space over $\mathbb{C}$, which we shall denote also by $\mathscr{V}_m$.

Exercise 10.19. Prove Proposition 10.18.

As with the vector space $\mathscr{K}$ of all knots, the vector space $\mathscr{V}$ of all Vassiliev invariants also admits a filtration.

Proposition 10.20. *There is a filtration on the space of Vassiliev invariants:*

$$\mathscr{V}_0 \leq \mathscr{V}_1 \leq \mathscr{V}_2 \leq \cdots . \tag{10.9}$$

Proof. To establish the result, it is sufficient to show that if $\theta \in \mathscr{V}_m$ then $\theta \in \mathscr{V}_{m+1}$. Let $K \in \mathscr{K}_{m+2}$. Then $K \in \mathscr{K}_{m+1}$ by the filtration (10.3). But $\theta \in \mathscr{V}_m$ implies that $\theta(K) = 0$ since $K \in \mathscr{K}_{m+1}$. Thus, $\theta(K) = 0$ for all $K \in \mathscr{K}_{m+2}$, so $\theta|_{\mathscr{K}_{m+2}} = 0$. Then $\theta \in \mathscr{V}_{m+1}$, thereby establishing the filtration. □

In general, to show that the degree of a knot invariant θ is m, we must show that $\theta(\mathscr{K}_{m+1}) = \{0\}$, that is, that θ vanishes on all linear combinations of knots with at least $m+1$ singular points. It suffices to show this for the generators, that is, to show that whenever K is a singular knot with at least $m+1$ singular points then $\theta(K) = 0$. But, by Lemma 10.11, $\mathscr{K}_{m+1}$ is generated by knots with exactly $m+1$ singular points so,

- to show that $\theta \in \mathscr{V}_m$, it suffices to show that $\theta(K) = 0$ for all singular knots K with exactly $m+1$ singular points.

We are now in a position to state two important relationships between the vector space $\mathscr{V}_m$ of degree m Vassiliev invariants and the dual of a particular quotient space associated with the vector space $\mathscr{K}$ of all knots.

Lemma 10.21.

$$\mathscr{V}_m \cong (\mathscr{K}/\mathscr{K}_{m+1})^\star .$$

Proof. By definition,

$$\mathscr{V}_m = \{\theta \in \mathscr{K}^\star : \theta(\mathscr{K}_{m+1}) = \{0\}\}.$$

But, from Lemma A.11,

$$\{\theta \in \mathscr{K}^\star : \theta(\mathscr{K}_{m+1}) = \{0\}\} \cong \hom(\mathscr{K}/\mathscr{K}_{m+1}, \mathbb{C}) = (\mathscr{K}/\mathscr{K}_{m+1})^\star,$$

and the result follows. □

Proposition 10.22.

$$\frac{\mathscr{V}_m}{\mathscr{V}_{m-1}} \cong \left(\frac{\mathscr{K}_m}{\mathscr{K}_{m+1}}\right)^\star. \tag{10.10}$$

Proof. The result would be fairly straightforward if we knew that the left-hand and the right-hand sides are finite dimensional, but do not (yet) know this. Thus, we begin by describing the elements of $\mathscr{V}_m/\mathscr{V}_{m-1}$, which are equivalence classes of elements of $\mathscr{V}_m$.

Let $[f], [g] \in \mathscr{V}_m/\mathscr{V}_{m-1}$ where $f, g \in \mathscr{V}_m$. Then, by Definition A.8, $[f] = [g]$ if and only if $f - g \in \mathscr{V}_{m-1}$. But

$$\mathscr{V}_{m-1} = \{\theta \in \mathscr{K}^\star : \theta(\mathscr{K}_m) = \{0\}\}, \tag{10.11}$$

so $f - g \in \mathscr{V}_{m-1}$ if and only if

$$(f - g)(\mathscr{K}_m) = \{0\}.$$

Therefore, $[f]$ is the equivalence class

$$[f] = \left\{\theta \in \mathscr{V}_m : \theta|_{\mathscr{K}_m} = f|_{\mathscr{K}_m}\right\}. \tag{10.12}$$

Then $[f] \in \mathscr{V}_m/\mathscr{V}_{m-1}$ may be identified with an element of $\mathscr{K}_m^\star$ as follows. For $K \in \mathscr{K}_m$, the action of $[f]$ is given by

$$[f](K) := f(K), \tag{10.13}$$

which is well-defined by (10.12). Furthermore, since $f \in \mathscr{V}_m$,

$$[f](\mathscr{K}_{m+1}) = f(\mathscr{K}_{m+1}) = \{0\},$$

and so $[f] \in \mathscr{V}_m/\mathscr{V}_{m-1}$ may be identified with an element of

$$\left\{\theta \in \mathscr{K}_m^\star : \theta(\mathscr{K}_{m+1}) = \{0\}\right\}. \tag{10.14}$$

Thus, we have defined a vector space morphism

$$\psi : \frac{\mathscr{V}_m}{\mathscr{V}_{m-1}} \to \left\{\theta \in \mathscr{K}_m^\star : \theta(\mathscr{K}_{m+1}) = \{0\}\right\},$$

where, for $K \in \mathscr{K}_m$, $\psi([f])(K) := f(K)$.

We show that ψ is both surjective and injective and hence is an isomorphism.
Surjectivity: Let $f \in \{\theta \in \mathscr{K}_m^\star : \theta(\mathscr{K}_{m+1}) = \{0\}\}$. Define $\widehat{f}$ by

$$\widehat{f}: \mathscr{K} \to \mathbb{C}: K \mapsto \begin{cases} f(K) & \text{if } K \in \mathscr{K}_m, \\ 0 & \text{otherwise.} \end{cases}$$

Since $f(\mathscr{K}_{m+1}) = \{0\}$, we have $\widehat{f} \in \mathscr{V}_m$. Let $[\widehat{f}\,] \in \mathscr{V}_m/\mathscr{V}_{m-1}$ be the equivalence class of $\widehat{f}$ in $\mathscr{V}_m/\mathscr{V}_{m-1}$. Then, for $K \in \mathscr{K}_m$, $\psi([\widehat{f}\,])(K) = \widehat{f}(K) = f(K)$, from which it follows that ψ is surjective.
Injectivity: Suppose that $\psi([f]) = \psi([g])$. Then $f(K) = g(K)$ for all $K \in \mathscr{K}_m$, so $(f - g)(\mathscr{K}_m) = \{0\}$ where $f, g \in \mathscr{K}^\star$. Then $f - g \in \mathscr{V}_{m-1}$ and so $[f] = [g]$, giving injectivity.

We have shown that ψ is an isomorphism, and so

$$\frac{\mathscr{V}_m}{\mathscr{V}_{m-1}} \cong \left\{\theta \in \mathscr{K}_m^\star : \theta(\mathscr{K}_{m+1}) = \{0\}\right\}.$$

By Lemma A.11, the set on the right-hand side is isomorphic to

$$\left(\frac{\mathscr{K}_m}{\mathscr{K}_{m+1}}\right)^\star,$$

and the result follows. □

Since we have established that Vassiliev invariants exist in each degree, and that at least some of these are non-trivial, it makes sense to investigate the structure of the vector space $\mathscr{V}_m$ of degree m Vassiliev invariants. A basic piece of information about $\mathscr{V}_m$ we can ask for is its dimension (or, indeed, if it is finite or infinite dimensional), and a basis for the vector space. We begin by calculating the dimensions of $\mathscr{V}_0$ (which was obtained in Exercise 10.14) and $\mathscr{V}_1$ from first principles as a prelude to studying the vector spaces $\mathscr{V}_m$ more systematically in the next chapter.

Proposition 10.23.

$$\dim(\mathscr{V}_0) = 1.$$

Proof. By Exercise 10.14, there is a degree 0 Vassiliev invariant (given by $[h^0]Q_{n,k}$). Let θ be one such invariant (it is in $\mathscr{V}_0$). In addition, let $K, K' \in \mathscr{K}$ be two knots which differ by a single crossing change. By the Vassiliev relation given in Definition 10.6, there is a singular knot $\mathfrak{b}$ with exactly one singular point such that $\mathfrak{b} = K - K'$. We have

$$\begin{aligned} 0 &= \theta(\mathfrak{b}) && (\theta \text{ is zero on } \mathscr{K}_1) \\ &= \theta(K - K') \\ &= \theta(K) - \theta(K') && (\theta \text{ is linear}) \end{aligned}$$

so $\theta(K) = \theta(K')$. Thus, a degree 0 Vassiliev invariant takes the same value on any two knots which differ by a single crossing change, and therefore by a finite sequence of crossing changes. But, by Exercise 1.19, any knot may be reduced to the unknot by a finite sequence of crossing changes so any pair of knots differ by a finite sequence of crossing changes. Thus, θ is constant on $\mathscr{K}$, and so $\theta \in \langle \kappa_1 \rangle$ so $\mathscr{V}_0 \leq \langle \kappa_1 \rangle$, where κ_1 is the constant invariant that takes the value 1 on every knot. But $\mathscr{V}_0$ is not the zero vector space. The result now follows. □

Proposition 10.24.

$$\dim(\mathscr{V}_1) = 1.$$

Proof. By Theorem 10.16, there exists a Vassiliev invariant of degree 1, so let $\theta \in \mathscr{V}_1$. Let $K, K' \in \mathscr{K}$ be knots that differ by exactly one crossing change and that have exactly one singular point. Then, by the Vassiliev relation, there is a knot $\mathfrak{b} \in \mathscr{K}$ with exactly two singular points such that $\mathfrak{b} = K - K'$. Then

$$\begin{aligned} 0 &= \theta(\mathfrak{b}) && (\theta \text{ is zero on } \mathscr{K}_2) \\ &= \theta(K - K') \\ &= \theta(K) - \theta(K') && (\theta \text{ is linear}). \end{aligned}$$

By the arguments of the proof of the previous proposition, θ takes the same value on all knots that differ in a finite number of crossing changes and have exactly one singular point. But, up to crossing changes, there is a unique knot with exactly one singular point, namely 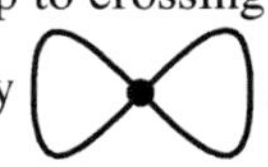. Thus,

$$= 0,$$

since the two knots on the right-hand side are in the same isotopy class and θ is constant on isotopy classes. Then $\theta(K) = 0$ for all $K \in \mathscr{K}$, and so $\theta \in \langle \kappa_1 \rangle$. However, from the filtration (10.9), $\mathscr{V}_0 \leq \mathscr{V}_1$, and we conclude that $\mathscr{V}_0 = \mathscr{V}_1$. The result then follows from Proposition 10.23. □

The argument used in the proof of this result is important and will be generalised in the next section to provide a connection between singular knots and Vassiliev invariants. Although the degrees 0 and 1 Vassiliev invariants are trivial, the Vassiliev invariants of degrees 2 and higher are not.

Before continuing, let us reconsider the question of determining the dimension of the vector space $\mathscr{V}_m$. Recall from (10.9) that we have a filtration

$$\mathscr{V}_0 \leq \mathscr{V}_1 \leq \mathscr{V}_2 \leq \cdots.$$

This means that all degree m Vassiliev invariants are also degree $m+k$ Vassiliev invariants for each $k \geq 1$. If we were to "count" the degree m Vassiliev invariants, by finding the dimension of $\mathcal{V}_m$, then we would also be including all the degree $m-1, \ldots, 0$ Vassiliev invariants. Instead, we should "count" the degree m Vassiliev invariants which are not Vassiliev invariants of lower degree. This task is better expressed in terms of determining the dimension of the quotient space $\mathcal{V}_m/\mathcal{V}_{m-1}$.

If $\dim(\mathcal{V}_m/\mathcal{V}_{m-1})$ were finite (we shall see shortly that it is), then we may recover $\dim(\mathcal{V}_m)$ inductively using (Theorem A.23)

$$\dim(\mathcal{V}_m) = \dim(\mathcal{V}_m/\mathcal{V}_{m-1}) + \dim(\mathcal{V}_{m-1}),$$

as a recurrence equation for $\dim(\mathcal{V}_m)$. This fact, among others, motivates the closer examination of $\mathcal{V}_m/\mathcal{V}_{m-1}$.

Chapter 11
Chord Diagrams

The vector space $\mathscr{V}$ of Vassiliev invariants is a space of functions and can be difficult to work with directly. We now introduce a vector space that is more manageable than $\mathscr{V}$, namely the combinatorial *vector space of chord diagrams*. Indeed, to understand Vassiliev invariants, we shall see it is enough to understand a quotient space of the vector space of chord diagrams by two particular combinatorial relations between chord diagrams, namely the 1T-relation and the 4T-relation.

A connection between the two vector spaces is established through a map from the space of knots to the space of chord diagrams. Formalising this map as a single-valued function requires the introduction of the vector space $\dot{\mathscr{K}}_m$ freely generated by the set of all singular knots of degree m where a singular knot is *not* identified through (10.2) with its Vassiliev resolution.

The arguments are quite dense, entailing technical issues of constructing vector spaces of combinatorial objects that are progressively easier to work with and, of course, isomorphisms between them. The combinatorial objects include *singular knots, chord diagrams, weight systems* and *Jacobi diagrams*, and will be introduced progressively in this and the following few chapters. While the local arguments are elementary, the overall structure and the ways in which the local arguments are used are subtle and quite intricate.

11.1 The Vassiliev Extension of a Knot Invariant

Definition 11.1 (Vector space of singular knots). Let $\dot{\mathscr{K}}$ denote the vector space of finite formal linear combinations of isotopy classes of singular oriented knots over $\mathbb{C}$. We call $\dot{\mathscr{K}}$ the *vector space of singular knots*.

That is, $\dot{\mathscr{K}}$ is the vector space over $\mathbb{C}$ with basis consisting of the isotopy classes of singular knots. For example, an element of $\dot{\mathscr{K}}$ is

D. M. Jackson and I. Moffatt, *An Introduction to Quantum and Vassiliev Knot Invariants*, CMS Books in Mathematics, https://doi.org/10.1007/978-3-030-05213-3_11

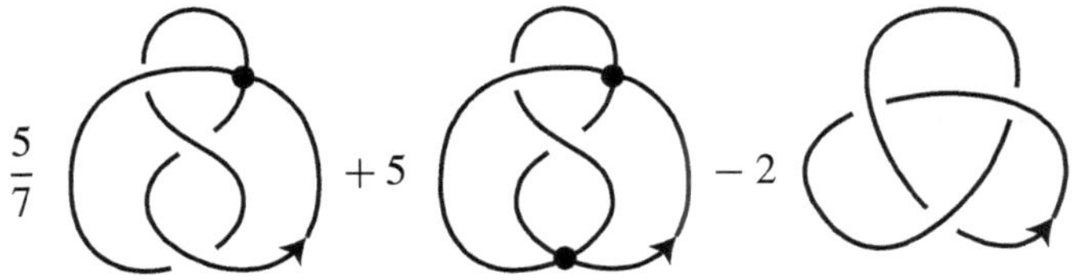

It is implicit in this definition that an element K of $\dot{\mathscr{K}}$ is *not* identified with its Vassiliev resolution. The same element K, regarded as an element of $\mathscr{K}$, *is* identified with its Vassiliev resolution. To emphasise that a singular knot K is to be regarded as an element of $\dot{\mathscr{K}}$, we shall rewrite K as $\dot{K}$.

We also note that the resolution operation described in Lemma 10.7, which expresses an oriented singular knot K as a linear combination of oriented knots $V_{\text{res}}(K)$, may be extended linearly to a map $V_{\text{res}}: \dot{\mathscr{K}} \to \mathscr{K}$. This is the *(Vassiliev) resolution map*.

Definition 11.2 (Vector space $\dot{\mathscr{K}}_m$). $\dot{\mathscr{K}}_m$ is the subspace of $\dot{\mathscr{K}}$ generated by all singular knots with exactly m singular points.

For example, the following figure shows an element of $\dot{\mathscr{K}}_3$.

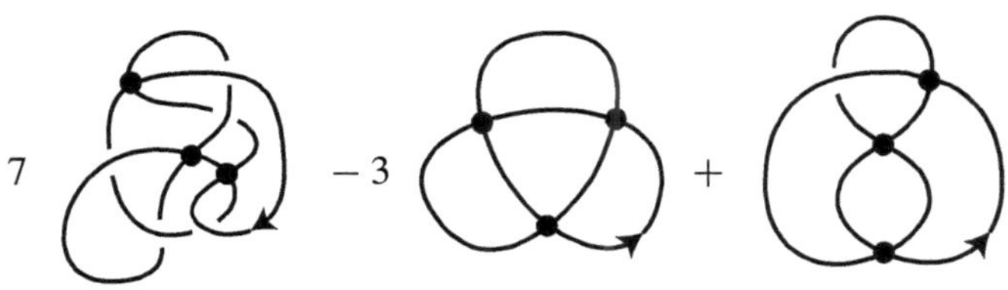

It follows immediately that $\dot{\mathscr{K}}$, while not admitting a filtration (compare with (10.3)), does admit a grading

$$\dot{\mathscr{K}} = \dot{\mathscr{K}}_0 \oplus \dot{\mathscr{K}}_1 \oplus \dot{\mathscr{K}}_2 \oplus \cdots . \tag{11.1}$$

This changes how we can work with $\dot{\mathscr{K}}$ as compared with $\mathscr{K}$.

An element of $\dot{\mathscr{K}}_0$ is a finite linear combination of knots, and an element of $(\dot{\mathscr{K}}_0)^\star$ is a $\mathbb{C}$-valued knot invariant. An element of $\dot{\mathscr{K}}$ is a finite linear combination of singular knots, and an element of $(\dot{\mathscr{K}})^\star$ is a $\mathbb{C}$-valued *singular knot* invariant.

Given a knot invariant $\theta: \mathscr{K} \to \mathbb{C}$, we may extend it to a singular knot invariant $\dot{\theta}: \dot{\mathscr{K}} \to \mathbb{C}$ by applying the Vassiliev relation iteratively to resolve all the singular points of $\dot{K}$ and then applying θ to the resulting element of $\mathscr{K}$. We refer to this process as the *Vassiliev extension* of a knot invariant θ. This is captured in the following definition.

Definition 11.3 (Vassiliev extension). Let $\theta\colon \mathscr{K} \to \mathbb{C}$ be a knot invariant. Then, the *Vassiliev extension* of θ is the singular knot invariant

$$\dot{\theta}\colon \dot{\mathscr{K}} \to \mathbb{C}$$

obtained by setting

$$\dot{\theta}(\dot{K}) := \theta\left(\sum_{A\subseteq S}(-1)^{|A|}\dot{K}_A\right)$$

for a singular knot $\dot{K} \in \dot{\mathscr{K}}$, extended linearly over $\dot{\mathscr{K}}$. Here, as in Lemma 10.7, S is the set of all singular points of $\dot{K}$, and $\dot{K}_A$ is the knot in which the singular points of $\dot{K}$ that are in A are resolved negatively and the remaining singular points are resolved positively.

For example, let $\theta\colon \mathscr{K} \to \mathbb{C}$ be a knot invariant. Then, from (10.1) and the linearity of θ, we have

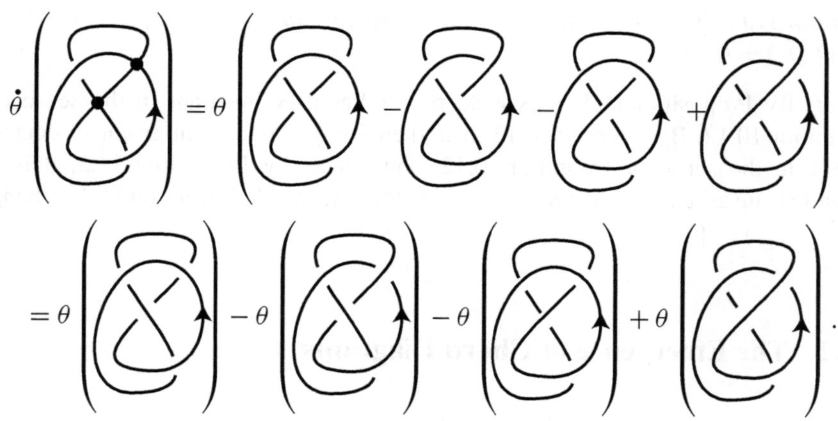

We now formulate a second definition of a Vassiliev invariant (the first was Definition 10.12) and then show that the two definitions are equivalent.

Definition 11.4 (Vassiliev invariant—2nd). A knot invariant $\theta\colon \mathscr{K} \to \mathbb{C}$ is a *Vassiliev invariant of degree m* if $\dot{\theta}\colon \dot{\mathscr{K}} \to \mathbb{C}$ satisfies $\dot{\theta}(\dot{\mathscr{K}}_{m+1}) = \{0\}$.

The next proposition establishes that the two definitions of a Vassiliev invariant are equivalent.

Proposition 11.5. *Definitions 10.12 and 11.4 are equivalent.*

Proof. Let θ be a Vassiliev invariant of degree m according to Definition 10.12. Then, $\theta\colon \mathscr{K} \to \mathbb{C}$ where

$$\theta(\mathscr{K}_{m+1}) = \{0\}. \tag{11.2}$$

Now, consider $\dot{K} \in \dot{\mathscr{K}}_{m+1}$. Then, from Definition 11.3,

$$\dot{\theta}(\dot{K}) := \theta\left(\sum_{A \subseteq S_K} (-1)^{|A|} K_A\right).$$

But, from (10.2), $\sum_{A \subseteq S_K} (-1)^{|A|} K_A = K$ where $K \in \mathscr{K}_{m+1}$. Then $\dot{\theta}(\dot{K}) = \theta(K) = 0$ from (11.2), so $\dot{\theta}(\dot{\mathscr{K}}_{m+1}) = \{0\}$, and therefore θ is a Vassiliev invariant according to Definition 11.4.

On the other hand, let θ be a Vassiliev invariant according to Definition 11.4. Then, $\theta\colon \mathscr{K} \to \mathbb{C}$ is such that $\dot{\theta}(\dot{\mathscr{K}}_{m+1}) = \{0\}$. Let $K \in \mathscr{K}_{m+1}$. Then, K is the Vassiliev resolution of some $\dot{K} \in \dot{\mathscr{K}}_{m+1}$. Then, by Definition 11.3, $\theta(K) = \dot{\theta}(\dot{K}) = 0$ since $\dot{K} \in \dot{\mathscr{K}}_{m+1}$. Thus, θ is a Vassiliev invariant according to Definition 10.12, concluding the proof. □

The next result is required because $\dot{\mathscr{K}}_{m+1} \not\subseteq \dot{\mathscr{K}}_m$, so we do not have a filtration for $\dot{\mathscr{K}}$ (as we have for $\mathscr{K}$ where the corresponding result is immediate).

Lemma 11.6. *If θ is a Vassiliev invariant of degree m and $\dot{K} \in \dot{\mathscr{K}}_{m+1+k}$, for $k \geq 0$, then $\dot{\theta}(\dot{K}) = 0$.*

Proof. By Proposition 11.5, θ is a degree m Vassiliev invariant in the sense of Definition 10.12. By Exercise 10.13, it is therefore a Vassiliev invariant of degree $m+k$ in the sense of Definition 10.12, and hence, by Proposition 11.5, it is a Vassiliev invariant of degree $m+k$ in the sense of Definition 11.4, giving $\dot{\theta}(\dot{\mathscr{K}}_{m+1+k}) = \{0\}$. □

11.2 The Emergence of Chord Diagrams

The following two lemmas show that Vassiliev invariants do not depend on the structure of non-singular crossings, that is, whether they are positive or negative, but do depend upon the order in which the singular points appear in a circular tour of the knot. The proof of this assertion is a straightforward generalisation of the method of the proof of Proposition 10.24.

Lemma 11.7. *Let θ be a degree m Vassiliev invariant and let $\dot{K}$ and $\dot{K}'$ be two singular knots in $\dot{\mathscr{K}}_m$, each with exactly m singular points. If $\dot{K}$ and $\dot{K}'$ differ by a finite sequence of crossing changes and isotopy, then $\dot{\theta}(\dot{K}) = \dot{\theta}(\dot{K}')$.*

Proof. We consider the case in which $\dot{K}$, $\dot{K}'$ are singular knots in $\dot{\mathscr{K}}_m$ that differ by only a single crossing change. Then, there exists a singular knot $\dot{\mathfrak{b}} \in \dot{\mathscr{K}}_{m+1}$ such that the application of the Vassiliev relation to a single singular point of $\mathfrak{b}$ gives $\dot{\mathfrak{b}} = \dot{K} - \dot{K}'$ where, without loss of generality, $\dot{K}$ has a positive crossing and $\dot{K}'$ has a negative crossing at the singular point. Then,

$$
\begin{aligned}
0 &= \dot{\theta}(\dot{\mathfrak{b}}) && (\text{since } \dot{\mathfrak{b}} \in \dot{\mathscr{K}}_{m+1}) \\
&= \dot{\theta}(\dot{K} - \dot{K}') \\
&= \dot{\theta}(\dot{K}) - \dot{\theta}(\dot{K}') && (\dot{\theta} \text{ is linear})
\end{aligned}
$$

Thus, $\dot{\theta}(\dot{K}) = \dot{\theta}(\dot{K}')$.

In the case in which $\dot{K}$ and $\dot{K}'$ differ by a finite sequence of crossing changes and isotopy, the result follows directly by applying the above argument a finite number of times, once for each crossing change, and noting that the value of θ does not change under isotopy. □

This lemma indicates that a degree m invariant does not depend upon the non-singular crossings in a knot with m singular points. The following lemma is, in effect, a rephrasing of the previous lemma. It indicates that the invariant is dependent only upon the *cyclic order* of the singular points.

Lemma 11.8. *Let θ be a degree m Vassiliev invariant and let $\dot{K} \in \dot{\mathscr{K}}_m$ be a singular knot with exactly m singular points. Then $\dot{\theta}(\dot{K})$ depends only upon the cyclic order in which the singular points of $\dot{K}$ appear in a cyclic tour of $\dot{K}$ in the direction of its orientation.*

Proof. The cyclic order of the singular points of a knot defines a singular knot modulo over- and under-crossings. □

We have shown the following.

Vassiliev invariants depend only upon the cyclic order of the singular points of singular knot.

That Vassiliev invariants depend only upon the cyclic order of the singular points of singular knot is the salient property of a singular knot as far as Vassiliev invariants are concerned. This information may be encoded in a more convenient form as follows.

Definition 11.9 (Chord diagram). A *chord diagram* consists of an oriented circle, called the *skeleton*, with a finite number of chords between points on the circle, and is considered up to orientation preserving diffeomorphisms of the circle. The *degree* of a chord diagram is the number of its chords.

For example, the sets of chord diagrams of degrees 1, 2 and 3 are, respectively,

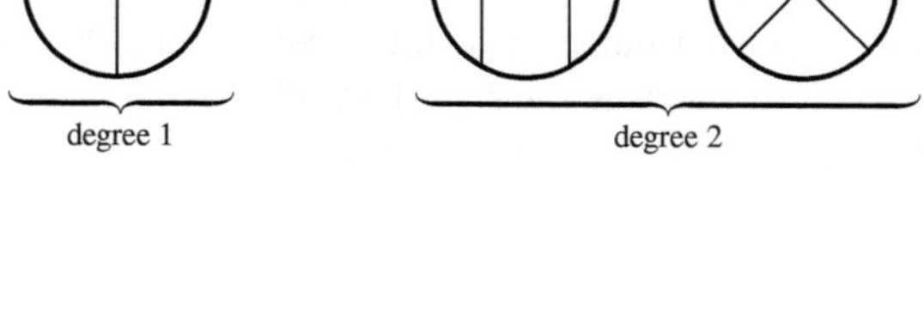

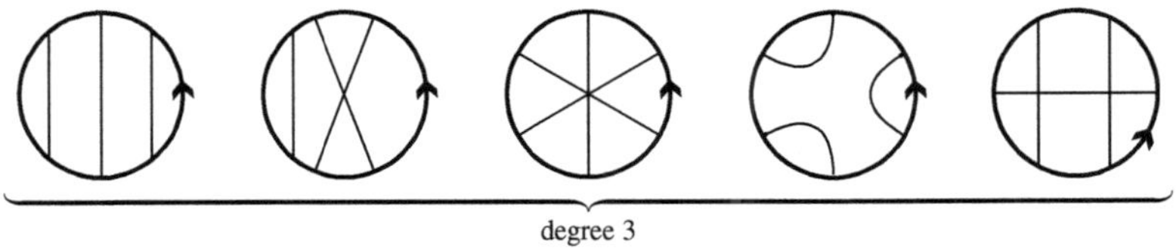

Convention 11.10 (Chord diagrams). Chord diagrams are drawn in the plane so that the chords are contained inside the skeleton.

For the purposes of this convention, it is to be recalled that, by the Jordan Curve Theorem, the skeleton of a chord diagram separates the plane into two regions, one of which contains the point at infinity while the other region does not. The latter is called the *interior* of the chord diagram.

Definition 11.11 (Vector spaces $\mathscr{C}$ and $\mathscr{C}_m$ of chord diagrams). $\mathscr{C}$ is the vector space over $\mathbb{C}$ that consists of all finite formal linear combinations of chord diagrams over $\mathbb{C}$, and $\mathscr{C}_m$ is the subspace of $\mathscr{C}$ generated by chord diagrams of degree exactly m.

A *base point* on a skeleton (or knot) is a point on a skeleton (or knot) that is used to specify a start and endpoints of a tour around the skeleton (or knot). We use the symbol "★" to mark a base point on a diagram.

A degree m chord diagram C is uniquely specified as a circular sequence in the following way. The labels $1, \ldots, m$ are assigned one to each chord of C. A point on the skeleton of C is selected as a base point, and a record is kept of the labels of chords as their ends are encountered in a tour of the skeleton in the direction of its orientation, starting from the base point. In the sequence of labels so formed, each element $1, \ldots, m$ appears exactly twice since each chord has two ends and is to be regarded as

1. a circular sequence, up to cyclic permutation of the order (to mod out the base point);
2. up to permutation of the labels $1, \ldots, m$ (to mod out the particular naming of the chords, since these are arbitrary).

It is clear that the set of all such circular sequences corresponds to the set of degree m chord diagrams, for each m. For example, the left-hand diagram of the following two chord diagrams

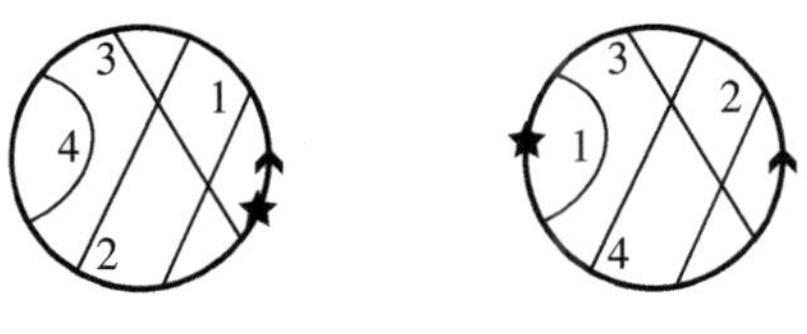

gives rise to a sequence $(1, 2, 3, 4, 4, 2, 1, 3)$ which we consider up to cyclic permutation of the order and permutation of the labels. Thus, the circular sequence $(1, 4, 2, 3, 2, 4, 3, 1)$ corresponds to the right-hand chord diagram which, by permutation of the labels and a cyclic shift, represents the same chord diagram.

Definition 11.12. Two chord diagrams are *equivalent* if, possibly after relabelling the chords, they determine the same circular sequences.

We have seen in Lemma 11.8 that the order that singular points appear in a tour of a knot diagram is important in the study of Vassiliev invariants. We may record this information in a chord diagram by defining a map ϕ_m from singular knots to chord diagrams.

Definition 11.13 (Map ϕ_m). Let $\dot{K} \in \dot{\mathscr{K}}_m$ be an oriented singular knot with exactly m singular points. Select any non-singular point on $\dot{K}$ as a base point, and label the singular points $1, \ldots, m$ arbitrarily. A tour of the knot from the base point in the direction of the orientation gives a sequence of the labels of the singular points in the order in which they are encountered. Each integer $1, \ldots, m$ appears exactly twice in the sequence.

Now, expunge the base point, so the sequence is now a circular sequence that is well-defined up to cyclic permutation and permutation of the labels. It is therefore precisely a chord diagram. We can extend ϕ_m to a map

$$\phi_m : \dot{\mathscr{K}}_m \to \mathscr{C}_m.$$

Clearly, ϕ_m does not depend upon the choice of base point and so is a well-defined map.

As an example of the application of ϕ_4, consider

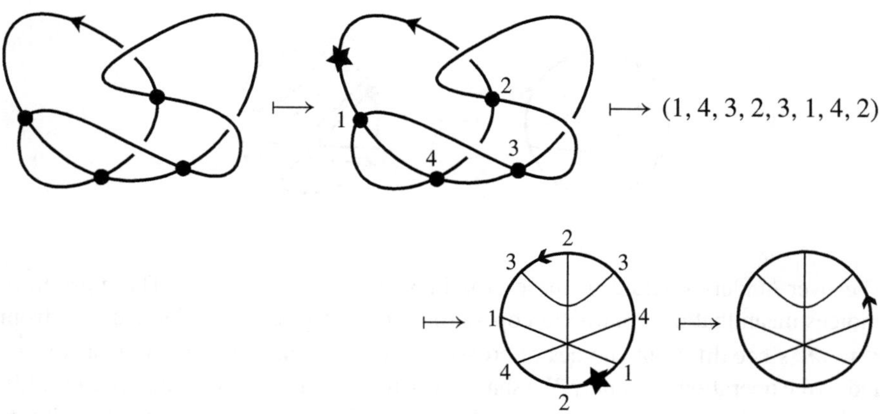

where the (oriented) singular knot on the left-hand side is mapped by ϕ_4 to the chord diagram on the right-hand side. The base points of the singular knot and the chord diagram are is marked by a "★" in the middle two diagrams.

As a second example, we have

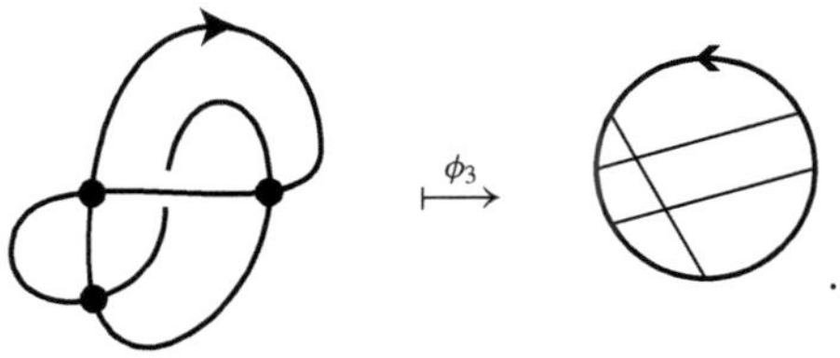

A knot diagram may be constructed from a chord diagram by *contracting the chords* using the operation that replaces a chord by the configuration shown on the right of the image below.

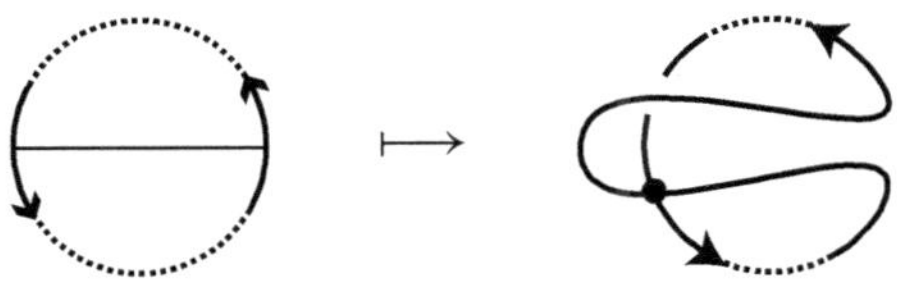

(11.3)

This transforms the skeleton of the chord diagram iteratively into a singular knot diagram. In this construction, each intersection of chords in the chord diagram will result in an intersection of arcs in the singular knot. An example of this operation is shown below.

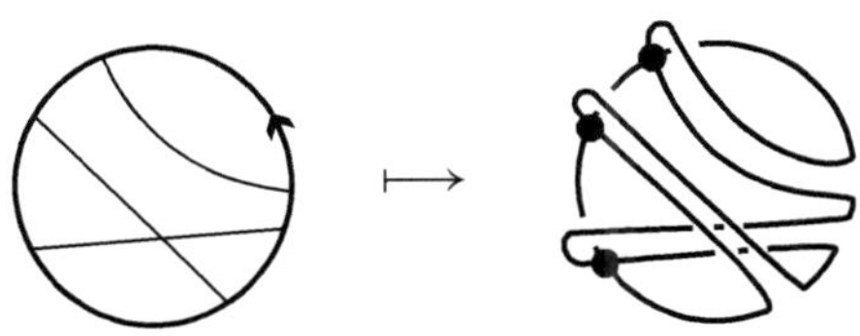

The over-/under-structure of these crossings is chosen arbitrarily. These arbitrary choices mean that the contraction operation *does not* give a well-defined map from $\mathscr{C}$ to $\dot{\mathscr{K}}$ since different choices of crossing may result in different elements of $\dot{\mathscr{K}}$, and so the operation is 1-many. We shall return to this point in detail later in Sect. 11.9.

An immediate consequence of the definition of ϕ_m and the contraction of chords is the following.

Proposition 11.14. *The map $\phi_m : \dot{\mathscr{K}}_m \to \mathscr{C}_m$ is surjective.*

Proof. Let C be a degree m chord diagram. A singular knot diagram $\dot{K}_C$ with exactly m singular points is formed from C by contracting each of the chords of C as in (11.3). Clearly, $\phi_m(\dot{K}_C) = C$, and so ϕ_m is surjective. □

Although ϕ_m is surjective, it is not injective, as the following exercise shows.

Exercise 11.15. Consider the following applications of ϕ_2 to the singular knots $\dot{K}$ and $\dot{K}'$:

$\dot{K} =$ 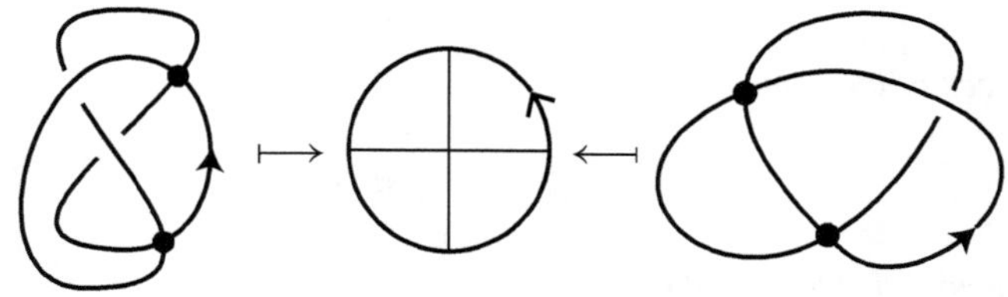 $= \dot{K}'$.

Let J be the Jones polynomial and $\dot{J}$ be its Vassiliev extension to singular knots. Show that $\dot{J}(\dot{K}) \neq \dot{J}(\dot{K}')$. Hence, conclude that ϕ_2 is not injective.

11.3 The Map $\alpha_m : \mathscr{V}_m \to \mathscr{C}_m^\star$

We now show that to understand Vassiliev invariants it is enough to understand the vector space $\mathscr{C}$ of chord diagrams. It is a discussion that will occupy the rest of this chapter. We construct a map $\alpha_m : \mathscr{V}_m \to \mathscr{C}_m^\star$ from the space of Vassiliev invariants to the dual of the space of chord diagrams. Chord diagrams may be used to rephrase both Lemmas 11.7 and 11.8 succinctly as follows.

Proposition 11.16. *Let $\theta \in \mathscr{V}_m$ be a degree m Vassiliev invariant and let $\dot{K}, \dot{K}' \in \dot{\mathscr{K}}_m$. Then*

$$\phi_m(\dot{K}) = \phi_m(\dot{K}') \implies \dot{\theta}(\dot{K}) = \dot{\theta}(\dot{K}'),$$

i.e. *the value of a degree m Vassiliev invariant of a singular knot with exactly m singular points depends only on the chord diagram of the singular knot.*

Proof. If $\phi_m(\dot{K}) = \phi_m(\dot{K}')$, then the cyclic order of the singular points of each of the two singular knots is the same and the knots differ only by crossing changes (and isotopy) of non-singular crossings. Therefore, by Lemma 11.8, any degree m Vassiliev invariant takes the same value on both knots. That is, $\dot{\theta}(\dot{K}) = \dot{\theta}(\dot{K}')$. □

Corollary 11.17. *Let $\theta \in \mathscr{V}_m$ be a degree m Vassiliev invariant. Then the following hold.*

1. *There exists a unique linear map $\omega_\theta : \mathscr{C}_m \to \mathbb{C}$ such that*

$$\dot{\theta} = \omega_\theta \circ \phi_m$$

or, equivalently, in terms of commutative diagrams,

$$\begin{array}{ccc} \dot{\mathscr{K}}_m & \xrightarrow{\phi_m} & \mathscr{C}_m \\ & \overset{\dot{\theta}}{\searrow} & \downarrow \exists!\, \omega_\theta \\ & & \mathbb{C} \end{array}$$

2. *ϕ_m induces a linear map*

$$\alpha_m : \mathscr{V}_m \to \mathscr{C}_m^\star : \theta \mapsto \omega_\theta.$$

3. *The element-wise action of ω_θ is*

$$\omega_\theta : C \mapsto \dot{\theta}(\dot{K})$$

for any $\dot{K}$ such that $\phi_m(\dot{K}) = C$.

Proof. For the first part, consider the map $\omega : \mathscr{C}_m \to \mathbb{C}$ defined as follows. Let $C \in \mathscr{C}_m$. Since ϕ_m is surjective by Proposition 11.14, there exists a singular knot $\dot{K} \in \dot{\mathscr{K}}_m$ such that $\phi_m(\dot{K}) = C$. Let

$$\omega : C \mapsto \dot{\theta}(\dot{K}),$$

and extend linearly to $\mathscr{C}_m$.

We next show that ω is well-defined; that is, it is independent of the choice of $\dot{K}$ used in its construction. Suppose that there is another singular knot $\dot{K}' \in \dot{\mathscr{K}}_m$ such that $\phi_m(\dot{K}') = C$. By Proposition 11.16, $\dot{\theta}(\dot{K}) = \dot{\theta}(\dot{K}')$, so the value of ω does not depend upon the choice of singular knot $\dot{K}$. Hence, ω is well-defined.

For all $\dot{K} \in \dot{\mathscr{K}}_m$, we have $(\omega \circ \phi_m)(\dot{K}) = \dot{\theta}(\dot{K})$ whence $\omega \circ \phi_m = \dot{\theta}$.

To show that there is a unique ω such that $\omega \circ \phi_m = \dot{\theta}$, suppose that there is another map $\varpi : \mathscr{C}_m \to \mathbb{C}$ such that $\varpi \circ \phi_m = \dot{\theta}$. Then, $\varpi(\phi_m(\dot{K})) = \omega(\phi_m(\dot{K}))$ for all $\dot{K} \in \dot{\mathscr{K}}_m$. But ϕ_m is surjective, so $\varpi(C) = \omega(C)$ for all $C \in \mathscr{C}_m$, whence $\varpi = \omega$. Thus, ω is the unique map with the prescribed property, and we therefore denote it by ω_θ. Thus,

$$\omega_\theta \circ \phi_m = \dot{\theta}. \tag{11.4}$$

For the second part, given θ and hence $\dot{\theta}$ there is a unique $\omega_\theta : \mathscr{C}_m \to \mathbb{C}$, that is, $\omega_\theta \in \mathscr{C}_m^\star$, induced by ϕ_m through (11.4). Thus, we can define α_m by

$$\alpha_m : \mathscr{V}_m \to \mathscr{C}_m^\star : \theta \mapsto \omega_\theta.$$

Now, consider $\theta_1, \theta_2 \in \mathscr{V}_m$. Then, $\alpha_m(\theta_1) = \omega_{\theta_1}$ and $\alpha_m(\theta_2) = \omega_{\theta_2}$. Also, $\alpha_m(\theta_1 + \theta_2) = \omega_{\theta_1+\theta_2}$. Let $\dot{K} \in \dot{\mathscr{K}}_m$ and $\phi_m(\dot{K}) = C$. Then,

$$\begin{aligned} \omega_{\theta_1+\theta_2}(C) &= \omega_{\theta_1+\theta_2} \circ \phi_m(\dot{K}) = (\dot{\theta}_1 + \dot{\theta}_2)(\dot{K}) && \text{(from (11.4))} \\ &= \dot{\theta}_1(\dot{K}) + \dot{\theta}_2(\dot{K}) = \omega_{\theta_1}(C) + \omega_{\theta_2}(C) \end{aligned}$$

for all $C \in \mathscr{C}_m$. Thus, $\omega_{\theta_1+\theta_2} = \omega_{\theta_1} + \omega_{\theta_2}$. Similarly, $\omega_{c\theta_1} = c\,\omega_{\theta_1}$ for all $c \in \mathbb{C}$. Thus, α_m is linear.

Finally, the element-wise action follows from the definition of ω_θ and its uniqueness. □

As an example of calculating the value of ω_θ, let c_3 be the coefficient of z^3 in the Alexander–Conway polynomial, $C(z)$ (Definition 2.30). We already know that c_3 is a Vassiliev invariant by Theorem 10.15. (Note that in this example C is the Alexander–Conway polynomial, rather than a chord diagram as it is in the rest of this chapter.) We determine ω_{c_3} () explicitly by using any preimage of  under ϕ_3. We have ϕ_3 () $=$ . Then, by repeated application of the Vassiliev relation, the Vassiliev resolution of the singular knot  is

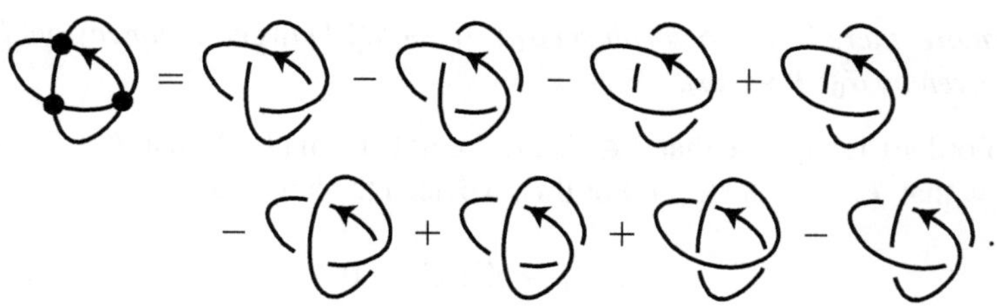

All but the first and last knot on the right-hand side of this resolution are equivalent to the unknot, so, making use of Exercise 2.31,

$\dot{C}$ () $= C$ ( $-$ ) $= (1+z^2) - (1-z^2) = 2z^2$

since three of the unknots in the resolution appear with plus signs and the three others with minus signs, thereby cancelling their contribution. Thus, $c_3 = 0$, and so

ω_{c_3} () $= 0.$

11.4 The Map $\overline{\alpha}_m : \mathscr{V}_m/\mathscr{V}_{m-1} \to \mathscr{C}_m^\star$

In order to gain a more detailed understanding of the space $\mathscr{V}_m$ of degree m Vassiliev invariants, we shall now refine the map α_m. This map is now to act on the quotient space $\mathscr{V}_m/\mathscr{V}_{m-1}$. In fact, we shall see that we may construct it as a map

$\overline{\alpha}_m : \mathscr{V}_m/\mathscr{V}_{m-1} \to \mathscr{C}_m^\star$ from the quotient space of Vassiliev invariants to the dual space of chord diagrams. The following lemma assists in this endeavour.

Lemma 11.18. *Let $\alpha_m : \mathscr{V}_m \to \mathscr{C}_m^\star$ be the map defined in Corollary 11.17, where $m > 0$. Then the following hold.*

1. We have

$$\ker(\alpha_m) = \mathscr{V}_{m-1}.$$

2. There is an injection

$$\overline{\alpha}_m : \frac{\mathscr{V}_m}{\mathscr{V}_{m-1}} \longrightarrow \mathscr{C}_m^\star.$$

3. The element-wise action of $\overline{\alpha}_m$ is

$$\overline{\alpha}_m : [\theta] \mapsto \omega_\theta,$$

where ω_θ is defined also in Corollary 11.17.

Furthermore, there is an isomorphism $\overline{\alpha}_0 : \mathscr{V}_0 \to \mathscr{C}_0^\star$ between 1-dimensional vector spaces given by $\overline{\alpha}_0 : \theta \mapsto \omega_\theta$.

Proof. For Part 1, suppose that $\theta \in \mathscr{V}_{m-1}$. We note from (10.9) that $\mathscr{V}_{m-1} \leq \mathscr{V}_m$. We first show that $\mathscr{V}_{m-1} \leq \ker(\alpha_m)$. For this, we need to show that

$$\omega_\theta : \mathscr{C}_m \to \mathbb{C} : C \mapsto 0.$$

To do this, let $C \in \mathscr{C}_m$. Then, there exists $\dot{K} \in \dot{\mathscr{K}}_m$ such that

$$\phi_m(\dot{K}) = C$$

since ϕ_m is surjective by Proposition 11.14. Then, $\omega_\theta(C) = \omega_\theta(\phi_m(\dot{K})) = \dot{\theta}(\dot{K})$ by Corollary 11.17. But $\dot{K} \in \dot{\mathscr{K}}_m$ and $\theta \in \mathscr{V}_{m-1}$ so, by Lemma 11.6, $\dot{\theta}(\dot{K}) = 0$. Thus, $\omega_\theta(C) = 0$ for all $C \in \mathscr{C}_m$. Thus,

$$\mathscr{V}_{m-1} \leq \ker(\alpha_m).$$

On the other hand, to prove that $\mathscr{V}_{m-1} \geq \ker(\alpha_m)$, we note that this is equivalent to the assertion

$$\alpha_m(\theta) = 0 \implies \theta \in \mathscr{V}_{m-1},$$

and thence to

$$\omega_\theta(C) = 0 \text{ for all } C \in \mathscr{C}_m \implies \dot{\theta}(\dot{\mathscr{K}}_m) = \{0\}.$$

We shall prove the latter assertion. Let $\theta \in \mathscr{V}_m$ be such that $\omega_\theta(C) = 0$ for all $C \in \mathscr{C}_m$. Then, $\omega_\theta(\phi_m(\dot{K})) = 0$ for all $\dot{K} \in \dot{\mathscr{K}}_m$, so $\dot{\theta}(\dot{K}) = \omega_\theta(\phi_m(\dot{K})) = 0$ for all $\dot{K} \in \mathscr{K}_m$, and thus $\dot{\theta}(\dot{\mathscr{K}}_m) = \{0\}$. Thus,

$$\mathscr{V}_{m-1} \geq \ker(\alpha_m).$$

It follows that $\mathscr{V}_{m-1} = \ker(\alpha_m)$.

For Part 2, we note that, by the First Isomorphism Theorem, Theorem A.9, α_m descends to an isomorphism

$$\overline{\alpha}_m : \frac{\mathscr{V}_m}{\ker(\alpha_m)} \longrightarrow \operatorname{Im}(\alpha_m).$$

But $\operatorname{Im}(\alpha_m) \leq \mathscr{C}_m^\star$ so there is an injection

$$\overline{\alpha}_m : \frac{\mathscr{V}_m}{\ker(\alpha_m)} \longrightarrow \mathscr{C}_m^\star.$$

completing the proof of Part 2.

For Part 3, the element-wise action of $\overline{\alpha}_m$ is given by

$$\overline{\alpha}_m : \frac{\mathscr{V}_m}{\mathscr{V}_{m-1}} \longrightarrow \mathscr{C}_m^\star : [\theta] \mapsto \alpha_m(\theta) = \omega_\theta$$

where $[\theta]$ is the equivalence class of θ.

The proof of the claim concerning $\overline{\alpha}_0$ is trivial. □

Although $\overline{\alpha}_m$ is injective, we shall see in Corollary 11.23 that it is not surjective.

11.5 The Finite Dimensionality of the Space $\mathscr{V}_m$

The existence of an injection $\overline{\alpha}_m : \mathscr{V}_m/\mathscr{V}_{m-1} \longrightarrow \mathscr{C}_m^\star$ has an important consequence.

Corollary 11.19. *The vector space $\mathscr{V}_m/\mathscr{V}_{m-1}$ is finite dimensional for each $m \geq 1$.*

Proof. There are only a finite number of chord diagrams of degree m, so $\mathscr{C}_m$ is finite dimensional. But $\dim\left(\mathscr{C}_m^\star\right) = \dim(\mathscr{C}_m)$ so the dual space $\mathscr{C}_m^\star$ is also finite dimensional for each m. Finally, since $\overline{\alpha}_m$ is injective, the quotient space $\mathscr{V}_m/\mathscr{V}_{m-1}$ is also finite dimensional, giving the result. □

We may now prove a result whose immediate crucial consequence is the finite dimensionality of the space of Vassiliev invariants of degree m.

Proposition 11.20. *Let*

$$\mathscr{C}_{\leq m} := \mathscr{C}_m \oplus \cdots \oplus \mathscr{C}_0$$

denote the subspace of $\mathscr{C}$ generated by all chord diagrams of degree at most m. Then, for $m \geq 0$, there is an injection

$$\beta_m : \mathscr{V}_m \longrightarrow \mathscr{C}_{\leq m}^\star.$$

Proof. By Theorem A.23, for $m \geq 1$ we can write

$$\mathscr{V}_m \cong \frac{\mathscr{V}_m}{\mathscr{V}_{m-1}} \oplus \mathscr{V}_{m-1} \tag{11.5}$$

Then, there is a map $\beta_m : \mathscr{V}_m \to \mathscr{C}^\star_{\leq m}$ given by

$$\begin{array}{ccccccccccccc}
\mathscr{V}_m & \cong & \frac{\mathscr{V}_m}{\mathscr{V}_{m-1}} & \oplus & \frac{\mathscr{V}_{m-1}}{\mathscr{V}_{m-2}} & \oplus & \cdots & \oplus & \frac{\mathscr{V}_1}{\mathscr{V}_0} & \oplus & \mathscr{V}_0 & & \text{(iterate (11.5))} \\
\beta_m \Big\downarrow & & \Big\downarrow \overline{\alpha}_m & & \Big\downarrow \overline{\alpha}_{m-1} & & \cdots & & \Big\downarrow \overline{\alpha}_1 & & \Big\downarrow \overline{\alpha}_0 & & \text{(Cor. 11.18)} \\
\mathscr{C}^\star_{\leq m} & \cong & \mathscr{C}^\star_m & \oplus & \mathscr{C}^\star_{m-1} & \oplus & \cdots & \oplus & \mathscr{C}^\star_1 & \oplus & \mathscr{C}^\star_0 & &
\end{array}$$

where the $\overline{\alpha}_i$ are from Lemma 11.18. That the morphism β_m on the left-hand side of the diagram is an injection follows since the direct sum of injective morphisms is injective (see Sect. A.6). The result follows. □

We have the following immediate corollary.

Corollary 11.21. *The vector space $\mathscr{V}_m$ of degree m Vassiliev invariants is finite dimensional for each m.*

Proof. The vector space $\mathscr{C}_m$ is finite dimensional for $m \geq 0$. Then, from Proposition 11.20, there is an injection $\beta_m : \mathscr{V}_m \longrightarrow \mathscr{C}^\star_{\leq m}$ so

$$\dim(\mathscr{V}_m) \leq \dim(\mathscr{C}^\star_{\leq m}) = \sum_{k=0}^{m} \dim(\mathscr{C}^\star_k) = \sum_{k=0}^{m} \dim(\mathscr{C}_k),$$

which is finite. The result follows. □

Before continuing, we highlight the significance of Proposition 11.20. It says that every Vassiliev invariant corresponds to a unique element of $\mathscr{C}^\star_{\leq m}$. Hence, by duality, every Vassiliev invariant corresponds to a unique element of $\mathscr{C}_{\leq m}$. Thus, we have the following:

To study Vassiliev invariants it is enough to study chord diagrams.

In terms of a mathematical approach, we have moved the theory of Vassiliev invariants from topology into pure combinatorics of diagrams.

11.6 A Quotient Space of Chord Diagrams

Lemma 11.18 provides an injective map

$$\overline{\alpha}_m : \frac{\mathscr{V}_m}{\mathscr{V}_{m-1}} \longrightarrow \mathscr{C}_m^\star.$$

The purpose of this section is to characterise the image of α_m in order to obtain an isomorphism

$$\overline{\alpha}_m : \frac{\mathscr{V}_m}{\mathscr{V}_{m-1}} \cong \frac{\mathscr{V}_m}{\ker(\alpha_m)} \longrightarrow \mathrm{Im}(\alpha_m)$$

using the First Isomorphism Theorem (Theorem A.9). The difficulty arises in finding a natural characterisation in terms of chord diagrams. The approach to this problem is to show that every element ω_θ of $\mathrm{Im}(\alpha_m)$ satisfies two particular relations (which implies that $\overline{\alpha}_m$ is *not* surjective). The relations are called the $1\mathsf{T}^\star$- and $4\mathsf{T}^\star$-relations. We then define a *weight system* to be a function in $\mathscr{C}_m^\star$ that satisfies both the $1\mathsf{T}^\star$- and $4\mathsf{T}^\star$-relations. Since weight systems are in $\mathrm{Im}(\alpha_m)$ by construction, this then will serve as a candidate for $\mathrm{Im}(\alpha_m)$.

The $1\mathsf{T}^\star$-relation arises, more or less, from the fact that Vassiliev invariants are invariant under the RI-move. (In fact, it is a little more general than this, for it arises from the move shown in (11.7) below.)

The $4\mathsf{T}^\star$-relation arises from the fact that the Vassiliev relation ensures that moving one singular point completely around another singular point as in $\mathfrak{a}_1$, to $\mathfrak{a}_4$, to $\mathfrak{a}_2$, to $\mathfrak{a}_3$ and back to $\mathfrak{a}_1$, where the $\mathfrak{a}_i$ are as in (11.8) and (11.9), does not change the resulting singular knot.

In the following Lemma, the two grey regions within the diagram

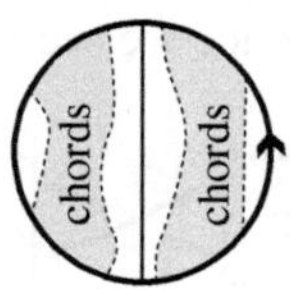

(11.6)

represent collections of chords that do not intersect the single chord that is drawn. Note that these collections may be empty of chords. The chord shown in the diagram is said to be an *isolated chord*. Thus, recalling Convention 11.10, the diagram in (11.6) represents any chord diagram that has a chord that does not intersect any other chord of the diagram.

Lemma 11.22. *Every $W \in \mathrm{Im}(\alpha_m) \leq \mathscr{C}_m^\star$ satisfies the* 1T**-relation*

1T* : W 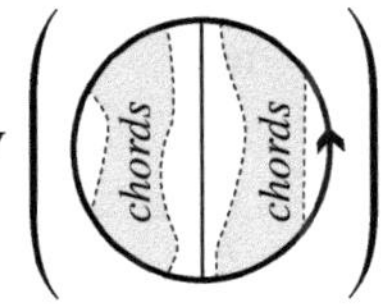 $= 0.$

Proof. Let $W \in \mathrm{Im}(\alpha_m)$. Then, from Corollary 11.17, $W = \omega_\theta$ for some Vassiliev invariant $\theta \in \mathscr{V}_m$, so

W 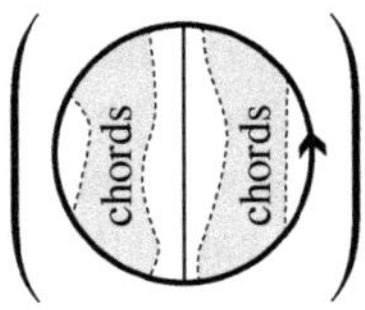 $= \omega_\theta$ 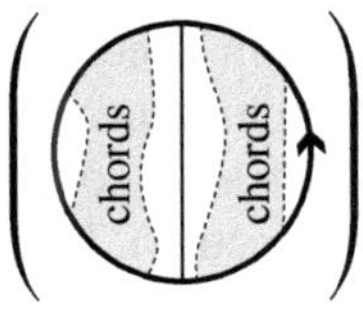 .

For  $\in \dot{\mathscr{K}}_m$ we have, by Definition 11.13,

ϕ_m  $=$ 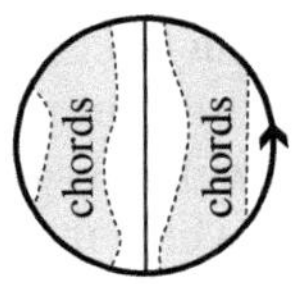

where, of course, the unspecified sets of chords arise from T and T'.

Applying the Vassiliev relation once we see that

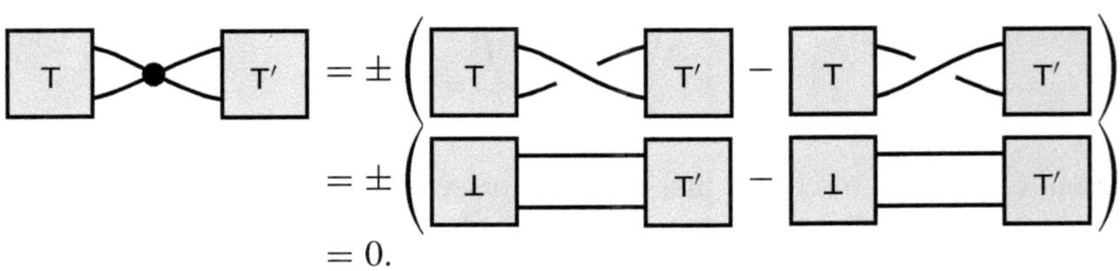

(11.7)

Thus applying $\dot{\theta}$ to this singular knot gives zero, and the result follows. □

We are now in a position to show that the map $\overline{\alpha}_m$ defined in Lemma 11.18 is not surjective.

Corollary 11.23. *The injective map*

$$\overline{\alpha}_m : \frac{\mathscr{V}_m}{\mathscr{V}_{m-1}} \longrightarrow \mathscr{C}_m^{\star}$$

is not surjective.

Proof. Consider the function f defined on the natural basis of $\mathscr{C}_m$, a finite dimensional space, by

$f : \mathscr{C}_m \to \mathbb{C} : C \mapsto \begin{cases} 1 \text{ if } C = \end{cases}$

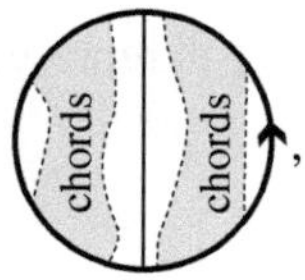

,

$\quad\quad 0$ otherwise,

extended linearly to $\mathscr{C}_m$. Then, clearly, f does not satisfy the $1\mathsf{T}^{\star}$-relation. □

In addition to satisfying the $1\mathsf{T}^{\star}$-relation, we shall show that every element in $\mathrm{Im}(\alpha_m)$ also satisfies a relation called the $4\mathsf{T}^{\star}$-relation. The $4\mathsf{T}^{\star}$-relation involves a linear combination of chord diagrams illustrated as

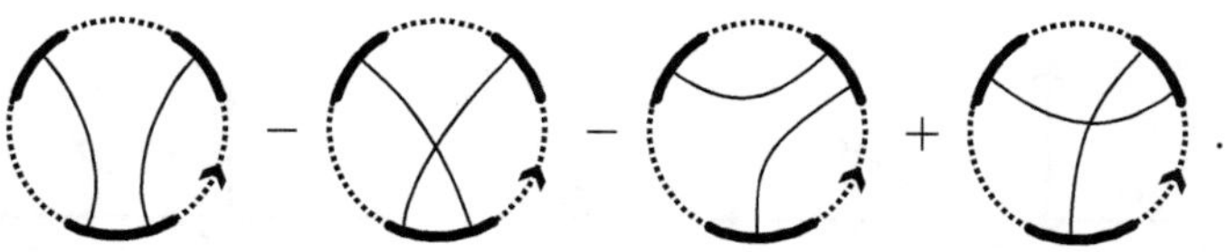

The bold arcs on the skeleton represent sections on the skeleton where there are no ends of chords except for those that are shown. There may be chords with ends on the non-bold sections of skeleton, and these chords may intersect the chords shown in the figure. However, all four of the diagrams should be identical except for the chords that are shown.

Lemma 11.24. *Every* $W \in \mathrm{Im}(\alpha_m) \leq \mathscr{C}_m^{\star}$ *satisfies the* $4\mathsf{T}^{\star}$*-relation*

$4\mathsf{T}^{\star}$: $\quad W \Bigg($

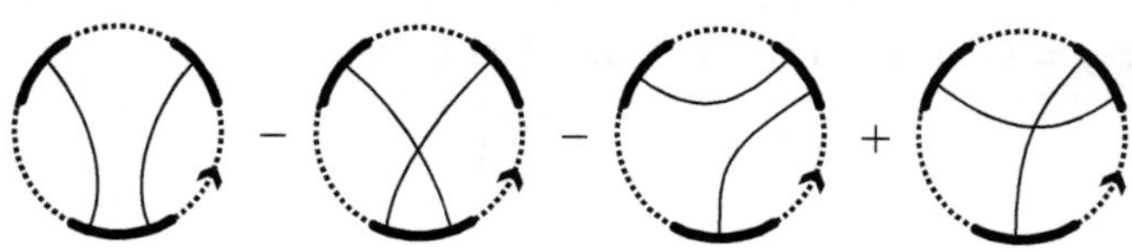

$\Bigg) = 0.$

Proof. Let $W \in \operatorname{Im}(\alpha_m)$. Then, from Lemma 11.18, $W = \omega_\theta$ for some $\theta \in \mathscr{V}_m$. Let

$C =$ 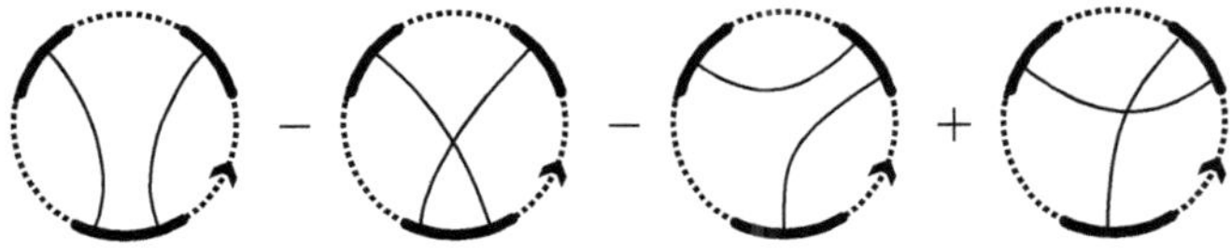 ,

and let $C_1, \ldots, C_4$ be the four chord diagrams in C as read from left to right.

Suppose that $\dot{\mathfrak{a}}_1 \in \dot{\mathscr{K}}_m$ is such that $\phi_m(\dot{\mathfrak{a}}_1) = C_1$ (such a $\dot{\mathfrak{a}}_1$ exists by Proposition 11.14). Then, we can represent $\dot{\mathfrak{a}}_1$ as

$\dot{\mathfrak{a}}_1 =$ 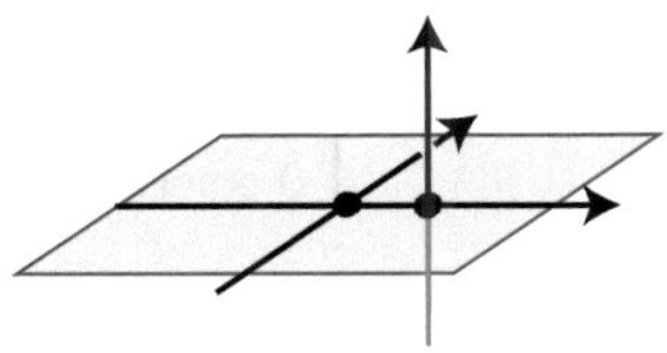 (11.8)

where the two singular points shown in the figure correspond to the two chords shown in the representation of C_1.

Let

$\dot{\mathfrak{a}}_2 =$ 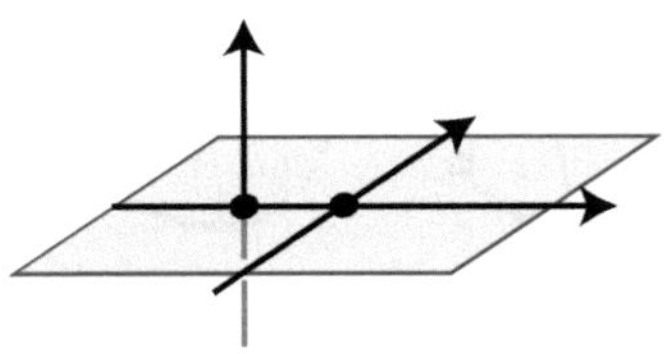 , $\dot{\mathfrak{a}}_3 =$ 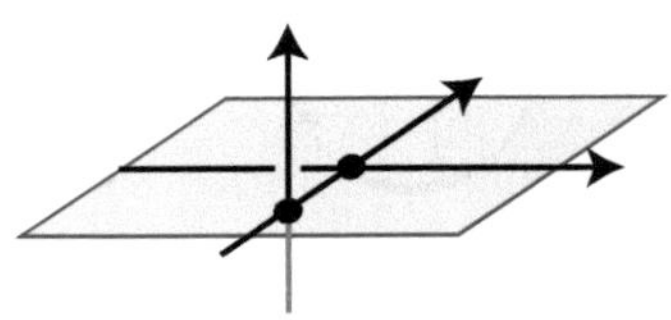 ,

$\dot{\mathfrak{a}}_4 =$ 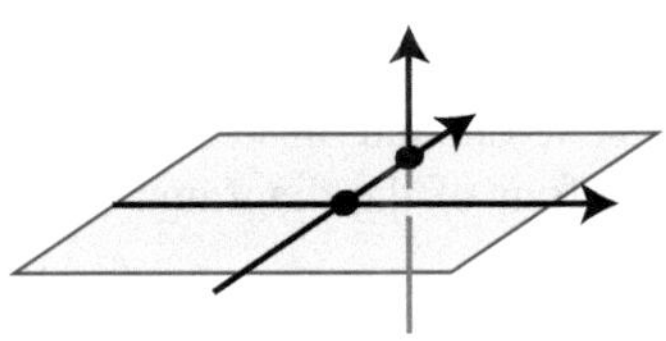 , (11.9)

be singular knots that are identical to $\dot{\mathfrak{a}}_1$ except for the region shown. Then, $\phi_m(\dot{\mathfrak{a}}_2) = C_2$, $\phi_m(\dot{\mathfrak{a}}_3) = C_3$ and $\phi_m(\dot{\mathfrak{a}}_4) = C_4$, and hence for

$$\dot{K} := \dot{\mathfrak{a}}_1 - \dot{\mathfrak{a}}_2 - \dot{\mathfrak{a}}_3 + \dot{\mathfrak{a}}_4 \in \dot{\mathscr{K}}_m$$

we have $\phi_m(\dot{K}) = C$. Then,

$$W(C) = \omega_\theta(C) = \dot{\theta}(\dot{K}). \tag{11.10}$$

Now, consider

$\mathfrak{b}_1 =$ 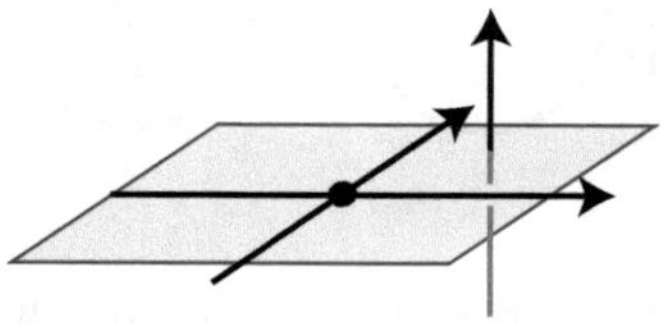 , $\mathfrak{b}_2 =$ 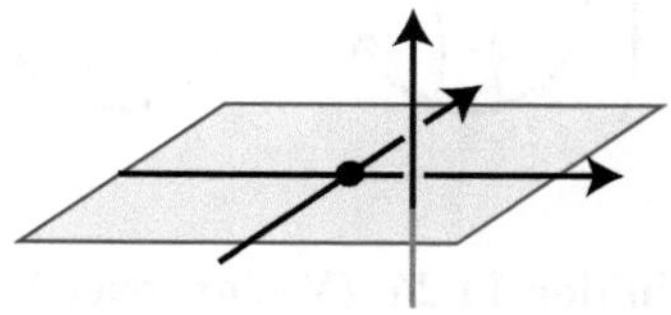 ,

$\mathfrak{b}_3 =$ 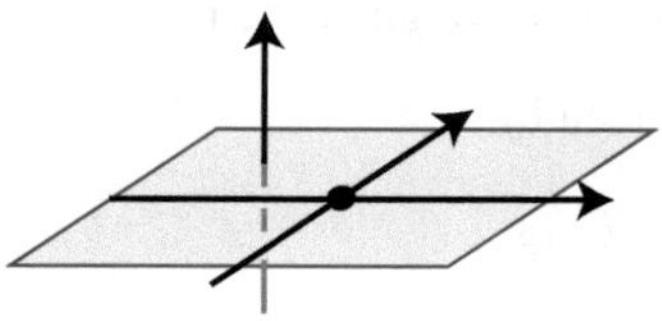 , $\mathfrak{b}_4 =$ 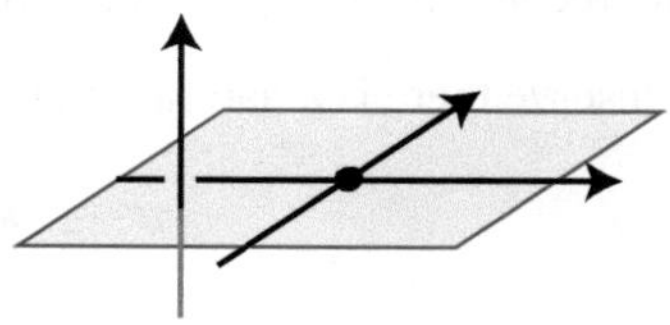 .

We have

$$\overset{\bullet}{\theta}(\overset{\bullet}{\mathfrak{a}}_1) = \overset{\bullet}{\theta}(\mathfrak{b}_1) - \overset{\bullet}{\theta}(\mathfrak{b}_2), \qquad \overset{\bullet}{\theta}(\overset{\bullet}{\mathfrak{a}}_2) = \overset{\bullet}{\theta}(\mathfrak{b}_3) - \overset{\bullet}{\theta}(\mathfrak{b}_4),$$

$$\overset{\bullet}{\theta}(\overset{\bullet}{\mathfrak{a}}_3) = \overset{\bullet}{\theta}(\mathfrak{b}_4) - \overset{\bullet}{\theta}(\mathfrak{b}_2), \qquad \overset{\bullet}{\theta}(\overset{\bullet}{\mathfrak{a}}_4) = \overset{\bullet}{\theta}(\mathfrak{b}_3) - \overset{\bullet}{\theta}(\mathfrak{b}_1),$$

where the diagrams shown are $\mathfrak{a}_1, \ldots, \mathfrak{a}_4$. It follows that

$$\begin{aligned}\overset{\bullet}{\theta}(\overset{\bullet}{K}) &= (\overset{\bullet}{\theta}(\mathfrak{b}_1) - \overset{\bullet}{\theta}(\mathfrak{b}_2)) - (\overset{\bullet}{\theta}(\mathfrak{b}_3) - \overset{\bullet}{\theta}(\mathfrak{b}_4)) - (\overset{\bullet}{\theta}(\mathfrak{b}_4) - \overset{\bullet}{\theta}(\mathfrak{b}_2)) + (\overset{\bullet}{\theta}(\mathfrak{b}_3) - \overset{\bullet}{\theta}(\mathfrak{b}_1))\\ &= 0,\end{aligned}$$

and therefore, from (11.10), that $W(C) = 0$, completing the proof. □

11.7 Weight Systems

Having seen that every linear map $W \in \operatorname{Im}(\alpha_m)$, where α_m was defined in Corollary 11.17, satisfies the $1\mathsf{T}^\star$- and $4\mathsf{T}^\star$-relations, we consider next the subspace $\mathscr{W}_m \leq \mathscr{C}_m^\star$ of all maps which satisfy these relations.

Definition 11.25 (Weight system). An *(unframed) weight system* of degree m is an element $f \in \mathscr{C}_m^\star$ which satisfies the *one term relation*, $1\mathsf{T}^\star$,

f 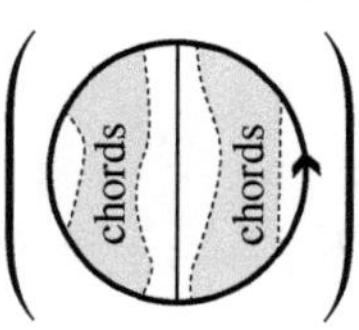

$= 0$

and the *four term relation*, $4\mathsf{T}^\star$,

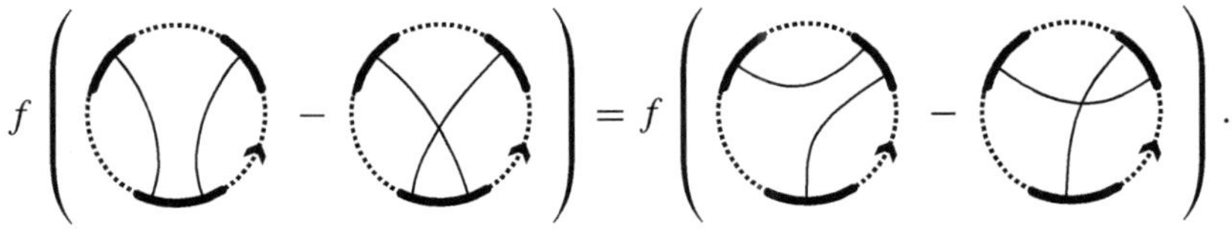

Definition 11.26 (Vector space $\mathscr{W}$). $\mathscr{W}_m \leq \mathscr{C}_m^\star$ is the vector space of weight systems of degree m, and $\mathscr{W} \leq \mathscr{C}^\star$ is the vector space of all weight systems.

Observe there is a natural grading on $\mathscr{W}$ given by

$$\mathscr{W} := \mathscr{W}_0 \oplus \mathscr{W}_1 \oplus \mathscr{W}_2 \oplus \cdots .$$

Example 11.27. As an example of a computation with the $1\mathsf{T}^\star$- and $4\mathsf{T}^\star$-relations in the vector space $\mathscr{W}_3$ of degree 3 weight systems, let f be a degree 3 weight system. Then,

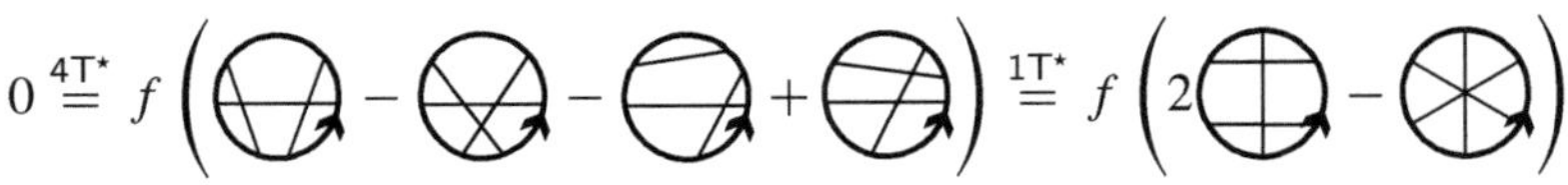

where the $1\mathsf{T}^\star$-relation has been applied to the third term of the middle member of the string of equalities and the linearity of f has been invoked throughout. This is the term that has an isolated chord so the $1\mathsf{T}^\star$-relation gives 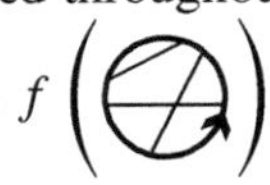 $= 0$. We therefore conclude that

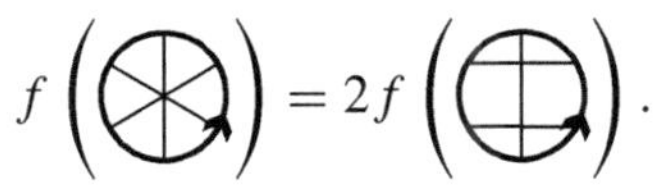 (11.11)

Although $\overline{\alpha}_m$ is injective by Lemma 11.18, it is not surjective by the Corollary 11.23. However, if we are able to determine the image of α_m, we may then restrict the codomain of α_m to obtain an isomorphism.

Lemma 11.28. *The map* $\overline{\alpha}_m : \mathscr{V}_m/\mathscr{V}_{m-1} \longrightarrow \mathscr{C}_m^\star$ *induces a well-defined injective linear map*

$$\overline{\alpha}_m : \frac{\mathscr{V}_m}{\mathscr{V}_{m-1}} \longrightarrow \mathscr{W}_m .$$

Proof. From Lemma 11.18(2),

$$\overline{\alpha}_m : \mathscr{V}_m/\mathscr{V}_{m-1} \longrightarrow \mathscr{C}_m^\star$$

is injective. Moreover, from Definition 11.25, $\mathscr{W}_m \leq \mathscr{C}_m^\star$, and from Lemmas 11.22 and 11.24, every $W \in \mathrm{Im}(\alpha_m) \leq \mathscr{C}_m^\star$ satisfies the $1\mathsf{T}^\star$- and $4\mathsf{T}^\star$-relations, so W is a weight system of degree m, whence $W \in \mathscr{W}_m$. Thus, $\mathrm{Im}(\alpha_m) \leq \mathscr{W}_m$, completing the proof. □

Proving $\overline{\alpha}_m$ is surjective requires much additional work, requiring something called the Kontsevich invariant, and so is deferred to Sect. 18.1 to avoid interrupting the present development. With this in mind, we shall assume surjectivity, and so the truth of Theorem 11.29, for the remainder of this chapter.

Theorem 11.29. *The map*

$$\overline{\alpha}_m : \frac{\mathscr{V}_m}{\mathscr{V}_{m-1}} \longrightarrow \mathscr{W}_m$$

is a vector space isomorphism.

The significance of Theorem 11.29 is that we have the following isomorphisms:

$$\begin{array}{ccccccccccc}
\mathscr{V}_m & \cong & \frac{\mathscr{V}_m}{\mathscr{V}_{m-1}} & \oplus & \frac{\mathscr{V}_{m-1}}{\mathscr{V}_{m-2}} & \oplus & \cdots & \oplus & \frac{\mathscr{V}_1}{\mathscr{V}_0} & \oplus & \mathscr{V}_0 \\
\beta_m \Big\downarrow & & \overline{\alpha}_m \Big\downarrow & & \overline{\alpha}_{m-1} \Big\downarrow & & \cdots & & \overline{\alpha}_1 \Big\downarrow & & \overline{\alpha}_0 \Big\downarrow \\
\mathscr{W}_{\leq m} & \cong & \mathscr{W}_m & \oplus & \mathscr{W}_{m-1} & \oplus & \cdots & \oplus & \mathscr{W}_1 & \oplus & \mathscr{W}_0
\end{array}$$

which establish the equivalence of Vassiliev invariants and weight systems,

$$\mathscr{V}_m \cong \mathscr{W}_{\leq m}.$$

We have shown that the following observation holds.

Studying Vassiliev invariants is equivalent to studying weight systems.

11.8 The Space $\mathscr{A}^c$ of Chord Diagrams

We briefly summarise our progress so far and discuss how to proceed. Our aim is to understand the space $\mathscr{V}_m$ of degree m Vassiliev invariants. We have

$$\mathscr{V}_m \cong \frac{\mathscr{V}_m}{\mathscr{V}_{m-1}} \oplus \frac{\mathscr{V}_{m-1}}{\mathscr{V}_{m-2}} \oplus \cdots \oplus \frac{\mathscr{V}_1}{\mathscr{V}_0} \oplus \mathscr{V}_0.$$

and that

$$\frac{\mathscr{V}_m}{\mathscr{V}_{m-1}} \cong \mathscr{W}_m,$$

so studying the space $\mathscr{V}_m$ is equivalent to studying the spaces

$$\mathscr{W}_i = \{f \in \mathscr{C}_i^\star : f \text{ satisfies } 1\mathsf{T}^\star \text{ and } 4\mathsf{T}^\star\},$$

for $0 \leq i \leq m$. To develop the theory of Vassiliev invariants, this is exactly what we need to do. However, to do this we need to revise our view of $\mathscr{W}_m$ in order to obtain a more amenable combinatorial model of it.

Each element f of $\mathscr{W}_m$ is a linear functional on chord diagrams, $f : \mathscr{C}_m \to \mathbb{C}$, that satisfies the $1\mathsf{T}^\star$ and $4\mathsf{T}^\star$-relations. Applying Lemma A.12 we see that, for suitable elements $1\mathsf{T}$ and $4\mathsf{T}$, which are readily read from Definition 11.25, we have

$$\mathscr{W}_m \cong \left(\frac{\mathscr{C}_m}{(1\mathsf{T}, 4\mathsf{T})}\right)^\star.$$

Thus to understand $\mathscr{W}_m$, it is enough to understand $(\mathscr{C}_m/(1\mathsf{T}, 4\mathsf{T}))^\star$, or just $\mathscr{C}_m/(1\mathsf{T}, 4\mathsf{T})$. Let us set this observation up formally.

Definition 11.30 **($1\mathsf{T}$-relation).** In the space $\mathscr{C}_m$ or $\mathscr{C}$ of chord diagrams, the $1\mathsf{T}$-*relation*, or *one term relation*, is

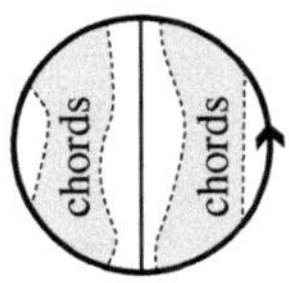

$= 0,$ (11.12)

and the left-hand side of this is called the $1\mathsf{T}$-*relator*.

As before, the two shaded regions of 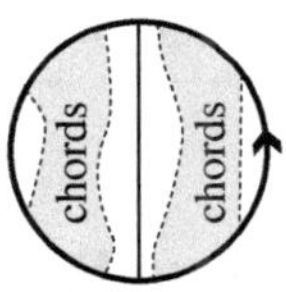

represent collections of chords that do not intersect the single chord that is drawn.

Definition 11.31 **($4\mathsf{T}$-relation).** In the space $\mathscr{C}_m$ or $\mathscr{C}$ of chord diagrams, the $4\mathsf{T}$-*relation*, or *four term relation*, is

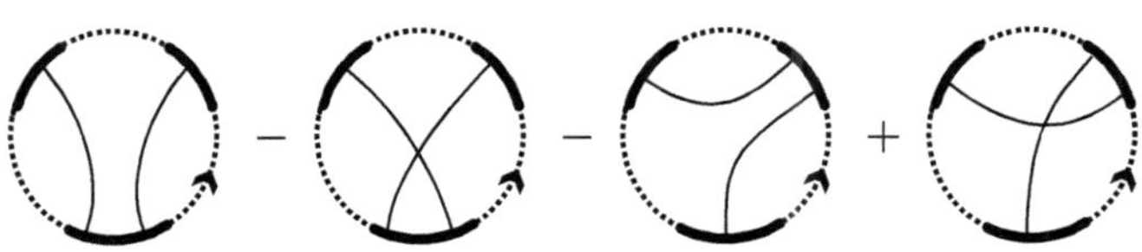
$= 0,$ (11.13)

and the left-hand side of this is called the $4\mathsf{T}$-*relator*.

As with the 4T*-relation, in Definition 11.31, the bold arcs on the skeleton represent sections on the skeleton where there are no ends of chords except for those shown. There may be chords with ends on the non-bold sections of the skeleton, and these chords may intersect the chords shown in (11.13). However, all four of the diagrams should be identical except for the chords that are shown.

Definition 11.32 (Vector spaces $\bar{\mathscr{A}}^c$ and $\bar{\mathscr{A}}^c_m$). The space $\bar{\mathscr{A}}^c$ of chord diagrams satisfying the 1T- and 4T-relations is

$$\bar{\mathscr{A}}^c := \frac{\mathscr{C}}{(1\mathsf{T}, 4\mathsf{T})}$$

where $(1\mathsf{T}, 4\mathsf{T})$ is the subspace of $\mathscr{C}$ generated by the 1T- and 4T-relators. The subspace of $\bar{\mathscr{A}}^c$ generated by degree m chord diagrams is denoted by $\bar{\mathscr{A}}^c_m$.

We have

$$\bar{\mathscr{A}}^c_m = \frac{\mathscr{C}_m}{(1\mathsf{T}, 4\mathsf{T})}.$$

By definition, an element of $\bar{\mathscr{A}}^c$ or $\bar{\mathscr{A}}^c_m$ is a finite formal $\mathbb{C}$-linear combination of equivalence classes of chord diagrams. It is more convenient instead to view the elements of these spaces in terms of representatives of the equivalence classes. That is, we often regard elements of $\bar{\mathscr{A}}^c$ or $\bar{\mathscr{A}}^c_m$ as finite formal $\mathbb{C}$-linear combinations of chord diagrams considered up to the 1T- and 4T-relations. Thus, two elements of $\bar{\mathscr{A}}^c$ or $\bar{\mathscr{A}}^c_m$ are considered to be equal if they are related to a finite sequence of the relations $(1\mathsf{T}) = 0$ and $(4\mathsf{T}) = 0$ (*i.e.* the relations in (11.12) and (11.13)). For example, the identities

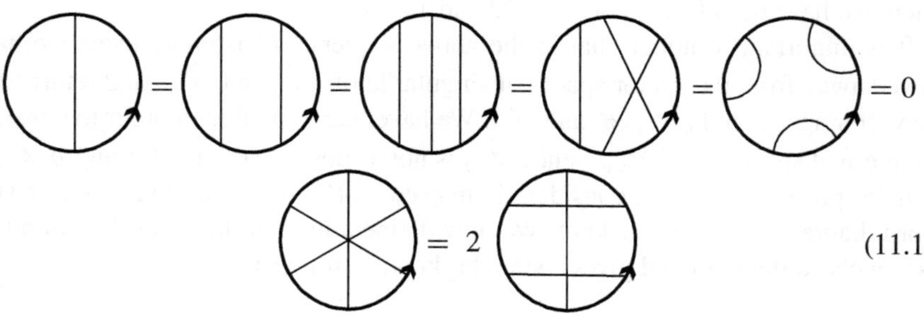

(11.14)

between chord diagrams hold. The first collection of identities follows from the 1T-relation since each of the chord diagrams has an isolated chord. The remaining identity follows from (11.11).

Observe that (11.14) immediately tells us that $\dim(\bar{\mathscr{A}}^c_0) \leq 1$, $\dim(\bar{\mathscr{A}}^c_1) = 0$, $\dim(\bar{\mathscr{A}}^c_2) \leq 1$ and $\dim(\bar{\mathscr{A}}^c_3) \leq 1$. In fact, these upper bounds are known to be sharp. The following table summarises the dimensions of $\bar{\mathscr{A}}^c_m$ for low values of m.

m	0	1	2	3	4	5	6	7	8	9	10	11	12
$\dim(\bar{\mathscr{A}}^c_m)$	1	0	1	1	3	4	9	14	27	44	80	132	232

Definition 11.25 and Lemma A.12 immediately give the following result.

Proposition 11.33. *Let $\mathscr{W}_m$ be the subspace of $\mathscr{C}^\star_m$ generated by the set of all degree m weight systems. Then*

$$\mathscr{W}_m \cong \left(\bar{\mathscr{A}}^c_m\right)^\star .$$

The significance of this result is as follows.

To study Vassiliev invariants it is enough to study the space $\bar{\mathscr{A}}^c$ of chord diagrams modulo the 1T- and 4T-relations.

11.9 Moving Between $\bar{\mathscr{A}}^c_m$ and $\mathscr{K}_m/\mathscr{K}_{m+1}$

We have isomorphisms $\mathscr{V}_m/\mathscr{V}_{m-1} \cong \mathscr{W}_m$ and $\mathscr{W}_m \cong \left(\bar{\mathscr{A}}^c_m\right)^\star$, from Theorem 11.29 and Proposition 11.33, respectively. We wish to work with the space $\bar{\mathscr{A}}^c_m$ of chord diagrams since they are amenable combinatorial objects. All of the spaces appearing here are finite dimensional so

$$\bar{\mathscr{A}}^c_m \cong \left(\bar{\mathscr{A}}^c_m\right)^{\star\star} \cong (\mathscr{W}_m)^\star \cong \left(\frac{\mathscr{V}_m}{\mathscr{V}_{m-1}}\right)^\star \cong \left(\frac{\mathscr{K}_m}{\mathscr{K}_{m+1}}\right)^{\star\star} \cong \frac{\mathscr{K}_m}{\mathscr{K}_{m+1}},$$

where we have used Propositions 10.22 and 11.33.

It is important to notice that in the above sequence of isomorphisms we have moved away from the vector spaces of singular knots, $\dot{\mathscr{K}}$ and $\dot{\mathscr{K}}_m$, and returned to the vector spaces of knots, $\mathscr{K}$ and $\mathscr{K}_m$. We have had to do this in order to consider the quotient spaces $\mathscr{K}_m/\mathscr{K}_{m+1}$ since $\dot{\mathscr{K}}_p$ is not a subspace of $\dot{\mathscr{K}}_q$ for any $p \neq q$.

In Proposition 11.14, we considered a map $\phi_m : \mathscr{K}_m \to \mathscr{C}_m$ which constructed the chord diagram of a singular knot. We now define a map in the other direction that takes a chord diagram and gives a singular knot, as follows.

Definition 11.34 (The map φ_m). Let

$$\varphi_m : \bar{\mathscr{A}}^c_m \to \frac{\mathscr{K}_m}{\mathscr{K}_{m+1}} \tag{11.15}$$

be the linear extension of the map which constructs a singular knot from a chord diagram by contracting its chords as in (11.3). Note that we are invoking Convention 10.9 here.

An example of the action of φ_3 on a chord diagram is given by

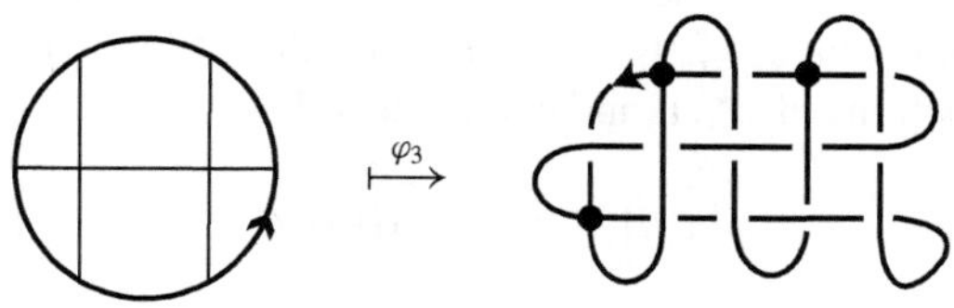

where the singular knot is considered as a representative of its equivalence class in $\mathscr{K}_3/\mathscr{K}_4$ and the chord diagram represents its equivalence class in $\bar{\mathscr{A}}_m^c$.

Lemma 11.35. *The linear map*

$$\varphi_m : \bar{\mathscr{A}}_m^c \to \mathscr{K}_m/\mathscr{K}_{m+1}$$

is well-defined.

Proof. First, we show that, in the definition of φ_m, it does not matter how we realise the singular knot in $\mathscr{K}_m/\mathscr{K}_{m+1}$. The argument for this is essentially the argument used in the proof of Proposition 11.16. Suppose C is a degree m chord diagram and that K and K' are two singular knots such that $K = \varphi_m(C)$ and $K' = \varphi_m(C)$. Then, K and K' both have exactly m singular points and differ by a finite number r of crossing changes. Then, there is a finite sequence $K = \mathfrak{b}_0, \mathfrak{b}_1, \ldots, \mathfrak{b}_r = K'$, for some nonnegative integer r, where $\mathfrak{b}_i$ and $\mathfrak{b}_{i+1}$ differ by a single crossing change, for $i = 0, \ldots, r-1$. Then, by Vassiliev's relation (Definition 10.6), for $i = 0, \ldots, r-1$, we have $\mathfrak{b}_i - \mathfrak{b}_{i+1} = \mathfrak{c}_i$ where $\mathfrak{c}_i$ is the knot obtained from $\mathfrak{b}_i$ by replacing, by a singular point, the point at which $\mathfrak{b}_i$ and $\mathfrak{b}_{i+1}$ differ by a crossing. Now, $\mathfrak{c}_0 \in \mathscr{K}_{m+1}$ so $\mathfrak{c}_0 = 0$ in $\mathscr{K}_m/\mathscr{K}_{m+1}$, in which case $\mathfrak{b}_0 = \mathfrak{b}_1$ in $\mathscr{K}_m/\mathscr{K}_{m+1}$. Iterating this process $r-1$ times, we have $\mathfrak{b}_0 = \mathfrak{b}_r$, whence $K = K'$, in $\mathscr{K}_m/\mathscr{K}_{m+1}$, so φ_m is well-defined as a map from $\mathscr{C}_m$.

It remains to show that φ_m is well-defined as a map from $\bar{\mathscr{A}}_m^c$; that is, it respects the 1T- and 4T-relations. The argument for this is similar to the proof of Lemma 11.28 and is therefore excluded. □

Exercise 11.36. Complete the proof of Lemma 11.35 by showing that the map φ_m satisfies the 1T- and 4T-relations.

We next show that φ_m and $\overline{\alpha}_m$ are dual to each other. In doing so, care must be taken about identifications of dual spaces. For this, we note the following.

Remark 11.37. We use the identifications

$$\frac{\mathscr{V}_m}{\mathscr{V}_{m-1}} \cong \left(\frac{\mathscr{K}_m}{\mathscr{K}_{m+1}}\right)^\star \quad \text{and} \quad \mathscr{W}_m \cong \left(\bar{\mathscr{A}}_m^c\right)^\star$$

from Propositions 10.22 and 11.33.

Lemma 11.38. *Under the identifications of Remark 11.37 we have* $\varphi_m^\star = \overline{\alpha}_m$.

Proof. For $\theta \in \mathscr{V}_m$, let $[\theta]$ be the class in $\mathscr{V}_m/\mathscr{V}_{m+1}$ containing θ. To establish the result, we must show that $\varphi_m^\star([\theta]) = \overline{\alpha}_m([\theta])$ for all $\theta \in \mathscr{V}_m$. This we do by using the element-wise actions of $\varphi_m^\star$, θ and $\overline{\alpha}_m$ to show that

$$\varphi_m^\star([\theta])(C) = \overline{\alpha}_m([\theta])(C) \tag{11.16}$$

for any degree m chord diagram C.

We consider the map $\overline{\alpha}_m$ first. Corollary 11.17 shows that $\alpha_m : \mathscr{V}_m \to \mathscr{W}_m$ has the property that, for each θ, the map $\alpha_m(\theta) = \omega_\theta$ is the unique map for which $\dot{\theta} = \omega_\theta \circ \phi_m$. Also, recall from Lemma 11.18 that

$$\overline{\alpha}_m : \frac{\mathscr{V}_m}{\mathscr{V}_{m-1}} \to \mathscr{W}_m : [\theta] \mapsto \omega_\theta$$

where $[\theta]$ is the class of $\mathscr{V}_m/\mathscr{V}_{m-1}$ containing θ.

To describe the action of $\overline{\alpha}_m([\theta])$, let C be any degree m chord diagram. Following (11.3) and Proposition 11.14, a singular knot $\dot{K} \in \dot{\mathscr{K}}_m$ may be constructed by contracting the chords of C with the result that $\phi_m(\dot{K}) = C$. Then, $\omega_\theta(C) = \omega_\theta(\phi_m(\dot{K})) = \dot{\theta}(\dot{K}) = \theta(K)$ where $K \in \mathscr{K}_m$ is the Vassiliev resolution of K. Thus, $\overline{\alpha}_m([\theta])(C) = \theta(K)$, which is the right-hand side of (11.16).

Next, we describe the action of $\varphi^\star([\theta])$. Since

$$\varphi_m : \bar{\mathscr{A}}_m^c \to \frac{\mathscr{K}_m}{\mathscr{K}_{m+1}},$$

we have, by taking duals,

$$\varphi_m^\star : \left(\frac{\mathscr{K}_m}{\mathscr{K}_{m+1}}\right)^\star \to (\bar{\mathscr{A}}_m^c)^\star : [f] \mapsto [f] \circ \varphi_m$$

and thence, by Propositions 10.22 and 11.33 (or Remark 11.37),

$$\varphi_m^\star : \frac{\mathscr{V}_m}{\mathscr{V}_{m-1}} \to \mathscr{W}_m.$$

To describe the action of $\varphi_m^\star$ on $[\theta] \in \mathscr{V}_m/\mathscr{V}_{m-1}$, let

$$F : \frac{\mathscr{V}_m}{\mathscr{V}_{m-1}} \longrightarrow \left(\frac{\mathscr{K}_m}{\mathscr{K}_{m+1}}\right)^\star \quad \text{and} \quad G : (\bar{\mathscr{A}}_m^c)^\star \longrightarrow \mathscr{W}_m$$

denote the isomorphisms constructed in Propositions 10.22 and 11.33 (*i.e.* those used in Remark 11.37). Then,

$$\varphi_m^\star([\theta]) = G \circ F([\theta]) \circ \varphi_m.$$

Then, as described in the proof of Proposition 10.22,

$$F([\theta])\colon \frac{\mathscr{K}_m}{\mathscr{K}_{m+1}} \to \mathbb{C}\colon [K] \mapsto \theta([K]).$$

Moreover, for $W \in \left(\mathscr{A}_m^c\right)^\star$ and $[C] \in \mathscr{A}_m^c$, the map G is described by

$$G(W)\colon C \mapsto W([C]).$$

Let $C \in \mathscr{C}_m$ and let $\dot{K} \in \dot{\mathscr{K}}_m$ where $\dot{K} \in \dot{\mathscr{K}}_m$ is such that $\phi_m(\dot{K}) = C$, as above, and K is the Vassiliev resolution of $\dot{K}$. By the construction of $\dot{K}$, we know that $[K] = \varphi_m([C])$ where $[K]$ is the equivalence class of K in $\mathscr{K}_m/\mathscr{K}_{m-1}$ and $[C]$ is the equivalence class in $\bar{\mathscr{A}}_m^c$ of C. Then, $F([\theta])\,(\varphi_m([C])) = F([\theta])([K]) = \theta(K)$, so

$$F([\theta]) \circ \varphi_m\colon [C] \mapsto \theta(K)$$

and, by applying the map G, we see that

$$\varphi_m^\star([\theta])\colon C \mapsto \theta(K).$$

Thus, $\varphi_m^\star([\theta])(C) = [\theta](K)$, which is the left-hand side of (11.16).

Combining these results, we have, $\varphi_m^\star([\theta])(C) = \theta(K) = \overline{\alpha}_m([\theta])(C)$ for all $\theta \in \mathscr{V}_m$ and all $C \in \mathscr{C}_m$, so $\varphi_m^\star = \overline{\alpha}_m$, giving the result. □

Corollary 11.39. *The linear map* $\varphi_m : \bar{\mathscr{A}}_m^c \to \mathscr{K}_m/\mathscr{K}_{m+1}$ *is surjective.*

Proof. By Theorem 11.28, $\overline{\alpha}_m$ is injective. Lemma 11.38 and Theorem A.19 give that $\overline{\alpha}_m^\star = \varphi_m^{\star\star}$, and hence φ_m is surjective. □

We shall prove the following dual version of Theorem 11.29 in Sect. 18.1.

Theorem 11.40. *The map* $\varphi_m : \bar{\mathscr{A}}_m^c \to \mathscr{K}_m/\mathscr{K}_{m+1}$ *is a vector space isomorphism.*

This theorem and its dual statement, Theorem 11.29, form one of the most important results about Vassiliev invariants.

11.10 A Summary of the Approach and Results

The development of Vassiliev invariants we have just seen in Chaps. 10 and 11 is fairly intricate, so here we provide a brief review of its overarching ideas.

Our aim was to understand the space $\mathscr{V}_m$ of degree m Vassiliev invariants. We observed that $\mathscr{V}_m \cong (\mathscr{K}/\mathscr{K}_{m+1})^\star$, and so understanding $\mathscr{K}_m$ is an equivalent problem. There is a filtration on each of the spaces $\mathscr{V}$ and $\mathscr{K}$:

$$\mathscr{V}_0 \leq \mathscr{V}_1 \leq \mathscr{V}_2 \leq \cdots,$$

and

$$\mathscr{K}_0 \geq \mathscr{K}_1 \geq \mathscr{K}_2 \geq \mathscr{K}_3 \geq \cdots.$$

The existence of these filtrations enabled us to express the spaces as direct sums,

$$\mathscr{V}_m \cong \frac{\mathscr{V}_m}{\mathscr{V}_{m-1}} \oplus \frac{\mathscr{V}_{m-1}}{\mathscr{V}_{m-2}} \oplus \cdots \oplus \frac{\mathscr{V}_1}{\mathscr{V}_0} \oplus \mathscr{V}_0,$$

and

$$\mathscr{K}_m \cong \frac{\mathscr{K}_m}{\mathscr{K}_{m+1}} \oplus \frac{\mathscr{K}_{m-1}}{\mathscr{K}_m} \oplus \cdots \oplus \frac{\mathscr{K}_1}{\mathscr{K}_2} \oplus \frac{\mathscr{K}}{\mathscr{K}_1}.$$

This shifted the problem of understanding $\mathscr{V}_m$ and $\mathscr{K}_m$ to that of understanding the quotient spaces $\mathscr{V}_i/\mathscr{V}_{i-1}$ and $\mathscr{K}_i/\mathscr{K}_{i+1}$.

By passing through the space of singular knots, we were able to show (albeit citing a result from Part IV that we are yet to see for the injectivity) that $\mathscr{V}_m/\mathscr{V}_{m-1} \cong \mathscr{W}_m$ and $\bar{\mathscr{A}}_m^c \cong \mathscr{K}_m/\mathscr{K}_{m+1}$.

Thus, we obtained the following two diagrams in which all mappings are isomorphisms:

$$\begin{array}{ccccccccccccc}
\mathscr{V}_m & \cong & \frac{\mathscr{V}_m}{\mathscr{V}_{m-1}} & \oplus & \frac{\mathscr{V}_{m-1}}{\mathscr{V}_{m-2}} & \oplus & \cdots & \oplus & \frac{\mathscr{V}_1}{\mathscr{V}_0} & \oplus & \mathscr{V}_0 \\
\beta_m \Big\downarrow & & \overline{\alpha}_m \Big\downarrow & & \overline{\alpha}_{m-1} \Big\downarrow & & \cdots & & \overline{\alpha}_1 \Big\downarrow & & \overline{\alpha}_0 \Big\downarrow \\
\mathscr{W}_{\leq m} & \cong & \mathscr{W}_m & \oplus & \mathscr{W}_{m-1} & \oplus & \cdots & \oplus & \mathscr{W}_1 & \oplus & \mathscr{W}_0
\end{array}$$

and

$$\begin{array}{ccccccccccccc}
\frac{\mathscr{K}}{\mathscr{K}_{m+1}} & \cong & \frac{\mathscr{K}_m}{\mathscr{K}_{m+1}} & \oplus & \frac{\mathscr{K}_{m-1}}{\mathscr{K}_m} & \oplus & \cdots & \oplus & \frac{\mathscr{K}_1}{\mathscr{K}_2} & \oplus & \frac{\mathscr{K}}{\mathscr{K}_1} \\
\Big\uparrow & & \varphi_m \Big\uparrow & & \varphi_{m-1} \Big\uparrow & & \cdots & & \varphi_1 \Big\uparrow & & \varphi_0 \Big\uparrow \\
\bar{\mathscr{A}}^c_{\leq m} & \cong & \bar{\mathscr{A}}^c_m & \oplus & \bar{\mathscr{A}}^c_{m-1} & \oplus & \cdots & \oplus & \bar{\mathscr{A}}^c_1 & \oplus & \bar{\mathscr{A}}^c_0
\end{array}$$

Using that $\mathscr{V}_m/\mathscr{V}_{m-1} \cong (\mathscr{K}_m/\mathscr{K}_{m+1})^{\star}$ gives the following commutative diagram in which all mappings are isomorphisms.

$$\begin{array}{ccc}
\frac{\mathscr{K}}{\mathscr{K}_{m+1}} & \xleftrightarrow{\text{dual}} & \mathscr{V}_m \\
\Big\updownarrow & & \Big\updownarrow \\
\bar{\mathscr{A}}^c_{\leq m} & \xleftrightarrow{\text{dual}} & \mathscr{W}_{\leq m}
\end{array}$$

Thus, result of this work is summarised by the following statement.

> To understand the Vassiliev invariants, it is enough to understand the combinatorial vector space of chord diagrams $\bar{\mathscr{A}}^c$.

This is exactly how we shall progress in the theory, with much of the remainder of Part III devoted to the study of $\bar{\mathscr{A}}^c$. However, the reader should remember that in doing this, we are really studying the space of Vassiliev invariants.

The following diagram, where all the arrows are isomorphisms, summarises the isomorphisms in Chaps. 10 and 11. We have not yet met the isomorphism $\check{Z}^u_m$. This will be the topic of Part IV of the book.

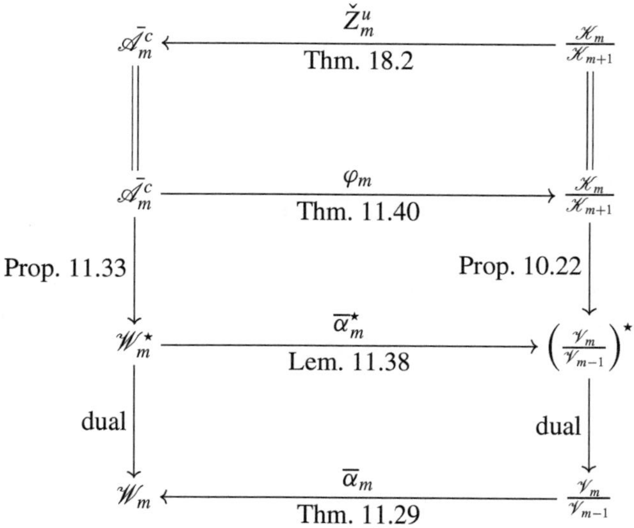

Chapter 12
Vassiliev Invariants of Framed Knots

We have seen earlier, in Chap. 9, that the natural setting for operator invariants is framed links, rather than links. Similarly, the natural setting for the theory of Vassiliev invariants is framed knots, rather than knots. We now develop the theory in this setting. For this, in this chapter, we provide a development for the theory of Vassiliev invariants of framed knots in a way that is parallel to the unframed case we have just seen in Chaps. 10 and 11. Again here the main point is to move from a vector space of framed knot invariants, $\mathscr{V}_m^f$, to a combinatorial vector space of chord diagrams $\mathscr{A}_m$.

The only real change in the theory (other than changing "knot" to "framed knot", and "RI" to "FRI") is that we no longer consider the 1T-relation in our spaces of chord diagrams. This makes the combinatorial analysis of these spaces more natural. Because of the similarity with the unframed case, we shall provide only a brief summary of the initial development of the theory. The reader should check the results by adapting the proofs of their unframed analogues from Chaps. 10 and 11.

Following the exposition of unframed Vassiliev invariants, we let $\mathscr{K}^f$ denote the vector space of finite formal linear combinations of isotopy classes of oriented framed knots over $\mathbb{C}$. We call $\mathscr{K}^f$ the *vector space of framed knots*.

Since the set of framed knots considered up to isotopy is equivalent to the set of framed knot diagrams considered up to the framed Reidemeister moves of Definition 3.3, we are at liberty to view $\mathscr{K}^f$ as a vector space of formal linear combinations of equivalence classes of oriented framed knot diagrams over $\mathbb{C}$. Since it simplifies the exposition, we shall generally think of $\mathscr{K}^f$ in this way.

As with the unframed case, we wish to consider quotient spaces of $\mathscr{K}^f$ generated by the Vassiliev relation

For this, we need to consider *singular framed knot diagrams*. As mathematical objects, these are defined as singular knot diagrams (Definition 10.4). Two

D. M. Jackson and I. Moffatt, *An Introduction to Quantum and Vassiliev Knot Invariants*, CMS Books in Mathematics, https://doi.org/10.1007/978-3-030-05213-3_12

singular framed knot diagrams are *equivalent* if they are related by a finite sequence of framed Reidemeister moves.

Let K be a singular framed knot diagram and S_K be its set of singular points. The *Vassiliev resolution*, $V_{\text{res}}(K)$, of K is the element

$$V_{\text{res}}(K) = \sum_{A \subseteq S_K} (-1)^{|A|} K_A$$

of $\mathscr{K}^f$ that results from using the Vassiliev relation to desingularize the diagram.

Again we adopt the following convention.

Convention 12.1. When working within the vector space $\mathscr{K}^f$, we shall make the identification

$$K \leftrightarrow V_{\text{res}}(K) \tag{12.1}$$

between a singular framed knot and its Vassiliev resolution. That is to say, K will be a linear combination of framed knot diagrams.

Definition 12.2 (Vector space $\mathscr{K}^f$). $\mathscr{K}_m^f$ is the (infinite dimensional) subspace of $\mathscr{K}^f$ spanned by the set of all oriented singular knots with (at least) m singular points.

The spaces $\mathscr{K}_m^f$, for $m \in \mathbb{N}_0$, provide a filtration on $\mathscr{K}^f$:

$$\mathscr{K}^f = \mathscr{K}_0^f \geq \mathscr{K}_1^f \geq \mathscr{K}_2^f \geq \mathscr{K}_3^f \geq \cdots .$$

Definition 12.3 (Vassiliev invariant — 1 st). A framed knot invariant θ is a *framed Vassiliev invariant of degree m* (or *finite-type invariant*) if it is a linear map $\theta : \mathscr{K}^f \to \mathbb{C}$ with the property that

$$\theta|_{\mathscr{K}_{m+1}^f} = 0,$$

where 0 is the zero map. The set of all framed Vassiliev invariants is denoted by $\mathscr{V}^f$, and the set of all framed Vassiliev invariants of degree m is denoted by $\mathscr{V}_m^f$.

The set $\mathscr{V}^f$ forms a vector space under the usual addition and scalar multiplication of functions in (10.8). Moreover,

$$\mathscr{V}_0^f \leq \mathscr{V}_1^f \leq \mathscr{V}_2^f \leq \cdots .$$

The spaces $\mathscr{V}^f$ and $\mathscr{K}^f$ are related *via* duality thus:

$$\mathscr{V}_m^f \cong \left(\mathscr{K}^f / \mathscr{K}_{m+1}^f\right)^{\star}.$$

Moreover,

$$\frac{\mathscr{V}_m^f}{\mathscr{V}_{m-1}^f} \cong \left(\frac{\mathscr{K}_m^f}{\mathscr{K}_{m+1}^f}\right)^{\star}. \tag{12.2}$$

Again, the theory of framed Vassiliev invariants is advanced by moving from the functional and topological spaces $\mathscr{V}_m$ and $\mathscr{K}_m$ to a combinatorial space of chord diagrams.

To move to the combinatorial spaces, we need to consider singular framed knots as objects in their own right, rather than just as a representation of an element of $\mathscr{K}^f$.

Definition 12.4 (Vector space $\dot{\mathscr{K}}^f$). We let $\dot{\mathscr{K}}^f$ denote the vector space of finite formal linear combinations of equivalence classes of framed singular oriented knot diagrams over $\mathbb{C}$. In addition, $\dot{\mathscr{K}}^f_m$ is the subspace of $\dot{\mathscr{K}}^f$ generated by all framed singular knot diagrams with exactly m singular points.

Note that $\dot{\mathscr{K}}^f$ is a graded vector space:

$$\dot{\mathscr{K}}^f = \dot{\mathscr{K}}^f_0 \oplus \dot{\mathscr{K}}^f_1 \oplus \dot{\mathscr{K}}^f_2 \oplus \cdots .$$

Any invariant of framed knots, θ, may be extended to an invariant of framed singular knots, $\dot{\theta}$, by considering its Vassiliev extension, just as in the unframed case of Definition 11.3. This observation gives a reformulation of the definition of a framed Vassiliev invariant.

Definition 12.5 (Framed Vassiliev invariant — 2nd). A framed knot invariant $\theta\colon \mathscr{K}^f \to \mathbb{C}$ is a *Vassiliev invariant of degree m* if $\dot{\theta}\colon \dot{\mathscr{K}}^f \to \mathbb{C}$ satisfies $\dot{\theta}(\dot{\mathscr{K}}_{m+1}) = \{0\}$.

Chord diagrams arise in the theory of framed Vassiliev invariants in the same way that they arise in the unframed case. The proof of Lemma 11.7 holds for framed knots, and it follows that if θ is a degree m framed Vassiliev invariant and if $\dot{K}, \dot{K}' \in \dot{\mathscr{K}}^f_m$, then, if $\dot{K}$ and $\dot{K}'$ differ by a finite sequence of crossing changes and isotopy, $\dot{\theta}(\dot{K}) = \dot{\theta}(\dot{K}')$. Thus the salient information about $\dot{K}$ and $\dot{K}'$ is the order in which the singular points appear along it, and this information is contained in a chord diagram. The map ϕ_m of Definition 11.3 applies to framed knot diagrams, giving a map

$$\phi_m : \dot{\mathscr{K}}^f_m \to \mathscr{C}_m .$$

The above observation on how degree m framed Vassiliev invariants detect crossing changes can be rephrased thus:

Proposition 12.6. *Let $\theta \in \mathscr{V}^f_m$ be a degree m Vassiliev invariant and let $\dot{K}, \dot{K}' \in \dot{\mathscr{K}}^f_m$. Then*

$$\phi_m(\dot{K}) = \phi_m(\dot{K}') \implies \dot{\theta}(\dot{K}) = \dot{\theta}(\dot{K}'),$$

i.e. *the value of a degree m Vassiliev invariant of a singular framed knot with exactly m singular points depends only on the chord diagram of the knot.*

The theory of framed Vassiliev invariants then progresses by formulating functions on framed knots into functions on chord diagrams. This leads us to *framed weight systems*.

Definition 12.7 (Framed weight system). A *framed weight system* of degree m is an element $f \in \mathscr{C}_m^\star$ which satisfies the *four-term relation*, $4\mathsf{T}^\star$,

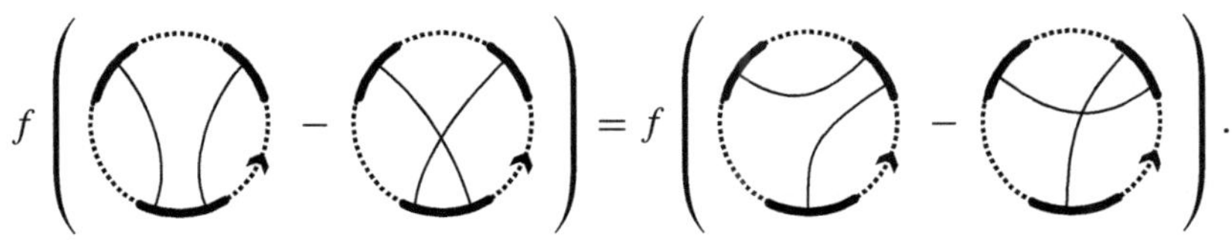

We let $\mathscr{W}_m^f$ be the complex vector space of framed weight systems of degree m, and $\mathscr{W}^f$ be the complex vector space of all framed weight systems.

The key result about weight systems is that they describe the quotient space $\mathscr{V}_m^f / \mathscr{V}_{m-1}^f$ of framed Vassiliev invariants. For this, we need the following maps. Following Lemma 11.18, let

$$\overline{\alpha}_m : \frac{\mathscr{V}_m^f}{\mathscr{V}_{m-1}^f} \to \mathscr{C}_m^\star$$

be defined by

$$\overline{\alpha}_m : [\theta] \mapsto \omega_\theta,$$

where ω_θ, as in Corollary 11.17, is defined for a degree m chord diagram C by $\omega_\theta(C) = \overset{\bullet}{\theta}(K)$ for any K such that $\phi_m(K) = C$. By adapting the proof of Lemma 11.28, we can obtain the following.

Lemma 12.8. *The map* $\overline{\alpha}_m : \mathscr{V}_m^f / \mathscr{V}_{m-1}^f \longrightarrow \mathscr{C}_m^\star$ *induces a well-defined injective linear map*

$$\overline{\alpha}_m : \frac{\mathscr{V}_m^f}{\mathscr{V}_{m-1}^f} \longrightarrow \mathscr{W}_m^f.$$

In fact, the following stronger result is true.

Theorem 12.9. *The map*

$$\overline{\alpha}_m : \frac{\mathscr{V}_m^f}{\mathscr{V}_{m-1}^f} \longrightarrow \mathscr{W}_m^f$$

is a vector space isomorphism.

We delay the proof of this theorem until Sect. 18.1, where it will follow from the theory of the Kontsevich invariant, but we assume its truth for the moment.

The significance of Theorem 12.9 is that we have the following commutative diagram, in which all vertical arrows are isomorphisms.

$$\begin{array}{ccccccccccc}
\mathscr{V}_m^f & \cong & \dfrac{\mathscr{V}_m^f}{\mathscr{V}_{m-1}^f} & \oplus & \dfrac{\mathscr{V}_{m-1}^f}{\mathscr{V}_{m-2}^f} & \oplus & \cdots & \oplus & \dfrac{\mathscr{V}_1^f}{\mathscr{V}_0^f} & \oplus & \mathscr{V}_0^f \\
\beta_m \big\downarrow & & \overline{\alpha}_m \big\downarrow & & \overline{\alpha}_{m-1} \big\downarrow & & \cdots & & \overline{\alpha}_0 \big\downarrow & & \overline{\alpha}_1 \big\downarrow \\
\mathscr{W}_{\leq m}^f & \cong & \mathscr{W}_m^f & \oplus & \mathscr{W}_{m-1}^f & \oplus & \cdots & \oplus & \mathscr{W}_1^f & \oplus & \mathscr{W}_0^f
\end{array}$$

Thus we have the equivalence of Vassiliev invariants and weight systems,

$$\mathscr{V}_m^f \cong \mathscr{W}_{\leq m}^f. \tag{12.3}$$

The conclusion of the discussion so far here is that, by (12.3), to study framed Vassiliev invariants, it is enough to study weight systems. In turn, to study weight systems, we may instead study their dual space, which is a vector space of chord diagrams.

Definition 12.10 (Vector space $\mathscr{A}^c$). The space $\mathscr{A}^c$ of chord diagrams modulo 4T-relation is

$$\mathscr{A}^c := \frac{\mathscr{C}}{(4\mathsf{T})}$$

where (4T) is the subspace of $\mathscr{C}$ generated by the 4T-relators,

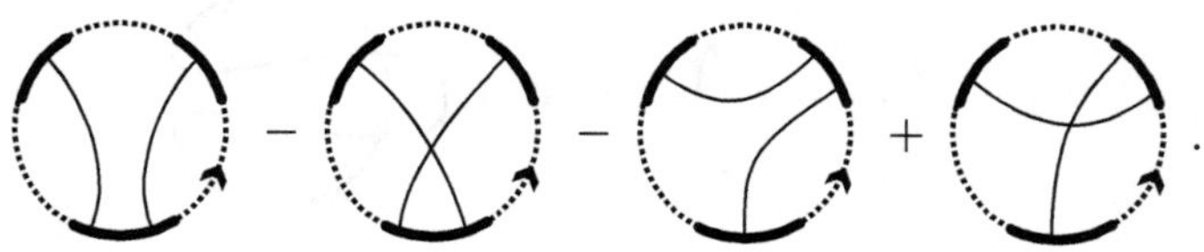

The subspace of $\mathscr{A}^c$ generated by degree m chord diagrams is denoted $\mathscr{A}_m^c$.

We have

$$\mathscr{A}_m^c = \frac{\mathscr{C}_m}{(4\mathsf{T})}.$$

The vector spaces $\mathscr{A}_m^c$ are finite dimensional (since there are only finitely many chord diagrams of a give degree). The next table summarises the dimensions of $\mathscr{A}_m^c$ for low values of m.

m	0	1	2	3	4	5	6	7	8	9	10	11	12
$\dim(\mathscr{A}_m^c)$	1	1	2	3	6	10	19	33	60	104	184	316	548

An application of Lemma A.12 relates $\mathscr{A}^c$ and $\mathscr{W}^f$ through duality:

$$\mathscr{W}_m^f \cong \left(\mathscr{A}_m^c\right)^\star. \tag{12.4}$$

Thus we may, equivalently, define a degree *m framed weight system* to be an element of the dual space $\left(\mathscr{A}_m^c\right)^\star$.

Since all of the spaces in question are finite dimensional (as there are finitely many chord diagrams of each degree), we have the following sequence of isomorphisms.

$$\mathscr{A}_m^c \cong \left(\mathscr{A}_m^c\right)^{\star\star} \cong \left(\mathscr{W}_m^f\right)^\star \cong \left(\frac{\mathscr{V}_m^f}{\mathscr{V}_{m-1}^f}\right)^\star \cong \left(\frac{\mathscr{K}_m^f}{\mathscr{K}_{m+1}^f}\right)^{\star\star} \cong \frac{\mathscr{K}_m^f}{\mathscr{K}_{m+1}^f},$$

Again, as in the unframed case, we may write down an explicit isomorphism from $\mathscr{A}_m^c$ to $\mathscr{K}_m^f / \mathscr{K}_{m+1}^f$. This is given by

$$\varphi_m : \mathscr{A}_m^c \to \frac{\mathscr{K}_m^f}{\mathscr{K}_{m+1}^f},$$

where φ_m is defined on a representative C of $[C] \in \mathscr{A}_m^c$ and C is a chord diagram, by the operation

$\varphi_m :$ 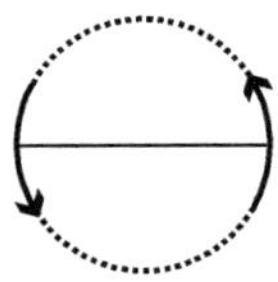 $\longmapsto$ 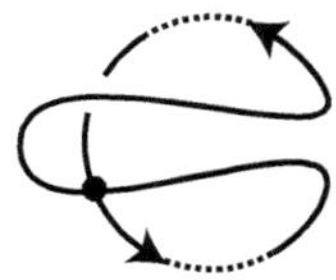 ,

and extended linearly. Here the operation indicated in the figure transforms the skeleton of the chord diagram iteratively into a knot diagram. In this construction, each intersection of chords in the chord diagram will result in four crossings in the singular knot. The over / under structure of these crossings should be chosen arbitrarily.

By following the approaches used to prove Lemmas 11.35 and 11.38, and Corollary 11.39, the following may be shown.

Lemma 12.11. *The map* $\varphi_m : \mathscr{A}_m^c \to \mathscr{K}_m^f / \mathscr{K}_{m+1}^f$ *is a well-defined, surjective linear map. Moreover,* $\varphi_m^\star = \overline{\alpha}_m$.

We shall prove the following dual result of Theorem 11.29 in Sect. 18.1 where it will follow from the theory of the Kontsevich invariant. The surjectivity of the map is due to Vassiliev. Its injectivity, whose proof we have yet to see, is a result of Kontsevich.

Theorem 12.12. *The map* $\varphi_m : \mathcal{A}^c_m \to \mathcal{K}^f_m / \mathcal{K}^f_{m+1}$ *is a vector space isomorphism.*

This discussion is summarised by the following diagram.

$$\begin{array}{ccccccccccc} \dfrac{\mathcal{K}^f}{\mathcal{K}^f_{m+1}} & \cong & \dfrac{\mathcal{K}^f_m}{\mathcal{K}^f_{m+1}} & \oplus & \dfrac{\mathcal{K}^f_{m-1}}{\mathcal{K}^f_m} & \oplus & \cdots & \oplus & \dfrac{\mathcal{K}^f_1}{\mathcal{K}^f_2} & \oplus & \dfrac{\mathcal{K}^f}{\mathcal{K}^f_1} \\ \uparrow & & \varphi_m \uparrow & & \varphi_{m-1} \uparrow & & \cdots & & \varphi_1 \uparrow & & \varphi_0 \uparrow \\ \mathcal{A}^c_{\leq m} & \cong & \mathcal{A}^c_m & \oplus & \mathcal{A}^c_{m-1} & \oplus & \cdots & \oplus & \mathcal{A}^c_1 & \oplus & \mathcal{A}^c_0 \end{array}$$

The relations between the various spaces described here are summarised in the following commutative diagrams in which all the mappings are isomorphisms.

$$\begin{array}{ccc} \dfrac{\mathcal{K}^f}{\mathcal{K}^f_{m+1}} & \xleftrightarrow{\text{dual}} & \mathcal{V}^f_m \\ \updownarrow & & \updownarrow \\ \mathcal{A}_{\leq m} & \xleftrightarrow{\text{dual}} & \mathcal{W}^f_{\leq m} \end{array}$$

$$\begin{array}{ccc} \mathcal{A}^c_m & \xleftarrow[\text{Thm. 18.1}]{\check{Z}_m} & \dfrac{\mathcal{K}^f_m}{\mathcal{K}^f_{m+1}} \\ \Vert & & \Vert \\ \mathcal{A}^c_m & \xrightarrow[\text{Thm. 12.12}]{\varphi_m} & \dfrac{\mathcal{K}^f_m}{\mathcal{K}^f_{m+1}} \\ (12.4) \downarrow & & \downarrow (12.2) \\ (\mathcal{W}^f_m)^\star & \xrightarrow[\text{Lem. 12.11}]{\overline{\alpha}^\star_m} & \left(\dfrac{\mathcal{V}^f_m}{\mathcal{V}^f_{m-1}}\right)^\star \\ \text{dual} \downarrow & & \downarrow \text{dual} \\ \mathcal{W}^f_m & \xleftarrow[\text{Thm. 12.9}]{\overline{\alpha}_m} & \dfrac{\mathcal{V}^f_m}{\mathcal{V}^f_{m-1}} \end{array}$$

Exercise 12.13. This section provided an overview of the fundamentals of the theory of framed Vassiliev invariants. Although we did not provide proofs of any results, all the proofs presented here may be obtained by adapting the corresponding theory for the unframed case. By adapting the corresponding results from Chaps. 10 and 11, verify that the claims made in the summary above are true.

Chapter 13
Jacobi Diagrams

13.1 Jacobi Diagrams

We shall now introduce a space $\mathscr{A}$ of *Jacobi diagrams* that is closely related to the space $\mathscr{A}^c$ of chord diagrams. In fact, this new space is isomorphic to $\mathscr{A}^c$, a result due to Bar-Natan that appeared in his seminal paper on Vassiliev invariants [13]. The introduction of Jacobi diagrams is properly motivated by Penrose's tensor calculus for universal enveloping algebras of Lie algebras (see [146]). However, here we shall take a combinatorial observation as our motivation, leaving the connection with Lie algebras to be discussed in Chap. 14.

The combinatorial observation is that the three expressions in

(13.1)

are equivalent through the 4T-relation. The evident rotational symmetry around the three points of attachment of the two chords to the skeleton may be encoded by representing each of the three expressions by a single trivalent (*i.e.* degree three) vertex:

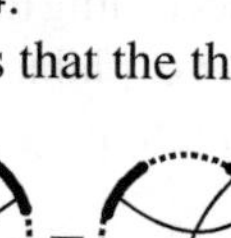

(13.2)

Thus, we are led to consider a space of unitrivalent graphs (*i.e.* graphs in which every vertex is of degree one or degree three) in which the degree one vertices lie on the skeleton. Moreover, we shall need to declare that every trivalent vertex may be replaced by chords to recover the original chord diagram:

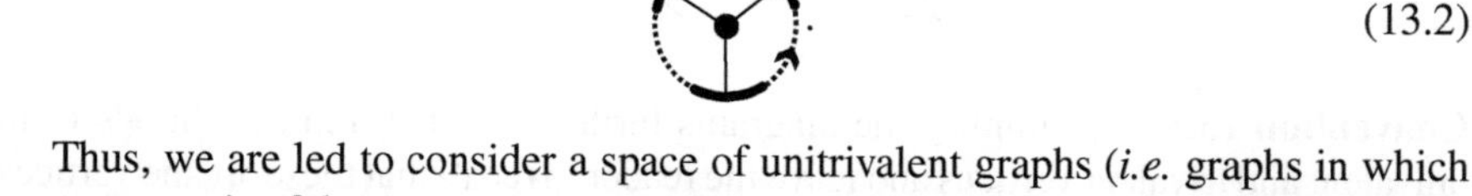

D. M. Jackson and I. Moffatt, *An Introduction to Quantum and Vassiliev Knot Invariants*,
CMS Books in Mathematics, https://doi.org/10.1007/978-3-030-05213-3_13

We now study these objects, and their relation to chord diagrams, in detail.

Definition 13.1 (Unitrivalent graph). A *unitrivalent graph* is a graph in $\mathbb{R}^3$ whose vertices (marked by •'s) are either of degree one, which are called *univalent vertices* or of degree three, which are called *trivalent vertices*. In addition, each trivalent vertex has a cyclic order on the half-edges incident with it.

The *degree* of a unitrivalent graph is half the number of its vertices.

To avoid cluttering the diagrams, we adopt the following convention.

Convention 13.2 (Cyclic order at trivalent vertices). If a cyclic order at a trivalent vertex is not specified in a drawing of a unitrivalent graph on the plane, then it is the cyclic order induced by the (anticlockwise) orientation of the plane.

Examples of unitrivalent graphs are

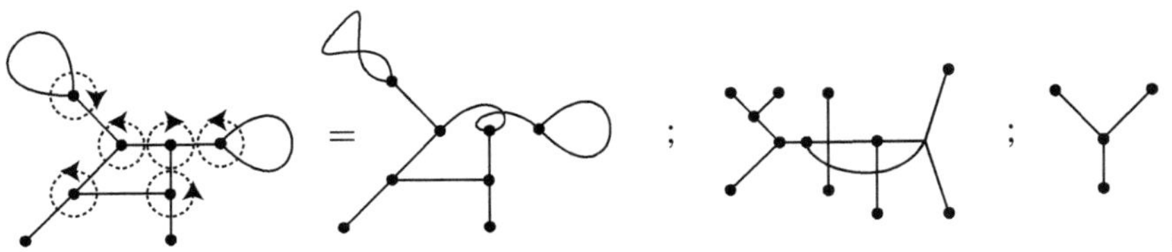

(13.3)

A cyclic orientation appears explicitly at each vertex of the leftmost diagram of (13.3). The second diagram is drawn according to Convention 13.2, so the orientations are to be assumed anticlockwise. This explains the adjustments made at two of the vertices that carried a clockwise orientation.

Exercise 13.3. Prove that the degree of a unitrivalent graph is an integer.

Remark 13.4. Points in the diagrams that appear to be "degree four vertices" are accidents of the drawing and are not vertices. Thus, for example,

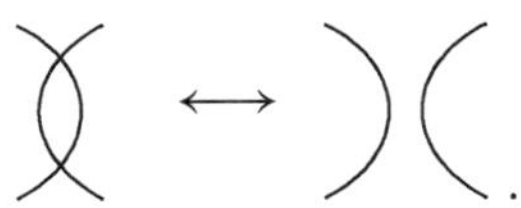

Convention 13.5. To simplify the diagrams further, we shall remove the •'s from univalent and trivalent vertices and leave the reader to recall that these are the vertices and that apparent 4-valent vertices are fictitious.

Definition 13.6 (Jacobi diagram). A *Jacobi diagram* is a connected diagram consisting of a single oriented circle, called the *skeleton*, and a unitrivalent graph. Moreover, each of the univalent vertices of the unitrivalent graph is identified with a point on the skeleton.

The *degree* of a Jacobi diagram is half the number of its vertices.

Examples of Jacobi diagrams are:

The leftmost Jacobi diagram is of degree five, the other two are of degree two. There is no cyclic order at the points where the univalent vertices meet the skeleton, so the two Jacobi diagrams on the right of the above display are equal.

Since a Jacobi diagram consists of a unitrivalent graph with additional structure, any drawing of it is simply a planar projection of it. Two drawings therefore represent the same Jacobi diagram if and only if they are projections of the same graph where the half-edges at corresponding trivalent vertices have the same cyclic ordering, and all the univalent vertices appear in the same cyclic order along the skeleton in the direction of its orientation.

A complete list of Jacobi diagrams of degrees zero, one and two is:

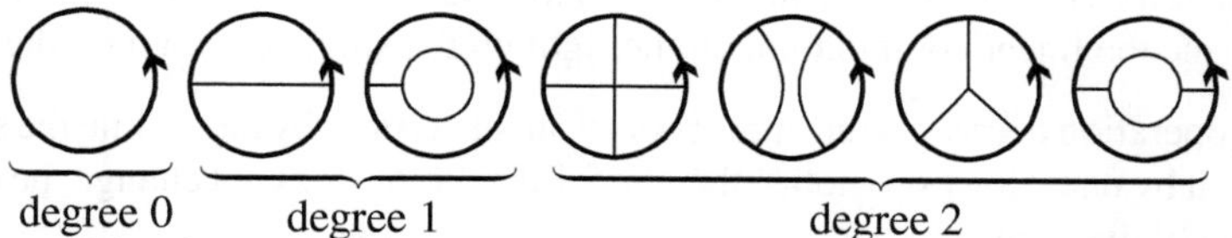

. (13.4)

The first of the degree 2 diagrams has a fictitious crossing that is an accident of the drawing (see Remark 13.4).

Convention 13.7. We shall generally draw Jacobi diagrams with a counterclockwise orientation of the skeleton.

Definition 13.8 (Vector spaces $\mathscr{D}$ and $\mathscr{D}_m$ of Jacobi diagrams). $\mathscr{D}$ is the vector space over $\mathbb{C}$ consisting of all finite formal linear combinations of Jacobi diagrams over $\mathbb{C}$, and $\mathscr{D}_m$ is the subspace of $\mathscr{D}$ generated by Jacobi diagrams of degree exactly m.

We have

$$\mathscr{D} := \bigoplus_{m=0}^{\infty} \mathscr{D}_m.$$

13.2 The STU-Relation and the Vector Space $\mathscr{A}$

Previously, in Chap. 12, we considered a quotient space $\mathscr{A}^c$ of the vector space $\mathscr{C}$ of chord diagrams by the 4T-relation. In an analogous way, we now consider a quotient space of the space of Jacobi diagrams by the following relation, called the STU-*relation*. It encapsulates the combinatorial observation given at the beginning of this chapter.

Definition 13.9 (STU-**relation**)**.** The STU-relation for Jacobi diagrams is

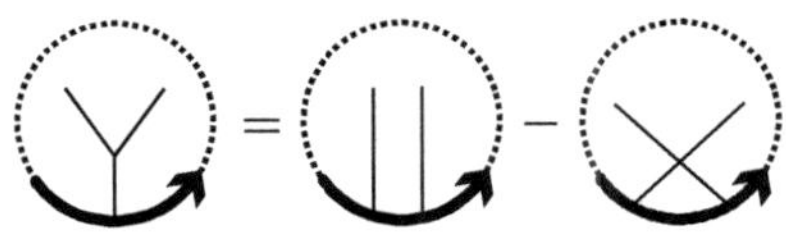

where the Jacobi diagrams are identical outside of the region shown, and where there are no vertices that meet the segment of the skeleton between the ends of the two edges on the bold arc in the middle diagram or the ends of the two edges on the bold arc in the diagram on the right.

The following observation is self-evident, but can easily be overlooked. There are two principal ways of applying the STU-relation: we can

- consider the edge joining the trivalent vertex to the skeleton in the diagram on the left-hand side of the equality, and then regard the relation as operating upon this (and therefore the incident trivalent vertex) to produce the expression on the right;
- or replace the linear combination on the right by the single diagram on the left.

The first operation decreases the number of trivalent vertices by one, while the second increases it by one. However, notice that the STU-relation does not change the degree of the Jacobi diagrams in it.

As instances of the use of the STU-relation, we consider the following examples. The first example is for a Jacobi diagram of degree 1, and the dotted region indicates where the STU-relation is being applied:

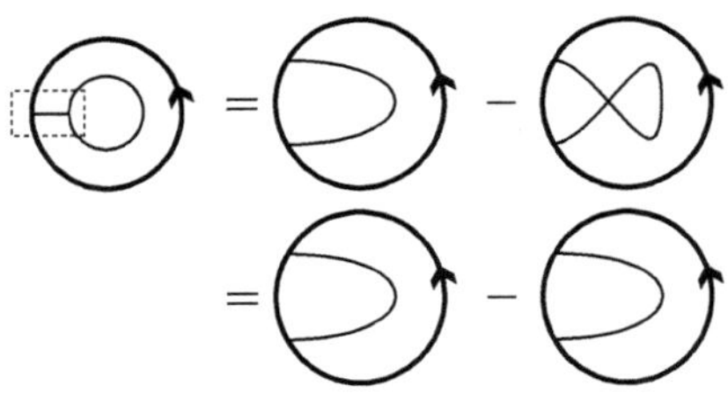

(STU)

('crossing' is fictitious)

$= 0.$

The second example is for a Jacobi diagram of degree 2. In this example, the STU-relation is used to eliminate trivalent vertices and thereby express a Jacobi diagram as a linear combination of chord diagrams.

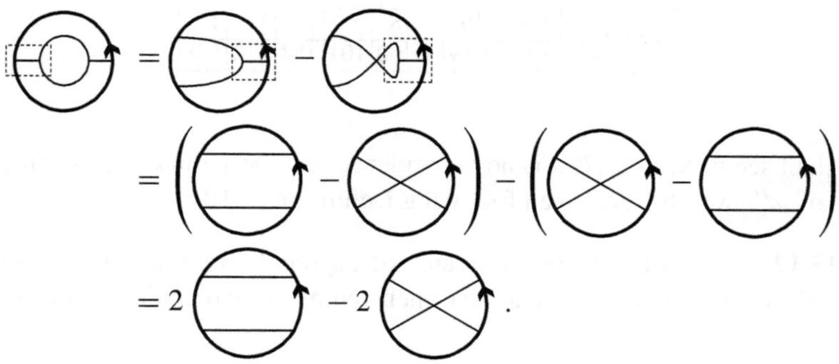

(13.5)

We may now introduce the vector space $\mathscr{A}$. This is the vector space of equivalence classes of all formal linear combinations of Jacobi diagrams where equivalence is given by the STU-relation.

Definition 13.10 (Vector spaces $\mathscr{A}$ and $\mathscr{A}_m$ of Jacobi diagrams). The space $\mathscr{A}$ of Jacobi diagrams satisfying the STU-relation is

$$\mathscr{A} := \frac{\mathscr{D}}{(\mathsf{STU})}$$

where (STU) denotes the subspace of $\mathscr{D}$ generated by the STU-relation. The subspace of $\mathscr{A}$ generated by degree m Jacobi diagrams is denoted by $\mathscr{A}_m$.

We have

$$\mathscr{A}_m = \frac{\mathscr{D}_m}{(\mathsf{STU})}.$$

Furthermore,

$$\mathscr{A} := \bigoplus_{m=0}^{\infty} \mathscr{A}_m.$$

Thus, two elements $[D]$ and $[D']$ of $\mathscr{A}$ are equal if and only if there is a finite sequence of applications of the STU-relation taking D to D'. Every Jacobi diagram D defines an element $[D]$ of $\mathscr{A}$. The element $[D]$ consists of all elements of $\mathscr{D}$ that may be obtained by sequences of applications of the STU-relation to D.

Convention 13.11. It is usual to represent an equivalence class of Jacobi diagrams $[D] \in \mathscr{A}$ by one of its members D. However, in this chapter we shall maintain a careful distinction between a Jacobi diagram D and its equivalence class $[D] \in \mathscr{A}$. Sometimes we specify the equivalence class by $[D]_{\mathsf{STU}}$ when there is potential for ambiguity. Furthermore, in this chapter we shall generally use the notation $D \stackrel{\mathsf{STU}}{=} D'$ to indicate that two elements of $\mathscr{D}$ are related by the STU-relation, in which case $[D] = [D']$.

The above examples of applications of the STU-relation, together with (13.4), show that $\dim(\mathscr{A}_1) \leq 1$ and $\dim(\mathscr{A}_2) \leq 2$. The dimensions of the spaces $\mathscr{A}_m$ for small values of m are shown in the following table.

m	0	1	2	3	4	5	6	7	8	9	10	11	12
$\dim(\mathscr{A}_m)$	1	1	2	3	6	10	19	33	60	104	184	316	548

As we shall see in Sect. 13.7, it is no coincidence that these dimensions coincide with those of $\mathscr{A}_m^c$, which were given following Definition 12.10.

Exercise 13.12. List all the Jacobi diagrams of degree 3. By applying the STU-relations to these diagrams, show that the dimension of $\mathscr{A}_3$ is bounded above by 3.

13.3 Identities from the STU-Relation

We recall from Definition 13.9 that the STU-relation states that

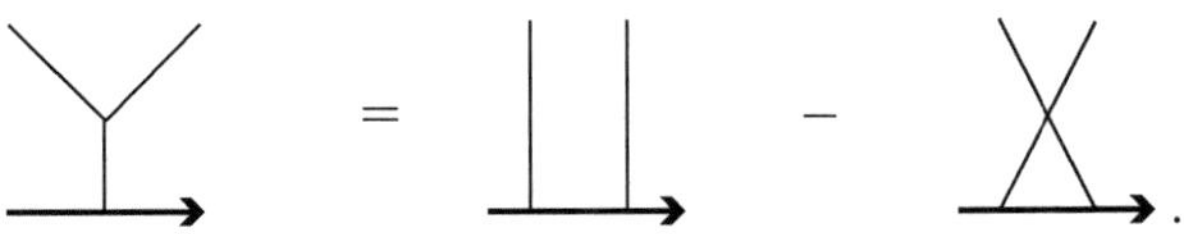

We shall prove several important identities for $\mathscr{A}$ by means of the STU-relation. We begin with an elementary identity.

From the STU-relation, we have

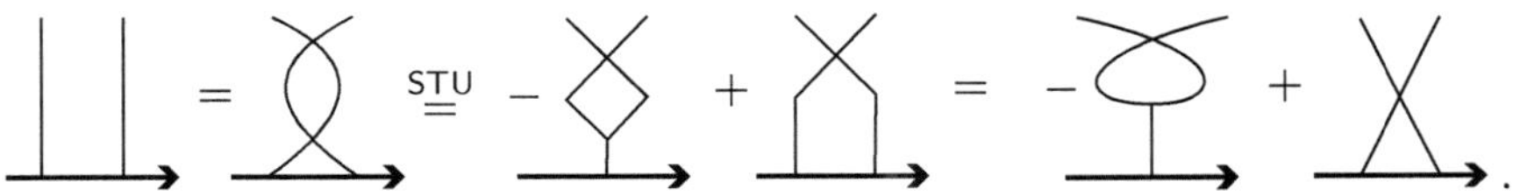

Then, substituting this expression for 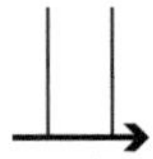 into the right-hand side of the STU-relation immediately gives an anti-symmetry relation

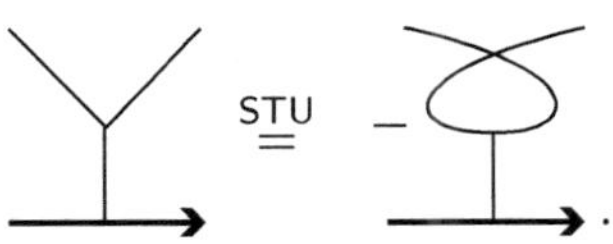

(13.6)

We shall shortly see (in Theorem 13.14) that this relation holds for all trivalent vertices, not just those adjacent to the skeleton.

Lemma 13.13. *The* STU*-relation implies the* 4T*-relation. In particular, the* 4T*-relation holds in* $\mathscr{A}$.

Proof. The proof is diagrammatic. Consider the first two terms of the 4T-relation given in (11.13). These terms are on the left-hand side of the next display. Then, by two applications of the STU-relation, we have

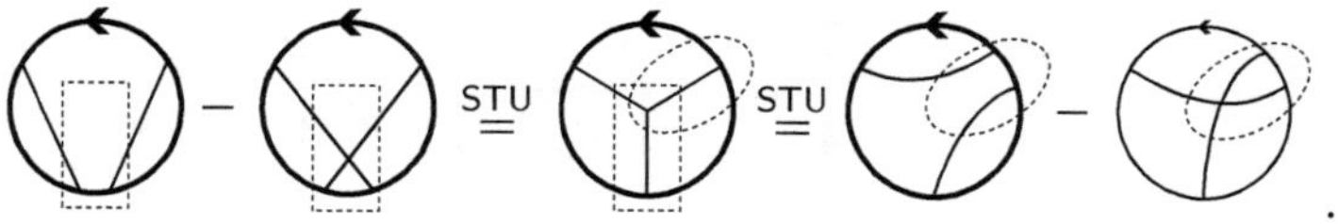

(13.7)

The first equality is obtained by the action of the STU-relation at the rectangular regions of the first two diagrams, to give the middle diagram whose rectangular region is the only region affected by the move. This move increases the number of trivalent vertices by one. The second equality is obtained by the action of the STU-relation applied to the middle diagram in the elliptical region. This move, which reduces the number of trivalent vertices by one, results in the linear combination of the rightmost two diagrams. The equality of the linear combination on the left and the linear combination on the right is precisely the 4T-relation. □

The AS-relation and IHX-relation in the theorem below will prove to be important in the subsequent development.

Definition 13.14 (AS- **and** IHX-**relations).** The AS- and IHX-relations for Jacobi diagrams are

1) AS: 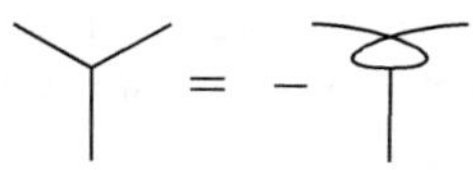 and 2) IHX: 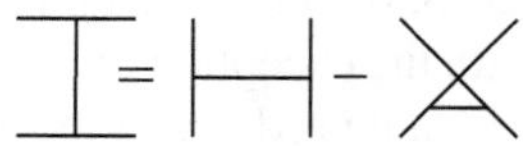

where the Jacobi diagrams are identical outside of the region shown.

As an example of the application of the AS-relation and the IHX-relation, consider

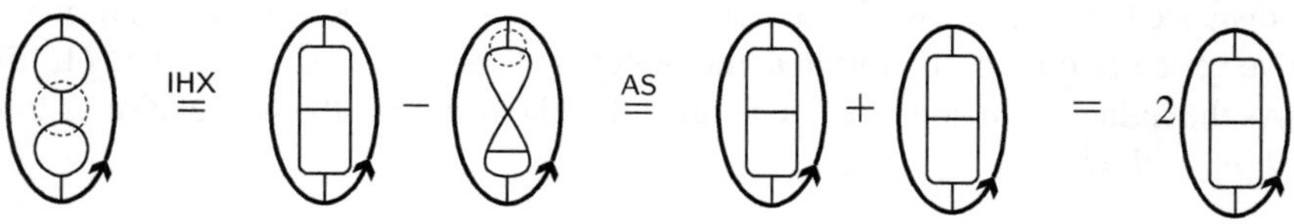

where the IHX-relation and AS-relation each act on the configuration in the dotted region of the diagram to their immediate left.

Recalling Convention 13.2 on the cyclic order at trivalent vertices, we note that we can represent the AS-relation as:

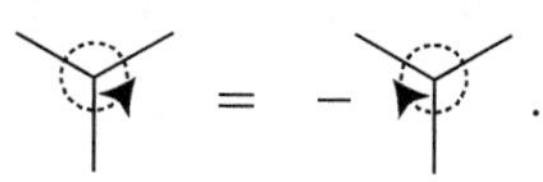

Theorem 13.15. *The* AS- *and* IHX-*relations hold in* $\mathscr{A}_m$ *and* $\mathscr{A}$.

Before proving Theorem 13.15, we introduce temporary notation for use in its proof.

1. We shall refer to the first of

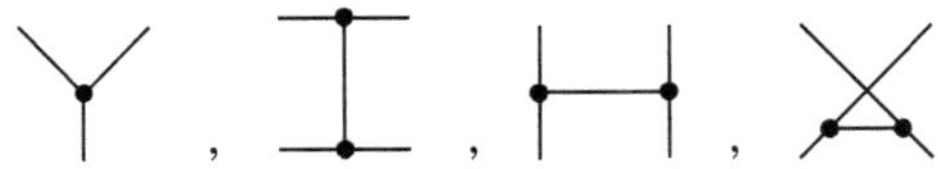

as an AS-*configuration*, and the remaining three as the IHX-*configurations*.

2. The "•"s are vertices that will be referred to as *active* vertices.

3. The *distance of a configuration* is the number of edges in a shortest path in a Jacobi diagram from the skeleton to an active vertex in the configuration. The distance is finite (since the Jacobi diagram is connected) and is greater than zero.

4. If a particular configuration, for example 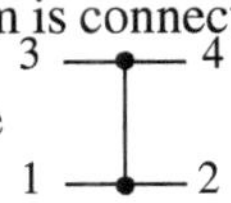 has distance k, then such a shortest path will be denoted by p_k.

5. The Jacobi diagram will be represented by 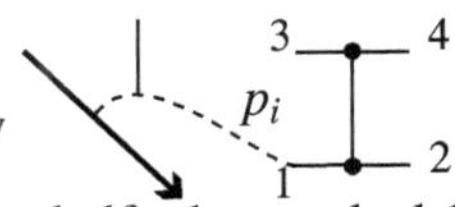 where the path p_i of i edges (shown as dotted) contains the half-edge marked 1 and where the rest of the Jacobi diagram has been suppressed (following our usual convention), with the exception of a half-edge at the trivalent vertex closest to the skeleton for it is at this edge that the STU-relation will be applied.

6. We shall show only an arc of the circular skeleton of the diagrams.

Proof. <u>*The* AS*−relation* </u>: As the base case, we consider a Jacobi diagram in which the distance for a particular AS-configuration is $k = 1$. Then there is an edge from the active vertex of this configuration to the skeleton. This is the case covered by (13.6).

As the inductive hypothesis, we assume that the result for the AS-relation is true for $k = n$. Now

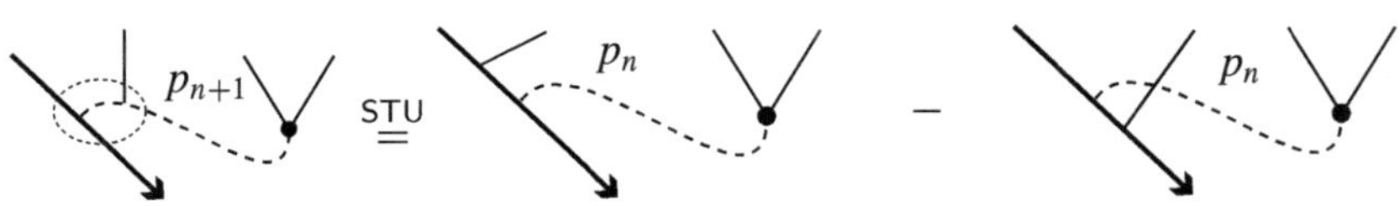

where the dotted ellipse indicates the configuration acted upon by the STU-relation.

Then, applying the inductive hypothesis to the right-hand side gives

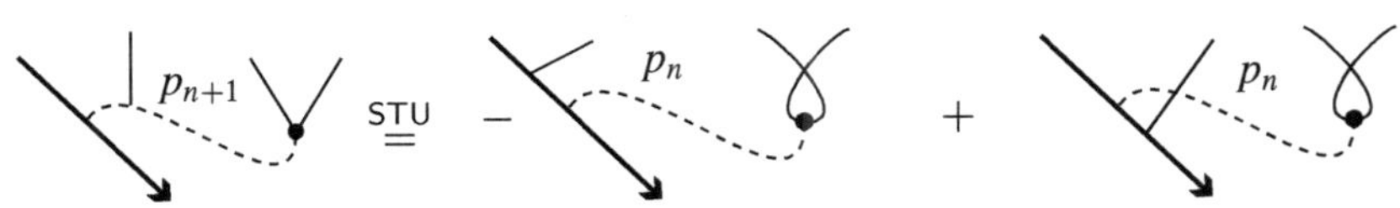

and so, again by the STU-relation,

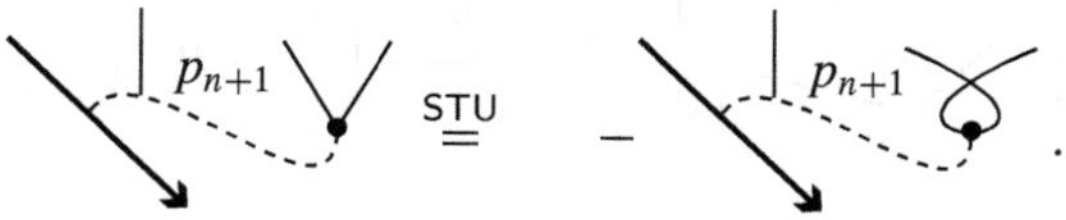

Thus, the result holds for the AS-relation.

The IHX*–relation* : For the inductive hypothesis, we shall assume that the result for the IHX-relation is true for $k = n$. With the exception of the base case, the proof of the inductive step is divided into four cases, one for each of the four edges of the I-configuration that is joined by the path p_{k+1} to the skeleton.

For each of the IHX-configurations in the IHX-relation

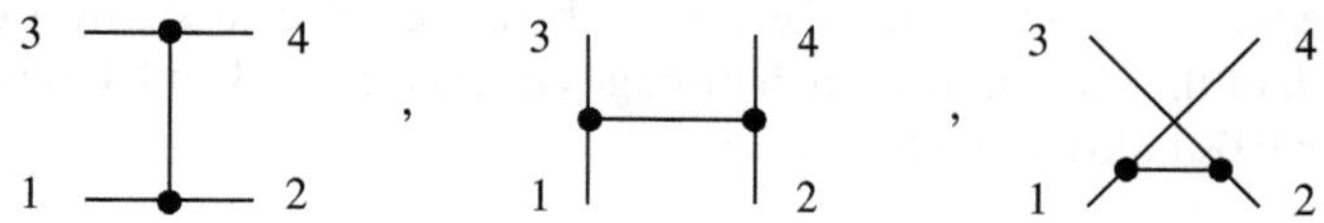

we shall consider shortest paths from the skeleton to it that contains the half-edge marked 1. A symmetry argument will then be used to dispatch more quickly the cases for the half-edges marked 2, 3 and 4.

Base case: As the base case, we consider

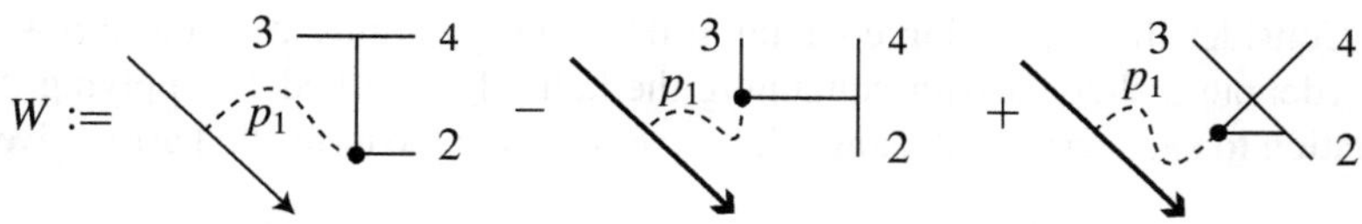

(13.8)

in which each Jacobi diagram has distance $k = 1$, (so p_1 consists of a single edge) We shall need the following diagrams that arise from these in the use of the STU-relation:

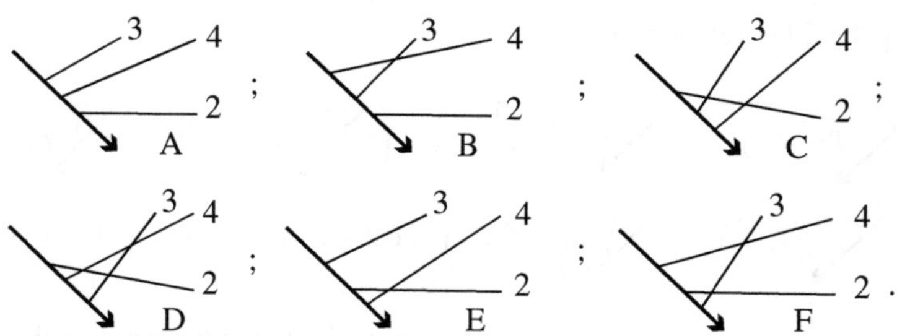

Then, applying the STU-relation to the first term in (13.8), we have

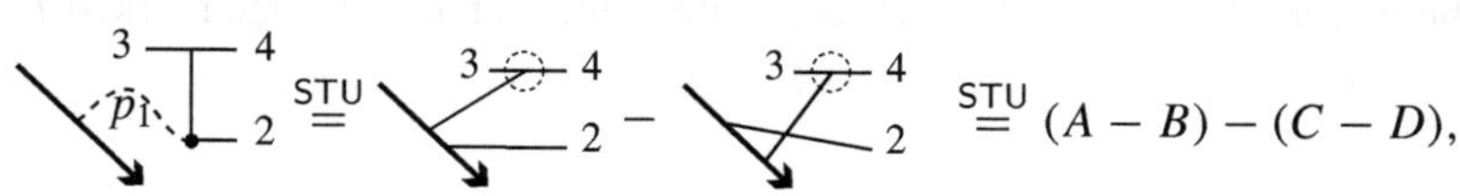

where a dotted circle in a diagram, here and in the next two cases, indicates the point at which the next STU-relation is applied.

Next, applying the STU-relation to the second term in (13.8), we have

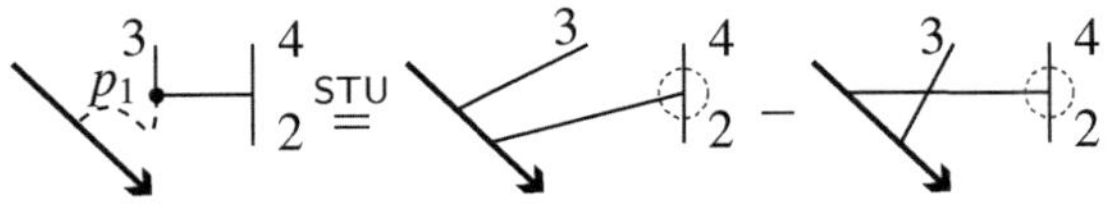

$\overset{\mathsf{STU}}{=} (A - E) - (F - D).$

Lastly, applying the STU-relation to the third term in (13.8), we have

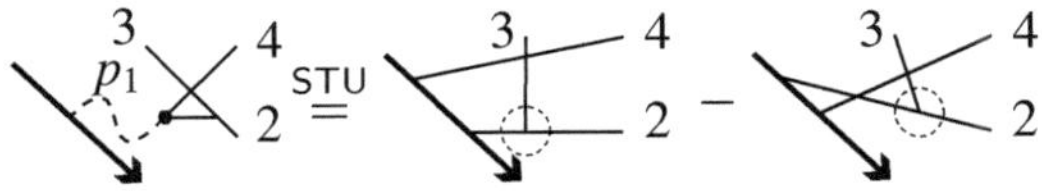

$\overset{\mathsf{STU}}{=} (B - F) - (E - C).$

Then, from (13.8), $W = (A - B - C + D) - (A - E - F + D) + (B - F - E + C) = 0$, so the result for IHX holds in the base case when the selected shortest edge-path contains the half-edge marked 1.

The arguments are similar for the cases where the selected shortest edge-path contains each of the remaining three half-edges marked by 2, 3 and 4 in turn. Thus, the result for IHX holds in the base case.

As the inductive hypothesis, we assume that the result for the IHX-relation is true for $k = n$. Within the induction, there are four case to consider depending upon the point of attachment of the shortest path p_{n+1}. Case 1 is the case in which 1 is the point of attachment. The remaining three cases are dealt with by reflecting or rotating the diagrams and then appealing to Case 1.

Case 1: Consider the expression containing IHX-configurations at distance $n + 1$. Let p_{n+1} denote a shortest path containing the half-edge marked 1. Applying the STU-relation to each term, and using $A', \ldots, F'$ to name the terms that arise, gives

$Y :=$ 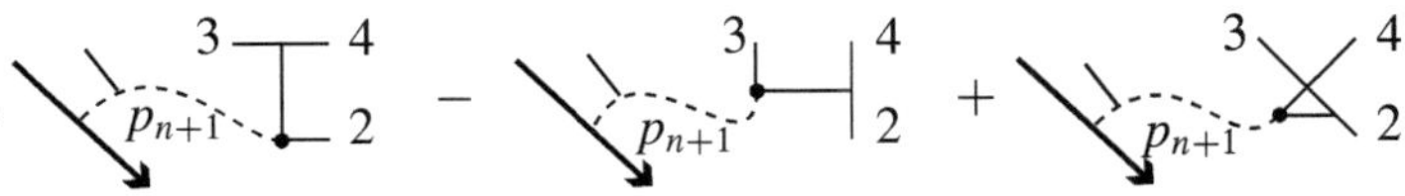

Applying the STU-relation gives

$Y =$ 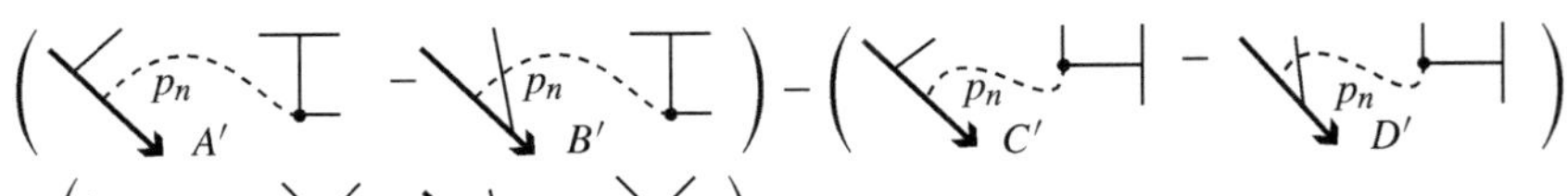

$+$ 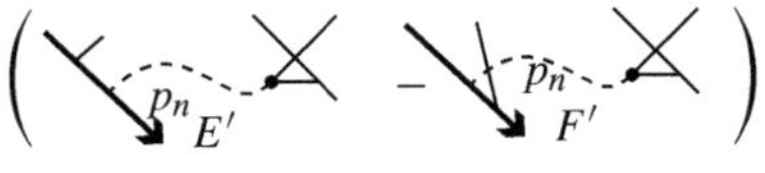

$= (A' - C' + E') - (B' - D' + F')$

upon rearranging the terms. Thus, $Y = 0$ since, under the inductive hypothesis, both $A' - C' + E'$ and $B' - D' + F'$ are 0 by the IHX-relation. The inductive step therefore holds for IHX when a shortest edge-path contains the half-edge marked 1.

We may express this case diagrammatically by

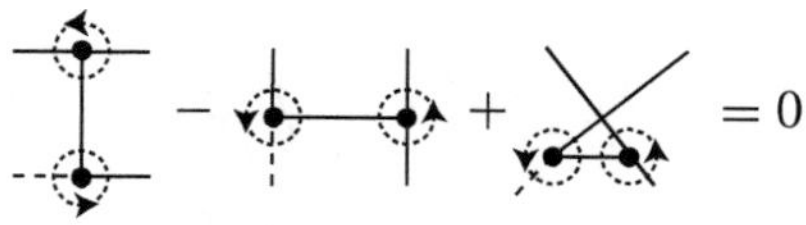 (13.9)

where the dotted half-edge is at the terminus of p_{n+1}, and where the orientation of the vertices is given explicitly. This is preparatory to the remaining three cases.
Case 2: By "flipping over" the drawing of each configuration of (13.9), we get 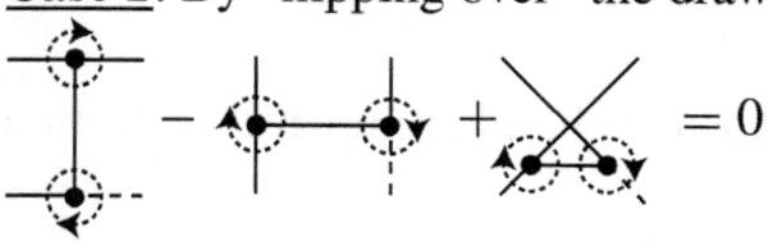 and then, by applying the AS-relation to each trivalent vertex, we obtain 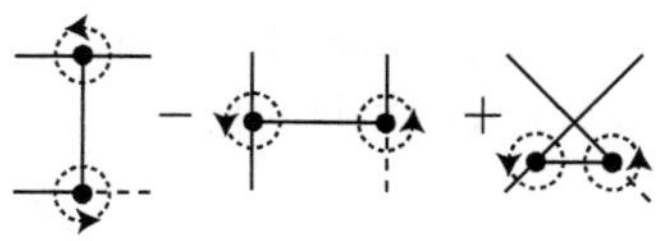 $= 0$. Thus, the inductive step holds in this case.

Case 3: Rotating each configuration of (13.9) through 180 degrees gives

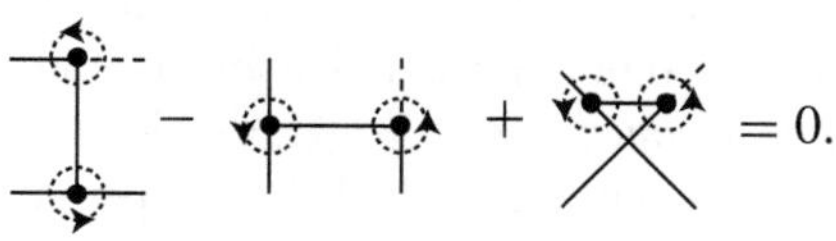

But 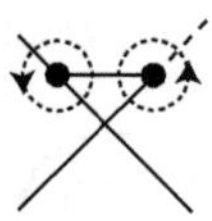 $\overset{\mathsf{AS}}{=} (-1)^2$ 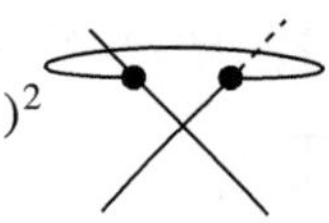 $=$ 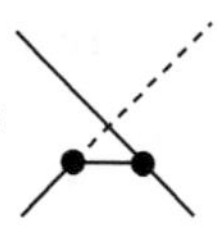 so the inductive step holds in this case.

Case 4: "Flipping over" the drawing of each configuration of (13.9), then applying the AS-relation once to each of the trivalent vertices, as in Case 2, gives 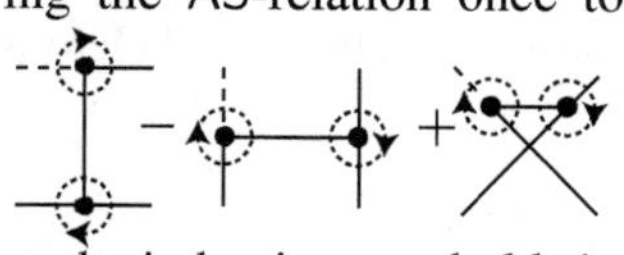 $= 0$. But $(-1)^2$ 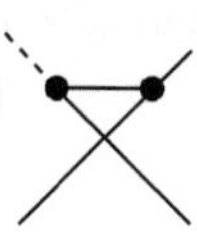 $=$ 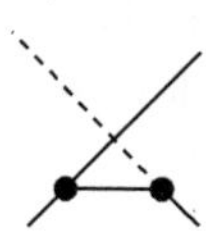, as in Case 3, so the inductive step holds in this case.

Thus, the inductive step holds for the case of IHX, so the result for the IHX-relation follows. This completes the proof. □

Proposition 13.16. *The following relations in $\mathscr{A}_m$ are equivalent forms of the* IHX-*relation.*

Jacobi form:

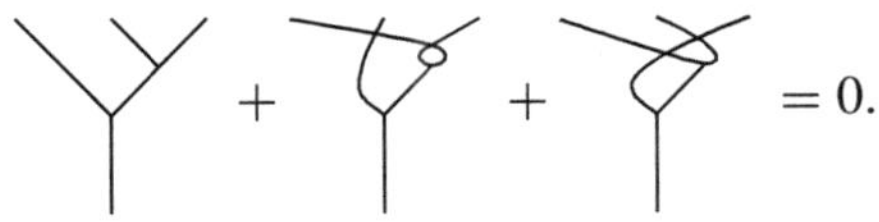

Symmetric form:

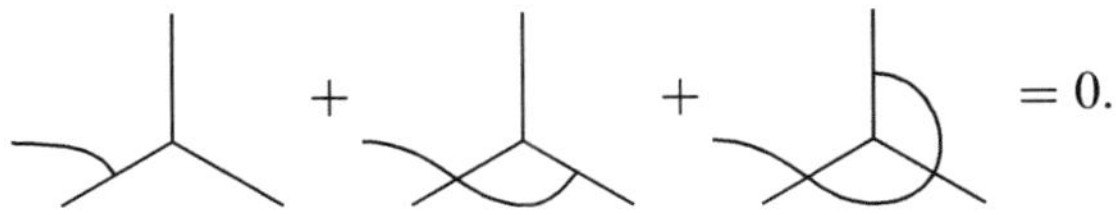

Exercise 13.17. Prove Proposition 13.16 by showing the identities follow from a rewriting of the IHX-relation.

(Note that the Jacobi form is so named because it is a diagrammatic version of the Jacobi relation in Lie algebras, as in [146].)

13.4 The Algebra Structure on $\mathscr{A}$

To turn the vector space $\mathscr{A}$ into an algebra, we must endow it with a product. The most obvious candidate for combining two Jacobi diagrams into a single Jacobi diagram is a form of *connect sum*, $\# : \mathscr{D} \otimes \mathscr{D} \to \mathscr{D}$ which will then induce a product $\# : \mathscr{A} \otimes \mathscr{A} \to \mathscr{A}$. That is, by cutting each skeleton open at some point to form a "gap", and then gluing the two cut skeletons together at their "gaps" to form a single skeleton. The following figure gives an example of the two stages of this construction:

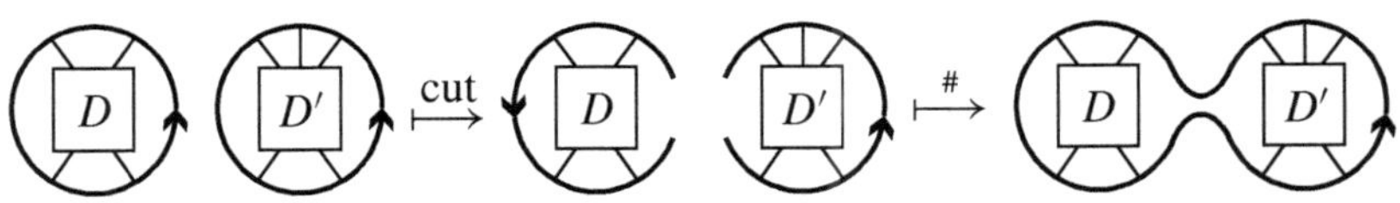

Immediately we run into a problem in defining such a product. The Jacobi diagram resulting from this process depends upon where we choose to cut the two skeletons open, as the following example demonstrates.

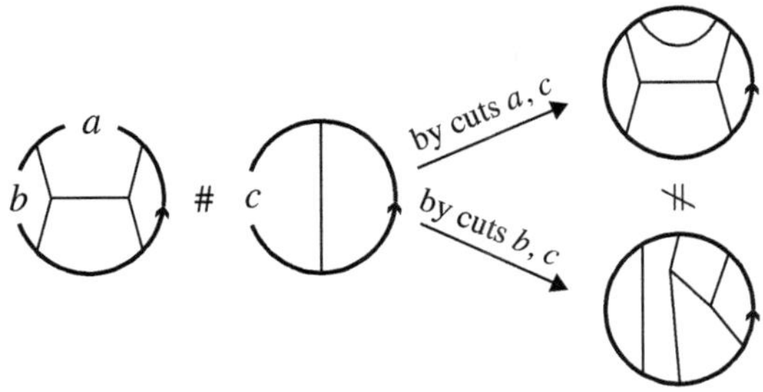

It therefore follows that the above procedure does not define a product $\mathscr{D} \otimes \mathscr{D} \to \mathscr{D}$.

However, we have not taken the STU-relation into account yet. We shall see that, although different choices of where to cut the skeletons open can result in different

Jacobi diagrams, such Jacobi diagrams are always related by the STU-relation. Thus, we do indeed have a product $\mathscr{A} \otimes \mathscr{A} \to \mathscr{A}$.

Exercise 13.18. Show that the two Jacobi diagrams on the right-hand side of the above figure are related by the STU-relation.

To proceed, we need to provide a formal definition of the product #. The first step in its formation is the choice of where to cut the skeletons. Thus, we need to consider Jacobi diagrams with a distinguished point, called a *base point*, on its skeleton. Such Jacobi diagrams are called *linear Jacobi diagrams*.

Definition 13.19 (Linearisation). A *linear Jacobi diagram* $(D, \mathfrak{l})$ consists of a Jacobi diagram D and a *base point* $\mathfrak{l}$ marked on its skeleton. The base point should not intersect any degree one vertices of the Jacobi diagram. In diagrams, the base point is denoted by the symbol "★".

A *linearisation* of a Jacobi diagram D is a linear Jacobi diagram $(D, \mathfrak{l})$ that results from choosing some base point of D.

Observe that the base point ★ of a linear Jacobi diagram provides a start and endpoint of the circular skeleton, and so we may represent the skeleton and base point of a linear Jacobi diagram as a directed line that starts and ends at the base point. We shall generally represent linear Jacobi diagrams in this way, as Jacobi diagrams with a skeleton that is a directed line. For example,

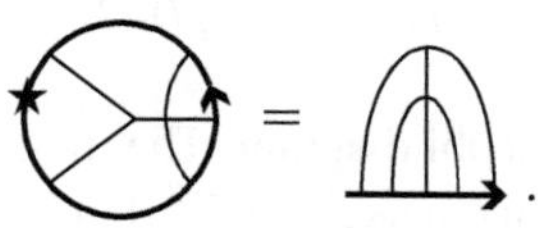

Definition 13.20 (Vector spaces $\mathscr{D}^l$ and $\mathscr{D}^l_m$ of linear Jacobi diagrams). $\mathscr{D}^l$ is the vector space over $\mathbb{C}$ that consists of all finite formal linear combinations of linear Jacobi diagrams over $\mathbb{C}$, and $\mathscr{D}^l_m$ is the subspace of $\mathscr{D}^l$ generated by linear Jacobi diagrams of degree exactly m.

We have,

$$\mathscr{D}^l = \bigoplus_{m=0}^{\infty} \mathscr{D}^l_m.$$

Let $D, D' \in \mathscr{D}^l_m$. Then we can define a product of these elements by concatenation:

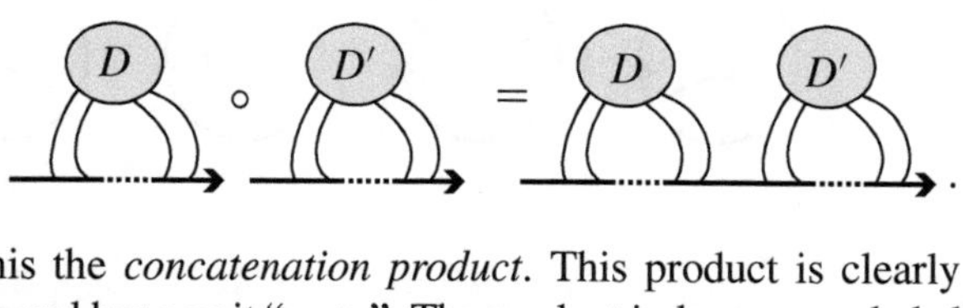

(13.10)

We call this the *concatenation product*. This product is clearly well-defined and associative and has a unit "⟶". The product induces a graded algebra structure on $\mathscr{D}^l_j$ since clearly

$$\circ\colon \mathscr{D}_i^l \otimes \mathscr{D}_j^l \to \mathscr{D}_{i+j}^l.$$

These observations are summarised in the following result.

Lemma 13.21. *$\mathscr{D}^l$ forms a graded, associative, non-commutative algebra, with unit, under concatenation product.*

To extend this product on $\mathscr{D}^l$ to a product on $\mathscr{A}$, given two Jacobi diagrams D and D', linearise them to get $(D, \mathfrak{l})$ and $(D', \mathfrak{l}')$, multiply them in $\mathscr{D}^l$ using the concatenation, and then forget the base point to obtain a Jacobi diagram. The problem with this construction is that, under it, different choices of base point in the linearisation may produce different Jacobi diagrams. However, we shall show that the resulting diagrams are related by the STU-relation and hence represent the same element in $\mathscr{A}$.

We shall state the result here but delay its proof until Sect. 13.6 where it will follow from more general considerations about linearised Jacobi diagrams.

Theorem 13.22. *The vector space $\mathscr{A}$ forms a graded commutative algebra with unit, where multiplication $\# : \mathscr{A} \otimes \mathscr{A} \to \mathscr{A}$ is defined as follows. Given Jacobi diagrams D and D' for each class, choose a linearisation $(D, \mathfrak{l})$ and $(D', \mathfrak{l}')$ of each diagram, form their product $(D, \mathfrak{l}) \circ (D', \mathfrak{l}')$ in $\mathscr{D}^l$ using concatenation, then obtain a Jacobi diagram $D\#D'$ by removing the linearisation. Then define $[D]_{\mathsf{STU}} \# [D']_{\mathsf{STU}}$ by*

$$[D]_{\mathsf{STU}} \# [D']_{\mathsf{STU}} := [D\#D']_{\mathsf{STU}}.$$

Exercise 13.23. Let D be a Jacobi diagram. The diagram D consists of a set Y of connected unitrivalent graphs and a skeleton $\mathbb{S}^1$. For $X \subseteq Y$, let D_X be the element of $\mathscr{D}$ obtained by deleting each of the unitrivalent graphs in $Y - X$ from D. Define an element $\Delta(D) \in \mathscr{D} \otimes \mathscr{D}$ by

$$\Delta(D) = \sum_{X \subseteq Y} D_X \otimes D_{Y-X} \tag{13.11}$$

For example,

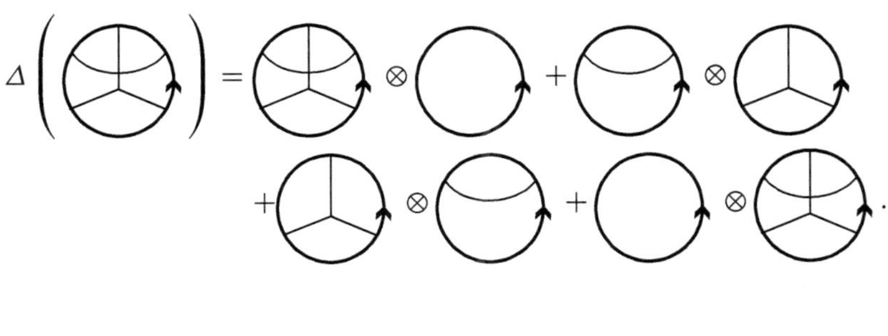

Show that $\Delta(D)$ induces a coassociative coproduct on $\mathscr{A}$ by

$$\Delta : \mathscr{A} \to \mathscr{A} \otimes \mathscr{A} : [D] \mapsto [\Delta(D)].$$

Hence show that, with this coproduct, $\mathscr{A}$ forms a commutative, cocommutative, graded bialgebra with unit and counit, where the counit is given by $\bar{\varepsilon}([D]) = 1$ if $[D] \in \mathscr{A}_0$, and $\bar{\varepsilon}([D]) = 0$ otherwise. In fact, it can be shown that $\mathscr{A}$ forms a Hopf algebra (see [13]).

13.5 Linearised Jacobi Diagrams and the Space $\mathscr{A}^l$

The algebra structure on $\mathscr{A}$ depends upon the properties of linearised Jacobi diagrams. In this section, we study linearised Jacobi diagrams in greater detail. In particular, we want to construct an analogue of the algebra $\mathscr{A} = \mathscr{D}/(\mathsf{STU})$ for the space $\mathscr{D}^l$ of formal linear combinations of linear Jacobi diagrams. To do so, it makes obvious sense to restrict the STU-relation so that it cannot be applied across the base point. In view of this, we define a new relation, the STU_l-relation.

Definition 13.24 (STU_l**-relation).** The STU_l-relation between Jacobi diagrams on a line segment is

STU_l :

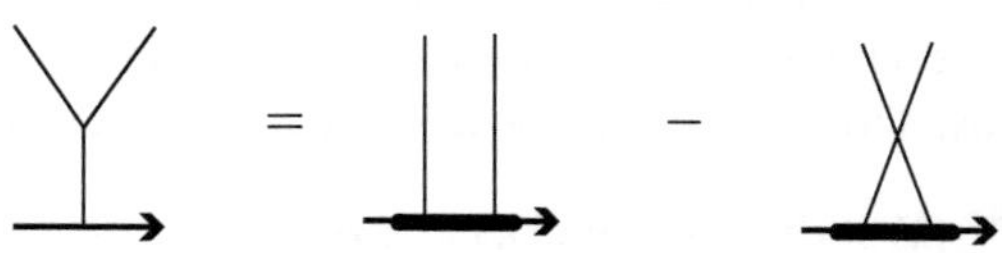

where the chord diagrams are identical outside the region shown, and where there are no other chords that meet the segment of the skeleton shown as bold in the figure.

We may now define our analogue of $\mathscr{A}$ for the space $\mathscr{D}^l$ freely generated by linear Jacobi diagrams.

Definition 13.25 (Vector spaces $\mathscr{A}^l$ and $\mathscr{A}^l_m$ of linear Jacobi diagrams). The space $\mathscr{A}^l$ of linear Jacobi diagrams satisfying the STU_l-relation is

$$\mathscr{A}^l := \frac{\mathscr{D}^l}{(\mathsf{STU}_l)}$$

where (STU_l) denotes the subspace of $\mathscr{D}^l$ generated by the STU_l-relation. The subspace of $\mathscr{A}^l$ generated by degree m linear Jacobi diagrams is denoted by $\mathscr{A}^l_m$.

We have

$$\mathscr{A}^l_m = \frac{\mathscr{D}^l_m}{(\mathsf{STU}_l)}.$$

Furthermore,

$$\mathscr{A}^l := \bigoplus_{m=0}^{\infty} \mathscr{A}^l_m.$$

The product on $\mathscr{D}^l$, given by concatenation (13.10), immediately induces a product on $\mathscr{A}^l$, and the following is readily seen to hold.

Theorem 13.26. *$\mathscr{A}^l$ endowed with concatenation product is a graded algebra with unit.*

Proof. The product is trivially well-defined, and the unit is clearly $[\longrightarrow]_{\mathsf{STU}_l}$. By inspection, the vector space $\mathscr{A}^l$ is graded since the STU_l-relation is grade preserving so any representative $D \in [D]_{\mathsf{STU}_l} \in \mathscr{A}^l_m$ has degree m. The algebra is graded since if $[D]_{\mathsf{STU}_l} \in \mathscr{A}^l_i$ and $[D']_{\mathsf{STU}_l} \in \mathscr{A}^l_j$, then $[D \circ D']_{\mathsf{STU}_l} \in \mathscr{A}^l_{i+j}$ since $[D \circ D']_{\mathsf{STU}_l}$ has degree $i + j$. □

The AS and IHX-relations of Theorem 13.15 also hold in $\mathscr{A}^l$. (This is since their proofs only considered an arc of the skeleton of the Jacobi diagram, and so the given proof of Theorem 13.15 is valid in $\mathscr{A}^l$.) The following two lemmas provide identities that are vital for understanding $\mathscr{A}^l$ and, as will become evident, $\mathscr{A}$.

The first identity says that univalent vertices may be moved along the linear skeleton of a Jacobi diagram which will in turn allow parts of a Jacobi diagram to be moved through each other. In this sense, part of a diagram may be "walked" along the skeleton past another part. The next result gives the essential step.

Lemma 13.27 (Walking Lemma 1). *Let B be part of a Jacobi diagram that contains no univalent vertices. Then*

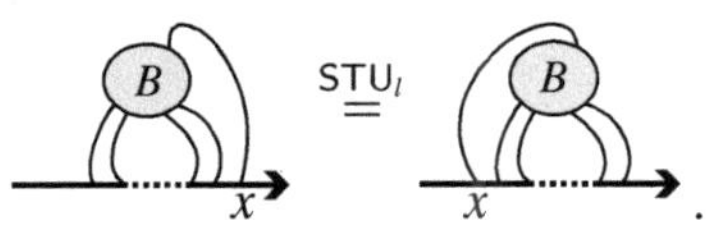

Proof. We consider the configuration $D :=$ 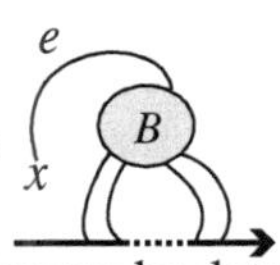

that is derived from the right-hand diagram in the enunciation of the lemma by detaching the univalent vertex x from the skeleton. In the construction to be described, the vertex x will be identified with other points on the skeleton in turn. Let e denote the edge terminated by x in the above diagram.

A set of weighted configurations is constructed from D as follows.

- First, the neighbourhood of each univalent and trivalent vertex of D is modified according to the rules

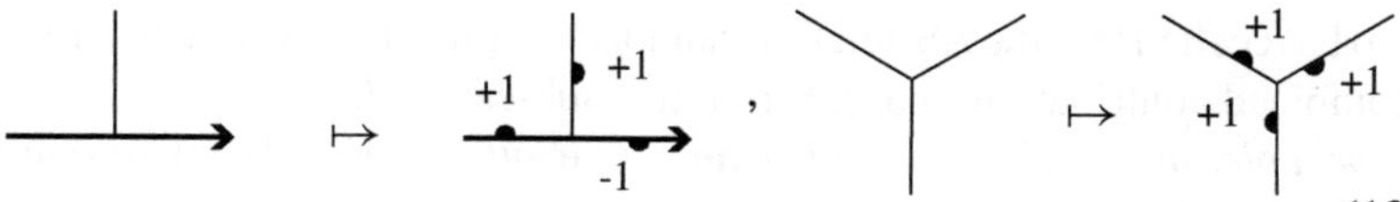

(13.12)

where the segment of the skeleton contains no other univalent vertices, the black semicircles are distributed, as shown, in the neighbourhood of each univalent and trivalent vertex, and these are weighted by $+1$ or -1 according to the patterns defined in (13.12); the black semicircles indicate points of attachment that may be occupied by x and are called *attachment points*.

- Second, x is assigned to each attachment point in turn, the resulting configuration is weighted by the weight of this point, and the black semicircles are then erased.
- Each configuration constructed in this way is a *signed Jacobi diagram* so there are as many of these as there are points of attachment associated with D. The resulting set of signed Jacobi diagrams is denoted in the proof by $\mathscr{M}_D$.

Let $S(D)$ denote the (weighted) sum of the elements in $\mathscr{M}_D$. Let B_1 and B_2 denote, respectively, the right-hand and left-hand diagrams in the statement of the lemma. To show that $B_1 \overset{\mathsf{STU}_l}{=} B_2$, we calculate $S(D)$ in two different ways:

1. *by summing over edges*: that is, by considering the pairs of attachment points on each edge;
2. *by summing over vertices*: that is, by considering the pair or triple of attachment points in the neighbourhood of each univalent and trivalent vertex;

and then equate the two results.

<u>*E: summing over edges*</u>: There are four sub-cases.

<u>*E1*</u>: *x occupies an attachment point at the ends of the skeleton (which is a line segment).* Then

$D_1 =$ 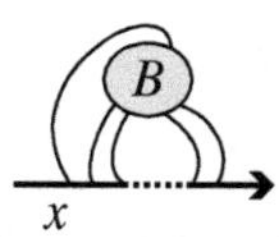

and $D_2 =$ 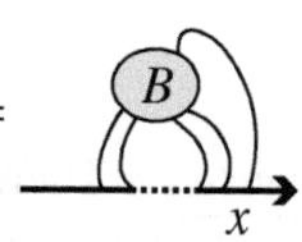

are clearly elements of $\mathscr{M}_D$ and therefore are terms in $S(D)$. The combined contribution to $S(D)$ from this sub-case is $D_1 - D_2$.

<u>*E2*</u>: *x occupies an attachment point on the skeleton between two univalent vertices.* The two contributions to $S(D)$ from e' are

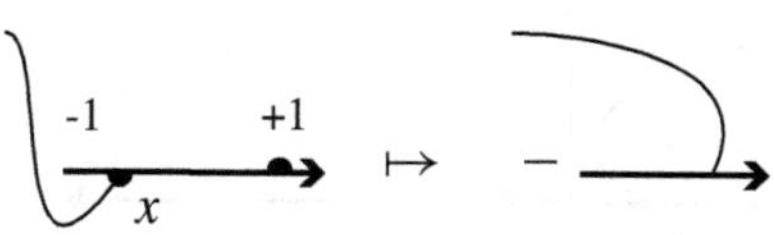

and

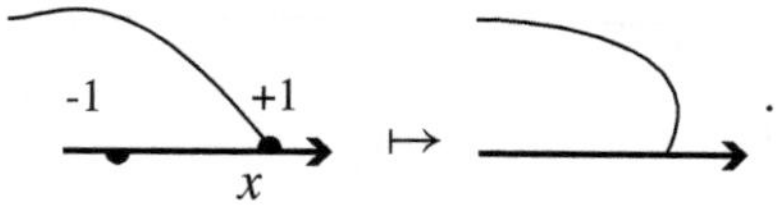

(It is to be recalled that the AS-relation cannot be applied to a univalent vertex.) Thus, the combined contribution to $S(D)$ from this sub-case is 0.

<u>E3</u>: *x occupies an attachment point on an edge e' other than e.* The two contributions to $S(D)$ from e' are

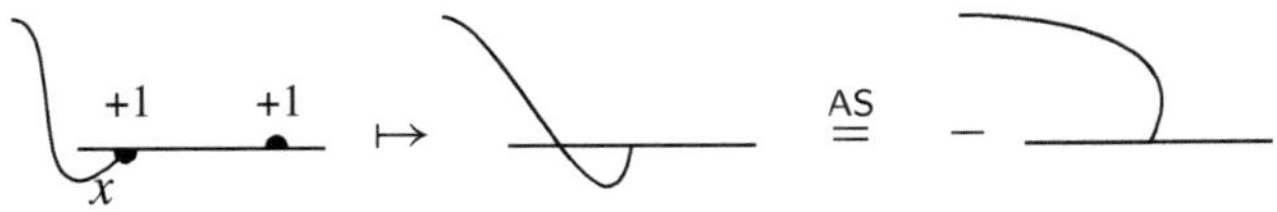

and

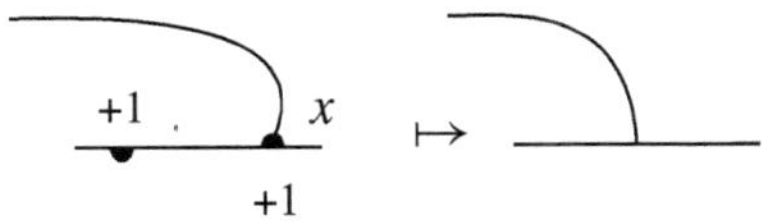

The combined contribution to $S(D)$ from this sub-case is 0.

<u>E4</u>: *x occupies the attachment point on the edge e.* Then

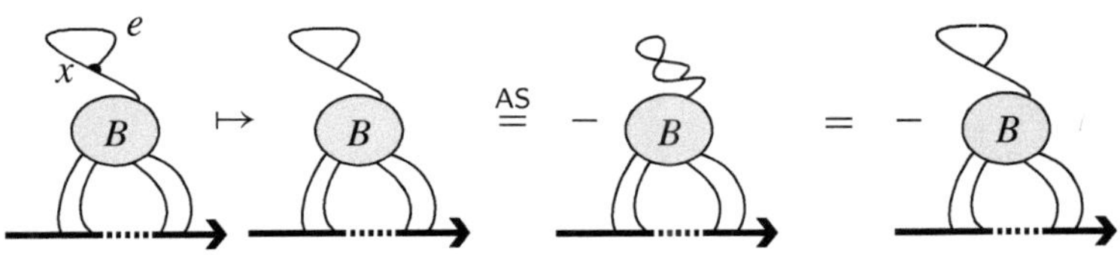

where the last equality follows by isomorphism of graphs, so this sub-case contributes 0 to $S(D)$.

Thus, the conclusion from these four sub-cases is that

$$S(D) = D_1 - D_2. \tag{13.13}$$

<u>*V: summing over vertices*</u>: There are two sub-cases.

<u>V1</u>: *x occupies an attachment point adjacent to a univalent vertex.* There are three contributions

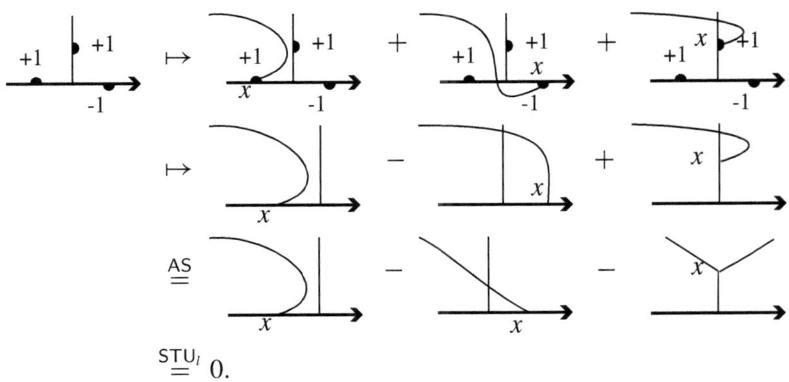

Thus, the contribution to $S(D)$ from this sub-case is 0.
<u>V2</u>: *x occupies an attachment point adjacent to a trivalent vertex.* There are three contributions

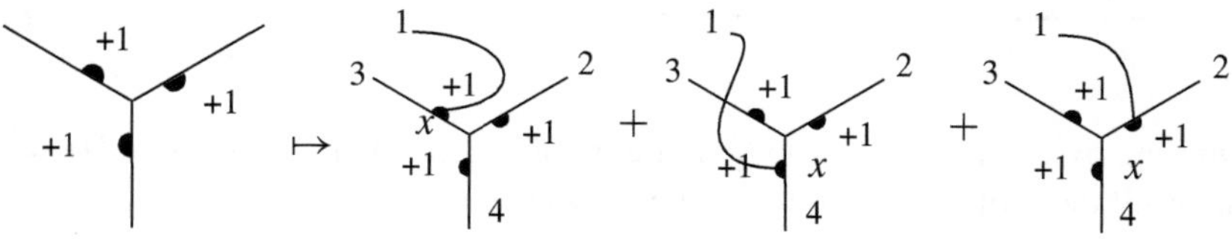

$$\overset{\mathsf{IHX}}{=} 0,$$

by Proposition 13.16. Thus, the contribution from this sub-case is 0, and the conclusion from these two sub-cases is therefore that

$$S(D) = 0. \tag{13.14}$$

The result now follows by equating the expressions (13.13) and (13.14) for $S(D)$ and recalling from Theorem 13.15 that the STU_l-relation implies both the AS- and the IHX-relations. □

The next result follows from the method of proof of the previous result.

Lemma 13.28 (Walking Lemma 2). *Let T and T' be two disjoint parts of a Jacobi diagram that contain no univalent vertices. Then*

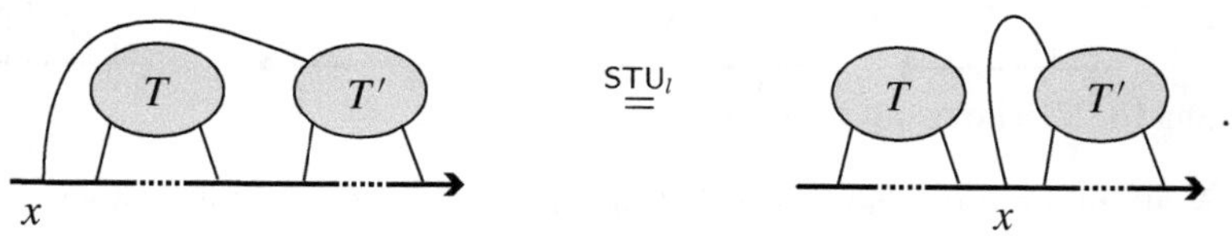

Proof. The weighting schemes for univalent vertices attached to T and for the segment of the skeleton between two adjacent univalent vertices are given in (13.12). Let b_1 and b_2 be the attachments points to the immediate left and immediate right of T. Consider the configuration

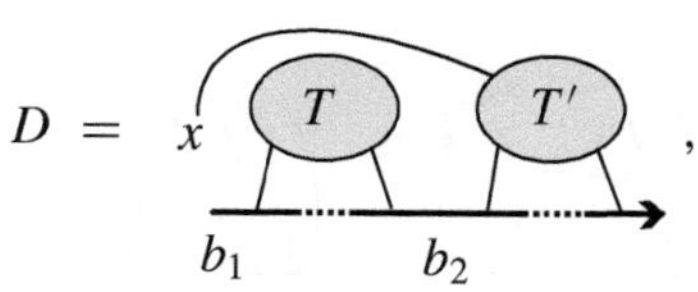

and let $S(D)$ be the signed sum of Jacobi diagrams obtained by allowing x to occupy each attachment point in turn. Then summing over edges as in Case E in the proof of Lemma 13.27 gives $S(D) = D_1 - D_2$ with

$D_1 =$ 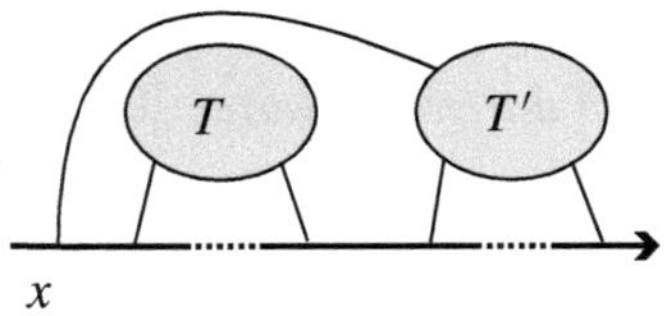

and $D_2 =$ 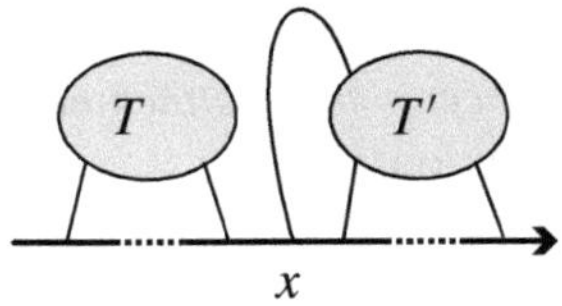

.

Summing over vertices as Case V in the proof of Lemma 13.27 gives $S(D) \overset{\mathsf{STU}_l}{=} 0$. The result then follows by equating these two expressions for $S(D)$. □

Corollary 13.29. *Let $(D, \mathfrak{l})$ and $(D, \mathfrak{l}')$ be two linearisations of a Jacobi diagram D. Then $(D, \mathfrak{l})$ and $(D, \mathfrak{l}')$ are related via the STU_l-relation.*

Proof. Let D be a Jacobi diagram and $\mathfrak{l}_1$ and $\mathfrak{l}_2$ be any base points defining the linearisations of D. It is sufficient to show that if the base points are separated by a univalent vertex on the skeleton, then the resulting Jacobi diagrams on a line segment are related by STU_l. Every Jacobi diagram D has the form

$D =$ 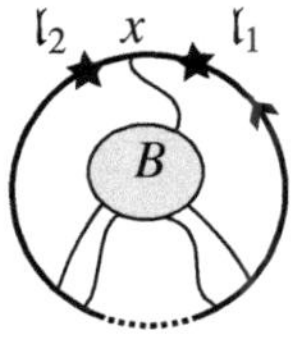

where the region B contains no univalent vertices. Opening at $\mathfrak{l}_1$ and $\mathfrak{l}_2$, respectively, gives 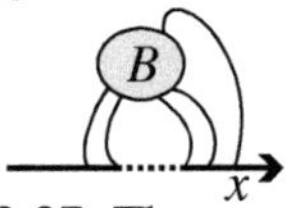 and 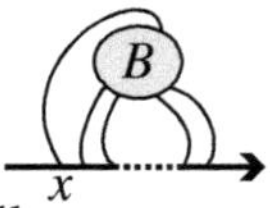. But 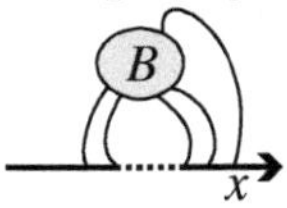 $\overset{\mathsf{STU}_l}{=}$ 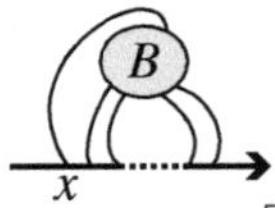, by Lemma 13.27. The result follows. □

The Walking Lemma also shows that the product on $\mathscr{A}^l$ is commutative.

Theorem 13.30. *$\mathscr{A}^l$ endowed with concatenation product is a graded, commutative algebra with unit.*

Proof. By Theorem 13.26, we know that $\mathscr{A}^l$ endowed with concatenation product is a graded algebra with unit. It remains to prove commutativity.

From Lemmas 13.27 and 13.28, we have

$D \circ D' =$ 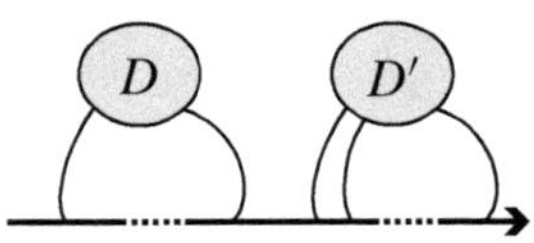 $\overset{\mathsf{STU}_l}{=}$ 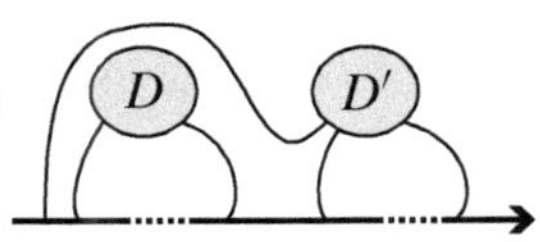

$\overset{\mathsf{STU}_l}{=} \cdots \overset{\mathsf{STU}_l}{=}$ 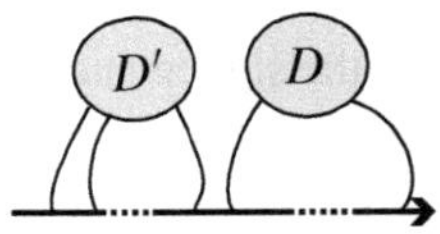 $= D' \circ D.$

This completes the proof. □

13.6 The Vector Spaces $\mathscr{A}$ and $\mathscr{A}^l$ are Isomorphic

The vector space $\mathscr{A}^l$ arose from the development of an algebra structure on $\mathscr{A}$. Here we will show that, in fact, these two spaces are isomorphic.

Clearly, there is a natural map

$$\lambda_m : \mathscr{D}^l_m \to \mathscr{D}_m$$

whose element-wise action on $(D, \mathfrak{l}) \in \mathscr{D}^l_m$ is to remove the linearisation $\mathfrak{l}$. This can be thought of as deleting the base point on a circular skeleton, or as identifying the endpoints of a linear skeleton of $(D, \mathfrak{l})$. For example,

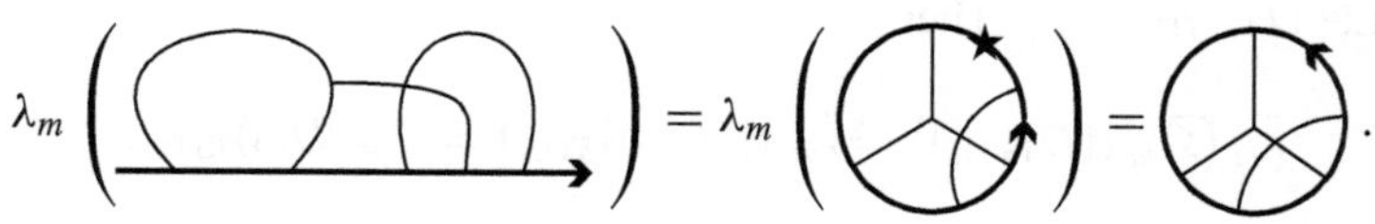

If p is the natural projection $\mathscr{D}_m \to \mathscr{D}_m/(\mathsf{STU})$, then, since λ_m is well-defined, so is

$$p \circ \lambda_m : \mathscr{D}^l_m \to \frac{\mathscr{D}_m}{(\mathsf{STU})}.$$

Lemma 13.31. *λ_m induces a linear map*

$$\overline{\lambda}_m : \mathscr{A}^l_m \to \mathscr{A}_m : [(D, \mathfrak{l})]_{\mathsf{STU}_l} \mapsto [\lambda_m(D, \mathfrak{l})]_{\mathsf{STU}} = [D]_{\mathsf{STU}}.$$

Proof. According to Lemma A.10, we need to show that $\lambda_m(\mathsf{STU}) = 0$. Now

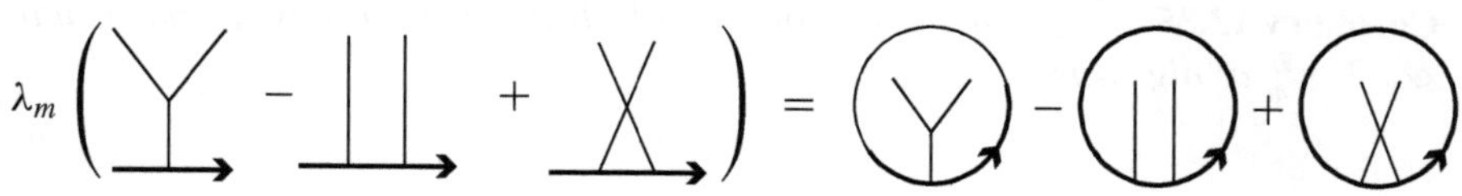

$$\stackrel{\mathsf{STU}}{=} 0.$$

□

Corollary 13.29 stated that for any Jacobi diagram D, any two linearisations are related by the STU_l-relation. This immediately gives the following.

Lemma 13.32. *There is a well-defined linear map*

$$\Omega_m : \mathscr{D}_m \to \mathscr{A}^l_m : D \mapsto [(D, \mathfrak{l})]_{\mathsf{STU}_l},$$

where $(D, \mathfrak{l})$ is any linearisation of D.

Lemma 13.33. *Ω_m induces a linear map*

$$\overline{\Omega}_m : \mathscr{A}_m \to \mathscr{A}^l_m : [D]_{\mathsf{STU}} \mapsto [(D, \mathfrak{l})]_{\mathsf{STU}_l}.$$

Proof. From Lemma 13.32, Ω_m is well-defined and it is easily seen that $\Omega_m(\mathsf{STU}) = \mathsf{STU}_l$. The result follows from Lemma A.10. □

Theorem 13.34. *$\overline{\Omega}_m$ and $\overline{\lambda}_m$ are inverse isomorphisms. In particular,*

$$\mathscr{A}_m \cong \mathscr{A}^l_m$$

as vector spaces.

Proof. Let $[D]_{\mathsf{STU}} \in \mathscr{A}_m$. Then

$$\overline{\lambda}_m \left(\overline{\Omega}_m \left([D]_{\mathsf{STU}}\right)\right) = \overline{\lambda}_m \left([(D, \mathfrak{l})]_{\mathsf{STU}_l}\right) = [\lambda_m(D, \mathfrak{l})]_{\mathsf{STU}}.$$

Moreover, $\lambda_m(D, \mathfrak{l}) = D$ since opening and closing the same diagram does not change the diagram. Thus, $\overline{\lambda}_m \circ \overline{\Omega}_m = \mathrm{id}$. A similar computation shows that $\overline{\Omega}_m \circ \overline{\lambda}_m = \mathrm{id}$. Hence, $\overline{\lambda}_m$ and $\overline{\Omega}$ are mutually inverse and hence isomorphisms. □

We are finally in a position to prove that the connected sum operation gives rise to an algebra structure on $\mathscr{A}$, as in Sect. 13.4.

Proof. (Theorem 13.22) The product of D and D' given in the statement of Theorem 13.22 is exactly a description of $\overline{\lambda}_m \left(\overline{\Omega}_m \left([D]_{\mathsf{STU}}\right) \cdot \overline{\Omega}_m \left([D']_{\mathsf{STU}}\right)\right)$, where the product is that of $\mathscr{A}^l$. Thus, the algebra structure on $\mathscr{A}$ is induced from that on $\mathscr{A}^l$. □

The following corollary is immediate.

Corollary 13.35. *$\overline{\Omega}_m$ and $\overline{\lambda}_m$ are inverse algebra isomorphisms. In particular, $\mathscr{A}_m \cong \mathscr{A}^l_m$ as algebras.*

13.7 The Vector Spaces $\mathscr{A}$ and $\mathscr{A}^c$ are Isomorphic

Recall that $\mathscr{C}$ is the vector space of chord diagrams (Definition 11.11), and $\mathscr{A}^c = \mathscr{C}/(4\mathsf{T})$ (Definition 12.10).

We begin with two observations about the appearance of chord diagrams in $\mathscr{A}$:

- by definition, every chord diagram is a Jacobi diagram, so *inclusion* gives a natural linear map $\iota : \mathscr{C} \to \mathscr{D}$;
- in the other direction, given any Jacobi diagram, we may repeatedly apply the STU-relation to express it as a linear combination of chord diagrams. We call such a rewriting a *resolution* of the Jacobi diagram. Specifically, by a *resolution* of a Jacobi diagram D, we mean any sequence of STU-relations that when applied to D results in a chord diagram.

For example,

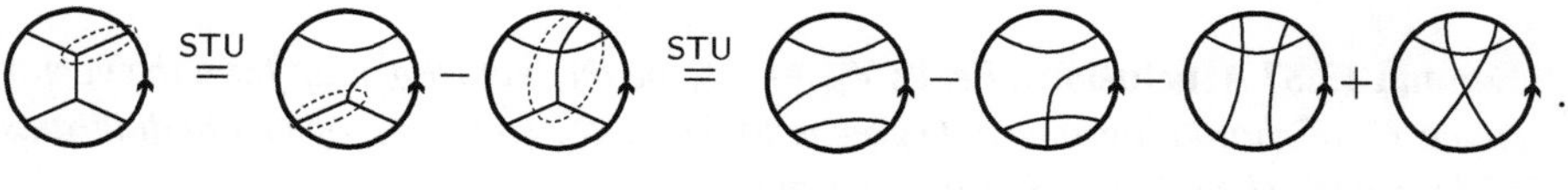

Our aim is to use inclusion and resolutions to define vector space isomorphisms between $\mathscr{A}^c = \mathscr{C}/(4\mathsf{T})$ and $\mathscr{A} = \mathscr{D}/(\mathsf{STU})$.

Before proceeding, we first decide what is needed to accomplish this.

- Starting with the inclusion map $\iota : \mathscr{C} \to \mathscr{D}$, and composing it with projection $p : \mathscr{D} \to \mathscr{D}/(\mathsf{STU})$ gives a linear map, $p \circ \iota : \mathscr{C} \to \mathscr{D}/(\mathsf{STU})$.
- To show this induces a well-defined map $\bar{\iota} : \mathscr{C}/(4\mathsf{T}) \to \mathscr{D}_m/(\mathsf{STU})$, by Lemma A.10 we must show that:

  if C and C' are elements of $\mathscr{C}$ related by the $4\mathsf{T}$-relation, then $p \circ \iota(C)$ and $p \circ \iota(C')$ are related by the STU-relation.

  Equivalently, we need to show if $C \overset{4\mathsf{T}}{=} C'$ then $C \overset{\mathsf{STU}}{=} C'$. We shall show this in Lemma 13.37, thus proving that inclusion induces a linear map $\bar{\iota} : \mathscr{A}^c \to \mathscr{A}$.

Defining a map from $\mathscr{A}$ to $\mathscr{A}^c$ using resolutions, on the other hand, is trickier since resolutions use STU-relations, and different ways of applying them to a Jacobi diagram result in different elements in $\mathscr{C}$. Thus, the resolution operation itself *is not* a well-defined map from $\mathscr{D}$ to $\mathscr{C}$. However,

- we shall show in Lemma 13.40 that, if different resolutions result in different elements of $\mathscr{C}$, then those elements are related by the $4\mathsf{T}$-relation.
- Thus, resolutions induce a linear map $\rho : \mathscr{D} \to \mathscr{C}/(4\mathsf{T})$.
- To obtain a linear map $\bar{\rho} : \mathscr{D}/(\mathsf{STU}) \to \mathscr{C}/(4\mathsf{T})$, we must show that if $D \overset{\mathsf{STU}}{=} D'$ then $\rho(D) \overset{4\mathsf{T}}{=} \rho(D')$, which we do in the proof of Theorem 13.36 at the end of this section, hence showing that resolutions induce a linear map $\bar{\rho} : \mathscr{A} \to \mathscr{A}^c$.

Furthermore, the two linear maps $\bar{\iota} : \mathscr{A}^c \to \mathscr{A}$ and $\bar{\rho} : \mathscr{A} \to \mathscr{A}^c$ define mutually inverse vector space isomorphisms. These observations are recorded in the following theorem.

Theorem 13.36. *The vector spaces $\mathscr{A}$ and $\mathscr{A}^c$ are isomorphic,*

$$\mathscr{A} \cong \mathscr{A}^c.$$

Moreover,

1. *inclusion induces a vector space isomorphism $\bar{\iota} : \mathscr{A}^c_m \to \mathscr{A}_m$;*
2. *resolution induces a vector space isomorphism $\bar{\rho} : \mathscr{A}_m \to \mathscr{A}^c_m$;*
3. *and these isomorphisms are inverses of each other.*

The remainder of this section is taken up with the proof of this result, through the following three lemmas.

Lemma 13.37 (Inclusion). *Let* $\iota : \mathscr{C}_m \to \mathscr{D}_m$ *be the inclusion map from the vector space* $\mathscr{C}_m$ *of chord diagrams of degree* m *to the vector space* $\mathscr{D}_m$ *of Jacobi diagrams of degree* m*. Then there exists a linear map*

$$\bar{\iota} : \mathscr{A}^c_m \to \mathscr{A}_m : [C]_{4\mathsf{T}} \mapsto [\iota(C)]_{\mathsf{STU}}.$$

Proof. The inclusion map ι may be composed with the natural projection map $p : \mathscr{D}_m \to \frac{\mathscr{D}_m}{(\mathsf{STU})} : D \mapsto [D]_{\mathsf{STU}}$ to give

$$p \circ \iota : \mathscr{C}_m \to \frac{\mathscr{D}_m}{(\mathsf{STU})} : D \mapsto [\iota(D)]_{\mathsf{STU}}.$$

Then, from Lemma 13.13, $(p \circ \iota)(4\mathsf{T}) = 0$. Thus, from Lemma A.10, $p \circ \iota$ induces a map $\bar{\iota}$.

$$\bar{\iota} : \frac{\mathscr{C}_m}{(4\mathsf{T})} \to \frac{\mathscr{D}_m}{(\mathsf{STU})} : [D]_{4\mathsf{T}} \mapsto [\iota(D)]_{\mathsf{STU}}.$$

The result now follows since $\mathscr{A}^c_m = \mathscr{C}_m/(4\mathsf{T})$ and $\mathscr{A}_m = \mathscr{D}_m/(\mathsf{STU})$. □

We next consider the use of resolutions to define a map $\overline{\rho} : \mathscr{A} \to \mathscr{A}^c$. To formalise our terminology, given a Jacobi diagram D of degree m with a trivalent vertex v, we define

- a <u>*resolution of D at v*</u> to be an application of the STU-relation that removes v;
- a <u>*resolution of D*</u> to be *any* sequence of applications of the STU-relation that when applied to D results in an element of $\mathscr{C}_m$.

If r is a resolution of D, then this resulting element of $\mathscr{C}_m$, which is a linear combination of chord diagrams of degree m, is denoted by $r(D)$. In general, different resolutions will result in different elements of $\mathscr{C}_m$. However, we shall now prove that these resolutions are equivalent modulo the 4T-relation and therefore each resolution of D results in the same element of $\mathscr{A}^c_m = \mathscr{C}_m/(4\mathsf{T})$. The proof of this result will follow from Lemmas 13.38 and 13.39.

Our proof that $\overline{\rho} : \mathscr{A} \to \mathscr{A}^c$ is well-defined will use induction on the number of trivalent vertices in a Jacobi diagram. In order to do this, we shall need to ensure that applications of the STU-relation do not result in too many trivalent vertices. We therefore introduce the notation that for any $D, D' \in \mathscr{D}$ such that each term has at most t trivalent vertices, by $D \overset{\mathsf{STU}_{\leq t}}{=} D'$ we mean that D and D' are related by a sequence of applications of the STU-relation with the property that no Jacobi diagram appearing in this sequence has more than t trivalent vertices.

Lemma 13.38. *Let $D, D' \in \mathscr{D}$ be such that no term of D or D' has more than $t \geq 0$ trivalent vertices. If $D \overset{4T,STU_{\leq t+1}}{=} D'$ then $D \overset{4T,STU_{\leq t}}{=} D'$.*

Proof. We consider sequences of applications of 4T- and STU-relations

$$D = D_1 \overset{4T,STU_{\leq t+1}}{=} D_2 \overset{4T,STU_{\leq t+1}}{=} \cdots \overset{4T,STU_{\leq t+1}}{=} D_n = D',$$

taking D to D' where each D_i is obtained from D_{i-1} by a single application of the 4T-relation or STU-relation. We assume all sequences of 4T- and STU-moves (*i.e.* applications of the 4T- or STU-relations) we write in this proof are of this type.

We proceed by induction on the total number τ of Jacobi diagrams with $t+1$ trivalent vertices in all of $D_1, \ldots, D_n$. Recall that each D_i is a linear combination of Jacobi diagrams.

For the base of the induction, if $\tau = 0$ in any sequence

$$D = D_1 \overset{4T,STU_{\leq t+1}}{=} \cdots \overset{4T,STU_{\leq t+1}}{=} D_n = D'$$

then every Jacobi diagram has at most t trivalent vertices, giving $D \overset{4T,STU_{\leq t}}{=} D'$.

For the inductive hypothesis, suppose that, for every sequence of 4T- and STU-moves $D = D_1 \overset{4T,STU_{\leq t+1}}{=} D_2 \overset{4T,STU_{\leq t+1}}{=} \cdots \overset{4T,STU_{\leq t+1}}{=} D_n = D'$ such that no more than $\tau = k$ Jacobi diagrams have $t+1$ trivalent vertices, we have $D \overset{4T,STU_{\leq t}}{=} D'$.

Then for the inductive step, suppose that in a sequence

$$D = D_1 \overset{4T,STU_{\leq t+1}}{=} \cdots \overset{4T,STU_{\leq t+1}}{=} D_n = D', \tag{13.15}$$

there are exactly $\tau = k+1 \geq 1$ Jacobi diagrams with $t+1$ trivalent vertices. Suppose that D_i is the first term in this sequence that contains a Jacobi diagram with $t+1$ trivalent vertices. Denote this Jacobi diagram by J. We will use this J throughout the proof. By examining how the 4T- and STU-moves act on J, it must be that a trivalent vertex v_1 is created in the move taking D_{i-1} to D_i. Then 4T-moves act on J until at some point a trivalent vertex v_2 of J is resolved (this is since no diagram has more than $t+1$ trivalent vertices). We shall focus upon J.

Claim: *The sequence*

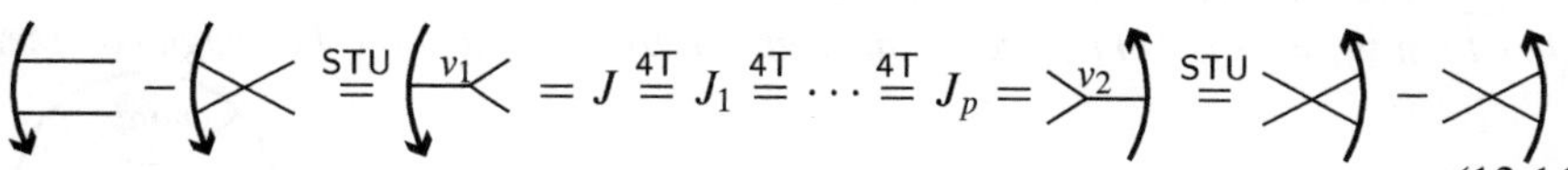

(13.16)

may be replaced by a sequence of 4T- and STU-moves with Jacobi diagrams that each have at most t trivalent vertices.

We assume the truth of the claim for the moment and explain how the lemma follows from it. First, observe that the sequence in (13.16) might not appear in (13.15) in that exact form, but some linear combination will appear in (13.15), possibly with some additional 4T- and STU-moves that act on components other than those in this linear combination. Then, by the claim, we can replace all the Jacobi diagrams that appear in this linear combination by sequence of 4T- and STU-moves with Jacobi diagrams that each have at most t trivalent vertices. The lemma then follows by induction.

It remains to prove the claim.

Proof of the claim. There are five cases to consider, according to how the two vertices v_1 and v_2 are related to each other. For these cases, let 1 be the edge connecting v_1 to the skeleton, and 2 be the edge connecting v_2 to the skeleton. The cases are:

<u>*Case 1*</u>: *v_1 and v_2 are different trivalent vertices. The Jacobi diagram then has the form $J =$* 

.

<u>*Case 2*</u>: *v_1 and v_2 are the same trivalent vertex, edges 1 and 2 are the same. The Jacobi diagram then has the form $J =$* 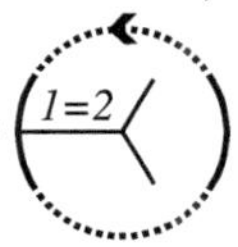

..

<u>*Case 3*</u>: *v_1 and v_2 are the same trivalent vertex, edges 1 and 2 are different, and there is an edge marked 3 that connects a trivalent vertex other than $v_1 = v_2$ to the skeleton. The Jacobi diagram then has the form $J =$* 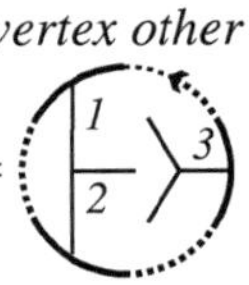

.

<u>*Case 4*</u>: *v_1 and v_2 are the same trivalent vertex, edges 1 and 2 are different, no vertex other than $v_1 = v_2$ is connected to the skeleton by an edge, but a third edge 3 connects $v_1 = v_2$ to the skeleton. The Jacobi diagram then has the form $J =$* 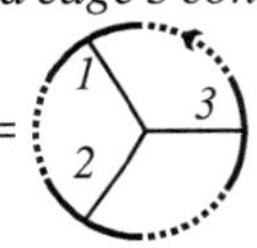

.

<u>*Case 5*</u>: *v_1 and v_2 are the same trivalent vertex, edges 1 and 2 are different, no vertex other than $v_1 = v_2$ is connected to the skeleton by an edge, no other edge connects $v_1 = v_2$ to the skeleton. The Jacobi diagram then has the form $J =$* 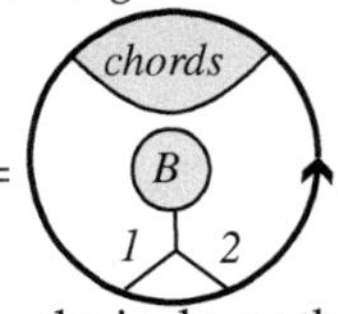

.

We analyse each of these cases in turn. Note that we can focus exclusively on the components on which the STU-relation acts, since the 4T-relation does not involve trivalent vertices.

For Case 1. In this case it must be that the sequence of STU-moves creating v_1 and resolving v_2 is

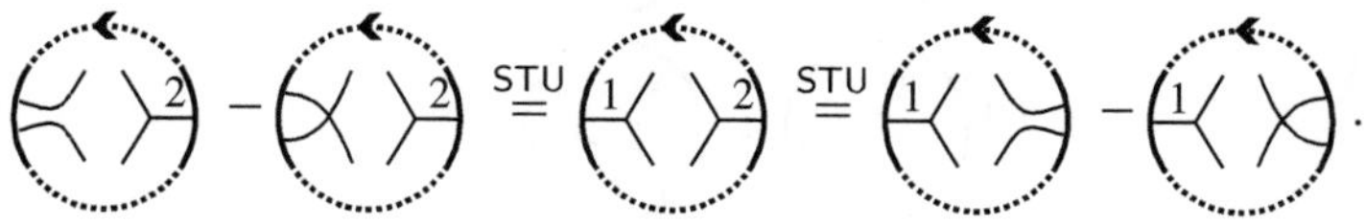

(13.17)

But we may apply the STU-relation in the following way to avoid creating v_1.

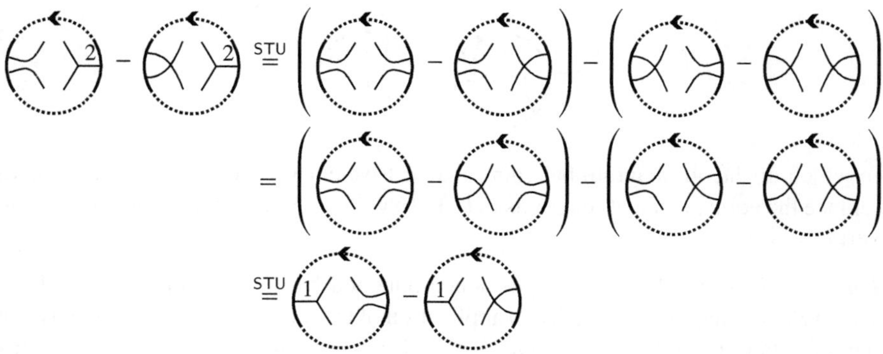

We may use this to replace the equivalence of diagrams in (13.16) with diagrams with having at most t trivalent vertices.

For Case 2. In this case, the STU-moves act like

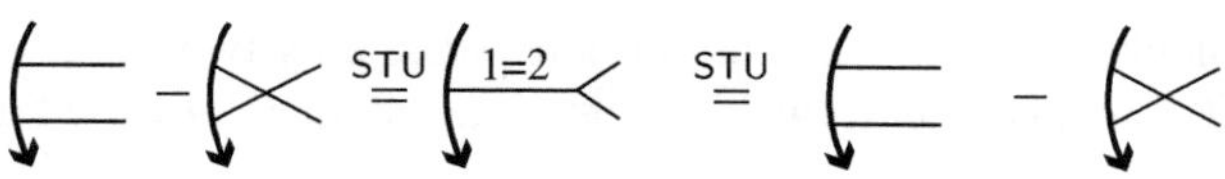

In this case, there is no need to create J and we may replace the occurrence of this configuration by the one in the left-hand side of (13.16).

For Case 3. First suppose that edges 1 and 2 are different and that there is third trivalent vertex adjacent to the skeleton. In this case, it must be that the sequence of STU-moves creating v_1 and resolving $v_2 = v_1$ is

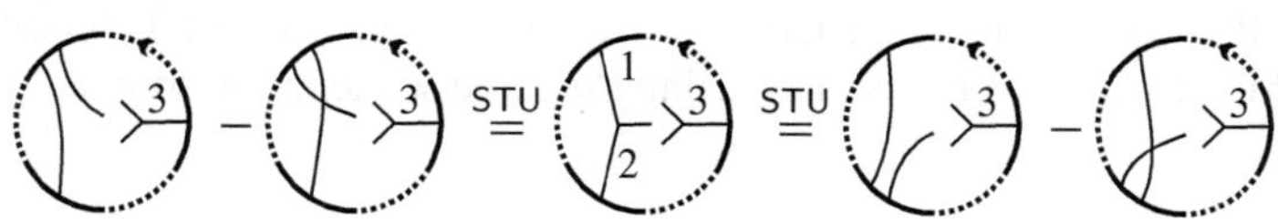

(13.18)

But by applying the STU-relation at the edge 3 first, we may achieve this equivalence by

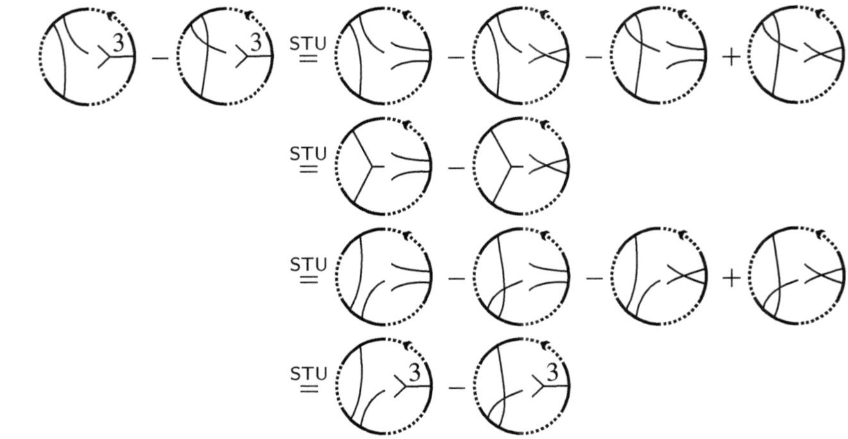

in which no Jacobi diagram has more than k trivalent vertices. We may use this to replace the equivalence of diagrams in (13.16) with diagrams with at most t trivalent vertices.

For Case 4. As in (13.7), the equivalence obtained by the two applications of the STU-relation may be attained by an application of the 4T-relation. Thus, we may use the 4T-relation to replace the equivalence of diagrams in (13.16). These diagrams have at most t trivalent vertices.

For Case 5. We may represent J as 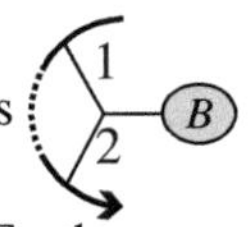 where all trivalent vertices are contained in the block denoted by B. Furthermore, since only two edges connect v_1 to the skeleton, J contains at least two trivalent vertices; thus, for Case 5 to occur, we must have $t \geq 1$.

The Jacobi diagram J is created by the STU-move

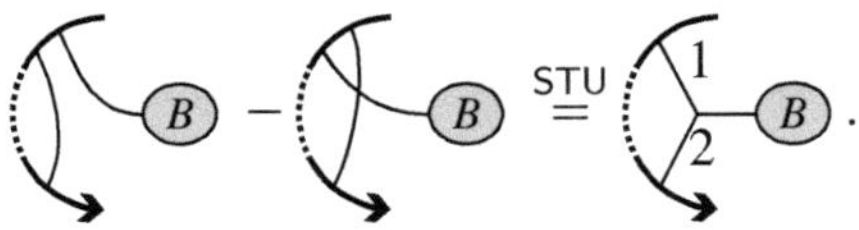

We focus on the expression on the left. We may use a sequence of STU-moves to resolve all of the trivalent vertices in B (using the same sequence in both diagrams) to write

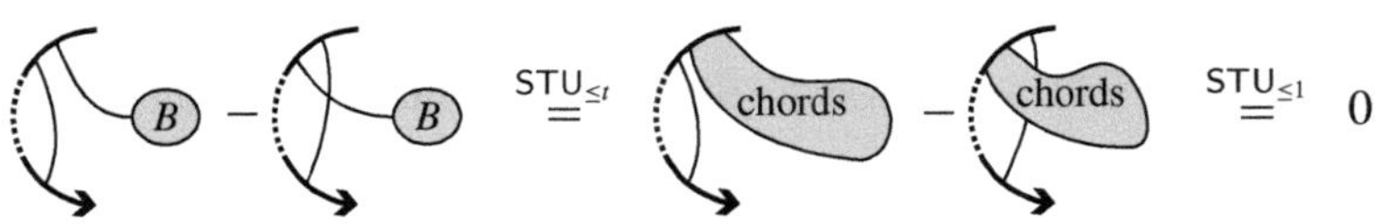

where the blocks of chords are identical, and the last equivalence is by an application of Lemma 13.28. Since $t \geq 1$ and none of the STU-moves involve diagrams with more than t trivalent vertices, we have that

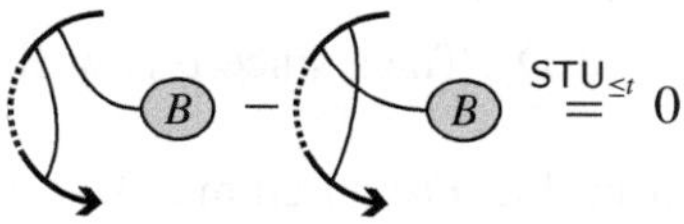

We may use this to replace the equivalence of diagrams in (13.16) with diagrams with at most t trivalent vertices.

This completes the proof of the claim. Hence, the proof of the lemma is complete. □

Lemma 13.39. *Let $D, D' \in \mathscr{D}$ be such that no term of D or D' has more than $t \geq 0$ trivalent vertices. If $D \overset{\mathsf{STU}}{=} D'$ then $D \overset{\mathsf{4T,STU}_{\leq t}}{=} D'$.*

Proof. Suppose that in the equivalence $D \overset{\mathsf{STU}}{=} D'$ the largest number of trivalent vertices in any Jacobi diagram is $t+k$ for some $k \geq 0$. If $k = 0$, we are done; otherwise, no term of D or D' has more than $t+k-1$ trivalent vertices, and $D \overset{\mathsf{4T,STU}_{\leq t+k}}{=} D'$. Then by Lemma 13.38, $D \overset{\mathsf{4T,STU}_{\leq t+k-1}}{=} D'$. Repeating this process gives that $D \overset{\mathsf{4T,STU}_{\leq t}}{=} D'$, as required. □

Lemma 13.40 (Resolutions). *The map*

$$\rho \colon \mathscr{D}_m \to \frac{\mathscr{C}_m}{(\mathsf{4T})} \colon D \mapsto [r(D)]_{\mathsf{4T}},$$

where $r(D)$ is any resolution of D, is well-defined.

Proof. Let $D \in \mathscr{D}_m$. To show that ρ is well-defined, we need to show that for any resolutions r_1 and r_2 of D we have $r_1(D) \overset{\mathsf{4T}}{=} r_2(D)$. Thus, if $r_1(D) = C_1$ and $r_2(D) = C_2$, we need to show that $C_1 \overset{\mathsf{4T}}{=} C_2$.

We have $D \overset{\mathsf{STU}}{=} C_1$ and $D \overset{\mathsf{STU}}{=} C_2$. Hence, $C_1 \overset{\mathsf{STU}}{=} C_2$. An application of Lemma 13.39 then gives that $C_1 \overset{\mathsf{4T}}{=} C_2$, as required. □

With the aid of the above lemmas, we may now prove that $\mathscr{A}_m^c \cong \mathscr{A}_m$.

Proof (Theorem 13.36). From Lemma 13.40, there is a well-defined linear map

$$\rho \colon \mathscr{D}_m \to \mathscr{A}_m^c \colon [D]_{\mathsf{STU}} \mapsto [r(D)]_{\mathsf{4T}},$$

where r is any resolution of the Jacobi diagram D.

To show that $\overline{\rho}$ is well-defined, we need to show that if $D_1 \overset{\mathsf{STU}}{=} D_2$, then $r_1(D_1) \overset{\mathsf{4T}}{=} r_2(D_2)$ for any resolutions r_1 and r_2 of D_1 and D_2. Suppose that $r_1(D_1) = C_1$ and $r_2(D_2) = C_2$, and suppose that r is a sequence of STU-moves taking D_2 to D_1, so that $r(D_2) = D_1$. Then $C_1 = r_1(r(D_2))$ and $C_2 = r_2(D_2)$, so C_1 and C_2 can both be obtained through resolutions of D_2. Then, since ρ is well-defined, $C_1 \overset{\mathsf{4T}}{=} C_2$. Thus, $\overline{\rho}$ is well-defined.

We now show that $\overline{\rho}$ is invertible. From Lemma 13.37, $\overline{\iota}([C]_{\mathsf{4T}}) = [C]_{\mathsf{STU}}$ where C is a chord diagram in $\mathscr{C}_m$, so

$$\begin{aligned}(\overline{\rho} \circ \overline{\iota})\,([C]_{\mathsf{4T}}) &= \overline{\rho}\,([C]_{\mathsf{STU}}) \\ &= [r(C)]_{\mathsf{4T}} \qquad \text{(Lem. 13.37; } r \text{ is any resolution of } C) \\ &= [C]_{\mathsf{4T}} \qquad \text{(since } C \text{ is already a chord diagram.)}\end{aligned}$$

Thus, $\overline{\rho} \circ \overline{\iota} = \mathrm{id}$. But $\mathscr{D}_m/(\mathsf{STU})$ and $\mathscr{C}_m/(\mathsf{4T})$ are finite dimensional since $\mathscr{D}_m$ and $\mathscr{C}_m$ are. Thus, ρ is invertible. This completes the proof. □

Chapter 14
Lie Algebra Weight Systems

We have seen the critical role the vector space $\mathscr{A}$ of Jacobi diagrams and weight systems, *i.e.* linear maps from $\mathscr{A}$ to $\mathbb{C}$, play in the theory of Vassiliev invariants. The purpose of this chapter is to construct weight systems in a canonical way from certain algebraic objects. This will be done by taking a Jacobi diagram or a chord diagram, and then constructing from it an element in an appropriate algebra. A representation of this algebra will provide an associated linear map, whose trace then produces a number in $\mathbb{C}$. We shall see in Sect. 18.2 that the weight systems emerging from this process are related to the theory of quantum invariants described in Part II of this book. This is important since it provides a way to study knot polynomials *via* combinatorics. The construction of these weight systems has three main steps.

1. We define a map $T_{\mathfrak{g}}$ that sends a Jacobi diagram D to an element $T_{\mathfrak{g}}(D)$ of the universal enveloping algebra $U(\mathfrak{g})$ of a Lie algebra $\mathfrak{g}$.
2. For a weight system, we need to obtain a number in $\mathbb{C}$ from $T_{\mathfrak{g}}(D)$. This is done by choosing a representation ρ of the universal enveloping algebra $U(\mathfrak{g})$.
3. The image of $T_{\mathfrak{g}}(D)$ under ρ is a linear map, which we may view as a matrix over $\mathbb{C}$. We obtain a complex number by taking the trace of this matrix.

This construction gives a weight system which we shall denote by $W_{\mathfrak{g},\rho}$.

As we shall describe presently, the element $T_{\mathfrak{g}}(D)$ is obtained by first decomposing D into elementary fragments, next by associating an algebraic object with each elementary fragment, and finally by using the decomposition of D to construct an element of $U(\mathfrak{g})$ from these algebraic objects.

Various choices are made in the computations of $T_{\mathfrak{g}}(D)$ and therefore of $W_{\mathfrak{g},\rho}(D)$. That the computations are independent of these choices, *i.e.*, that $T_{\mathfrak{g}}$ and $W_{\mathfrak{g},\rho}$ give well-defined maps, is non-trivial, requiring a good amount of work. It is essential to this well-definedness that $T_{\mathfrak{g}}(D)$ be evaluated in a universal enveloping algebra $U(\mathfrak{g})$ of a Lie algebra $\mathfrak{g}$.

The Lie algebra weight systems, including their combinatorial descriptions, constructed in this chapter are due to Bar-Natan [11–13].

D. M. Jackson and I. Moffatt, *An Introduction to Quantum and Vassiliev Knot Invariants*, CMS Books in Mathematics, https://doi.org/10.1007/978-3-030-05213-3_14

14.1 Introduction to Lie Algebra Weight Systems

In this section, we give an overview of the construction of the map $T_{\mathfrak{g}}$ that sends a Jacobi diagram D to an element $T_{\mathfrak{g}}(D)$ of the universal enveloping algebra $U(\mathfrak{g})$. We describe the key points of this construction and give an *informal* overview of how Lie algebras appear in this theory. The next section contains a formal treatment of map $T_{\mathfrak{g}}$, showing, in particular, that it is well-defined. The later parts of this chapter will focus on how the map $T_{\mathfrak{g}}$ gives rise to weight systems through representations of $U(\mathfrak{g})$.

Let us proceed with this programme. Suppose we are given:

- a Jacobi diagram D;
- a metrized Lie algebra $\mathfrak{g}$ over $\mathbb{C}$ with a basis $\{e_i : i = 1, \dots n\}$. (We shall explain the term "metrized Lie algebra" shortly.)

In the following three subsections, we break up the description of $T_{\mathfrak{g}}$ into the three steps:

1. decomposing a Jacobi diagram,
2. understanding metrized Lie algebras, and
3. the construction of $T_{\mathfrak{g}}(D)$.

Decomposing a Jacobi Diagram

We may "cut" the Jacobi diagram D into "elementary fragments" that consist of an arc or a trivalent vertex as shown:

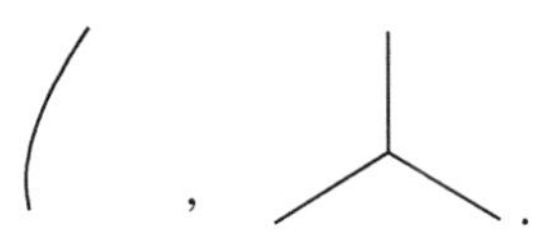

(14.1)

The ends of these elementary fragments may, or may not, lie on the skeleton of D. For example,

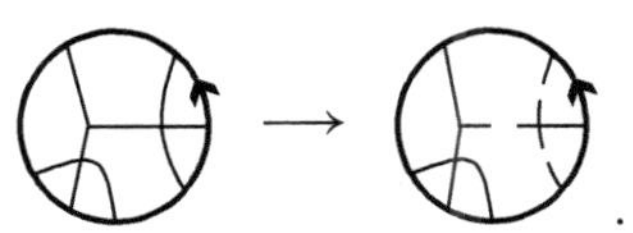

There are, of course, infinitely many ways to choose such a decomposition. For example, a Jacobi diagram and three of its decompositions are shown below.

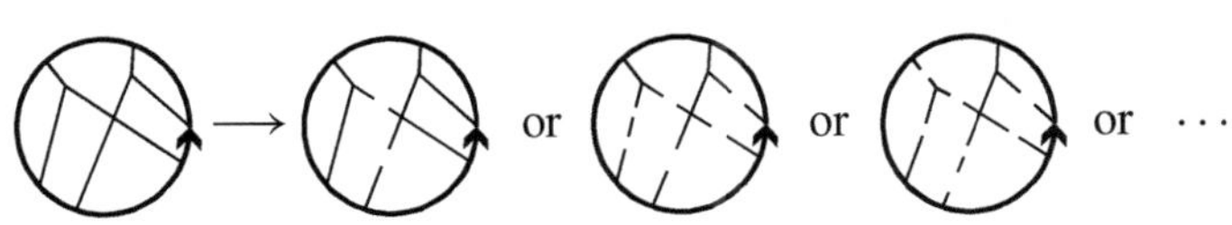

We use such a decomposition of D to compute $T_{\mathfrak{g}}(D)$, and we need to ensure that the value of $T_{\mathfrak{g}}(D)$ is independent of the choice of decomposition we make. We shall see that our use of Lie algebras ensures this independence.

Understanding the Lie Algebras

The map $T_{\mathfrak{g}}(D)$ takes its values in the universal enveloping algebra of a metrized Lie algebra $\mathfrak{g}$. We met Lie algebras and their universal enveloping algebras in Sect. 8.4. A *metrized Lie algebra* consists of a Lie algebra and a symmetric, non-degenerate, ad-invariant bilinear form $\langle\cdot,\cdot\rangle$. We shall explain these terms next.

Let $\mathfrak{g}$ be a finite dimensional Lie algebra over $\mathbb{C}$ with a basis $\{e_i : i = 1, \ldots n\}$. The Lie algebra $\mathfrak{g}$ is equipped with a Lie bracket $[\cdot,\cdot]$ which may be described by its action on a basis by

$$[\cdot,\cdot] : V \otimes V \to V : e_i \otimes e_j \mapsto \sum_k \gamma_{i,j}{}^k e_k. \tag{14.2}$$

We recall that the scalars $\gamma_{i,j}{}^k \in \mathbb{C}$ are called *structure constants*.

Suppose that $\mathfrak{g}$ is equipped with a *bilinear form*, which is a map $\langle\cdot,\cdot\rangle : \mathfrak{g} \otimes \mathfrak{g} \to \mathbb{C}$, and that the action of $\langle\cdot,\cdot\rangle$ on the basis of $\mathfrak{g}$ is

$$\langle e_i, e_j \rangle = h_{i,j}, \tag{14.3}$$

for some scalars $h_{i,j} \in \mathbb{C}$. The bilinear form $\langle\cdot,\cdot\rangle$ is said to be *symmetric* if $\langle e_i, e_j\rangle = \langle e_j, e_i\rangle$, or equivalently $h_{i,j} = h_{j,i}$, for all i, j.

The definition of the term "non-degenerate bilinear form" makes use of the facts, among others, that $\mathrm{Hom}(U, V) \cong U^\star \otimes V$ and $V^{\star\star} \cong V$ for finite dimensional vector spaces U and V (see Sect. A.9). Using the isomorphisms, we have

$$\begin{aligned}\langle\cdot,\cdot\rangle \in \mathrm{Hom}(\mathfrak{g}\otimes\mathfrak{g}, \mathbb{C}) &\cong (\mathfrak{g}\otimes\mathfrak{g})^\star \otimes \mathbb{C} \cong \mathfrak{g}^\star \otimes \mathfrak{g}^\star \otimes \mathbb{C} \\ &\cong \mathfrak{g}^\star \otimes \mathfrak{g}^\star \cong \mathrm{Hom}(\mathfrak{g}^{\star\star}, \mathfrak{g}^\star) \cong \mathrm{Hom}(\mathfrak{g}, \mathfrak{g}^\star).\end{aligned} \tag{14.4}$$

Thus, we may regard the bilinear form $\langle\cdot,\cdot\rangle$ as a linear map

$$\langle\cdot,\cdot\rangle : \mathfrak{g} \to \mathfrak{g}^\star : e_i \mapsto \sum_j h_{i,j} e^j. \tag{14.5}$$

This view of $\langle\cdot,\cdot\rangle$ will probably surprise a reader meeting it for the first time.

Exercise 14.1. Verify that the action of $\langle\cdot,\cdot\rangle : \mathfrak{g} \to \mathfrak{g}^\star$ is given by $x \mapsto \langle\cdot, x\rangle$, where $x \in \mathfrak{g}$.

A bilinear form $\langle\cdot,\cdot\rangle$ is said to be *non-degenerate* if, when regarded as a linear map $\langle\cdot,\cdot\rangle \in \mathrm{Hom}(\mathfrak{g}, \mathfrak{g}^\star)$, it is a vector space isomorphism. We denote the inverse map of $\langle\cdot,\cdot\rangle$ by $\langle\cdot,\cdot\rangle^{-1}$ and denote its action on dual basis $\{e^i : i = 1,\ldots,n\}$ of $\mathfrak{g}$ by

$$\langle\cdot,\cdot\rangle^{-1} : V^\star \to V : e^i \mapsto \sum_j h^{i,j} e_j. \tag{14.6}$$

Exercise 14.2. Verify that the following identity relating the $h_{i,j}$ and the $h^{i,j}$ holds:

$$\sum_q h_{p,q}\, h^{q,r} = \delta_{p,r}. \tag{14.7}$$

The final term needed for the definition of a metrized Lie algebra is ad-invariant. Ad-invariance provides a compatibility between the Lie bracket $[\cdot,\cdot]$ and the bilinear form $\langle\cdot,\cdot\rangle$. The bilinear form $\langle\cdot,\cdot\rangle$ is said to be *ad-invariant* if

$$\langle [x,y], z\rangle = \langle x, [y,z]\rangle, \qquad \text{for all } x, y, z \in \mathfrak{g}. \tag{14.8}$$

Definition 14.3 (Metrized Lie algebra). A *metrized Lie algebra* consists of a finite dimensional Lie algebra $\mathfrak{g}$ together with a symmetric, non-degenerate, ad-invariant bilinear form.

Exercise 14.4. Consider the Lie algebra $\mathfrak{sl}_2(\mathbb{C})$ from Example 8.42, of 2×2 matrices with trace zero, and Lie bracket given by $[A, B] = A \cdot B - B \cdot A$. As in Exercise 8.45, this Lie algebra has a basis $\{X, Y, H\}$ where

$$X = \begin{bmatrix} 0 & 1 \\ 0 & 0 \end{bmatrix}, \qquad Y = \begin{bmatrix} 0 & 0 \\ 1 & 0 \end{bmatrix}, \qquad H = \begin{bmatrix} 1 & 0 \\ 0 & -1 \end{bmatrix}.$$

Define a bilinear form $\langle\cdot,\cdot\rangle : \mathfrak{sl}_2(\mathbb{C}) \otimes \mathfrak{sl}_2(\mathbb{C}) \to \mathbb{C}$ by $\langle A, B\rangle = \mathrm{Tr}\,(A \cdot B)$. Verify that $\langle\cdot,\cdot\rangle$ is symmetric, non-degenerate and ad-invariant. Conclude that $\mathfrak{sl}_2(\mathbb{C})$ together with $\langle\cdot,\cdot\rangle$ is a metrized Lie algebra.

Universal enveloping algebras of Lie algebras were introduced in Definition 8.46. For a Lie algebra $\mathfrak{g}$ with basis $\{e_1, e_2, \ldots, e_n\}$ and structure constants ${\gamma_{i,j}}^k$, so $[e_i, e_j] = \sum_{i=1}^n {\gamma_{i,j}}^k e_k$, the universal enveloping algebra, $U(\mathfrak{g})$, of $\mathfrak{g}$ was defined as the associative unital algebra generated by elements $E_1, E_2, \ldots, E_n$ and subject to the relations $E_i E_j - E_j E_i = \sum_{i=1}^n {\gamma_{i,j}}^k E_k$, for each i, j. Here it is convenient to use a different formulation of universal enveloping algebras.

The *tensor algebra*, $\mathfrak{T}(\mathfrak{g})$, of a Lie algebra $\mathfrak{g}$ is defined to be the algebra over $\bigoplus_{r\geq 0} \mathfrak{g}^{\otimes r}$ with multiplication given by

$$(x_1 \otimes \ldots \otimes x_p) \otimes (y_1 \otimes \ldots \otimes y_q) \mapsto x_1 \otimes \ldots \otimes x_p \otimes y_1 \otimes \ldots \otimes y_q.$$

The elements of $\mathfrak{T}(\mathfrak{g})$ consist of finite linear combinations of tensor products of elements of $\mathfrak{g}$. It is easily seen that $\mathfrak{T}(\mathfrak{g})$ is an associative unital algebra (see Exercise 8.7).

The *universal enveloping algebra*, $U(\mathfrak{g})$, of $\mathfrak{g}$ may be defined as the quotient of the tensor algebra $\mathfrak{T}(\mathfrak{g})$ by the ideal generated by the relation $[x, y] = x \otimes y - y \otimes x$,

$$U(\mathfrak{g}) := \frac{\mathfrak{T}(\mathfrak{g})}{[x, y] = x \otimes y - y \otimes x} = \frac{\bigoplus_{r \geq 0} \mathfrak{g}^{\otimes r}}{[x, y] = x \otimes y - y \otimes x}. \tag{14.9}$$

Exercise 14.5. Verify that the constructions of $U(\mathfrak{g})$ given here and in Definition 8.46 result in isomorphic algebras.

The Construction of $T_{\mathfrak{g}}(D)$

We may now describe the action of the map $T_{\mathfrak{g}} : \mathscr{A} \to U(\mathfrak{g})$. For this suppose that, as above, $\mathfrak{g}$ is a finite dimensional metrized Lie algebra over $\mathbb{C}$ with a basis $\{e_i : i = 1, \dots n\}$ and bilinear form $\langle \cdot, \cdot \rangle$. Also as above, suppose that the element-wise actions of $[\cdot, \cdot]$, $\langle \cdot, \cdot \rangle$, and $\langle \cdot, \cdot \rangle^{-1}$ are, respectively,

$$[\cdot, \cdot] : e_i \otimes e_j \mapsto \sum_k \gamma_{i,j}{}^k e_k\,, \quad \langle \cdot, \cdot \rangle : e_i \mapsto \sum_j h_{i,j} e^j\,, \quad \langle \cdot, \cdot \rangle^{-1} : e^i \mapsto \sum_j h^{i,j} e_j.$$

Define an element $S \in \mathfrak{g} \otimes \mathfrak{g} \otimes \mathfrak{g}$ by

$$S := - \sum_{i,j,a,b,c} \gamma_{i,j}{}^c\, h^{i,a} h^{j,b}\, e_a \otimes e_b \otimes e_c,$$

and let $S^{a,b,c}$ denote the coefficient of $e_a \otimes e_b \otimes e_c$ in S, so

$$S^{a,b,c} := - \sum_{i,j} \gamma_{i,j}{}^c\, h^{i,a} h^{j,b}.$$

For a Jacobi diagram D, $T_{\mathfrak{g}}(D)$ is constructed as follows.

1. Choose a decomposition of D into elementary fragments and choose a base point, $\bigstar$, on its skeleton.
2. Label the end of each arc of an elementary piece with a distinct (indexing) element.
3. Form an expression by

   a. Assigning a tensor to each labelled elementary piece using the scheme

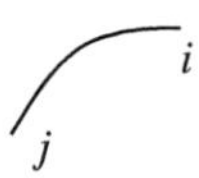

$\longmapsto h^{i,j}\, e_i \otimes e_j$ and
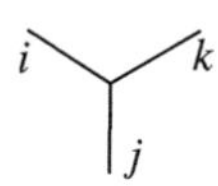

$\longmapsto S^{i,j,k}\, e_i \otimes e_j \otimes e_k.$ (14.10)

   b. Wherever a Jacobi diagram is "cut" in forming the elementary decomposition, apply the bilinear form $\langle \cdot, \cdot \rangle$ to the corresponding basis elements, thus

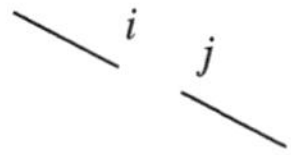

$$\longmapsto \quad \langle e_i, e_j \rangle = h_{i,j}. \tag{14.11}$$

c. What remains is a scalar made up of the $h^{i,j}$, $h_{i,j}$ and $S^{i,j,k}$ and basis elements e_i that come from arc ends that lie on the skeleton. Order these basis elements by reading along the skeleton from the base point, and tensor them in this order. Then place the scalar expression in front of this tensor.

4. Take the sum of this expression with the indices ranging from $1, \dots, n$ (*i.e.* over all of the basis elements of the Lie algebra).
5. The resulting element is in the tensor algebra $\mathfrak{T}(\mathfrak{g})$. Project it into $U(\mathfrak{g})$ (see (14.9)).

Example 14.6. Consider the Jacobi diagram shown below on the left, with the decomposition and choice of base point shown in the middle, and the labelling shown on the right. We have drawn the skeleton with the base point as a straight line in rightmost figure for clarity.

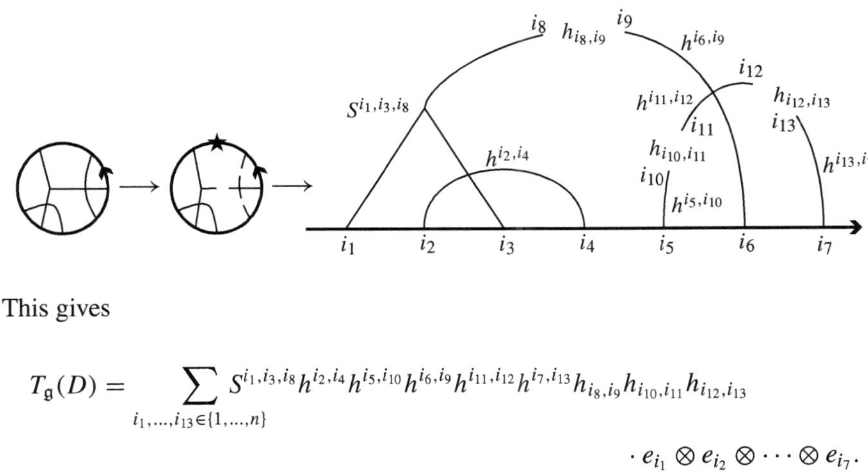

This gives

$$T_{\mathfrak{g}}(D) = \sum_{i_1,\dots,i_{13} \in \{1,\dots,n\}} S^{i_1,i_3,i_8} h^{i_2,i_4} h^{i_5,i_{10}} h^{i_6,i_9} h^{i_{11},i_{12}} h^{i_7,i_{13}} h_{i_8,i_9} h_{i_{10},i_{11}} h_{i_{12},i_{13}}$$
$$\cdot e_{i_1} \otimes e_{i_2} \otimes \cdots \otimes e_{i_7}.$$

Example 14.7.

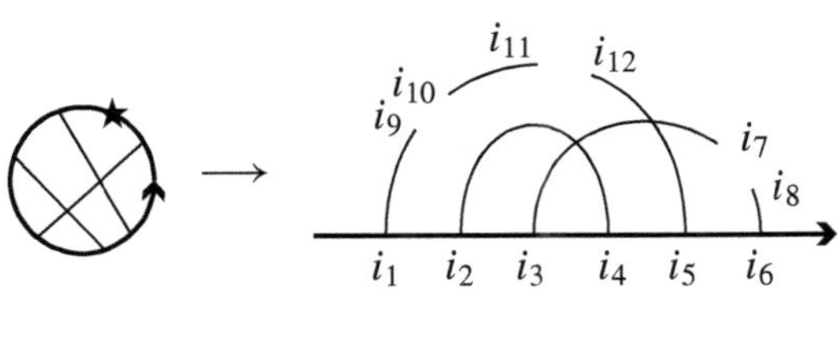

$$T_{\mathfrak{g}}(D) = \sum_{i_1,\dots,i_{12}\in\{1,2,\dots,n\}} h^{i_1,i_9} h^{i_2,i_4} h^{i_3,i_7} h^{i_5,i_{12}} h^{i_6,i_8} h^{i_{10},i_{11}} h_{i_7,i_8} h_{i_9,i_{10}} h_{i_{11}i_{12}} \cdot e_{i_1} \otimes e_{i_2} \otimes e_{i_3} \otimes e_{i_4} \otimes e_{i_5} \otimes e_{i_6}.$$

Exercise 14.8. Consider the chord diagram and the two of its decompositions shown below. By using the fact that the $h_{i,j} = h_{j,i}$ and $h^{i,j} = h^{j,i}$ satisfy the identity in Exercise 14.2 to show that constructing $T_{\mathfrak{g}}(D)$ from either decomposition results in equal elements of $U(\mathfrak{g})$.

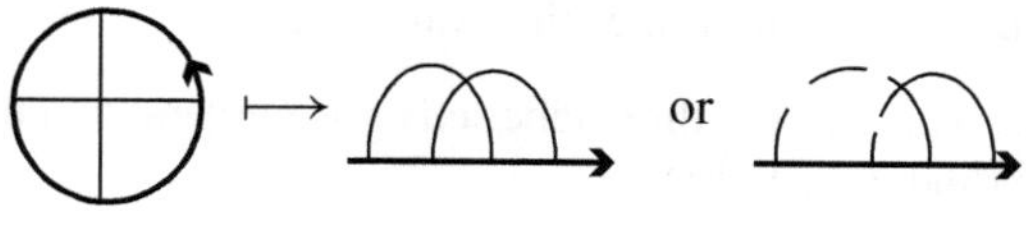

Exercise 14.9. The STU-relation gives the equality 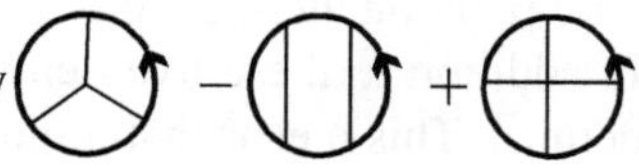 $= 0$ in $\mathscr{A}$. Using the decompositions

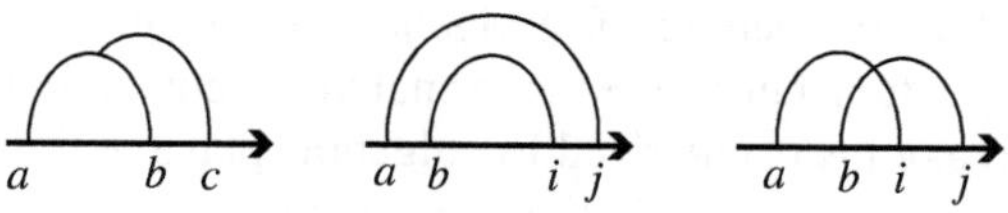

verify that

$$T_{\mathfrak{g}}\left(\right.$$  $$\left.\right) = -\sum_{i,j,a,b,c} \gamma_{i,j}{}^{c}\, h^{i,a} h^{j,b}\, e_a \otimes e_b \otimes e_c,$$

$$T_{\mathfrak{g}}\left(\right.$$  $$\left.\right) = \sum_{i,j,a,b} h^{i,b} h^{a,j}\, e_a \otimes e_b \otimes e_i \otimes e_j,$$

$$T_{\mathfrak{g}}\left(\right.$$  $$\left.\right) = \sum_{i,j,a,b} h^{i,a} h^{b,j}\, e_a \otimes e_b \otimes e_i \otimes e_j.$$

Hence, use the fact that $T_{\mathfrak{g}}$ evaluates in $U(\mathfrak{g})$ to verify that

$$T_{\mathfrak{g}}\left(\right.$$ 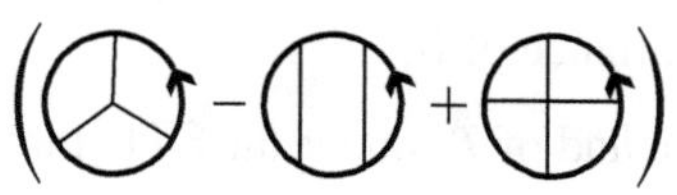 $$\left.\right) = 0.$$

Independence of Choices Made in the Construction

The reader will have noticed that several choices needed to be made in the computation of $T_{\mathfrak{g}}(D)$. There were choices made for the following:

- decomposition of a Jacobi diagram;
- order of labels of each cut in the decomposition

  (*i.e.* in (14.11), which of the two labelled ends of the arc we take as the first index in $h_{i,j}$);

- order of the labels on each arc in the decomposition

  (*i.e.* in (14.10), which of the two labelled ends of the arc we take as the first index in $h^{i,j}$);

- linearisation of the cyclic order at each trivalent vertex

  (*i.e.* in (14.10), which of the three labelled ends incident to a trivalent vertex we take as the first index in $S^{i,j,k}$);

- base point on the skeleton.
- In addition to these five items, but separate from them, we need $T_{\mathfrak{g}}$ to be well-defined. This means that it must be invariant under the STU-relation.

The requirement of independence of $T_{\mathfrak{g}}$ under these six items explains the use of enveloping algebras of metrized Lie algebras. We examine this independence formally in the next section, but before the formal treatment it is helpful to understand the role each of the axioms of metrized Lie algebras plays in this independence. In short, following the order of items in the above listing in turn, the *independence of the choice* of:

- decomposition:

  arises from the fact that the $h_{i,j}$ and $h^{i,j}$ are coefficients from inverse maps $\langle \cdot, \cdot \rangle$ and $\langle \cdot, \cdot \rangle^{-1}$ (hence the non-degeneracy of $\langle \cdot, \cdot \rangle$ is required). This means that the contributions from any "extra cuts" cancel.

- order of labels of each cut in the decomposition:

  arises from the fact that $\langle \cdot, \cdot \rangle$ is symmetric and so $h_{i,j} = h_{j,i}$.

- order of the labels on each arc in the decomposition:

  follows since $\langle \cdot, \cdot \rangle$ is symmetric, therefore $\langle \cdot, \cdot \rangle^{-1}$ is, and so $h^{i,j} = h^{j,i}$.

- linearisation of the cyclic order at each trivalent vertex:

  arises from the ad-invariance of $\langle \cdot, \cdot \rangle$.

- base point on the skeleton and of $T_{\mathfrak{g}}$ under the STU-move:

  uses the fact that $T_{\mathfrak{g}}$ is evaluated in the universal enveloping algebra $U(\mathfrak{g})$.

14.2 The Universal $\mathfrak{g}$-Weight System

In the previous section, we described the construction of a map $T_{\mathfrak{g}} : \mathscr{A} \to U(\mathfrak{g})$, where $\mathfrak{g}$ is a metrized Lie algebra. We shall now formalise this construction and show that it is independent of the several choices we made in the decomposition of the Jacobi or chord diagram, and that it defines a map from $\mathscr{A}$ to $U(\mathfrak{g})$.

For convenience, we collect the associated algebraic notation here that we shall need. Throughout this section, we let $\mathfrak{g}$ denote a metrized Lie algebra with finite basis $\{e_i\}_{i\in I}$. This means that the Lie algebra $\mathfrak{g}$ is equipped with a bilinear form $\langle\cdot,\cdot\rangle$ that is symmetric (*i.e.* $\langle x, y\rangle = \langle y, x\rangle$ for all $x, y \in \mathfrak{g}$), ad-invariant (*i.e.* $\langle [x, y], z\rangle = \langle x, [y, z]\rangle$ for all $x, y, z \in \mathfrak{g}$) and non-degenerate (*i.e.* $\langle\cdot,\cdot\rangle$ is invertible when considered as a map $\mathfrak{g} \to \mathfrak{g}^\star$).

We recall the following from Sect. 14.1. The action, from (14.2), of the Lie bracket on basis elements is

$$\left[e_i, e_j\right] = \sum_k \gamma_{i,j}{}^k e_k,$$

and the action, from (14.3), of $\langle\cdot,\cdot\rangle$ is

$$\left\langle e_i, e_j\right\rangle = h_{i,j}.$$

From (14.4), *via* the isomorphisms $\mathrm{Hom}(\mathfrak{g}\otimes\mathfrak{g},\mathbb{C}) \cong (\mathfrak{g}\otimes\mathfrak{g})^\star\otimes\mathbb{C} \cong \mathfrak{g}^\star\otimes\mathfrak{g}^\star \cong \mathrm{Hom}(\mathfrak{g}^{\star\star},\mathfrak{g}^\star) \cong \mathrm{Hom}(\mathfrak{g},\mathfrak{g}^\star)$, we may consider $\langle\cdot,\cdot\rangle$ as a map

$$\langle\cdot,\cdot\rangle : \mathfrak{g} \to \mathfrak{g}^\star : e_i \mapsto \sum_j h_{i,j} e^j.$$

By non-degeneracy, this map is invertible. We denote the action of this inverse, $\langle\cdot,\cdot\rangle^{-1}$, by

$$\langle\cdot,\cdot\rangle^{-1} : \mathfrak{g}^\star \to \mathfrak{g} : e^i \mapsto \sum_j h^{i,j} e_j.$$

We define S by

$$S = -\sum_{i,j,a,b,c} \gamma_{i,j}{}^c h^{i,a} h^{j,b} e_a \otimes e_b \otimes e_c,$$

and let $S^{a,b,c}$ denote the coefficient of $e_a \otimes e_b \otimes e_c$ in S:

$$S^{a,b,c} := -\sum_{i,j} \gamma_{i,j}{}^c h^{i,a} h^{j,b}. \tag{14.12}$$

Finally, we realise the universal enveloping algebra of $\mathfrak{g}$ as the quotient space of the tensor algebra $\mathfrak{T}(\mathfrak{g})$:

$$U(\mathfrak{g}) = \frac{\mathfrak{T}(\mathfrak{g})}{[x, y] = x\otimes y - y\otimes x}. \tag{14.13}$$

Thus, projecting from $\mathfrak{T}(\mathfrak{g})$ to $U(\mathfrak{g})$ can be thought of as imposing the relation $[x, y] = x \otimes y - y \otimes x$ on $\mathfrak{T}(\mathfrak{g})$.

We defined $T_{\mathfrak{g}} : \mathscr{D} \to \mathfrak{T}(\mathfrak{g})$ in the previous section by means of a diagrammatic construction that involves the decomposition of a Jacobi diagram into elementary fragments. The following definitions and notation are needed to make this precise.

The Construction of $T_{\mathfrak{g}}(D)$

Definition 14.10 (Fragment, decomposition, colouring, linearisation). Let D be a Jacobi diagram.

1. An *elementary fragment* of D is a graph of the form

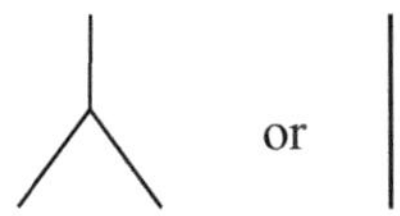

   where the univalent vertices may lie on the skeleton.
2. An *elementary decomposition* $\mathfrak{d}$ of D is a separation of D into elementary fragments only. We write $\mathfrak{f} \in \mathfrak{d}$ if $\mathfrak{f}$ is an elementary fragment in $\mathfrak{d}$.
3. A $\mathfrak{g}$-*colouring* ϑ of an elementary decomposition $\mathfrak{d}$ of D is an assignment of an index $i \in I$, where I is the index set for the basis $\{e_i\}_{i \in I}$ of $\mathfrak{g}$, to the univalent ends of edges of fragments.
4. A *linearisation* of the cyclic order at a trivalent vertex of D is obtained by distinguishing an edge incident with the vertex. A "★" is used to indicate the start of the linear order.
5. A *linearisation* of the skeleton of D is obtained by distinguishing a point, denoted by "★", on the skeleton. A linearisation of a skeleton gives a linear order of the degree one vertices of a Jacobi diagram.
6. A diagram $D \in \mathscr{D}$ endowed with a linearisation of the skeleton $\mathfrak{l}$, a linearisation of the cyclic order at each trivalent vertex ℓ, a decomposition $\mathfrak{d}$ and a $\mathfrak{g}$-colouring ϑ is denoted by

$$(D, \mathfrak{l}, \ell, \mathfrak{d}, \vartheta)$$

   and a diagram endowed with a linearisation of the skeleton $\mathfrak{l}$, a linearisation of the cyclic order at each trivalent vertex ℓ, a decomposition $\mathfrak{d}$ is denoted by

$$(D, \mathfrak{l}, \ell, \mathfrak{d}).$$

7. The *distinguished fragments* of $(D, \mathfrak{l}, \ell, \mathfrak{d}, \vartheta)$ are

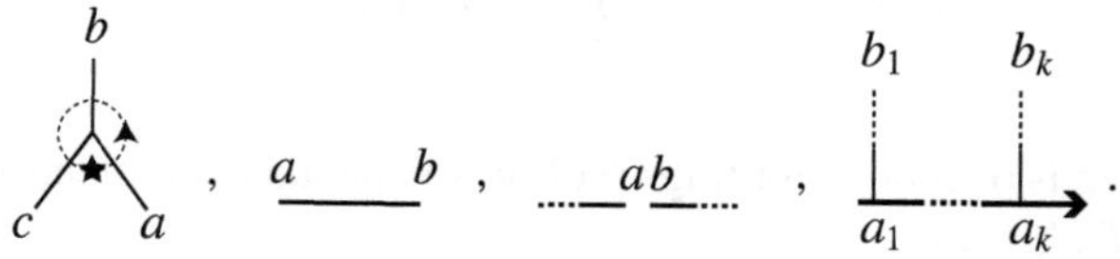

We shall usually represent the linearisation of a skeleton by drawing it as a straight line which should be understood to be starting and ending at the point ★. (See Sect. 13.5 for a discussion of linearised Jacobi diagrams.)

We shall call $(D, \mathfrak{l}, \ell, \mathfrak{d})$ a *fragmentary decomposition* of D, and $(D, \mathfrak{l}, \ell, \mathfrak{d}, \vartheta)$ a $\mathfrak{g}$-*coloured fragmentary decomposition* of D.

A unique tensor

$$T_{\mathfrak{g}}(D, \mathfrak{l}, \ell, \mathfrak{d}, \vartheta) \in \mathfrak{T}(\mathfrak{g})$$

may be assigned to a $\mathfrak{g}$-coloured fragmentary decomposition $(D, \mathfrak{l}, \ell, \mathfrak{d}, \vartheta)$ by assigning element-wise actions to each of the four types of distinguished fragments by the following rules. For this purpose, let

$T_{\mathfrak{g}}:$ 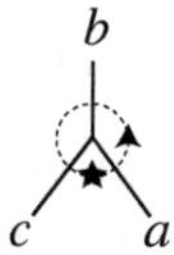

$\longmapsto S^{a,b,c} := -\sum_{s,t\in I} \gamma_{s,t}{}^{c}\, h^{s,a} h^{t,b},$

$T_{\mathfrak{g}}:$ 

$\longmapsto h^{a,b},$ (14.14)

$T_{\mathfrak{g}}:$ 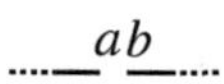

$\longmapsto h_{a,b} = \langle e_a, e_b \rangle,$

$T_{\mathfrak{g}}:$ 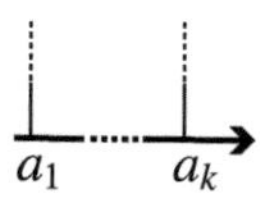

$\longmapsto e_{a_1} \otimes \cdots \otimes e_{a_k},$

and

$$T_{\mathfrak{g}}(D, \mathfrak{l}, \ell, \mathfrak{d}, \vartheta) := \prod_{\mathfrak{f}\in\mathfrak{d}} T_{\mathfrak{g}}(\mathfrak{f}, \mathfrak{l}, \ell, \mathfrak{d}, \vartheta) \tag{14.15}$$

where the product is over the set of all distinguished fragments $\mathfrak{f}$ in the decomposition $\mathfrak{d}$ of D. The only choice in this construction is the order in which the product of the $T_{\mathfrak{g}}(\mathfrak{f}, \ell, \mathfrak{d}, \vartheta)$ is taken, but since all fragments map to scalars, with the exception of a (single) tensor factor arising from the skeleton (the fourth case in (14.14)), the product is independent of this choice.

Now let

$$T_{\mathfrak{g}}(D, \mathfrak{l}, \ell, \mathfrak{d}) := \sum_{\vartheta} T_{\mathfrak{g}}(D, \mathfrak{l}, \ell, \mathfrak{d}, \vartheta) \in \mathfrak{T}(\mathfrak{g})$$

where the sum is over all colourings ϑ of the elementary fragments of the decomposition $\mathfrak{d}$ of $(D, \mathfrak{l}, \ell, \mathfrak{d})$.

Notice that the construction of $T_{\mathfrak{g}}(D, \mathfrak{l}, \ell, \mathfrak{d})$ is exactly our construction of $T_{\mathfrak{g}}(D)$ from in Sect. 14.1, as illustrated in Example 14.6, but with a specified decomposition.

Independence from the Choices

Recall that our ultimate aim in this section is to define a map from $\mathscr{A}$ to $U(\mathfrak{g})$ *via* the fragmentary decompositions of Definition 14.10. In particular, given an element $[D]_{\mathsf{STU}}$ in $\mathscr{A}$, we wish to obtain an element of $U(\mathfrak{g})$ by choosing a Jacobi diagram D in the class $[D]_{\mathsf{STU}}$, choosing a fragmentary decomposition $(D, \mathfrak{l}, \ell, \mathfrak{d})$ of it, then calculating $T_{\mathfrak{g}}(D, \mathfrak{l}, \ell, \mathfrak{d})$. However, in the process just described, we have had to make several choices in constructing $T_{\mathfrak{g}}(D, \mathfrak{l}, \ell, \mathfrak{d})$. In order to obtain a well-defined map from $\mathscr{A}$ to $U(\mathfrak{g})$, we need to show that $T_{\mathfrak{g}}(D, \mathfrak{l}, \ell, \mathfrak{d})$ is independent of all of these choices. The following theorem shows that this is indeed the case.

Theorem 14.11. *Let* $\mathfrak{g}$ *be a metrized Lie algebra,* $\pi : \mathfrak{T}(\mathfrak{g}) \to U(\mathfrak{g})$ *be the projection map, and* $[D]_{\mathsf{STU}} \in \mathscr{A}$. *Let* D *be a representative of* $[D]_{\mathsf{STU}}$ *and* $(D, \mathfrak{l}, \ell, \mathfrak{d})$ *be a* fragmentary decomposition *of* D. *Then the value of*

$$\pi \circ T_{\mathfrak{g}}(D, \mathfrak{l}, \ell, \mathfrak{d}) \in U(\mathfrak{g})$$

is independent of the choice of

1. *linearisation of the cyclic order at a trivalent vertices,* ℓ*;*
2. *elementary decomposition,* $\mathfrak{d}$*;*
3. *choice of representative* D*; and*
4. *linearisation of the skeleton,* $\mathfrak{l}$.

Hence, $\pi \circ T_{\mathfrak{g}}$ *defines a map from* $\mathscr{A}$ *to* $U(\mathfrak{g})$.

Assuming the validity of Theorem 14.11 for the moment, we may make the following definition.

Definition 14.12 (Universal $\mathfrak{g}$ weight system). Let $\mathfrak{g}$ be a metrized Lie algebra, $\langle \cdot, \cdot \rangle$ denote its bilinear form, and $\pi : \mathfrak{T}(\mathfrak{g}) \to U(\mathfrak{g})$ be the projection map. Then $T_{\mathfrak{g}} : \mathscr{A} \to U(\mathfrak{g})$ denotes the map $\pi \circ T_{\mathfrak{g}}(D, \mathfrak{l}, \ell, \mathfrak{d})$ defined by Theorem 14.11. We call the map $T_{\mathfrak{g}}$ the *unframed universal* $(\mathfrak{g}, \langle \cdot, \cdot \rangle)$*-weight system.*

It remains to prove Theorem 14.11. The rest of this section will be devoted to its proof, which will follow from a series of four lemmas that establishes each of the four assertions in turn.

Lemma 14.13. *Let $\mathfrak{g}$ be a metrized Lie algebra. Then $T_{\mathfrak{g}}(D, \mathfrak{l}, \ell, \mathfrak{d}) \in \mathfrak{T}(\mathfrak{g})$ is independent of the linearisation of the cyclic order at a trivalent vertices, ℓ.*

Proof. We must show that if ℓ_1 and ℓ_2 are linearisations of the cyclic order at a trivalent vertex, then

$$T_{\mathfrak{g}}(D, \mathfrak{l}, \ell_1, \mathfrak{d}) = T_{\mathfrak{g}}(D, \mathfrak{l}, \ell_2, \mathfrak{d}).$$

It is sufficient to consider a single trivalent vertex. This is equivalent to showing that

$$\sum_{a,b,c} S^{a,b,c} = \sum_{a,b,c} S^{b,c,a}. \tag{14.16}$$

By ad-invariance, we have

$$\langle [e_i, e_j], e_k \rangle = \langle e_i, [e_j, e_k] \rangle .$$

Then

$$\sum_l \gamma_{i,j}{}^l \langle e_l, e_k \rangle = \sum_l \gamma_{j,k}{}^l \langle e_i, e_l \rangle$$

so

$$\sum_l \gamma_{i,j}{}^l h_{l,k} = \sum_l \gamma_{j,k}{}^l h_{i,l}. \tag{14.17}$$

But $\langle [\cdot, \cdot], \cdot \rangle \in \mathrm{Hom}\,(\mathfrak{g} \otimes \mathfrak{g} \otimes \mathfrak{g}, \mathbb{C}) \cong \mathfrak{g}^\star \otimes \mathfrak{g}^\star \otimes \mathfrak{g}^\star$. Under this isomorphism (14.17) becomes

$$\sum_{i,j,k,l} \gamma_{i,j}{}^l h_{l,k}\, e^i \otimes e^j \otimes e^k = \sum_{i,j,k,l} \gamma_{j,k}{}^l h_{i,l}\, e^i \otimes e^j \otimes e^k. \tag{14.18}$$

Elements in $\mathfrak{g} \otimes \mathfrak{g} \otimes \mathfrak{g}$ may be obtained by applying the isomorphism

$$h^{-1} \colon \mathfrak{g}^\star \to \mathfrak{g} \colon e^s \mapsto \sum_t h^{s,t}\, e_t$$

to (14.18) to obtain

$$\sum_{\substack{i,j,k,l \\ a,b,c}} \gamma_{i,j}{}^l h_{l,k} h^{i,a} h^{j,b} h^{k,c}\, e_a \otimes e_b \otimes e_c = \sum_{\substack{i,j,k,l \\ a,b,c}} \gamma_{j,k}{}^l h_{i,l} h^{i,a} h^{j,b} h^{k,c}\, e_a \otimes e_b \otimes e_c. \tag{14.19}$$

Comparing the coefficients of $e_a \otimes e_b \otimes e_c$ throughout (14.19) gives

$$\sum_{i,j,k,l} \gamma_{i,j}{}^l h_{l,k} h^{i,a} h^{j,b} h^{k,c} = \sum_{i,j,k,l} \gamma_{j,k}{}^l h_{i,l} h^{i,a} h^{j,b} h^{k,c}.$$

But by (14.7)

$$\sum_t h_{s,t}\, h^{t,u} = \delta_{s,u}, \tag{14.20}$$

and recalling that $h_{i,l} = h_{l,i}$, the above equality becomes

$$\sum_{i,j} \gamma_{i,j}{}^{c} h^{i,a} h^{j,b} = \sum_{j,k} \gamma_{j,k}{}^{a}\, h^{j,b} h^{k,c}$$

so from (14.12)

$$S^{a,b,c} = S^{b,c,a}$$

which establishes (14.16), and the result now follows. □

Lemma 14.14. *Let $\mathfrak{g}$ be a metrized Lie algebra. Then $T_{\mathfrak{g}}(D, \mathfrak{l}, \ell, \mathfrak{d}) \in \mathfrak{T}(\mathfrak{g})$ is independent of the choice of the decomposition $\mathfrak{d}$ of D.*

Proof. It suffices to show that if $\mathfrak{d}_1$ and $\mathfrak{d}_2$ are decompositions of D then $T_{\mathfrak{g}}(D, \mathfrak{l}, \ell, \mathfrak{d}_1)$ equals $T_{\mathfrak{g}}(D, \mathfrak{l}, \ell, \mathfrak{d}_2)$. We observe that the decompositions are related by a finite sequence of moves of the two forms

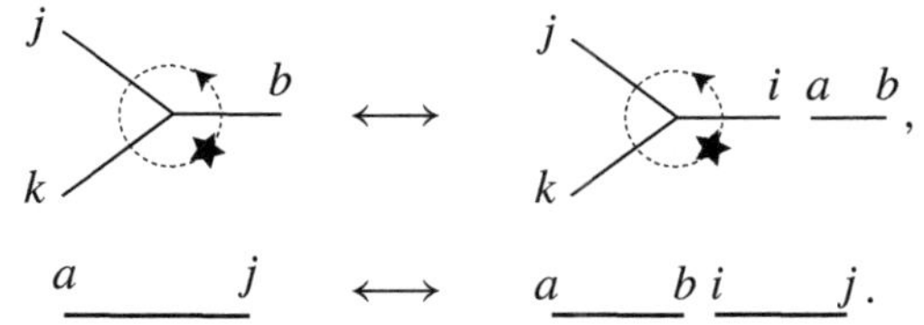

where a colouring has been applied to the constituents on the right-hand side of the two moves.

By Lemma 14.13, $T_{\mathfrak{g}}(D, \mathfrak{l}, \ell, \mathfrak{d})$ is independent of the linearisation ℓ, so it suffices to consider a single application of each of the two moves and therefore to show for each move that the contribution to $\sum_{\vartheta} T_{\mathfrak{g}}(D, \mathfrak{l}, \ell, \mathfrak{d}, \vartheta)$ from the left-hand side is equal to the contribution from the right-hand side.

For the first move, we need to show that

$$S^{b,j,k} = \sum_{a,i} S^{i,j,k}\, h_{i,a}\, h^{a,b}.$$

But this follows by (14.7).

For the second move, we need to show that

$$h^{a,j} = \sum_{i,b} h^{a,b}\, h_{b,i}\, h^{i,j}.$$

Again this follows by (14.7).

The result now follows. □

We now want to show that $\pi \circ T_{\mathfrak{g}}(D, \mathfrak{l}, \ell, \mathfrak{d})$ is unchanged by an application of the STU-relation. However, we need to be particularly careful in showing this since we have not yet shown independence of choice of linearisation of the skeleton. Thus, we shall need to make sure that our application of the STU-relation does not "cross" the base point. Accordingly, we need a version of the STU-relation that behaves well with respect to the linearisation. Thus, we use the STU_l-*relation* on Jacobi diagrams with a linearised skeleton from Definition 13.24:

Lemma 14.15. *Let $\mathfrak{g}$ be a metrized Lie algebra, $\pi : \mathfrak{T}(\mathfrak{g}) \to U(\mathfrak{g})$ be the projection map, $(D, \mathfrak{l})$ and $(D', \mathfrak{l}')$ be Jacobi diagrams with a linearised skeletons such that $(D', \mathfrak{l}')$ is obtained from $(D, \mathfrak{l})$ by STU_l-moves. Then $\pi \circ T_{\mathfrak{g}}(D, \mathfrak{l}, \ell, \mathfrak{d}) = \pi \circ T_{\mathfrak{g}}(D', \mathfrak{l}', \ell', \mathfrak{d}')$.*

Proof. It is enough to show that $\pi \circ T_{\mathfrak{g}}(D, \mathfrak{l}, \ell, \mathfrak{d}) = \pi \circ T_{\mathfrak{g}}(D', \mathfrak{l}', \ell', \mathfrak{d}')$ when $(D, \mathfrak{l})$ and $(D', \mathfrak{l}')$ are related by a single STU_l-move. By the previous two lemmas, we know that the value of $T_{\mathfrak{g}}$ is independent of the choice of decomposition and linearisation of trivalent vertices and so we may choose these for our convenience.

Thus, we need to show that, for some choices of decomposition and linearisation of trivalent vertices,

$\pi \circ T_{\mathfrak{g}}$ 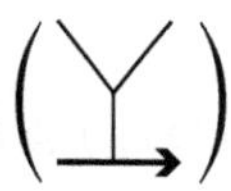 $= \pi \circ T_{\mathfrak{g}}$ 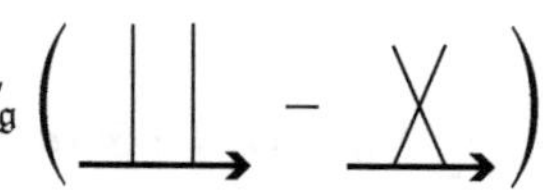 . (14.21)

In $U(\mathfrak{g})$, we have

$$\sum_k \gamma_{i,j}{}^k e_k = [e_i, e_j] = e_i \otimes e_j - e_j \otimes e_i. \tag{14.22}$$

Then the contribution of the fragment $\mathfrak{f} :=$  in D to $\pi \circ T_{\mathfrak{g}}$ 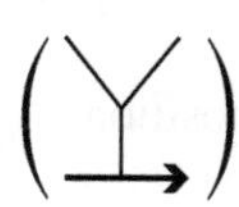 can be computed as

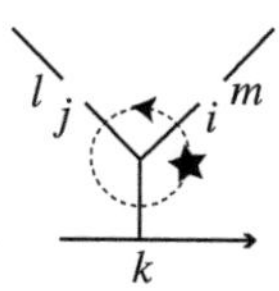

$$\mapsto S^{i,j,k} \langle e_l, e_j \rangle \langle e_m, e_i \rangle \, e_k = S^{i,j,k} h_{l,j} h_{m,i} \, e_k$$

by means of (14.14). We shall denote the contribution to $\pi \circ T_{\mathfrak{g}}$ 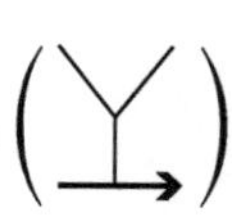 from D temporarily by X. Then

$$X = \sum_{i,j,k,l,m} S^{i,j,k} h_{l,j} h_{m,i} \, K_{D-\mathfrak{f}} \left(L_{D-\mathfrak{f}} \otimes e_k \otimes R_{D-\mathfrak{f}} \right)$$

where $L_{D-\mathfrak{f}}$ and $R_{D-\mathfrak{f}}$ are, respectively, tensor factors arising from the univalent vertices to the left and the right of the univalent vertex in the fragment $\mathfrak{f} =$ 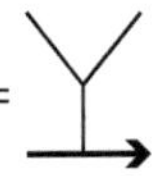, and $K_{D-\mathfrak{f}}$ is a scalar factor arising from the rest of D. Then

$$\begin{aligned} X &= - \sum_{a,b,i,j,k,l,m} \gamma_{a,b}{}^{k} h^{a,i} h^{b,j} h_{l,j} h_{m,i} K_{D-\mathfrak{f}} \, (L_{D-\mathfrak{f}} \otimes e_k \otimes R_{D-\mathfrak{f}}) \\ &= - \sum_{a,b,k,l,m} \gamma_{a,b}{}^{k} \delta_{a,m} \, \delta_{b,l} K_{D-\mathfrak{f}} \, (L_{D-\mathfrak{f}} \otimes e_k \otimes R_{D-\mathfrak{f}}) && \text{Eq. (14.7) twice} \\ &= - \sum_{k,l,m} \gamma_{m,l}{}^{k} K_{D-\mathfrak{f}} \, (L_{D-\mathfrak{f}} \otimes e_k \otimes R_{D-\mathfrak{f}}) \\ &= - \sum_{l,m} K_{D-\mathfrak{f}} \, (L_{D-\mathfrak{f}} \otimes [e_m, e_l] \otimes R_{D-\mathfrak{f}}) && \text{Eq. (14.22)} \\ &= - \sum_{l,m} K_{D-\mathfrak{f}} \, (L_{D-\mathfrak{f}} \otimes (e_m \otimes e_l - e_l \otimes e_m) \otimes R_{D-\mathfrak{f}}). \end{aligned}$$

Thus, the contribution from D to $\pi \circ T_{\mathfrak{g}}$ 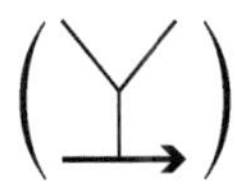 is

$$- \sum_{l,m} K_{D-\mathfrak{f}} \, (L_{D-\mathfrak{f}} \otimes (e_m \otimes e_l - e_l \otimes e_m) \otimes R_{D-\mathfrak{f}}). \tag{14.23}$$

On the other hand, the contributions of the fragments $\mathfrak{f}_1 =$  and $\mathfrak{f}_2 =$  of D to $\pi \circ T_{\mathfrak{g}}$ 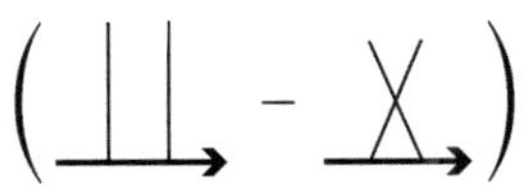 may be computed through the decompositions

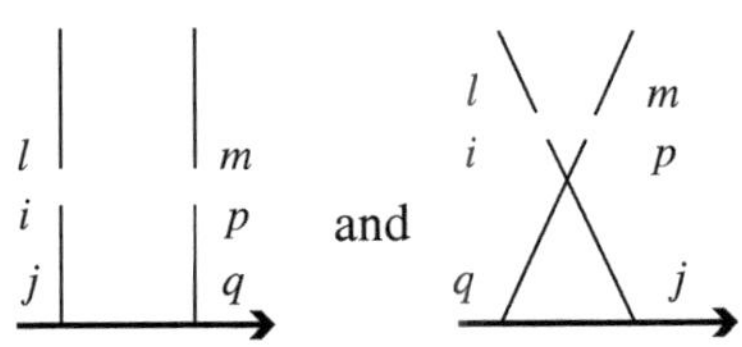

respectively.

Let Y temporarily denote the sum of these contributions. Then proceeding as before, we have

$$\begin{aligned}
Y = & \sum_{i,j,l,m,p,q} K_{D-\mathfrak{f}_1} h^{i,j} h^{p,q} \langle e_i, e_l \rangle \langle e_m, e_p \rangle \left(L_{D-\mathfrak{f}_1} \otimes e_j \otimes e_q \otimes R_{D-\mathfrak{f}_1} \right) \\
& - \sum_{i,j,l,m,p,q} K_{D-\mathfrak{f}_2} h^{i,j} h^{p,q} \langle e_i, e_l \rangle \langle e_m, e_p \rangle \left(L_{D-\mathfrak{f}_2} \otimes e_q \otimes e_j \otimes R_{D-\mathfrak{f}_2} \right) \\
= & \sum_{i,j,l,m,p,q} K_{D-\mathfrak{f}_1} h^{i,j} h^{p,q} h_{i,l} h_{m,p} \left(L_{D-\mathfrak{f}_1} \otimes (e_j \otimes e_q - e_q \otimes e_j) \otimes R_{D-\mathfrak{f}_1} \right) \\
= & \sum_{j,l,m,q} K_{D-\mathfrak{f}_1} \delta_{l,j} \delta_{m,q} \left(L_{D-\mathfrak{f}_1} \otimes (e_j \otimes e_q - e_q \otimes e_j) \otimes R_{D-\mathfrak{f}_1} \right) \\
= & \sum_{l,m} K_{D-\mathfrak{f}_1} \left(L_{D-\mathfrak{f}_1} \otimes (e_l \otimes e_m - e_m \otimes e_l) \otimes R_{D-\mathfrak{f}_1} \right).
\end{aligned}$$

Thus, the contribution from D to $\pi \circ T_{\mathfrak{g}}$ 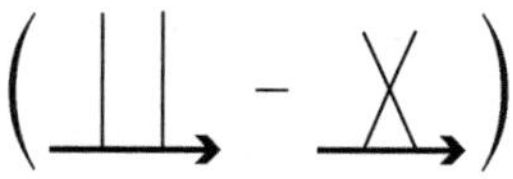 is

$$\sum_{j,l,m,q} K_{D-\mathfrak{f}_1} \left(L_{D-\mathfrak{f}_1} \otimes (e_l \otimes e_m - e_m \otimes e_l) \otimes R_{D-\mathfrak{f}_1} \right). \tag{14.24}$$

We see that (14.23) and (14.24) are equal, and therefore that (14.21) holds, completing the proof. □

Lemma 14.16. *Let $\mathfrak{g}$ be a metrized Lie algebra, $\pi : \mathfrak{T}(\mathfrak{g}) \to U(\mathfrak{g})$ be the projection map, and D be a Jacobi diagram. Let $\mathfrak{l}$ and $\mathfrak{l}'$ be two linearisations of the skeleton of D. Then $\pi \circ T_{\mathfrak{g}}(D, \mathfrak{l}, \ell, \mathfrak{d}) = \pi \circ T_{\mathfrak{g}}(D, \mathfrak{l}', \ell', \mathfrak{d}')$.*

Proof. Consider the linearised Jacobi diagrams $(D, \mathfrak{l})$ and $(D, \mathfrak{l}')$. By Corollary 13.29 we have that $(D, \mathfrak{l}')$ may be obtained from $(D, \mathfrak{l})$ by applications of the STU_l-relation. That $\pi \circ T_{\mathfrak{g}}(D, \mathfrak{l}, \ell, \mathfrak{d}) = \pi \circ T_{\mathfrak{g}}(D, \mathfrak{l}', \ell', \mathfrak{d}')$ follows from an application of Lemma 14.15. □

Proof (Theorem 14.11*).* Parts (1)–(4) of Theorem 14.11 follow immediately from Lemmas 14.13–14.16, respectively. □

For referencing later, we summarise the construction of $T_{\mathfrak{g}}$ and of Definition 14.12 below.

Definition 14.17 (Universal $\mathfrak{g}$-weight system). Let $\mathfrak{g}$ be a metrized Lie algebra and $\langle \cdot, \cdot \rangle$ denote its bilinear form. Then $T_{\mathfrak{g}} : \mathscr{A} \to U(\mathfrak{g})$ denotes the *unframed universal* $(\mathfrak{g}, \langle \cdot, \cdot \rangle)$*-weight system.* $T_{\mathfrak{g}}(D)$ is computed from any fragmentary decomposition of D by the following association to obtain an element of $\mathfrak{T}(\mathfrak{g})$, then projecting into $U(\mathfrak{g})$:

$T_{\mathfrak{g}}:$ 

$\longmapsto S^{a,b,c} := - \sum_{s,t \in I} \gamma_{s,t}{}^{c} h^{s,a} h^{t,b},$

$T_{\mathfrak{g}}:$ 

$\longmapsto h^{a,b},$

$T_{\mathfrak{g}}:$ 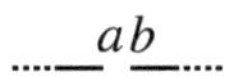

$\longmapsto h_{a,b} = \langle e_a, e_b \rangle ,$

$T_{\mathfrak{g}}:$ 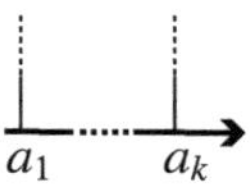

$\longmapsto e_{a_1} \otimes \cdots \otimes e_{a_k}.$

14.3 Lie Algebra Weight Systems

We have just constructed a map $T_{\mathfrak{g}} : \mathscr{A} \to U(\mathfrak{g})$ where $\mathfrak{g}$ is a metrized Lie algebra. Since we are studying Vassiliev invariants, our interest is in weight systems, *i.e.* maps from $\mathscr{A}$ to $\mathbb{C}$. There is a canonical way to obtain weight systems from $T_{\mathfrak{g}}$ as follows. If we are given a metrized Lie algebra $\mathfrak{g}$ together with a representation ρ of its universal enveloping algebra, $\rho\colon U(\mathfrak{g}) \longrightarrow \mathrm{End}(V)$, where V is some complex vector space, we may obtain a weight system $W_{\mathfrak{g},\rho}$ through the composition

$$W_{\mathfrak{g},\rho}\colon D \overset{T_{\mathfrak{g}}}{\longmapsto} T_{\mathfrak{g}}(D) \overset{\rho}{\longmapsto} \rho \circ T_{\mathfrak{g}}(D) \overset{\mathrm{Tr}}{\longmapsto} \mathrm{Tr}\left(\rho \circ T_{\mathfrak{g}}(D)\right) \in \mathbb{C}, \qquad (14.25)$$

where Tr denotes the trace. We call a weight system obtained in this way a *Lie algebra weight system.*

An important fact is that we can obtain a Lie algebra weight system from the classical Lie algebras $\mathfrak{sl}_n$, $\mathfrak{gl}_n$, $\mathfrak{so}_n$, $\mathfrak{sp}_{2n}$ and any of their representations. Every weight system corresponds to a Vassiliev invariant, and the knot invariants coming from classical Lie algebras are important since, as we shall see, they relate to the Reshetikhin–Turaev invariants of Part II. Here we shall focus upon one of the simplest cases, the Lie algebra $\mathfrak{sl}_2$ and its standard two-dimensional representation. In this section, we shall find a combinatorial description of this Lie algebra weight system. Later, in Sect. 18.2, we shall identify the knot invariants corresponding to this weight system.

Recall from Definition 8.44 that $\mathfrak{sl}_2$ has basis $\{X, Y, H\}$ with Lie bracket given by

$$[X, Y] = H, \quad [H, Y] = -2Y, \quad [H, X] = 2X.$$

Then from Definition 8.47 and (8.19), its universal enveloping algebra $U(\mathfrak{sl}_2)$ is the tensor algebra of $\mathfrak{sl}_2$ with relations

$$X \otimes Y - Y \otimes X = H, \quad H \otimes Y - Y \otimes H = -2Y, \quad H \otimes X - X \otimes H = 2X.$$

Let ρ denote the standard two-dimensional representation of $U(\mathfrak{sl}_2)$. It is given by

$$\rho(X) = \begin{bmatrix} 0 & 1 \\ 0 & 0 \end{bmatrix}; \quad \rho(Y) = \begin{bmatrix} 0 & 0 \\ 1 & 0 \end{bmatrix}; \quad \rho(H) = \begin{bmatrix} 1 & 0 \\ 0 & -1 \end{bmatrix}. \tag{14.26}$$

Note that since we are viewing $U(\mathfrak{sl}_2)$ as a tensor algebra, the multiplication is $\otimes$ and so $\rho(x \otimes y) = \rho(x) \cdot \rho(y)$ (and not $\rho(x) \otimes \rho(y)$).

We may define a bilinear form $\langle \cdot, \cdot \rangle : \mathfrak{sl}_2 \otimes \mathfrak{sl}_2 \to \mathfrak{sl}_2$ by

$$\langle x, y \rangle = \mathrm{Tr}\left(\rho(x) \cdot \rho(y)\right).$$

With this, $\mathfrak{sl}_2$ forms a metrized Lie algebra (see Exercise 14.4).

From (14.4), we may view the bilinear form as an element of $\mathfrak{sl}_2^\star \otimes \mathfrak{sl}_2^\star$, writing it as

$$\langle \cdot, \cdot \rangle = \sum_{A,B \in \{X,Y,H\}} h_{A,B}\, A^* \otimes B^* = h_{X,X} X^* \otimes X^* + h_{X,Y} X^* \otimes Y^* + \cdots + h_{H,H} H^* \otimes H^*,$$

where the coefficients of the nine terms in the expression are

$$h_{X,X} = \langle X, X \rangle = 0, \qquad h_{Y,Y} = \langle Y, Y \rangle = 0, \quad h_{H,H} = \langle H, H \rangle = 2,$$

$$h_{X,Y} = h_{Y,X} = \langle Y, X \rangle = 1, \quad h_{X,H} = h_{H,X} = \langle X, H \rangle = \langle H, X \rangle = 0,$$

$$h_{Y,H} = h_{H,Y} = \langle H, Y \rangle = 0.$$

Then, viewing $\langle \cdot, \cdot \rangle$ as a map (see (14.5)) from $\mathfrak{g}$ to $\mathfrak{g}^\star$ we have

$$X \overset{\langle \cdot,\cdot \rangle}{\longmapsto} Y^\star, \quad Y \overset{\langle \cdot,\cdot \rangle}{\longmapsto} X^\star, \quad H \overset{\langle \cdot,\cdot \rangle}{\longmapsto} 2H^\star.$$

This map is invertible (see (14.6)) with inverse $\langle \cdot, \cdot \rangle^{-1} : U(\mathfrak{g})^\star \to U(\mathfrak{g})$ given by

$$X^\star \overset{\langle \cdot,\cdot \rangle^{-1}}{\longmapsto} Y, \quad Y^\star \overset{\langle \cdot,\cdot \rangle^{-1}}{\longmapsto} X, \quad H^\star \overset{\langle \cdot,\cdot \rangle^{-1}}{\longmapsto} \tfrac{1}{2} H.$$

Thus, $h^{X,Y} = h^{Y,X} = 1$, $h^{H,H} = \frac{1}{2}$, $h^{X,X} = h^{Y,Y} = h^{X,H} = h^{H,X} = h^{Y,H} = h^{H,Y} = 0$.

We may now compute $W_{\mathfrak{sl}_2,\rho}$ for a number of Jacobi diagrams, following Example 14.6 for the construction of $T_{\mathfrak{sl}_2}(D)$.

Example 14.18.

$T_{\mathfrak{sl}_2}\Big($  $\Big) = \sum_{i,j\in\{X,Y,H\}} h^{i,j}\, i \otimes j = X \otimes Y + Y \otimes X + \frac{1}{2}\, H \otimes H.$

Applying the representation ρ of $U(\mathfrak{sl}_2)$ to this expression gives

$$\rho(X)\cdot\rho(Y) + \rho(Y)\cdot\rho(X) + \tfrac{1}{2}\rho(H)\cdot\rho(H) = \begin{bmatrix} 1 & 0 \\ 0 & 0 \end{bmatrix} + \begin{bmatrix} 0 & 0 \\ 0 & 1 \end{bmatrix} + \tfrac{1}{2}\begin{bmatrix} 1 & 0 \\ 0 & 1 \end{bmatrix},$$

whose trace is equal to 3, so, from (14.25), $W_{\mathfrak{sl}_2,\rho}\Big($  $\Big) = 3$.

Example 14.19.

 $\overset{T_{\mathfrak{sl}_2}}{\longmapsto} \sum_{i,j,k,l\in\{X,Y,H\}} h^{i,k}h^{j,l}\, i \otimes j \otimes k \otimes l$

$$\begin{aligned}
&= X \otimes X \otimes Y \otimes Y + X \otimes Y \otimes Y \otimes X + \tfrac{1}{2}X \otimes H \otimes Y \otimes H + \\
&\quad Y \otimes X \otimes X \otimes Y + Y \otimes Y \otimes X \otimes X + \tfrac{1}{2}Y \otimes H \otimes X \otimes H + \\
&\quad \tfrac{1}{2}H \otimes X \otimes H \otimes Y + \tfrac{1}{2}H \otimes Y \otimes H \otimes X + \tfrac{1}{4}H \otimes H \otimes H \otimes H \\
&\overset{\mathrm{Tr}\circ\rho}{\longmapsto} (0 + 0 - \tfrac{1}{2}) + (0 + 0 - \tfrac{1}{2}) + (-\tfrac{1}{2} - \tfrac{1}{2} + \tfrac{1}{2})
\end{aligned}$$

so $W_{\mathfrak{sl}_2,\rho}\Big($ 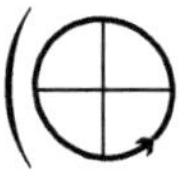 $\Big) = -\frac{3}{2}$.

Example 14.20. For $T_{\mathfrak{sl}_2}\Big($  $\Big)$, it is necessary to determine the structure constants $\gamma_{i,j}{}^k$ of the Lie algebra, which are defined by given by $[e_i, e_j] = \sum_k \gamma_{i,j}{}^k e_k$. We know that $[X, Y] = H$, $[H, Y] = -2Y$, $[H, X] = 2X$ and so

$$\begin{aligned}
&\gamma_{X,Y}{}^H = 1, \quad \gamma_{H,Y}{}^Y = -2, \; \gamma_{H,X}{}^X = 2, \\
&\gamma_{Y,X}{}^H = -1, \; \gamma_{Y,H}{}^Y = 2, \quad \gamma_{X,H}{}^X = -2,
\end{aligned}$$

and $\gamma_{i,j}{}^{k} = 0$ otherwise. Now

$T_{\mathfrak{sl}_2}\Big($  $\Big) = \sum_{i,j,k\in\{X,Y,H\}} S^{i,j,k}\, i \otimes j \otimes k,$

where $S^{i,j,k} := -\sum_{s,t\in\{X,Y,H\}} \gamma_{s,t}{}^{k}\, h^{s,i} h^{j,t}$. Then

$T_{\mathfrak{sl}_2}\Big($  $\Big)$

$$\begin{aligned}
&= (S^{H,X,Y} H \otimes X \otimes Y + S^{X,H,Y} X \otimes H \otimes Y)\\
&\quad + (S^{H,Y,X} H \otimes Y \otimes X + S^{Y,H,X} Y \otimes H \otimes X)\\
&\quad + (S^{X,Y,H} X \otimes Y \otimes H + S^{Y,X,H} Y \otimes X \otimes H)\\
&= (H \otimes X \otimes Y - X \otimes H \otimes Y)\\
&\quad + (-H \otimes Y \otimes X + Y \otimes H \otimes X)\\
&\quad + (X \otimes Y \otimes H - Y \otimes X \otimes H),
\end{aligned}$$

so

$\rho T_{\mathfrak{sl}_2}\Big($  $\Big)$

$$= \left(\begin{bmatrix}1&0\\0&0\end{bmatrix} - \begin{bmatrix}-1&0\\0&0\end{bmatrix}\right) + \left(-\begin{bmatrix}0&0\\0&-1\end{bmatrix} + \begin{bmatrix}0&0\\0&1\end{bmatrix}\right) + \left(\begin{bmatrix}1&0\\0&0\end{bmatrix} - \begin{bmatrix}0&0\\0&-1\end{bmatrix}\right).$$

Thus, $W_{\mathfrak{sl}_2,\rho}\Big($  $\Big) = 6$.

Exercise 14.21. Compute $W_{\mathfrak{sl}_2,\rho}\Big($  $\Big)$ directly as in the examples above, and by applying the STU-relation which gives

$W_{\mathfrak{sl}_2,\rho}\Big($  $\Big) = W_{\mathfrak{sl}_2,\rho}\Big($ 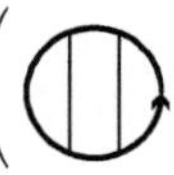 $\Big) - W_{\mathfrak{sl}_2,\rho}\Big($  $\Big).$

14.4 A Combinatorial Description of $W_{\mathfrak{sl}_2,\rho}$

Our aim is to find a combinatorial description of the weight system $W_{\mathfrak{sl}_2,\rho}$, where ρ is the standard two-dimensional representation of $\mathfrak{sl}_2(\mathbb{C})$. (This description is given in Theorem 14.24.) Since, from Theorem 13.36, $\mathscr{A} \cong \mathscr{A}^c$, we can always find a chord diagram to represent any Jacobi diagram, and so it is enough to find a combinatorial description of the action of $W_{\mathfrak{sl}_2,\rho}$ on chord diagrams. We proceed by examining how to determine $W_{\mathfrak{sl}_2,\rho}$ on a chord diagram.

A Preliminary Expression for $W_{\mathfrak{sl}_2,\rho}$

To determine $W_{\mathfrak{sl}_2,\rho}(C)$, we first choose a base point on the skeleton of the chord diagram C and label the end of each chord by $i_1, \ldots, i_{2n}$ with respect to the linear order on the skeleton given by the base point. For example,

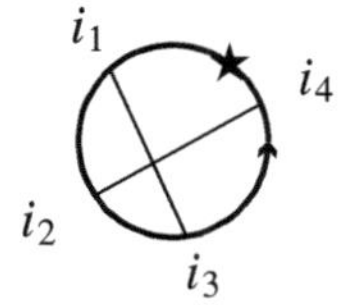

.

In the computation of $W_{\mathfrak{sl}_2,\rho}(C)$, each label i_k will be one of the generators X, Y, H of $\mathfrak{sl}_2$, and $W_{\mathfrak{sl}_2,\rho}(C)$ will be obtained as a sum over all ways of choosing $i_1, \ldots, i_{2n}$ to be elements of $\{X, Y, H\}$. Then by definition

$$W_{\mathfrak{sl}_2,\rho}(C) = \operatorname{Tr} \sum_{\substack{i_1,\ldots,i_{2n} \\ \in\{X,Y,H\}}} \left(\prod_{\substack{(i_p,i_q) \text{ label} \\ \text{ends of a chord}}} h^{i_p,i_q} \right) \rho(i_1)\rho(i_2)\cdots\rho(i_{2n}),$$

where the $\rho(i_k)$ are one of the representing matrices

$$\rho(X) = \begin{bmatrix} 0 & 1 \\ 0 & 0 \end{bmatrix}, \quad \rho(Y) = \begin{bmatrix} 0 & 0 \\ 1 & 0 \end{bmatrix}, \quad \rho(H) = \begin{bmatrix} 1 & 0 \\ 0 & -1 \end{bmatrix}.$$

Now let $\rho(i_j)_{a,b}$ denote the (a, b)-entry of the matrix $\rho(i_j)$. Then we may write the trace as a sum of matrix elements to get

$$\begin{aligned} W_{\mathfrak{sl}_2,\rho}(C) &= \sum_{\substack{i_1,\ldots,i_{2n} \\ \in\{X,Y,H\}}} \left(\prod_{\substack{(i_p,i_q) \text{ label} \\ \text{ends of a chord}}} h^{i_p,i_q} \right) \cdot \operatorname{Tr}\left(\rho(i_1)\rho(i_2)\cdots\rho(i_{2n})\right). \qquad (14.27) \\ &= \sum_{\substack{i_1,\ldots,i_{2n} \\ \in\{X,Y,H\}}} \left(\prod_{\substack{(i_p,i_q) \text{ label} \\ \text{ends of a chord}}} h^{i_p,i_q} \right) \sum_{\substack{a_1,a_2,\ldots,a_{2n} \\ \in\{1,2\}}} \rho(i_1)_{a_1,a_2}\rho(i_2)_{a_2,a_3}\cdots\rho(i_{2n})_{a_{2n},a_1}. \end{aligned}$$

For example,

$$W_{\mathfrak{sl}_2,\rho}\left(\quad \right) = \operatorname{Tr} \sum_{\substack{i_1,\ldots,i_4 \\ \in\{X,Y,H\}}} h^{i_1,i_3}\, h^{i_2,i_4}\, \rho(i_1)\rho(i_2)\rho(i_3)\rho(i_4)$$

$$= \sum_{\substack{i_1,\dots,i_4 \\ \in\{X,Y,H\}}} h^{i_1,i_3} h^{i_2,i_4} \operatorname{Tr}(\rho(i_1)\rho(i_2)\rho(i_3)\rho(i_4))$$

$$= \sum_{\substack{i_1,\dots,i_4 \\ \in\{X,Y,H\}}} h^{i_1,i_3} h^{i_2,i_4} \sum_{\substack{a_1,a_2,a_3,a_4 \\ \in\{1,2\}}} \rho(i_1)_{a_1,a_2}\rho(i_2)_{a_2,a_3}\rho(i_3)_{a_3,a_4}\rho(i_4)_{a_4,a_1}.$$

To evaluate $W_{\mathfrak{sl}_2,\rho}(C)$, we may eliminate all terms in (14.27) that are necessarily zero. We have $h^{X,Y} = 1$, $h^{Y,X} = 1$ and $h^{H,H} = \frac{1}{2}$ and h^{i_p,i_q} is zero otherwise. This has two consequences as follows.

Summing Over Allowed Labellings

The first consequence is that we need only to consider summands in (14.27) in which the chords are labelled as one of

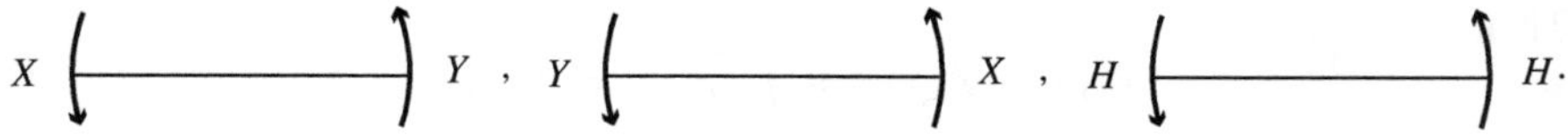

We shall call any labelling in which each chord is labelled by any one of these three types an *allowed labelling*. Thus, we may restrict the first sum in (14.27) to allowed labellings.

The second consequence is that, for allowed labellings,

$$\prod_{\substack{(i_p,i_q)\text{ label} \\ \text{ends of a chord}}} h^{i_p,i_q} = \left(\tfrac{1}{2}\right)^{\#HH\text{-chords}},$$

where $\#HH$-chords denotes the number of chords that have both ends labelled by H. Thus,

$$W_{\mathfrak{sl}_2,\rho}(C) = \sum_{\substack{i_1,\dots,i_{2n} \\ \in\{X,Y,H\} \\ \text{allowed labelling}}} \left(\tfrac{1}{2}\right)^{\#HH\text{-chords}} \sum_{\substack{a_1,a_2,\dots,a_{2n} \\ \in\{1,2\}}} \rho(i_1)_{a_1,a_2}\rho(i_2)_{a_2,a_3}\cdots\rho(i_{2n})_{a_{2n},a_1}. \tag{14.28}$$

We now examine when $\rho(i_1)_{a_1,a_2}\rho(i_2)_{a_2,a_3}\cdots\rho(i_{2n})_{a_{2n},a_1}$ can be nonzero. For this, consider a chord in C labelled as

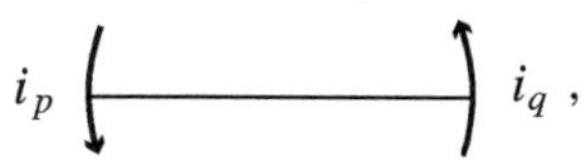

which gives rise to two terms $\cdots\rho(i_p)_{\alpha,\beta}\cdots\rho(i_q)_{\gamma,\delta}\cdots$ in the summand. We know i_p is one of X, Y or H, and so we know the zero entries of $\rho(i_p)$, and similarly for

$\rho(i_q)$. Thus, we may examine under what conditions both $\rho(i_p)_{\alpha,\beta}$ and $\rho(i_q)_{\gamma,\delta}$ are nonzero. There are three cases:

Case 1: Because we are considering allowed labellings, we have

$$\rho(i_p) = \rho(X) \iff \rho(i_q) = \rho(Y).$$

Then

$$\rho(i_p)_{\alpha,\beta} \neq 0 \implies \rho(i_p)_{\alpha,\beta} = \rho(i_p)_{1,2}$$

and

$$\rho(i_q)_{\gamma,\delta} \neq 0 \implies \rho(i_q)_{\gamma,\delta} = \rho(i_q)_{2,1}.$$

and both terms contribute a 1.

Case 2: Because we are considering allowed labellings, we have

$$\rho(i_p) = \rho(Y) \iff \rho(i_q) = \rho(X).$$

Then

$$\rho(i_p)_{\alpha,\beta} \neq 0 \implies \rho(i_p)_{\alpha,\beta} = \rho(i_p)_{2,1}$$

and

$$\rho(i_q)_{\gamma,\delta} \neq 0 \implies \rho(i_q)_{\gamma,\delta} = \rho(i_q)_{1,2}.$$

and both terms contribute a 1.

Case 3: Because we are considering allowed labellings, we have

$$\rho(i_p) = \rho(H) \iff \rho(i_q) = \rho(H).$$

Then

$$\rho(i_p)_{\alpha,\beta} \neq 0 \implies \rho(i_p)_{\alpha,\beta} = \rho(i_p)_{1,1} \text{ or } \rho(i_p)_{2,2}$$

and

$$\rho(i_q)_{\gamma,\delta} \neq 0 \implies \rho(i_q)_{\gamma,\delta} = \rho(i_q)_{1,1} \text{ or } \rho(i_q)_{2,2}.$$

and the terms contribute a 1 or a -1.

We may use this information to obtain a combinatorial description of

$$\sum_{\substack{a_1,a_2,\ldots,a_{2n} \\ \in\{1,2\}}} \rho(i_1)_{a_1,a_2}\rho(i_2)_{a_2,a_3}\cdots\rho(i_{2n})_{a_{2n},a_1}, \tag{14.29}$$

which is from the right-hand side of (14.28), as follows.

For each labelled chord i_p 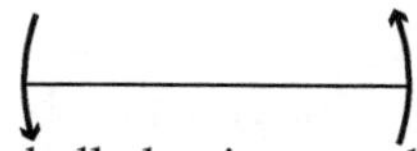 i_q, decorate the skeleton near the ends of each chord with two labelled points, to obtain

$\rho(i_p)_{\alpha,\beta}$ 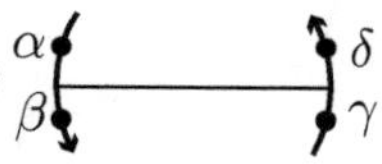

$\rho(i_q)_{\gamma,\delta}$

where the two labelled points correspond to the entries of the matrices $\rho(i_p)$ and $\rho(i_q)$ as indicated in the figure. There is a correspondence between a labelling of these points and an expression $\rho(i_1)_{a_1,b_1}\rho(i_2)_{a_2,b_2}\cdots\rho(i_{2n})_{a_{2n},b_{2n}}$ as shown in

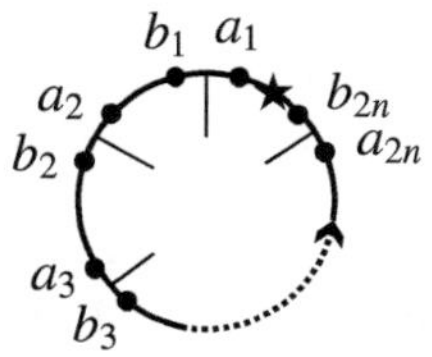

$$\longleftrightarrow \quad \rho(i_1)_{a_1,b_1}\rho(i_2)_{a_2,b_2}\cdots\rho(i_{2n})_{a_{2n},b_{2n}}.$$

Whenever the indices b_i and a_{i+1} match, we shall record this by connecting them with a bold line. In (14.29), some of the indices match up so the correspondence becomes

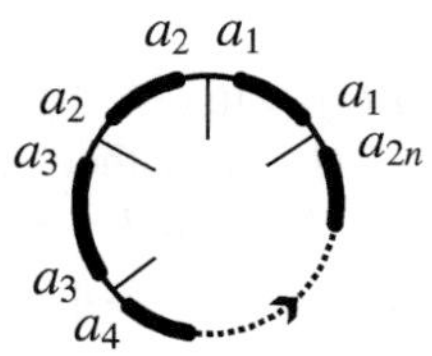

$$\longleftrightarrow \quad \rho(i_1)_{a_1,a_2}\rho(i_2)_{a_2,a_3}\cdots\rho(i_{2n})_{a_{2n},a_1}.$$

We sum over all $a_1, a_2, \ldots, a_{2n} \in \{1, 2\}$, which means that each a_j in the figure is either valued as 1 or 2.

Calculations with Decorated Chords

Returning to the decorated chord

$\rho(i_p)_{\alpha,\beta}$ 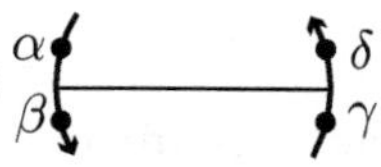

$\rho(i_q)_{\gamma,\delta}$,

we determine the contributions it makes to (14.29) for all possible values of i_p, i_q, α, β, γ, δ. There are three cases to consider, and these are specified by the three allowed labellings of the chord. We require both $\rho(i_p)_{\alpha,\beta}$ and $\rho(i_q)_{\gamma,\delta}$ to be nonzero.

Case 1: If $\rho(i_p) = X$ and $\rho(i_q) = Y$, then $\alpha = \delta = 1$ and $\beta = \gamma = 2$ (otherwise one of the entries of the matrix is zero). This case may be recorded by

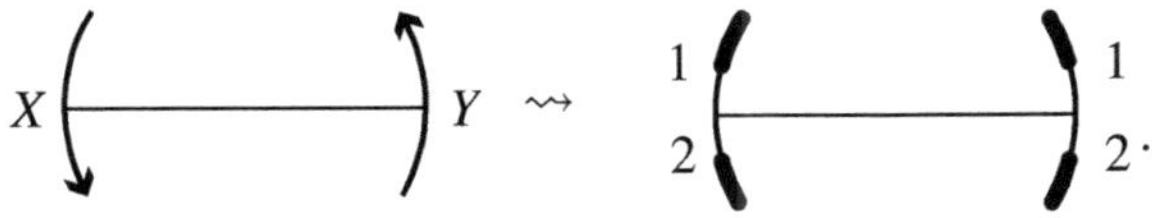

We may represent the equalities that $\alpha = \delta = 1$ and $\beta = \gamma = 2$ combinatorially by smoothing the skeleton along the chord as shown

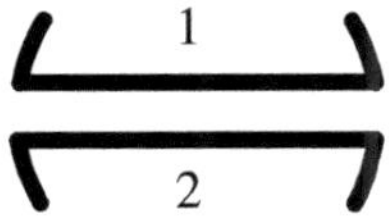

In this case, the total contribution of $\rho(i_p)_{\alpha,\beta}$ and $\rho(i_q)_{\gamma,\delta}$ to (14.29) is 1.

Case 2: Similarly, if $\rho(i_p) = Y$ and $\rho(i_q) = X$, then $\alpha = \delta = 2$ and $\beta = \gamma = 1$ (otherwise one of the entries of the matrix is zero). This case may be recorded by

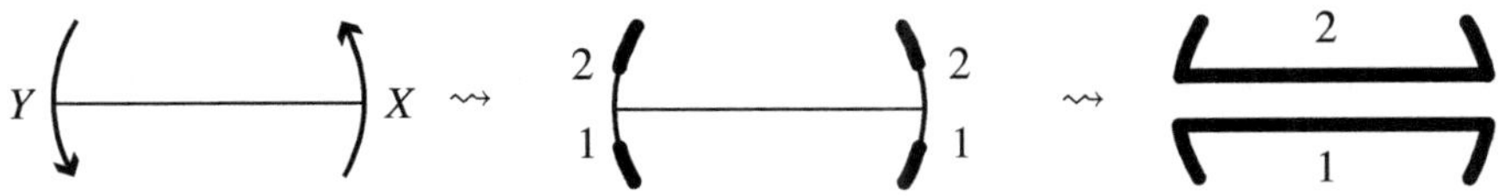

In this case, the total contribution of $\rho(i_p)_{\alpha,\beta}$ and $\rho(i_q)_{\gamma,\delta}$ to (14.29) is 1.

Case 3: This is the case in which $\rho(i_p) = \rho(i_q) = H$. Then there are four ways that $\rho(i_p)_{\alpha,\beta}$ and $\rho(i_q)_{\gamma,\delta}$ can be nonzero. Either $\alpha = \beta = 1$ and $\gamma = \delta = 2$, or $\alpha = \beta = 2$ and $\gamma = \delta = 1$, or $\alpha = \beta = \gamma = \delta = 1$, or $\alpha = \beta = \gamma = \delta = 2$. These four situations may be recorded by

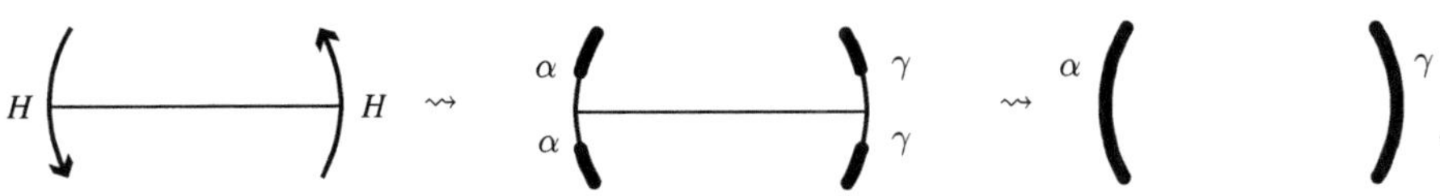

We see that the total contribution of $\rho(i_p)_{\alpha,\beta}$ and $\rho(i_q)_{\gamma,\delta}$ to (14.29) is 1 unless $\alpha = \beta \neq \gamma = \delta$ in which case it is -1.

Example 14.22. As an example, consider the term in $W_{\mathfrak{sl}_2,\rho}$  given by the labelling 

. Its contribution is, from (14.28),

$$\frac{1}{2} \sum_{i,j,k,l \in \{1,2\}} \rho(H)_{i,j}\, \rho(X)_{j,k}\, \rho(H)_{k,l} \rho(Y)_{l,i}.$$

In this sum, the only nonzero summand is $\rho(H)_{1,1}\rho(X)_{1,2}\rho(H)_{2,2}\rho(Y)_{2,1} = 1 \cdot 1 \cdot (-1) \cdot 1$. This summand is represented diagrammatically as

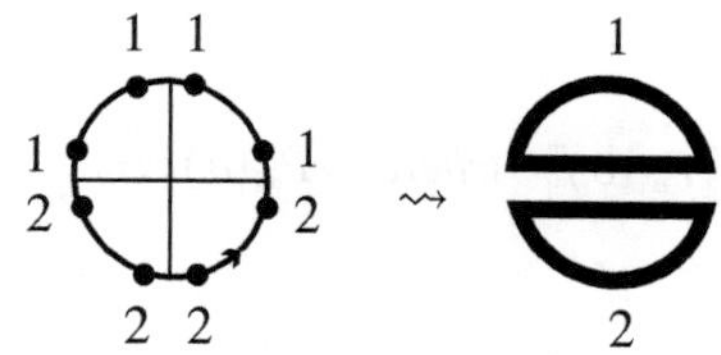

The total contribution of the labelling to $W_{\mathfrak{sl}_2,\rho}(C)$ is then $-\frac{1}{2}$.

A State Model for $W_{\mathfrak{sl}_2,\rho}$

The object we have just constructed is a *state model* for $W_{\mathfrak{sl}_2,\rho}$. We define a *state* σ of a chord diagram C as the configuration obtained by the local replacement of each chord by one of the following configurations:

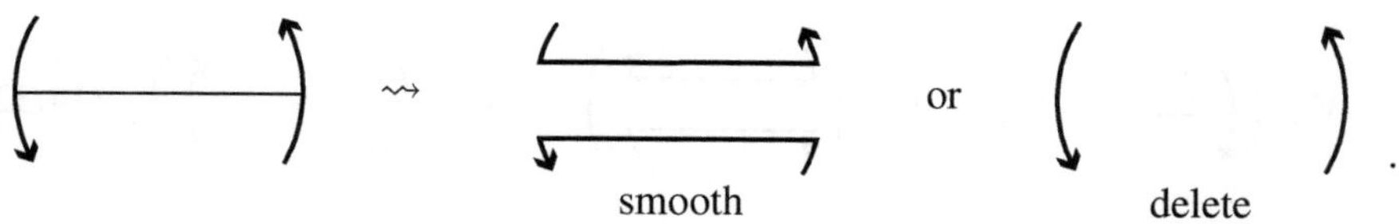

A $(1, 2)$-colouring of a state is an assignment of 1 or 2 to each of its closed curves. We shall say that a $(1, 2)$-colouring of a state is *permissible* if the two arcs that result from smoothing a chord have different colours.

To summarise the combinatorial construction, we have shown that there is a 1-1 correspondence

$$\left\{\begin{matrix}\text{non-zero terms in}\\ \mathrm{Tr}(\rho(i_1)\cdots\rho(i_{2n}))\end{matrix}\right\} \overset{1-1}{\longleftrightarrow} \left\{\begin{matrix}\text{permissible } (1,2)\text{-}\\ \text{colourings of states}\end{matrix}\right\}.$$

where it is to be recalled that the trace term on the left comes from (14.27).

Let $del(\sigma)$ denote the total number of deleted chords in the state. Consider a $(1, 2)$-coloured state. There are a number of pairs of arcs in the state that correspond to the ends of deleted chords. These pairs of arcs will either have the same colour or different colours in the colouring. Let

- $del_=(\sigma)$ denote the number of such pairs with the same colour
- $del_{\neq}(\sigma)$ denote the number of pairs with the different colours,

and let

$$\xi(\sigma) := \left(\tfrac{1}{2}\right)^{del(\sigma)} (-1)^{del_{\neq}(\sigma)}. \tag{14.30}$$

Then from our construction and (14.28), we have the following theorem.

Theorem 14.23.

$$W_{\mathfrak{sl}_2,\rho}(C) = \sum_{\text{states } \sigma} [\![P_{\text{all}}(\sigma)]\!] \quad \textit{where} \quad [\![P_{\text{all}}(\sigma)]\!] := \sum_{\substack{\text{permissible} \\ \{1,2\}\text{-colourings}}} \xi(\sigma) \tag{14.31}$$

A Skein Relation for $W_{\mathfrak{sl}_2}$

The expression in (14.31) for $W_{\mathfrak{g},\rho}(C)$ may be simplified further by observing that the right-hand side of (14.31) is entirely combinatorial and may be applied to chord diagrams with more than one skeleton component. We shall use this fact to obtain a skein relation for $W_{\mathfrak{sl}_2,\rho}(C)$.

Theorem 14.24. *There is a well-defined function* $W_{\mathfrak{sl}_2}$ *on chord diagrams whose skeleton is a union of disjoint oriented circles defined by the relation*

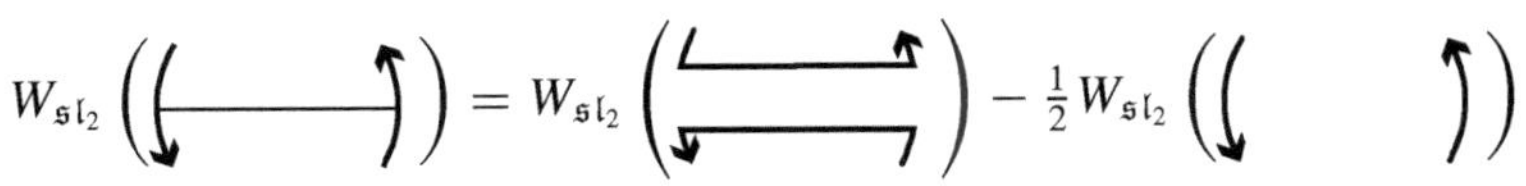

(14.32)

together with the initial condition on chord diagrams of n skeleton components and no chords

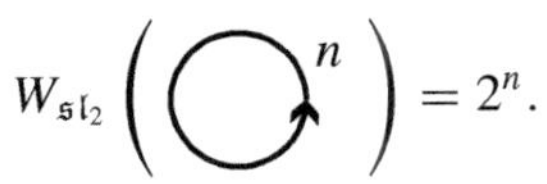

(14.33)

Moreover, for a chord diagram C,

$$W_{\mathfrak{sl}_2}(C) = W_{\mathfrak{sl}_2,\rho}(C).$$

Before proving the theorem, we illustrate it with two examples.

Example 14.25.

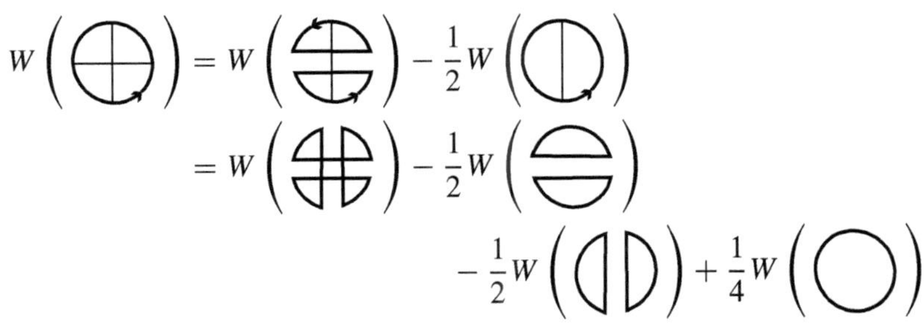

$$= 2 - 2 - 2 + \frac{1}{2} = -\frac{3}{2}.$$

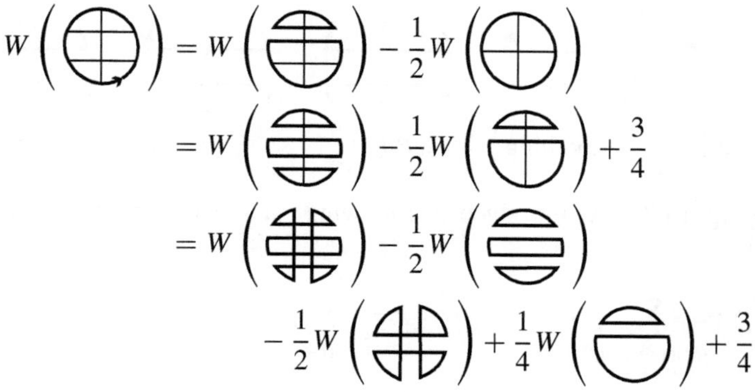

$$= 4 - 4 - 1 + 1 + \frac{3}{4} = \frac{3}{4}.$$

Exercise 14.26. Since $W_{\mathfrak{sl}_2}(C) = W_{\mathfrak{sl}_2,\rho}(C)$, Theorem 14.24 shows that $W_{\mathfrak{sl}_2}$ satisfies the 4T-relation for chord diagrams on one skeleton component. Prove that it satisfies the 4T-relation for chord diagrams with more than one skeleton component.

Proof (Theorem 14.24*).* Let C be a chord diagram, possibly with more than one circular skeleton component. Define a function P taking chord diagrams to $\mathbb{C}$ by the state sum

$$P(C) := \sum_{\text{states } \sigma} [\![P_{\text{all}}(\sigma)]\!].$$

Observe that, by (14.31), $P(C) = W_{\mathfrak{sl}_2,\rho}(C)$ when C has one skeleton component. We show that P satisfies the identities (14.32) and (14.33), which gives $W_{\mathfrak{sl}_2} = P$.

Suppose that C contains some chords and choose any chord c of C. We can partition the terms in the defining sum of $P(C)$ according to whether c is deleted or smoothed:

$$P(C) := \sum_{\substack{\text{states } \sigma \\ \text{with } c \text{ smoothed}}} [\![P_{\text{all}}(\sigma)]\!] + \sum_{\substack{\text{states } \sigma \\ \text{with } c \text{ deleted}}} [\![P_{\text{all}}(\sigma)]\!]. \tag{14.34}$$

We rewrite the two sums on the right-hand side.

Let C_s be the chord diagram obtained by smoothing c. For each state σ of C_s, the set $P_s(\sigma)$ of permissible $\{1, 2\}$-colourings of states of C_s may be partitioned into the sets:

- $P_s(\sigma, =)$, of those in which arcs on the skeleton that were created by the smoothing of the chord c have the same colour, and
- $P_s(\sigma, \neq)$, of those in which they have a different colour.

These different situations are indicated in the following figure.

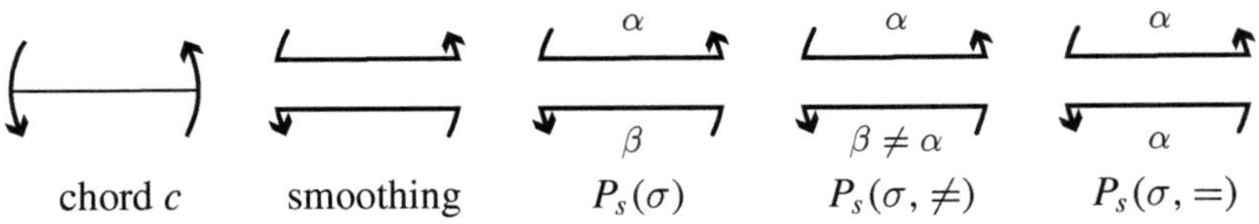

Recalling (14.30), we introduce the following notation to simplify the expressions that arise.

$$[\![P_s^{\neq}(\sigma)]\!] := \sum_{\substack{\text{colourings in}\\ P_s(\sigma,\neq)}} \xi(\sigma)\,,\quad [\![P_s^{=}(\sigma)]\!] := \sum_{\substack{\text{colourings in}\\ P_s(\sigma,=)}} \xi(\sigma)\,,\quad [\![P_s(\sigma)]\!] := \sum_{\substack{\text{colourings in}\\ P_s(\sigma)}} \xi(\sigma).$$

Then, for the first sum on the right-hand side of (14.34),

$$\sum_{\substack{\text{states } \sigma \text{ of } C\\ \text{with } c \text{ smoothed}}} [\![P_{\text{all}}(\sigma)]\!] = \sum_{\substack{\text{states}\\ \sigma \text{ of } C_s}} [\![P_s^{\neq}(\sigma)]\!] = \sum_{\substack{\text{states}\\ \sigma \text{ of } C_s}} [\![P_s(\sigma)]\!] - \sum_{\substack{\text{states}\\ \sigma \text{ of } C_s}} [\![P_s^{=}(\sigma)]\!]. \tag{14.35}$$

Now let C_d be the chord diagram obtained by deleting c. For each state σ of C_s, the set $P_d(\sigma)$ of permissible $\{1,2\}$-colourings of states of C_d may be partitioned into two sets:

- $P_d(\sigma,=)$, of those in which arcs that were at the ends of the chord c have the same colour,
- $P_d(\sigma,\neq)$, of those in which arcs have a different colour.

These different situations are listed in the following:

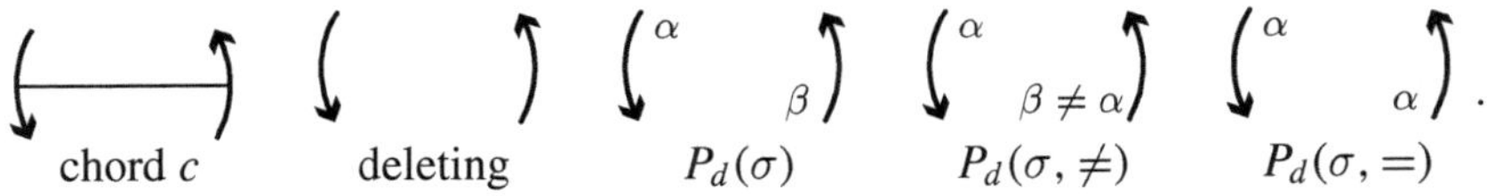

Then, for the second sum on the right-hand side of (14.34),

$$\sum_{\substack{\text{states } \sigma \text{ of } C\\ \text{with } c \text{ deleted}}} [\![P_{\text{all}}(\sigma)]\!] = \frac{1}{2}\left(\sum_{\substack{\text{states}\\ \sigma \text{ of } C_d}} [\![P_d^{=}(\sigma)]\!] + (-1) \sum_{\substack{\text{states}\\ \sigma \text{ of } C_d}} [\![P_d^{\neq}(\sigma)]\!], \right) \tag{14.36}$$

where the factor of $\frac{1}{2}$ arises to account for the differences in $del(\sigma)$ when changing from states of C to states of C_d, and the -1 because of the difference in $del_{\neq}(\sigma)$, and $\xi(\sigma)$ is defined in (14.30).

Substituting (14.35) and (14.36) into (14.34) and rearranging gives

$$P(C) = \sum_{\substack{\text{states}\\ \sigma \text{ of } C_s}} [\![P_s(\sigma)]\!] - \sum_{\substack{\text{states}\\ \sigma \text{ of } C_s}} [\![P_s^{=}(\sigma)]\!] + \tfrac{1}{2} \sum_{\substack{\text{states}\\ \sigma \text{ of } C_d}} [\![P_d^{=}(\sigma)]\!] - \tfrac{1}{2} \sum_{\substack{\text{states}\\ \sigma \text{ of } C_d}} [\![P_d^{\neq}(\sigma)]\!]. \tag{14.37}$$

The summations $\sum_{\substack{\text{states}\\ \sigma \text{ of } C_s}} [\![P_s^{=}(\sigma)]\!]$ and $\sum_{\substack{\text{states}\\ \sigma \text{ of } C_d}} [\![P_d^{=}(\sigma)]\!]$ are equal since any state of C_d with a colouring in $P_d(\sigma, =)$ may be transformed to a state of C_s with a colouring in $P_s(\sigma, =)$, and *vice versa* using

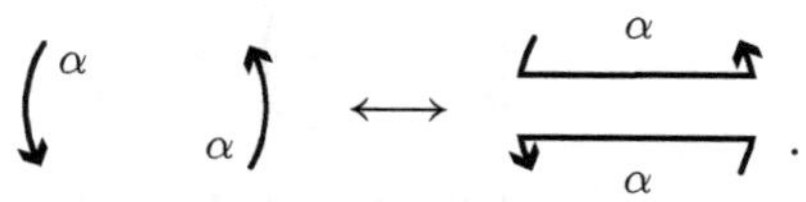

$P_s(\sigma, =)$ $P_s(\sigma, =)$

Thus, from (14.37), we have

$$P(C) = \sum_{\substack{\text{states}\\ \sigma \text{ of } C_s}} [\![P_s(\sigma)]\!] - \tfrac{1}{2} \sum_{\substack{\text{states}\\ \sigma \text{ of } C_d}} [\![P_d(\sigma)]\!].$$

It is readily seen that this is equation is precisely

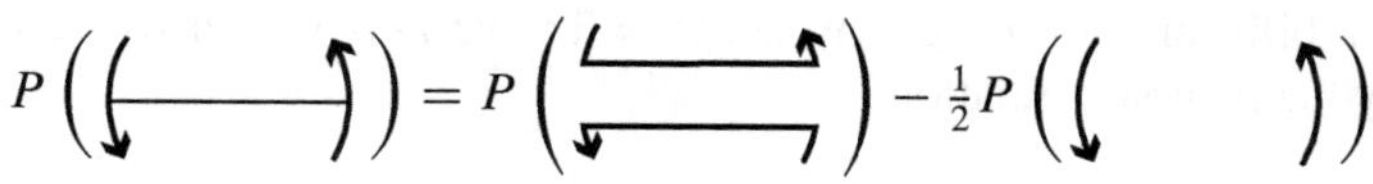

It is easily checked that

$= 2^n$. (14.38)

Thus, we have shown that $P(C)$ satisfies the defining relations (14.32) and (14.33) of $W_{\mathfrak{sl}_2}$ and so $P = W_{\mathfrak{sl}_2}$. Since $P(C)$ was defined as a sum, it follows that the recursion relation for P, and therefore $W_{\mathfrak{sl}_2}$ is independent of the order that the deletions and smoothings are applied to the chords. Thus, $W_{\mathfrak{sl}_2}$ is well-defined. Finally, $W_{\mathfrak{sl}_2}(C) = W_{\mathfrak{sl}_2,\rho}(C)$ when C has one skeleton component since the defining sum of $P(C)$ coincides with (14.31). □

We shall return to the recursive definition of $W_{\mathfrak{sl}_2,\rho}$ provided by Theorem 14.24 later in Sect. 18.2.

Exercise 14.27. Define an invariant of chord diagrams as follows. Let C be a chord diagram possibly with more than one circular skeleton component. Define $Q(C)$ by

$$Q(C) := \sum_{\sigma:\, chords \to \{1,-1/2\}} \left(\prod_{c \text{ a chord}} \sigma(c) \right) 2^{|\sigma(c)|} \tag{14.39}$$

where the sum is over all colourings of the chords of C with the elements 1 and $-\frac{1}{2}$, the product is over all chords of C, and $|\sigma(c)|$ is obtained by smoothing all 1 coloured chords, and deleting all $-\frac{1}{2}$ coloured chords then counting the number of skeleton components that remain. This is indicated by

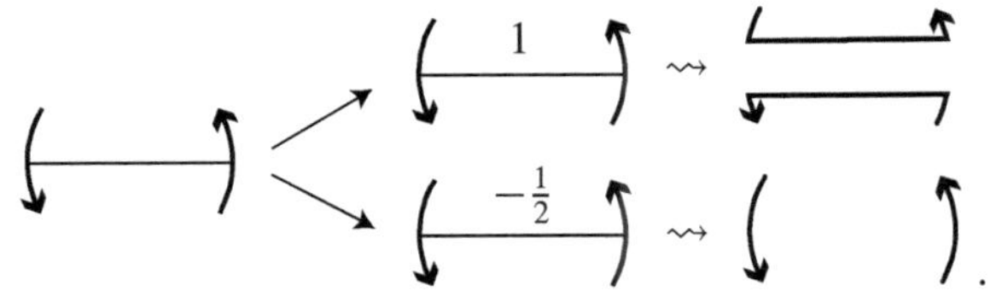

(14.40)

Prove that Q satisfies the recursive definition of $W_{\mathfrak{sl}_2}$ given in Theorem 14.24 and hence provides another combinatorial model for $W_{\mathfrak{sl}_2,\rho}$.

Exercise 14.28. Let C be a chord diagram with n chords that is specified by the circular sequence (as in Sect. 11.2) by $112233\cdots nn$. Show that $W_{\mathfrak{sl}_2,\rho}(C) \neq 0$. Hence, conclude that $\mathscr{A}_n$ has dimension at least 1 for all n. What does this tell you about $\mathscr{V}_n^f$?

Exercise 14.29. Consider the universal enveloping algebra $U(\mathfrak{gl}_n)$ where $\mathfrak{gl}_n$ has basis $\{E_{i,j} : 1 \leq i, j \leq n\}$. Let ρ be the standard n-dimensional representation of $\mathfrak{gl}_n$ in which $\rho(E_{i,j})$ consists of the matrix with (i, j)-entry 1 and all other entries 0. Define a bilinear form on $\mathfrak{gl}_n$ by $\langle x, y\rangle = \mathrm{Tr}\,(\rho(x)\cdot\rho(y))$. Prove that $W_{\mathfrak{gl}_n,\rho}$ is defined by the recursion relation

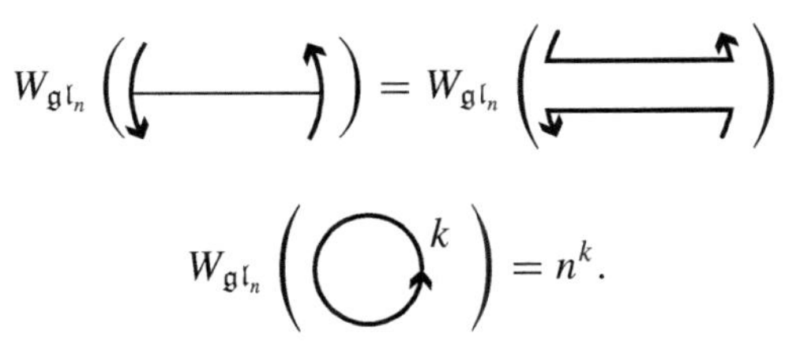

Exercise 14.30. Consider the universal enveloping algebra $U(\mathfrak{sl}_n)$ where $\mathfrak{sl}_n$ has basis $\{E_{i,j} : 1 \leq i, j \leq n, i \neq j\} \cup \{H_i : 1 \leq i \leq n-1\}$. Let ρ be the standard n-dimensional representation of $\mathfrak{sl}_n$ in which $\rho(E_{i,j})$ consists of the matrix with (i, j)-entry 1 and all other entries 0; and $\rho(H_i)$ has (i, i)-entry 1, $(i+1, i+1)$-entry -1, and all other entries 0. Define a bilinear form on $\mathfrak{sl}_n$ by $\langle x, y\rangle = \mathrm{Tr}\,(\rho(x)\cdot\rho(y))$. Prove that $W_{\mathfrak{sl}_n,\rho}$ is defined by the recursion relation

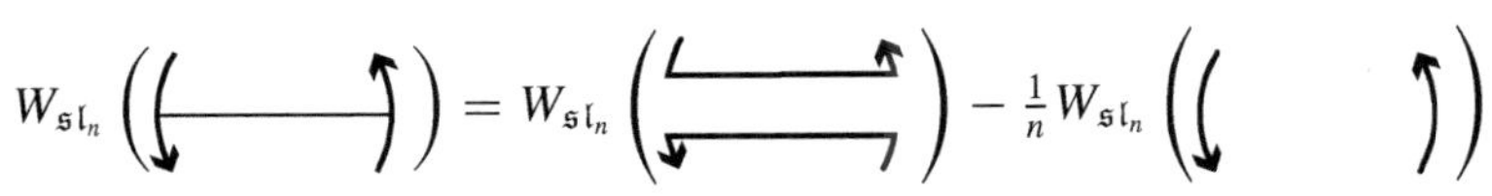

(14.41)

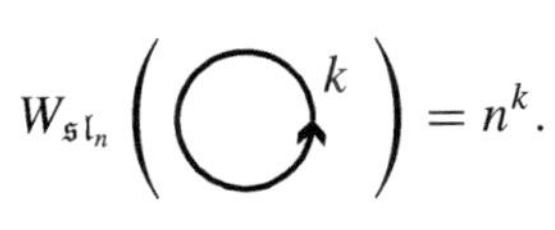

Part IV
The Kontsevich Invariant

Chapter 15
q-Tangles

In this part, we give an overview of the Kontsevich invariant, $\check{Z}$, and shall see that it is a universal object for both Vassiliev invariants and quantum invariants. This means that, in some sense, it contains *all* of these invariants.

It was introduced by Kontsevich in [102] as a universal Vassiliev invariant. It has its roots in Chern–Simons theory and the Knizhnik–Zamolodchikov equation, and the work of Khono and of Drinfel'd. It appeared in Kontsevich's paper as an iterated integral. Here, however, we will approach the Kontsevich invariant through the combinatorial extension to framed oriented tangles of Le and Murakami [116]. Since our focus is on the application of the invariant, we keep our exposition in this part very streamlined, quoting key results as needed. We refer a reader interested in additional background and approaches to the Kontsevich invariant to, for example, the excellent expositions in [37, 90, 144]. An interested reader can find some other approaches to the Kontsevich invariant in [6, 16, 32, 149].

15.1 Parenthesisations and Tangles

A *parenthesized word* on the alphabet $\{+, -\}$ is a finite word in the symbols "+" and "−", together with a (binary) *parenthesization*, or "bracketing", of it. For example, $(-+)-$, $-(-((-+)-))$ and $(((++)-)+)(+-)$ are instances of parenthesised words.

More formally, parenthesised words on the alphabet $\{+, -\}$ can be defined recursively by defining the empty word, denoted by ( ) or e, $(+)$, and $(-)$ as parenthesised words. Then for any parenthesised words π and π', we define $(+(\pi))$, $(-(\pi))$, $((\pi)+)$, $((\pi)-)$ and $((\pi)(\pi'))$ to be parenthesised words.

We impose a convention that redundant brackets are removed from parenthesised words. For example, this means that we remove any brackets around singletons or pairs, writing, *e.g.* "+" rather than "(+)", or "−+" rather than "(−+)". It also

D. M. Jackson and I. Moffatt, *An Introduction to Quantum and Vassiliev Knot Invariants*, CMS Books in Mathematics, https://doi.org/10.1007/978-3-030-05213-3_15

means that parenthesised words are not enclosed in brackets, so, for example, we write $+(-+)$ rather than $(+(-+))$.

With this convention, all the parenthesised words on the alphabet $\{+, -\}$ of length at most two are

$$(\,),\ +,\ -,\ ++,\ +-,\ -+,\ --\,.$$

The parenthesised words of length three are

$$\begin{gathered} +(++),\ +(+-),\ +(-+),\ +(--),\ -(++),\ -(+-),\ -(-+),\ \ -(--)\\ (++)+,\ (++)-,\ (+-)+,\ (+-)-,\ (-+)+,\ (-+)-,\ (--)+,\ (--)-\,. \end{gathered}$$

The 5×2^4 parenthesised words of length four arise from all possible ways of substituting either $+$ or $-$ for the $\cdot$ in $\cdot(\cdot(\cdot\cdot))$, $\cdot((\cdot\cdot)\cdot)$, $(\cdot\cdot)(\cdot\cdot)$, $(\cdot(\cdot\cdot))\cdot$, $((\cdot\cdot)\cdot)\cdot$.

A *q-tangle* (or *quasi-tangle*) consists of an oriented tangle together with two parenthesised words on the alphabet $\{+, -\}$ associated with the top and bottom of the tangle. Each endpoint of a component on $\mathbb{R} \otimes \{0\}$ and $\mathbb{R} \otimes \{1\}$ is associated with a symbol "+" if it is directed upwards and "−" if it is directed downwards. This associates a word with the bottom of the tangle on $\mathbb{R} \otimes \{0\}$ and the top of the tangle on $\mathbb{R} \otimes \{1\}$. A q-tangle consists of the tangle together with parenthesisations of these words.

For example,

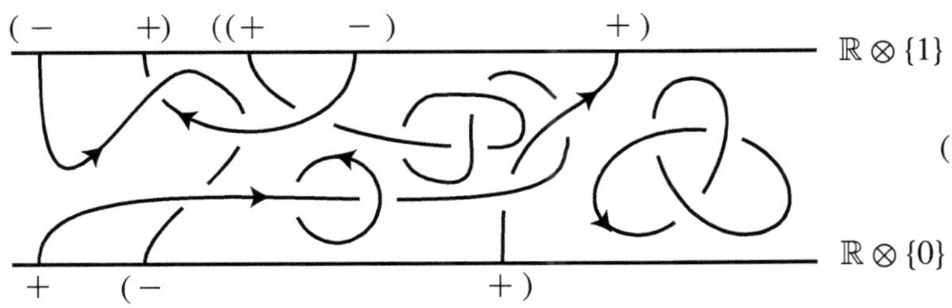

(15.1)

is a parenthesised q-tangle with six components. The parenthesisation at the top is $(-+)((+-)+)$ and the one at the bottom is $+(-+)$.

Definition 15.1 ((Framed) q-tangle). A *(framed) q-tangle* is a (framed) oriented tangle such that each boundary point of the tangle is assigned the symbol "+" if the strand is oriented upwards at this boundary point: otherwise, it is assigned the symbol "−". This defines words associated with $\mathbb{R} \otimes \{0\}$ and $\mathbb{R} \otimes \{1\}$, and a parenthesisation is assigned to each of these words.

The *underlying (framed) tangle* of a (framed) q-tangle is obtained by forgetting the parenthesised words.

A *diagram of a (framed) q-tangle* consists of a diagram for the underlying (framed) tangle endowed with the parenthesised words of the (framed) q-tangle on the boundary points of the (framed) tangle diagram. We shall henceforth identify (framed) q-tangles with their diagrams.

Although the parenthesisations of the words on the top and bottom of a q-tangle can differ, the words themselves cannot (since they are determined by the tangle). Accordingly in diagrams of q-tangles, we generally omit the signs $+$ and $-$ and parenthesise the endpoints of the tangles instead, for example, writing

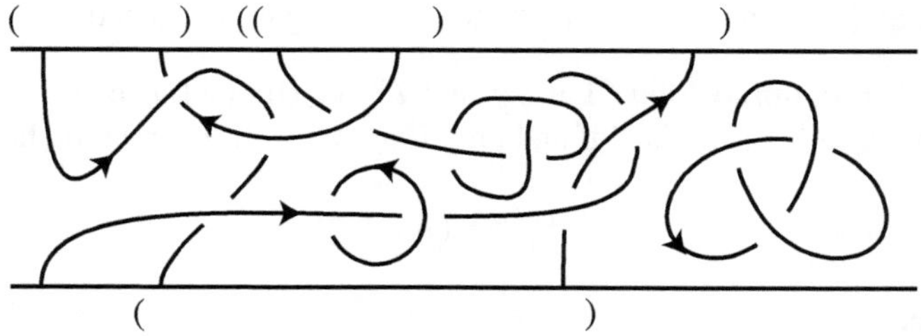

for the q-tangle in (15.1). Clearly, no information is lost by using this notational convention. Furthermore, sometimes we emphasise the parenthesisation by drawing the ends of the tangle close together when they are close in the parenthesisation (as (15.2) below).

We generally use a letter such as T to denote a q-tangle. However, at times we want to emphasise the parenthesised words and use the notation $T_{w_0}^{w_1}$ to denote a q-tangle consisting of an oriented tangle T with parenthesised words w_0 and w_1 on $\mathbb{R} \times \{0\}$ and $\mathbb{R} \times \{1\}$, respectively. Sometimes we apply this notation to q-tangle diagrams, for example, writing the q-tangle shown in (15.1) as

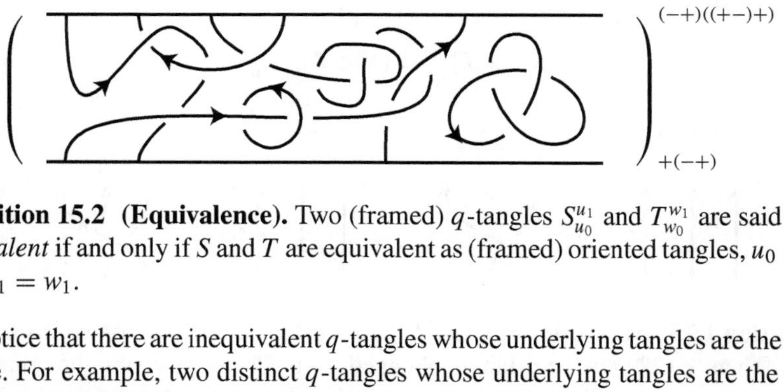

Definition 15.2 (Equivalence). Two (framed) q-tangles $S_{u_0}^{u_1}$ and $T_{w_0}^{w_1}$ are said to be *equivalent* if and only if S and T are equivalent as (framed) oriented tangles, $u_0 = w_0$ and $u_1 = w_1$.

Notice that there are inequivalent q-tangles whose underlying tangles are the same tangle. For example, two distinct q-tangles whose underlying tangles are the same trivial tangle with three components are

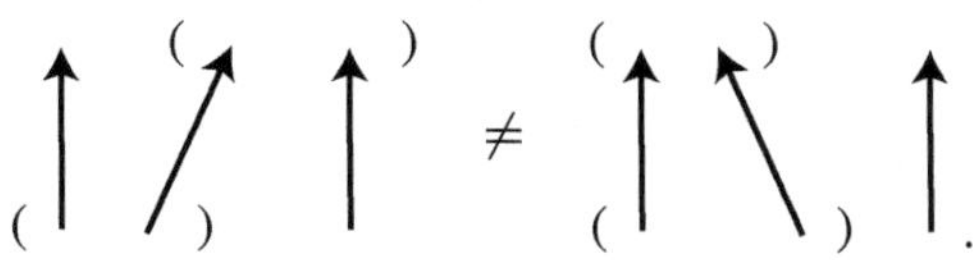

(15.2)

15.2 Three Operations on q-Tangles

The composition of two q-tangles is obtained by composing (*i.e.* "stacking") the two underlying tangles subject to the obvious compatibility relation that the parenthesised words agree at the boundary line where the two q-tangles are joined.

Definition 15.3 (Composition). Let $S_{u_0}^{u_1}$ and $T_{w_0}^{w_1}$ be (framed) q-tangles. Then, the *composition* $T_{w_0}^{w_1} \circ S_{u_0}^{u_1}$ is defined if and only if $u_1 = w_0$. If this relation holds, then

$$T_{w_0}^{w_1} \circ S_{u_0}^{u_1} := (T \circ S)_{u_0}^{w_1}.$$

For example,

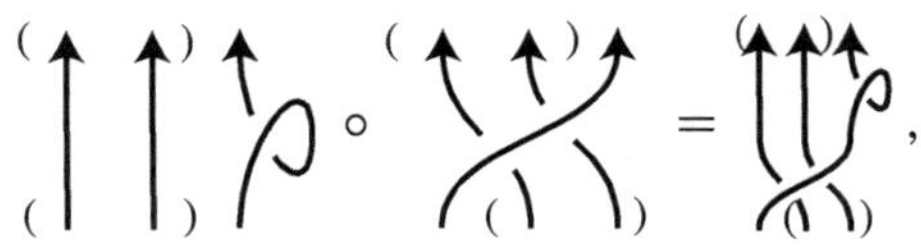

but

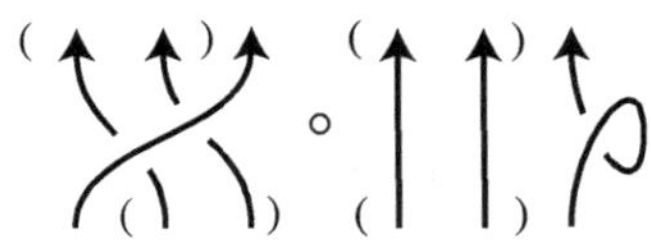

is not defined.

It is implicit in the definition of composition that the "internal" parenthesised words u_1 and w_0, of the original q-tangles, are forgotten.

Definition 15.4 (Tensor product). The *tensor product* of two (framed) q-tangles $S_{u_0}^{u_1}$ and $T_{w_0}^{w_1}$ is the (framed) q-tangle

$$S_{u_0}^{u_1} \otimes T_{w_0}^{w_1} := (S \otimes T)_{(u_0)(w_0)}^{(u_1)(w_1)}.$$

As an example,

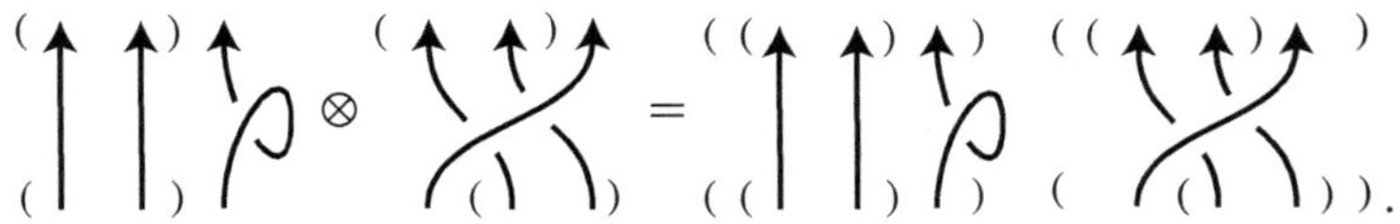

Definition 15.5 (Operation s_a). Let $T_{w_0}^{w_1}$ be a q-tangle and a be one of its components. Then, $s_a(T_{w_0}^{w_1})$ is the q-tangle obtained from $T_{w_0}^{w_1}$ by reversing the orientation of the component labelled by a, and by replacing the symbol $+$ by $-$, or the symbol $-$ by $+$ at the positions in w_0 and w_1 corresponding to the component a.

As an example,

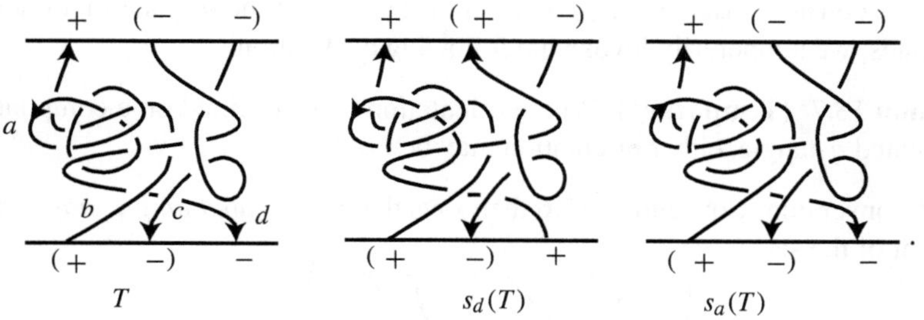

(15.3)

15.3 Decomposing q-Tangles

Our ultimate aim is to construct an invariant $\check{Z}$ of framed q-tangles. Moreover, we want to do this by proceeding in a way similar to the construction of operator invariants in Chap. 7. There we constructed an invariant Q by using $\otimes$ and $\circ$ to decompose a link diagram into elementary tangles,

then specified what value Q takes on each of these, and finally extended Q to all tangles through $\otimes$ and $\circ$.

Defining an invariant of q-tangles by an analogous process, however, is more involved as the parenthesisations introduce their own subtleties. The main issue is that we cannot use the tensor product $\otimes$ to build our q-tangles since we cannot obtain all q-tangles through tensor products of the above elementary tangles. For example, the q-tangle 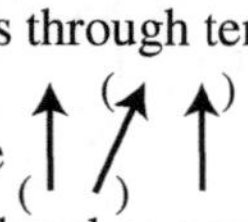 cannot be realised as a tensor of copies of  and will need to be considered as one of our elementary q-tangles. Thus, we need a larger set of elementary q-tangles than we did elementary tangles.

One difference between our previous work with tangles and our present work with q-tangles is that we are no longer going to insist that crossings involve two upward-oriented strands (recall this convention for tangles was introduced in Sect. 7.1.3). Relaxing this condition on the strands of a crossing is certainly conceptually more natural, but it also turns out in the present setting to be more natural mathematically as well.

Convention 15.6 Henceforth, by the term *oriented tangle diagram* we mean an element of the set

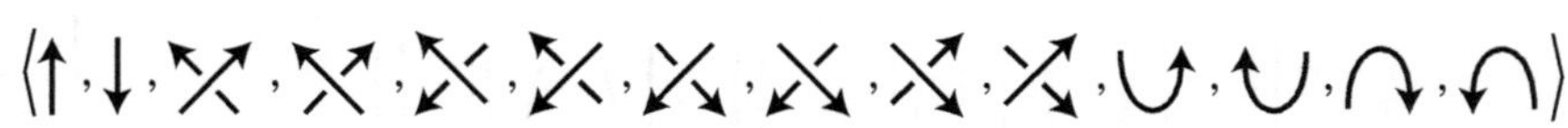

(Recall the notation $\langle\cdot\rangle$ for tangle generation from Definition 7.7, which means that the tangles are generated from elements of the set through the use of $\circ$ and $\otimes$.)

Since we consider oriented tangles which differ only in the orientation of some of its strands, we temporarily invoke the following convention.

Convention 15.7 (Temporary). Where orientation is unspecified on a component of an oriented q-tangle, either orientation may be used.

With this convention, Convention 15.6 then says that an *oriented tangle diagram* is an element of the set

Our interest here is in framed q-tangles. From Definition 15.2, two framed q-tangles $S_{u_0}^{u_1}$ and $T_{w_0}^{w_1}$ are equivalent if and only if $u_1 = w_1, u_0 = w_0$ and their underlying framed tangles are equivalent. The following theorem then arises from Theorem 7.8, giving a set of Turaev moves for framed q-tangles.

Theorem 15.8 *The framed q-tangles $S_{u_0}^{u_1}$ and $T_{w_0}^{w_1}$ are equivalent if and only if $u_1 = w_1, u_0 = w_0$ and their diagrams are related by a finite sequence of the following moves* qFT_0 *to* qFT_5.

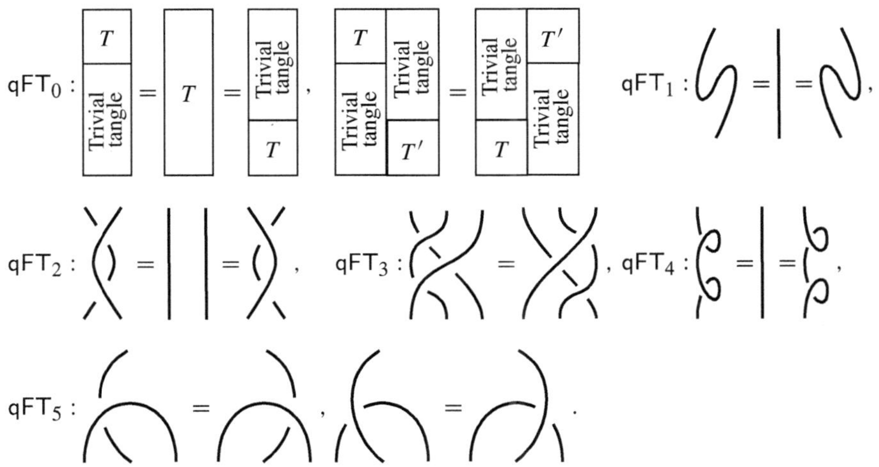

It is clear (and has already been seen) that every oriented tangle can be written as a composite of the following four families of tangles:

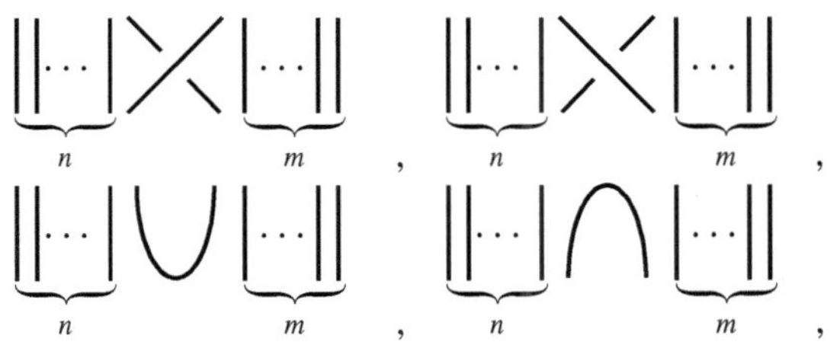

(15.4)

where we are invoking Convention 15.7, and m and n are nonnegative integers.

We may then obtain four families of *elementary q-tangles* by parenthesising each of these tangles subject to the following conditions:

- The tangles in (15.4) involving crossings must have identical parenthesisations at the top and bottom.
- In the tangles in (15.4) involving minima, the two ends of a strand at the top involved in the cup must be parenthesised as $(-+)$ or $(+-)$, as appropriate, and the parenthesisation at the bottom can be obtained from the one at the top by deleting this $(-+)$ or $(+-)$.
- In the tangles in (15.4) involving maxima, the two ends of a strand at the bottom involved in the cap must be parenthesised as $(-+)$ or $(+-)$, as appropriate, and the parenthesisation at the top can be obtained from the one at the bottom by deleting this $(-+)$ or $(+-)$.

For example,

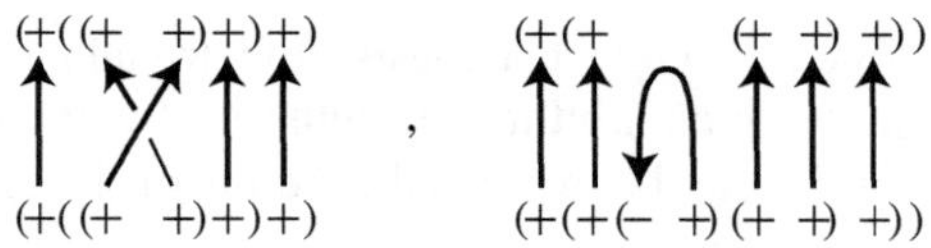

are valid elementary q-tangles, but

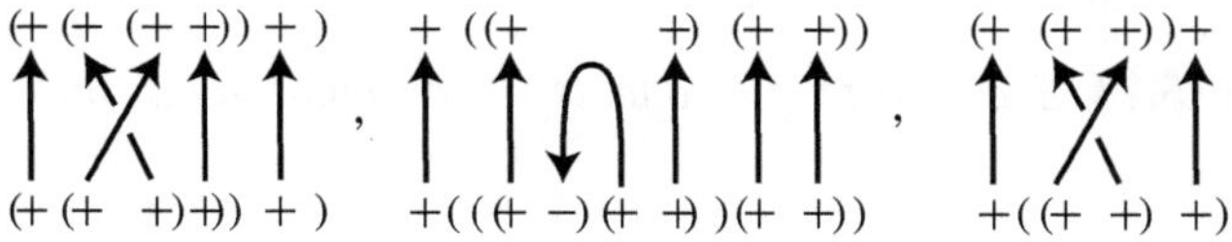

are not.

Parenthesisations should be thought of as corresponding to "closeness" and that only close stands can cross or be "created" or "annihilated" in a cup or cap.

In addition to the elementary tangles involving crossings, caps and cups as above, we also need to be able to change parenthesisations at the ends of the q-tangles. We do this by including in our set of elementary q-tangles, q-tangles that consist of a trivial tangle with different parenthesisations at the top and bottom. Initially, it might appear that we need to allow all possible combinations of parenthesisations at each end of the tangles, but this is not the case.

We include as *elementary q-tangles* the q-tangles of the form

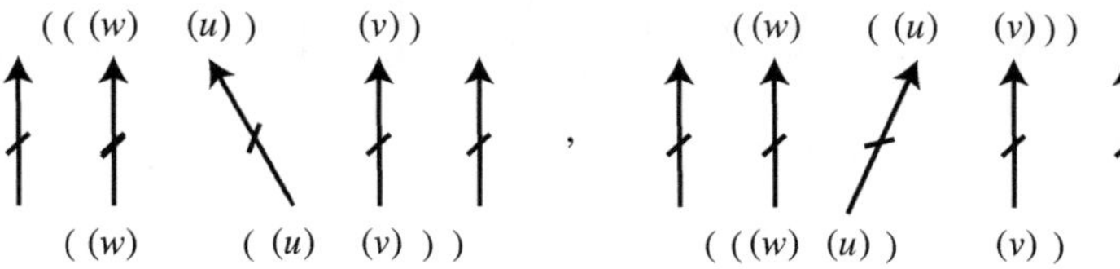

(15.5)

in which:

- A *striation* (*i.e.* the diagonal mark) on a strand indicates that it may be replicated a nonnegative number of times.
- The u, v and w are parenthesisations of a word in $+$ and $-$.
- Outside the displayed portion of the bracketing, the parenthesisations are the same.

For example,

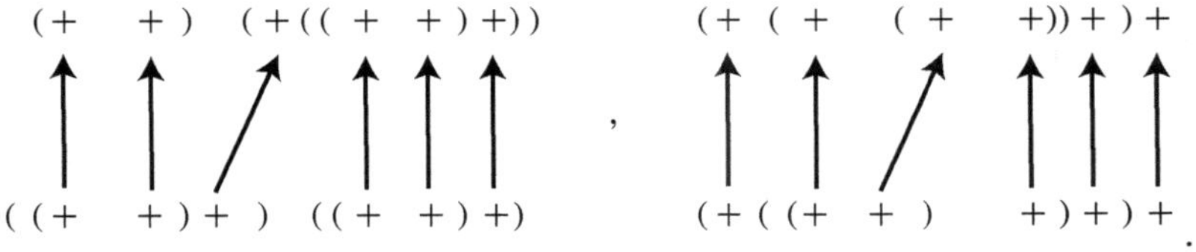

are valid elementary q-tangles.

Exercise 15.9 Let T_u^w be a q-tangle that consists of a trivial tangle (*i.e.* one with no crossings, caps or cups) with an arbitrary parenthesised word u at the bottom and w at the top. Show that T_u^w can be expressed in terms of the elementary q-tangles shown in (15.5). Hence, prove that every q-tangle can be expressed as a composite of elementary q-tangles.

Exercise 15.9 gives the following.

Proposition 15.10 *Every q-tangle admits a decomposition into elementary q-tangles.*

For convenience, we shall now formalise the above definition of the elementary q-tangles. Again we make use of striations to denote multiple parallel components of a q-tangle, writing  to denote a nonnegative number, n, of copies of that strand.

For a q-tangle T_a^b, we use $\alpha_w\left(k, T_a^b\right)$ to denote the q-tangle obtained by replacing the kth component $\uparrow_k$ of $\left(\uparrow_1 \cdots \uparrow_n\right)_w^w$ with T and by replacing the element $+$ or $-$ in the kth position of w with (b) at the top and (a) at the bottom.

For example, using e to denote the empty word,

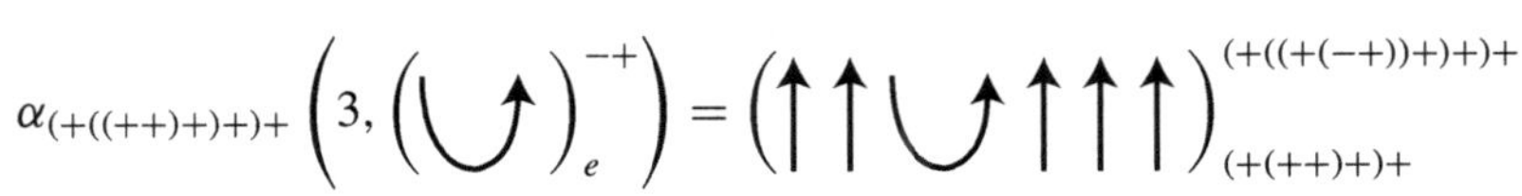

Definition 15.11 (Basic q-tangles). Let w be a parenthesised word with underlying word in $\{+, -\}$, let e denote the empty word, and let

$(E)^b_a \in$

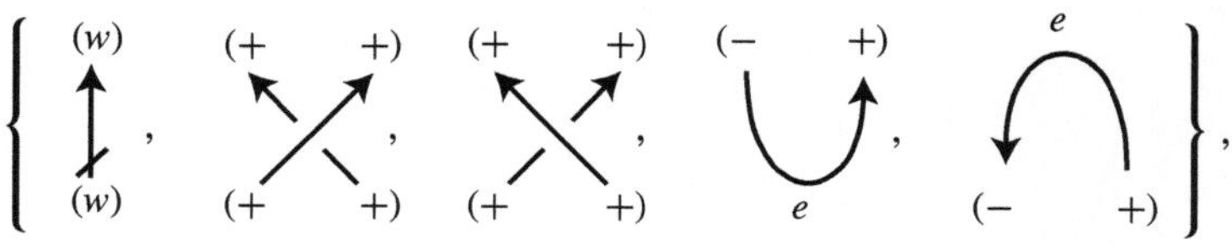

be a q-tangle. A q-tangle $\alpha_w\left(k, (E)^b_a\right)$ or a q-tangle of the form given in (15.5) is said to be a *basic q-tangle.*

Definition 15.12 (Elementary q-tangle). An *elementary q-tangle* is a q-tangle obtained from a basic q-tangle by a finite number of applications of s_i (sign reversal of i-th component) for any finite number of permissible values of i.

A basic q-tangle is an elementary q-tangle with a particular orientation. At times, it is convenient to work with basic q-tangles since, by making use of s_i, we can consider fewer types of q-tangles than we would have to were we to consider elementary q-tangles in general. We will return to q-tangles in Chap. 17 after discussion of Jacobi diagrams.

Chapter 16
Jacobi Diagrams on a 1-Manifold

The ultimate aim in this part of the book is the construction of an invariant $\check{Z}$ of framed q-tangles. In this chapter, we consider the space where $\check{Z}$ takes its values. Ultimately, we want $\check{Z}$ to map a knot to a space of chord diagrams $\mathscr{A}^c$ (actually we need it to take values in a space $\widehat{\mathscr{A}^c}$ of infinite formal linear combinations of chord diagrams). However, since we shall define $\check{Z}$ on q-tangles, rather than knots alone, we shall need to generalise the definition of $\mathscr{A}^c$ so that it can have a skeleton consisting of:

- circles: these correspond to the closed components in the q-tangle;
- lines: these correspond to the components of the q-tangle that are lines.

16.1 Chord Diagrams on X

Definition 16.1 (Chord diagrams with skeleton X). Let X be a compact, oriented, 1-manifold (*i.e.* X is a disjoint union of oriented copies of $\mathbb{S}^1$ and oriented copies of the unit interval $I := [0, 1]$). A *chord diagram with skeleton* X consists of X and a collection of chords with ends on distinct points of X. The skeleton X is considered up to orientation preserving homeomorphisms.

We shall use the following notation and terminology:

- the *degree* of a chord diagram is the number of chords it possesses; and
- $\mathscr{C}(X)$ denotes the vector space consisting of all finite formal linear combinations of chord diagrams on X with coefficients in $\mathbb{C}$;
- $\mathscr{C}_m(X)$ denotes the subspace of $\mathscr{C}(X)$ generated by all degree m chord diagrams on X.

D. M. Jackson and I. Moffatt, *An Introduction to Quantum and Vassiliev Knot Invariants*,
CMS Books in Mathematics, https://doi.org/10.1007/978-3-030-05213-3_16

As with earlier appearances of chord diagrams, our interest here is not in the vector space $\mathscr{C}(X)$ itself, but rather in this space modulo 4T-relation. To translate the 4T-relation to the present context consider

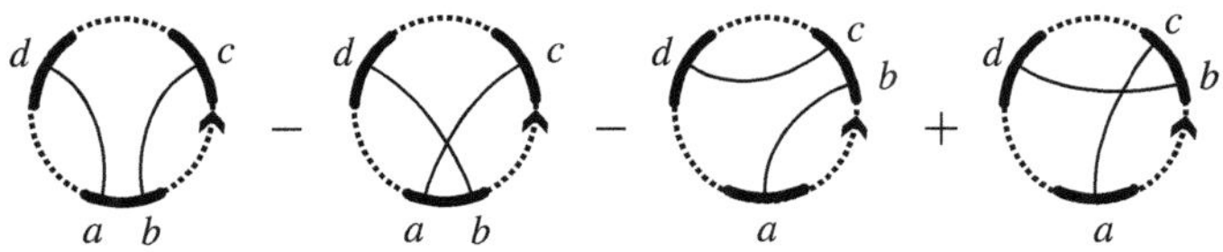

$$= 0. \qquad (16.1)$$

This may be depicted as

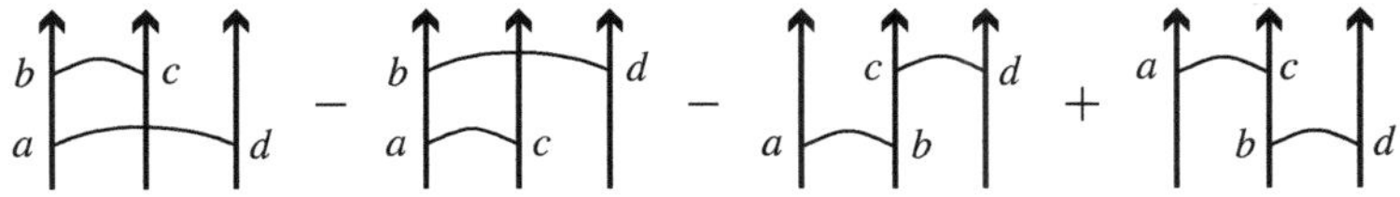

$$= 0 \quad (16.2)$$

by representing the skeletal arcs as directed arrows, then marking the points of contact of chords with these directed arcs by a, b, c and d, and finally joining these points by lines to represent the chords. Thus, deleting the labels a, b, c and d we obtain, in the present setting, the 4T-*relation*

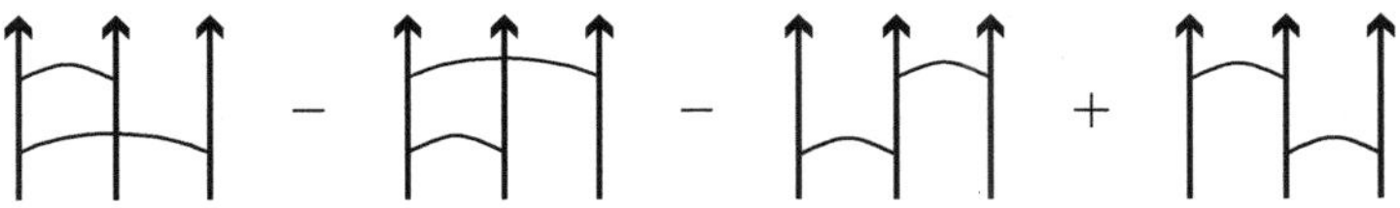

$$= 0 \quad (16.3)$$

where the arrows are distinct skeletal arcs in the manifold X, and possibly arcs of the same component. The skeletal arcs may arise from copies of an interval or copies of $\mathbb{S}^1$, and there are no other chords on these local arcs. The four chord diagrams are identical outside of the region shown.

Definition 16.2 (Vector space $\mathscr{A}^c(X)$). The vector spaces $\mathscr{A}^c_m(X)$ and $\mathscr{A}^c(X)$ are defined as

$$\mathscr{A}^c_m(X) := \frac{\mathscr{C}_m(X)}{(4\mathsf{T})} \quad \text{and} \quad \mathscr{A}^c(X) := \frac{\mathscr{C}(X)}{(4\mathsf{T})}.$$

The following is an example of the application of the 4T-relation in $\mathscr{A}^c(X)$. For clarity, the chords not involved in the application of the 4T-relation are shown as dashed lines. The thickened portions of the skeleton indicate the portions of it that are involved in the 4T-relation, with the convention that there or no other chords incident with these thickened portions.

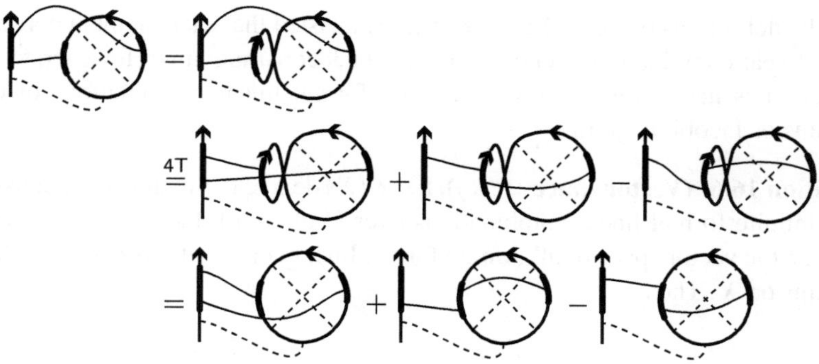

Proposition 16.3. $\mathscr{A}_m^c = \mathscr{A}_m^c(\mathbb{S}^1)$.

Proof. Since $\mathscr{C}_m(\mathbb{S}^1) = \mathscr{C}_m$, we need only to show that the 4T-relations in the two spaces coincide. This was done in (16.1) and (16.2). □

It has already be shown (in Theorem 13.36) that the vector space $\mathscr{A}^c = \mathscr{A}^c(\mathbb{S}^1)$ is isomorphic to the vector space $\mathscr{A} = \mathscr{A}(\mathbb{S}^1)$ of Jacobi diagrams on the circle $\mathbb{S}^1$ modulo the STU-relation. The analogous property holds for the space $\mathscr{A}^c(X)$.

Definition 16.4 (Jacobi diagram on X). Let X be a compact, oriented, 1-manifold. A *Jacobi diagram on X* consists of a unitrivalent graph, with oriented trivalent vertices, such that all univalent vertices lie on the *skeleton* X. The *degree* of a Jacobi diagram on X is half the number of vertices of the diagram.

Let $\mathscr{D}(X)$ be the vector space of all finite formal linear combinations over $\mathbb{C}$ of Jacobi diagrams on X, and let $\mathscr{D}_m(X)$ be the subspace of $\mathscr{D}(X)$ generated by degree m Jacobi diagrams.

The STU-relation on $\mathscr{A}(X)$ or $\mathscr{A}_m(X)$ is

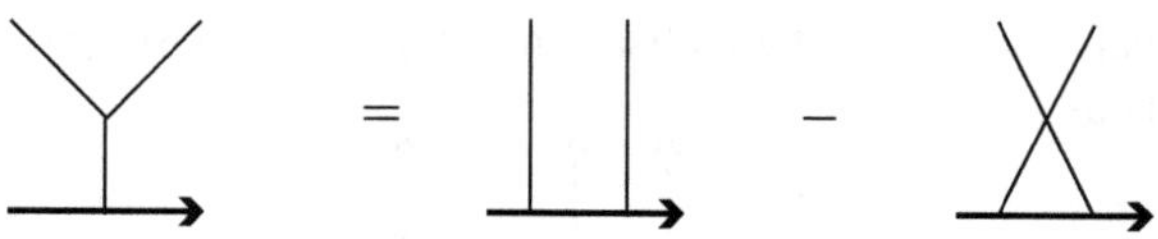

where the $\longrightarrow$ is an arc of X (which may be part of a circular skeleton component), and no other univalent vertices lie on these arcs.

Definition 16.5 (Vector space $\mathscr{A}(X)$). The vector spaces $\mathscr{A}_m(X)$ and $\mathscr{A}(X)$ are defined as

$$\mathscr{A}_m(X) := \frac{\mathscr{D}_m(X)}{(\mathsf{STU})} \quad \text{and} \quad \mathscr{A}(X) := \frac{\mathscr{D}(X)}{(\mathsf{STU})}.$$

Many properties of $\mathscr{A} = \mathscr{A}(\mathbb{S}^1)$ hold in $\mathscr{A}(X)$. For example, the IHX- and AS-relations hold in $\mathscr{A}(X)$ since the proof of Theorem 13.15, which gave these relations in $\mathscr{A}$, did not require the fact that the skeleton was $\mathbb{S}^1$. As another such example, we shall see later, in Theorem 16.19, that $\mathscr{A}^c(X) \cong \mathscr{A}(X)$.

In the definitions of $\mathscr{A}^c(X)$ and $\mathscr{A}(X)$, we insisted that each element consists of a *finite* linear combination of chord diagrams or Jacobi diagrams. However, to make progress it is necessary to consider *infinite* formal linear combinations of chord diagrams or Jacobi diagrams.

Definition 16.6 (Vector space $\widehat{\mathscr{A}}(X)$). Let $\widehat{\mathscr{C}}(X)$ denote the vector space consisting of all infinite formal linear combinations over $\mathbb{C}$ of chord diagrams on X, and let $\widehat{\mathscr{D}}(X)$ be the vector space of all infinite formal linear combinations over $\mathbb{C}$ of Jacobi diagrams on X. Then

$$\widehat{\mathscr{A}^c}(X) := \frac{\widehat{\mathscr{C}}(X)}{(\mathsf{4T})} \quad \text{and} \quad \widehat{\mathscr{A}}(X) := \frac{\widehat{\mathscr{D}}(X)}{(\mathsf{STU})}.$$

For example, the following element R will be important to us:

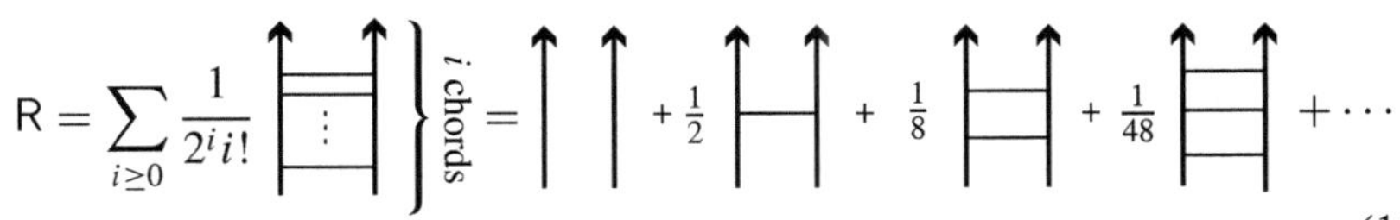

(16.4)

Since the sum has an infinite number of terms, R is an element of $\widehat{\mathscr{A}^c}(\uparrow\uparrow)$ but not of $\mathscr{A}^c(\uparrow\uparrow)$.

16.2 Composition and Tensor Products in $\mathscr{A}(X_n)$

For a unit interval I, let X_n be the realisation of $\bigsqcup_{i=1}^{n} I$ consisting of the n oriented unit intervals. Thus,

$$X_n = \underbrace{\uparrow\ \uparrow\ \cdots\ \uparrow}_{n}. \tag{16.5}$$

At times, we shall need to specify the components of X_n. For this, we index the components from left to right with the numbers 1 through n, thus

$$X_n = \underset{1}{\uparrow}\ \underset{2}{\uparrow}\ \cdots\ \underset{n}{\uparrow}. \tag{16.6}$$

Convention 16.7 We shall use the diagrammatic convention that 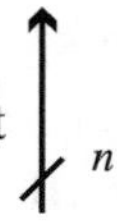 denotes X_n. At times, we shall omit the integer n.

There is a natural product on Jacobi diagrams on X_n given by stacking the skeleton components.

Definition 16.8 (Composition of Jacobi diagrams). Let $D, D' \in \mathscr{A}(X_n)$. Then composition is defined by

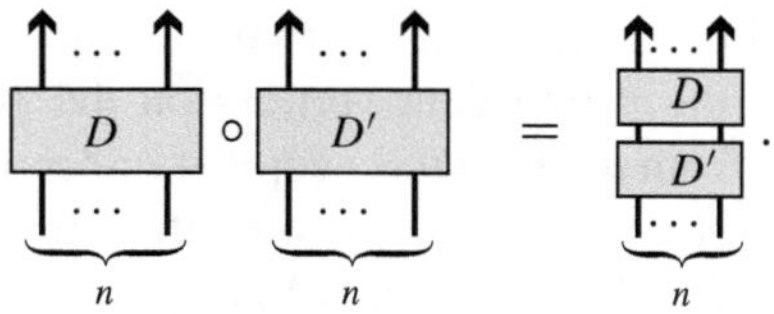

To simplify notation, we shall often drop the symbol $\circ$ and use concatenation for composition, for example, writing $D'D$ for $D \circ D'$.

As an example of composition, we express the above element $\mathsf{R} \in \widehat{\mathscr{A}^c}\left(\uparrow\uparrow\right)$ from (16.4) in a more compact form. We let $H :=$  $\in \widehat{\mathscr{A}^c_2}\left(\uparrow\uparrow\right)$; that is, H consists of two skeletal components joined by a single chord. Then

$$\mathsf{R} := \exp\left(\tfrac{1}{2}H\right) = \sum_{i\geq 0}\frac{1}{2^i i!}\left(H\right)^i = \sum_{i\geq 0}\frac{1}{2^i i!}\ \text{(diagram with } i \text{ chords)} \tag{16.7}$$

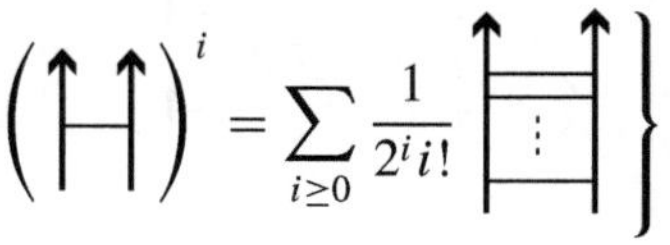

illustrates the use of composition.

Similarly, we may form new Jacobi diagrams from old ones by placing them next to each other.

Definition 16.9 (Tensor product of Jacobi diagrams). The *tensor product* of Jacobi diagrams D and D', with $D \in \mathscr{A}(X_n)$ and $D' \in \mathscr{A}(X_m)$, is

$$D \otimes D' \in \mathscr{A}(X_{n+m})$$

where $D \otimes D' := D \sqcup D'$. That is, diagrammatically,

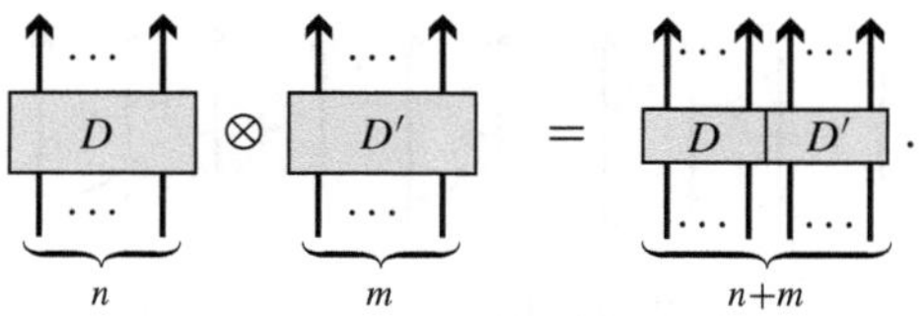

16.3 The ε_a, s_a and ∇_a Operators on Jacobi Diagrams

Definition 16.10 (Operators ε_a, s_a and ∇_a). Let D be a chord diagram or Jacobi diagram on X. Suppose that the skeleton components in X have been labelled arbitrarily, and that a is the label of such a component. Then

1. $\varepsilon_a(D) := \begin{cases} 0 & \text{if D has vertices on a,} \\ D - a & \text{otherwise.} \end{cases}$
2. $s_a(D) := (-1)^{m_a} D'$ where D' is obtained from D by reversing the orientation of the component a, and m_a is the number of univalent vertices on a.
3. $\nabla_a(D)$ is the Jacobi diagram obtained from D by the following construction. Let $v_1, \ldots, v_n$ label the n univalent vertices on the (skeleton) component a. Detach the n chords from a and make a copy of a, denoted by a', with points v_i' corresponding to v_i for $i = 1, \ldots, n$. Now reattach the chords to either v_i or to v_i' (but not both), for $i = 1, \ldots, n$. Then $\nabla_a(D)$ is the formal sum of all of the 2^n diagrams that can be formed in this way.

We also denote by ε_a, s_a and ∇_a the linear extensions of these operations to $\mathscr{A}_m^c(X)$, $\mathscr{A}^c(X)$, $\widehat{\mathscr{A}^c}(X)$, $\mathscr{A}_m(X)$, $\mathscr{A}(X)$ and $\widehat{\mathscr{A}}(X)$.

For example, if the skeleton components are labelled from left to right with 1, 2, or 1, 2, 3, as appropriate, then

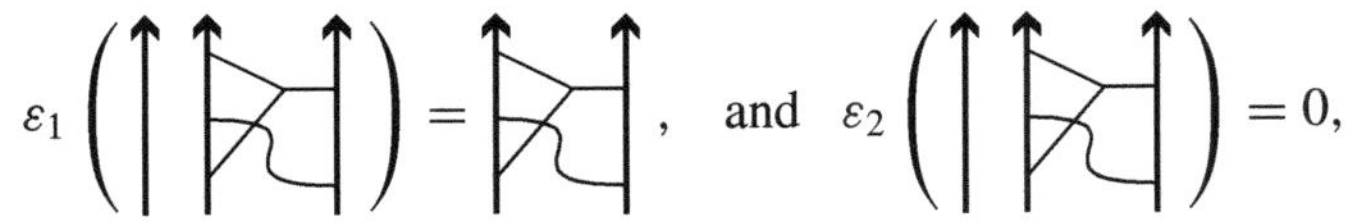

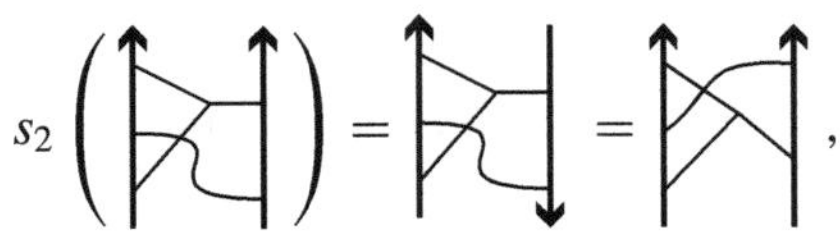

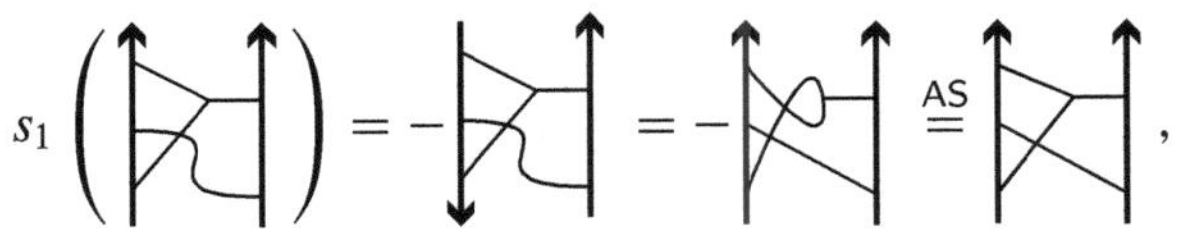

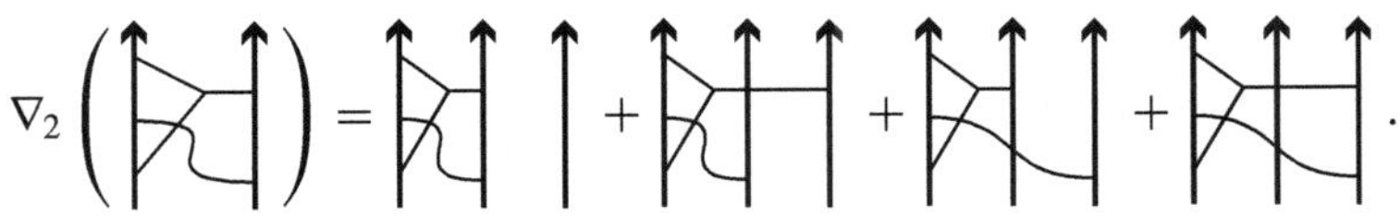

The index of ε, s and ∇ will usually be suppressed when they are applied to a Jacobi diagram with only one skeleton component.

When we consider Jacobi diagrams on the 1-manifold X_n, as in (16.6), we can avoid using the indices by making use of the tensor product of maps:

$$\nabla_i := \mathrm{id}^{\otimes(i-1)} \otimes \nabla \otimes \mathrm{id}^{\otimes(n-i)}.$$

For example,

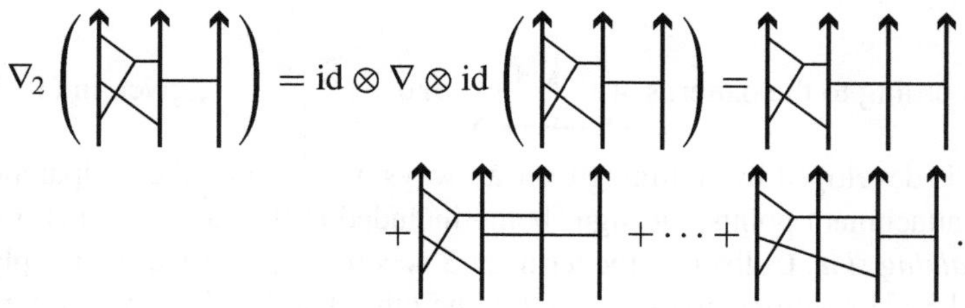

Exercise 16.11. Verify that $(\nabla \otimes \mathrm{id}) \circ \nabla = (\mathrm{id} \otimes \nabla) \circ \nabla$.

Later we shall need to apply ∇ repeatedly to a component. For this, we define

$$\nabla^{(n)} := (\nabla \otimes \mathrm{id}^{\otimes(n-1)}) \circ \cdots \circ (\nabla \otimes \mathrm{id}) \circ \nabla. \tag{16.8}$$

As an example,

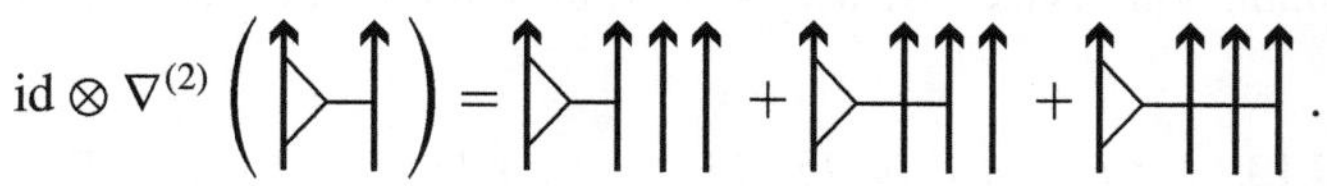

We also can specify the component of X_n that $\nabla^{(n)}$ acts on by writing

$$\nabla_i^{(n)} := \mathrm{id}^{\otimes(i-1)} \otimes \nabla^{(n)} \otimes \mathrm{id}^{\otimes(n-i)}. \tag{16.9}$$

Exercise 16.12. Verify that

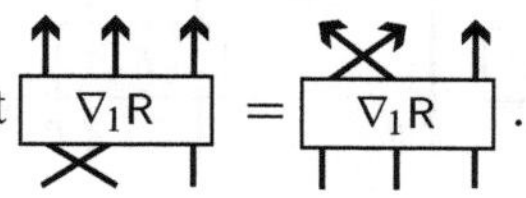

16.4 The Walking Lemma and Some Commutativity Results

We now examine $\widehat{\mathscr{A}}(X_n)$ in more detail, by looking at an extension of the Walking Lemma, (Lemmas 13.27 and 13.28) and using it to prove some commutation results for $\widehat{\mathscr{A}}(X_n)$.

The key idea in the proof the Walking Lemma was to move a *peripatetic vertex* x (*i.e.* the vertex that "travels") in 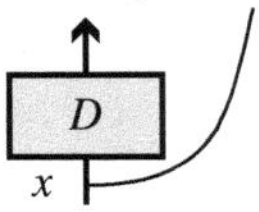

past D so that the diagram becomes 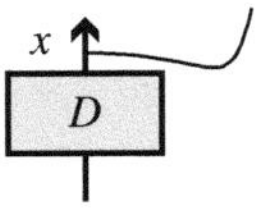

. In doing so, the attachment points at every vertex of D are marked by signs according to the patterns 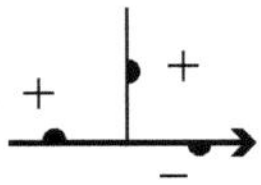

and 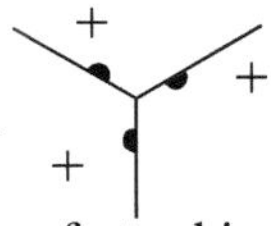

, given in (13.12), and a sum S is developed by summing over all ways of attaching the peripatetic vertex x to the attachment points, the signs being included in the sum. We shall refer to S as the *walking sum*. Collecting the terms in S according to vertices and applying the IHX- and STU-relations gives $S = 0$. On the other hand, collecting the terms of S according to edges gives $S =$ 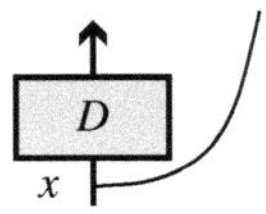

$-$ 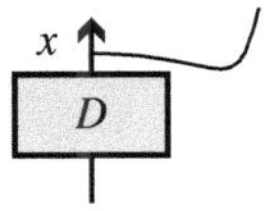

, so we can conclude that

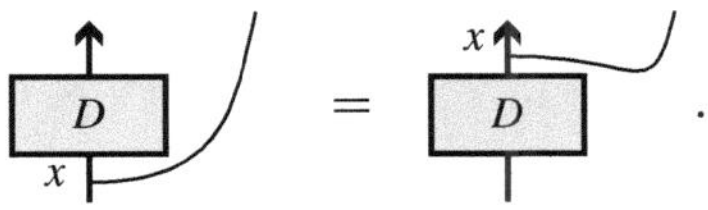

(16.10)

It is this summation device that will be used in some form in each of the following lemmas.

Lemma 16.13. *Let D and E be Jacobi diagrams $D \in \mathscr{A}(X_1)$ and $E \in \mathscr{A}(X_{n+1})$. Then in $\mathscr{A}(X_{n+1})$ and $\widehat{\mathscr{A}}(X_{n+1})$ we have*

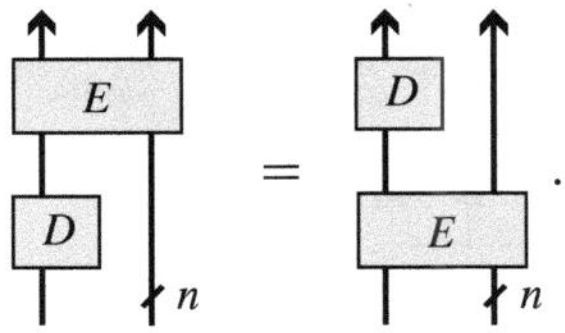

(16.11)

More generally, the following holds:

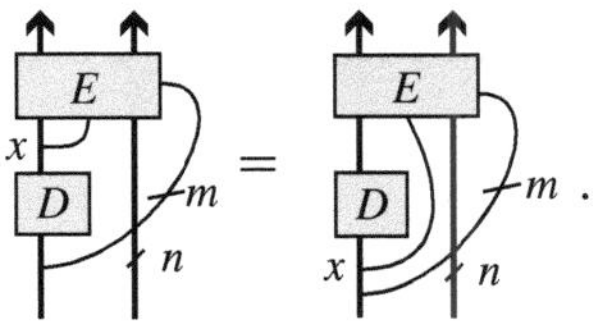

(16.12)

Moreover, these identities hold in $\mathscr{A}(X)$ if some of the shown skeleton components are parts of skeleton components.

Proof. When $n = 0$, the lemma is a trivial application of Lemma 13.28. If $n > 0$, the proof of the lemma is very similar to that of Lemma 13.28, so we provide only a sketch. There are two cases.
Case 1: If E does not have a vertex on the first skeleton component, then the result follows trivially.
Case 2: It therefore may be assumed that E has a vertex on the first skeleton component. Each vertex of E on the first skeleton component must be moved over D, which may be done inductively as follows. Assume that $p \geq 0$ vertices have been moved over D so the diagram on the left in (16.12) has the form 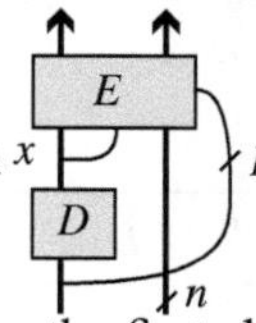

where x is the univalent vertex between D and E that is closest to D on the first skeleton component. It is now enough to show that x may be "moved over" D to obtain 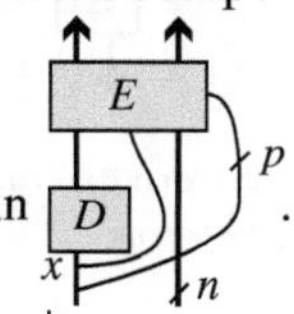

.

To move the vertex x over D, attachment points with patterns 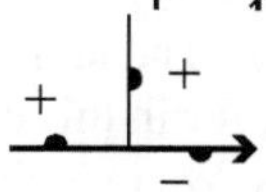

and 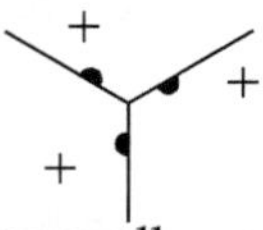

are added to each vertex of D. A walking sum S is obtained by summing over all ways of attaching x to the attachment points, with signed of the summands determined by the signs of the attachment points that were used. Then, collecting terms in S by vertices gives $S = 0$, while collecting terms of S by edges gives

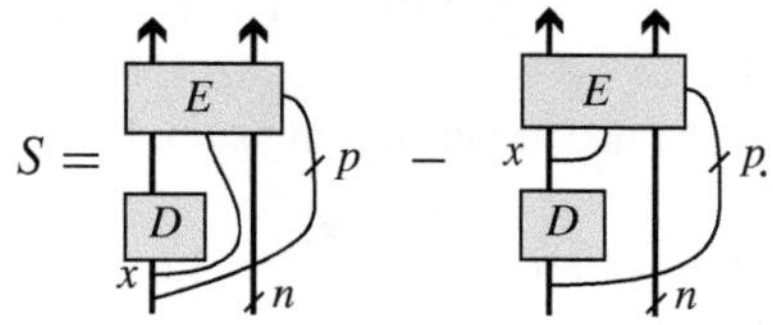

This completes the proof. □

For notational convenience, let

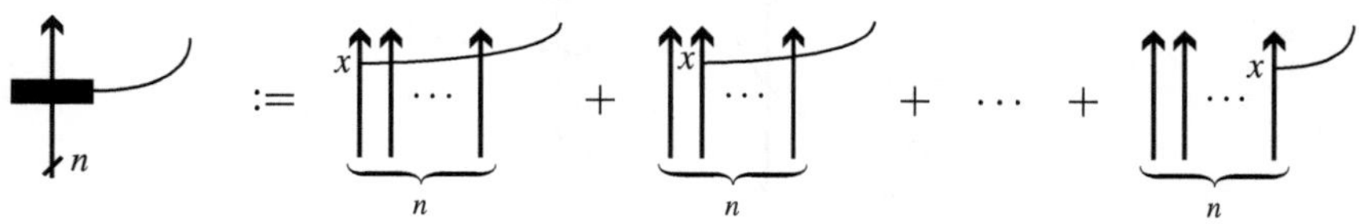

where the peripatetic vertex x visits each skeleton component, and the sum has n terms. In this notation, note that if $D =$ 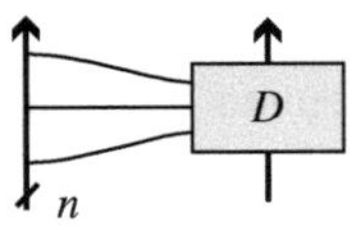

then $\nabla_1^{(n)}(D) =$ 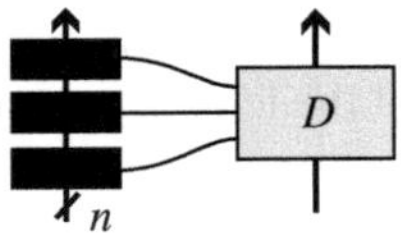

.

The next lemma is a generalisation of Lemma 16.13 to more skeleton components. It has almost the same proof as that lemma.

Lemma 16.14. *Let* $D \in \mathscr{A}(X_{m+1})$ *and* $E \in \mathscr{A}(X_n)$*. Then, in* $\mathscr{A}(X_{m+n})$ *and* $\widehat{\mathscr{A}}(X_{m+n})$*, we have*

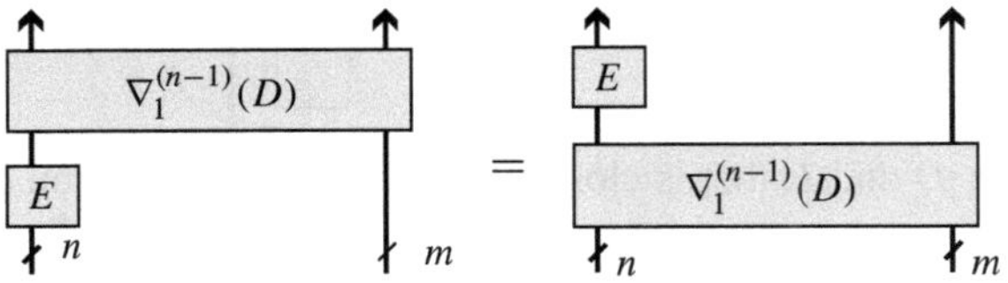

for $m, n \geq 0$.

Moreover, this holds if some of the shown skeleton components form arcs of larger a skeleton.

Proof. The lemma follows trivially if D has no vertices on its first skeleton component, for in this case $\nabla^{(n)} \otimes \mathrm{id}^{\otimes m}(D)$ has no vertices on the first n skeleton components. We may therefore assume that D has vertices on its first skeleton component. We show that

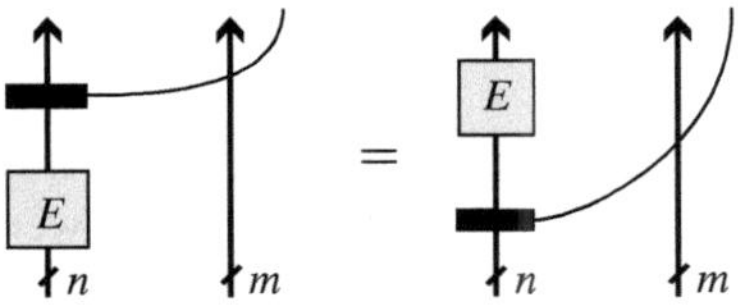

(16.13)

where the peripatetic vertex is not part of E. The lemma then follows by repeated application of this identity.

To prove (16.13), we make use of the walking sum. For this, the signed attachment points are given by 

and 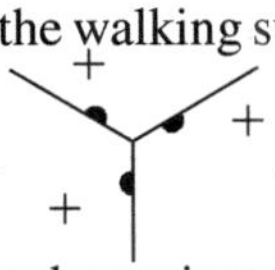

. Let S be the walking sum, *i.e.* the signed sum over all ways of attaching the peripatetic vertex to the signed attachment points. Then, collecting terms of S by edges gives

$S =$ 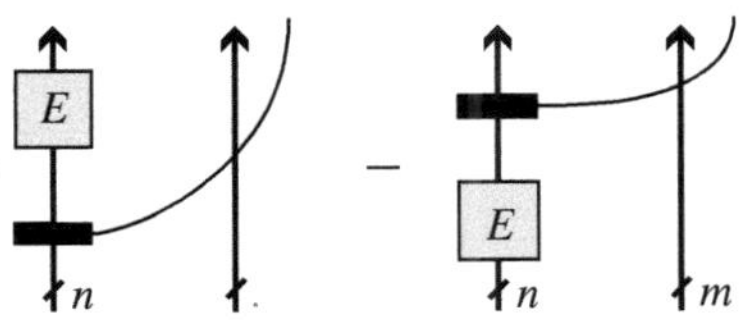

.

On the other hand, collecting the terms of S according to vertices gives $S = 0$ so (16.13) holds, and the lemma therefore follows. □

Corollary 16.15. *Let $D \in \mathscr{A}(X_1)$ or $\widehat{\mathscr{A}}(X_1)$. Then $\nabla^{(n)}(D)$ is central in $\mathscr{A}(X_{n+1})$ and $\widehat{\mathscr{A}}(X_{n+1})$.*

Proof. This is immediate from Lemma 16.14 with $m = 0$. □

The following lemma provides a class of central elements in $\mathscr{A}(X_2)$ and $\widehat{\mathscr{A}}(X_2)$.

Lemma 16.16. *For $F \in \mathscr{A}(X_2)$,*

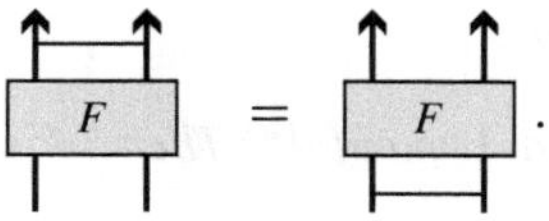

Moreover, this holds if some of the shown skeleton components are parts of circular skeleton components.

Proof. An application of Lemma 16.14 with $D =$  and $E =$ 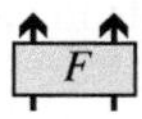

gives

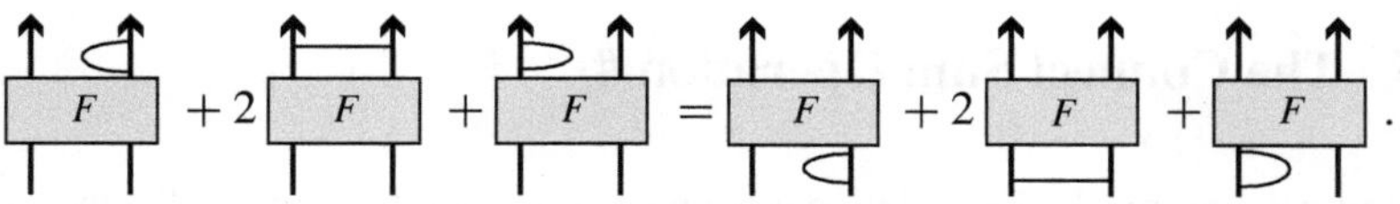

In this identity, the first term on the left-hand side equals the first term on the right-hand side by Lemma 16.13. Also by that lemma, the third term on the left-hand side equals the third term on the right-hand. The result follows. □

An immediate corollary concerning the element R from (16.4) is the following.

Corollary 16.17. *Let $F \in \mathscr{A}(X_2)$. Then*

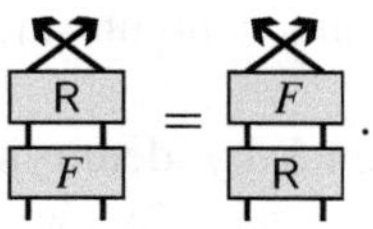

That is, R *is central in $\widehat{\mathscr{A}}(X_2)$.*

We record the following identity for later use.

Lemma 16.18. *Let $D \in \mathscr{A}(X_n)$. Then*

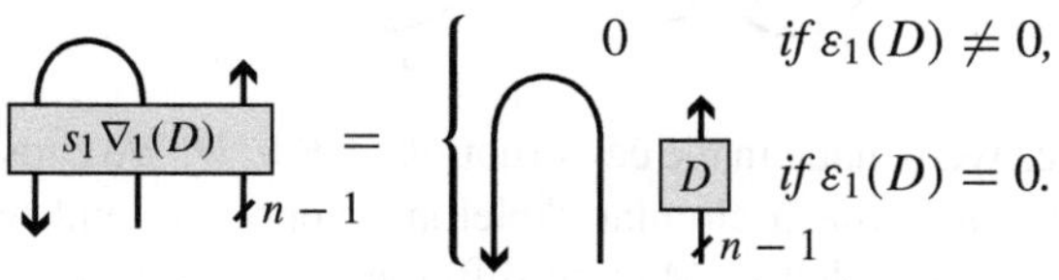

Proof. If $\varepsilon_1(D) = 0$, then $\nabla_1(D)$ has no vertices on the first or second components, so the lemma holds in this case. On the other hand, if $\varepsilon_1(D) \neq 0$, then

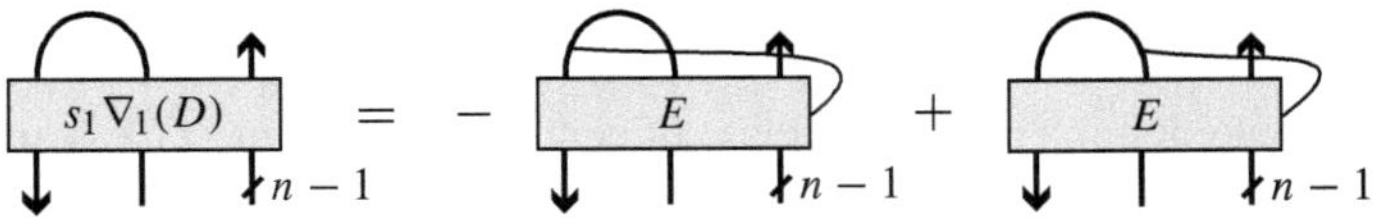

for some block E. But the two diagrams on the right-hand side of the equation are identical and the result follows. □

Theorem 16.19. *Let X be a 1-manifold. Then $\mathscr{A}^c(X) \cong \mathscr{A}(X)$ and $\widehat{\mathscr{A}}^c(X) \cong \widehat{\mathscr{A}}(X)$ as vector spaces.*

Proof. The proof is identical to the proof of Theorem 13.36 since there we used segments of the skeleton with no assumption that they were on a copy of $\mathbb{S}^1$. It therefore holds in each of the spaces. □

16.5 The Connect Sum Operation $\#_i$

We may use Lemma 16.13 to define a connect sum operation on $\mathscr{A}(X)$ and $\widehat{\mathscr{A}}(X)$. For this, let E be a Jacobi diagram on X. Suppose that the skeleton components in X have been labelled arbitrarily, and that a is the label of such a component. Further suppose that D is a Jacobi diagram on a single skeleton component. We define a Jacobi diagram $E\#_a D$ as follows:

1. If the skeleton of D is a circle, then choose a linearisation of it and regard D as being in $\mathscr{A}(\uparrow)$ or $\widehat{\mathscr{A}}(\uparrow)$.
2. Choose a directed arc that contains no univalent vertices on the component of the X that is labelled by a.
3. Form a new Jacobi diagram on X by identifying the (linearised) skeleton component of D with the chosen arc on X in a way that preserves the directions.

Two examples of this operation are shown below. The choice in the examples is precisely where the operation is acting at the points on the skeleton marked with a ★.

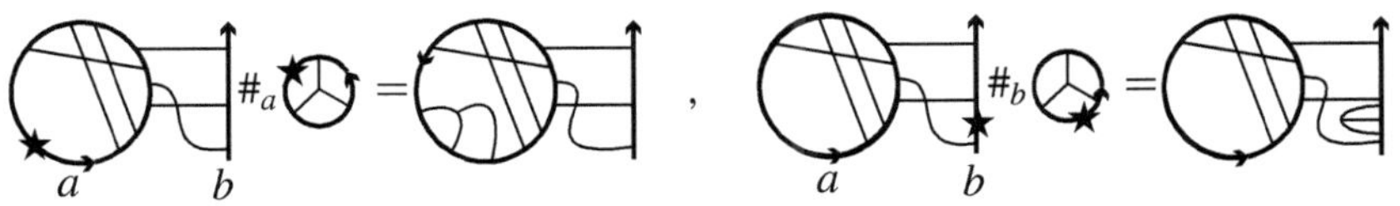

Various choices were made in the construction of $E\,\#_a D$. We chose a linearisation of D in the case that a was a circular skeleton component, and we chose the arc on the component X labelled a. Making different choices in the construction can result in different Jacobi diagrams. Thus, $E\,\#_a D$ is not a well-defined operation in

$\mathscr{D}(X)$. However, different Jacobi diagrams that result from different choices will be equivalent up to the STU-relation. Thus, $\#_a$ gives a well-defined operation in $\mathscr{A}(X)$ or $\widehat{\mathscr{A}}(X)$, as the following result proves.

Theorem 16.20. *Let E be a Jacobi diagram on X and suppose that a component of X is labelled a. Further suppose that D is a Jacobi diagram on a single skeleton component. Then as an element of $\mathscr{A}(X)$ or $\widehat{\mathscr{A}}(X)$, $E \#_a D$ is independent of all choices made in its construction.*

Exercise 16.21. Use Corollary 13.29 and Lemma 16.13 to prove Theorem 16.20.

16.6 Associators

We conclude this chapter with the definition of a special element Φ of $\widehat{\mathscr{A}^c}(X_3)$, called an *associator*, from [57, 59]. In the current setting of constructing a map $\check{Z}$ from q-tangles to $\widehat{\mathscr{A}^c}(X)$, associators will be associated with changes in parenthesization in q-tangles as in Eq. (15.2).

Definition 16.22 (Associator). An *associator* Φ, also denoted by

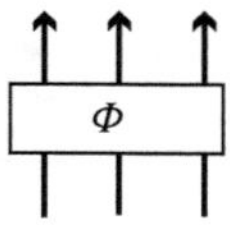

,

is a non-trivial invertible element of $\widehat{\mathscr{A}^c}(X_3)$ such that

(1) $\varepsilon_1(\Phi) = \varepsilon_2(\Phi) = \varepsilon_3(\Phi) =$ 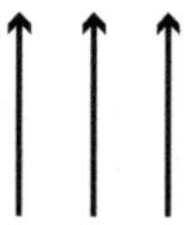 , (2) $\Phi^{-1} =$ 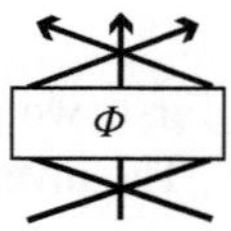 ,

(3) 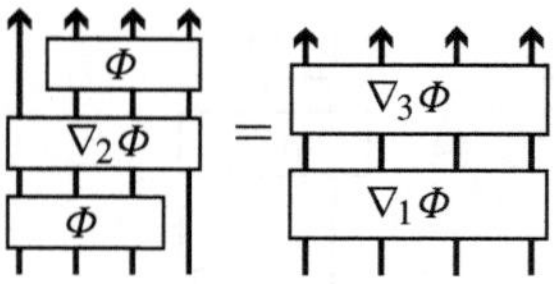 , (4) 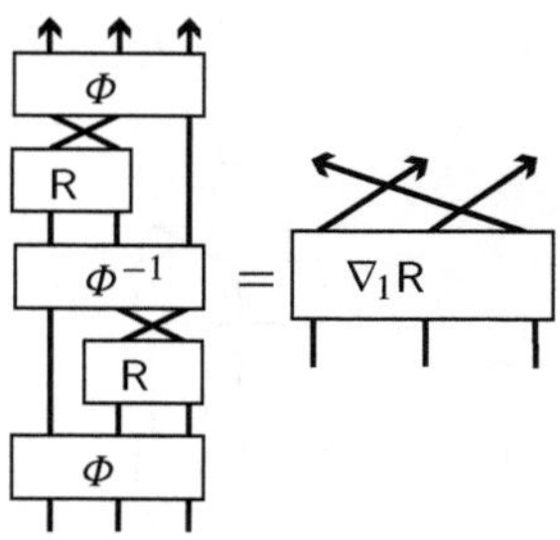 .

where the element $\mathsf{R} = \exp\left(\frac{1}{2}\right.$ 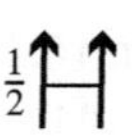 $\left.\right)$ is from (16.7). The relations (3) and (4) are, respectively, the *pentagon relation* and the *(positive) hexagon relation*.

We shall now deduce some properties of associators that we shall need later.

Lemma 16.23. $\Phi =$  *+ (terms in Jacobi diagrams of degree* ≥ 2*).*

Exercise 16.24. Use Condition (1) of Definition 16.22 to prove Lemma 16.23.

Lemma 16.25. *Let* $D \in \widehat{\mathscr{A}}(X_1)$. *Then*

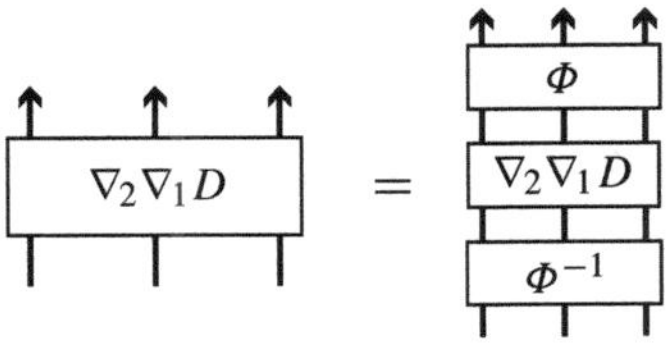

Proof. The result follows since $\nabla^{(n)}(\mathrm{D})$, and so $\nabla_2\nabla_1 D$, is central from Corollary 16.15. □

Lemma 16.26 (Negative hexagon relation). *For the element* $\mathsf{R} = \exp\left(\tfrac{1}{2}\right.$ 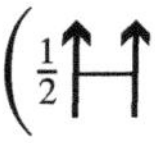 $\left.\right)$ *from (16.7),*

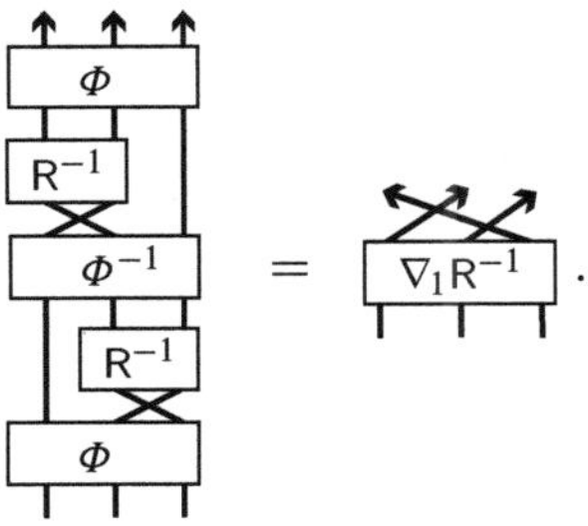

Proof. First, we invert the hexagon relation given in (4) of Definition 16.22, as follows. The inverse of the left-hand side of the hexagon relation is

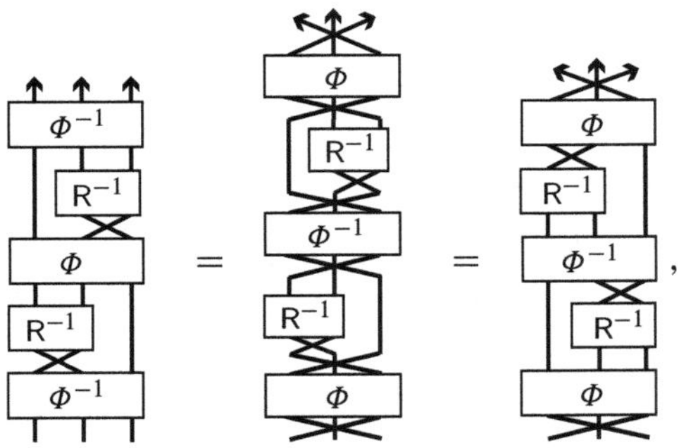

where the first equality is from (2) of Definition 16.22 and the second by redrawing the diagram in a simpler way. To find the inverse of the right-hand side of the hexagon relation, observe that $\nabla_1(D \circ D') = \nabla_1(D) \circ \nabla_1(D')$ where D and D' are Jacobi diagrams, so $(\nabla_1(\mathsf{R}))^{-1} = \nabla_1(\mathsf{R}^{-1})$. Thus, the inverse of the right-hand side of the hexagon relation is 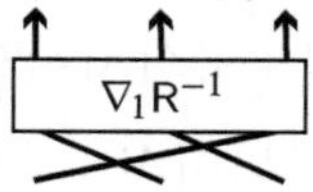 and therefore, equating these two inverses,

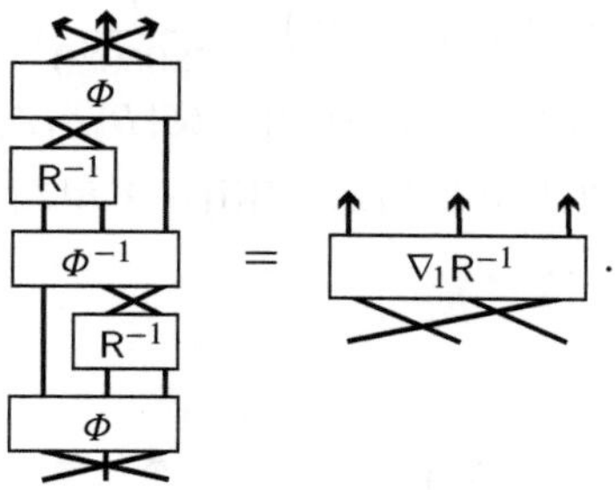

(16.14)

Conjugating the left-hand side of this relation by  and using the symmetry of R^{-1} gives precisely the left-hand side of the lemma. To conjugate the right-hand side of (16.14), we note that

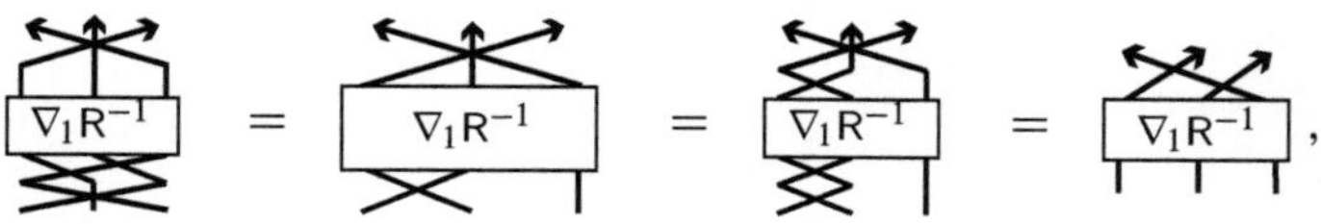

where the second equality is by the symmetry of $\nabla_1(\mathsf{R}^{-1})$. But this is the right-hand side of the lemma, so the result follows. □

One issue that we have not addressed so far is that of the existence of associators. After all, we could greatly simplify our proofs of the above properties if associators did not exist—the above statements would be empty statements and so trivially true! In fact, associators do exist and so the theory presented here is meaningful.

Theorem 16.27. *Associators exist. Moreover, there exists an associator with rational coefficients.*

We omit the (analytic) proof of Theorem 16.27 and refer the reader to Drinfel'd's original work [56, 57] or the expositions in [90, 144] for details. An alternative inductive approach to the construction of associators was given by Bar-Natan in [16]. The implementation of this inductive approach resulted in the following expression for an associator.

$$\begin{aligned}\log \Phi =&\frac{[A,B]}{48}-\frac{8}{11520}\big([A[A[A,B]]]+[A[B[A,B]]]\big)\\&+\frac{1}{5806080}(96[A[A[A[A[A,B]]]]]+4[A[A[A[B[A,B]]]]]\\&+65[A[A[B[B[A,B]]]]]+68[A[B[A[A[A,B]]]]]\\&+4[A[B[A[B[A,B]]]]]\big)-\frac{1}{5806080}(96[B[B[B[B[B,A]]]]]\\&+4[B[B[B[A[B,A]]]]]+65[B[B[A[A[B,A]]]]]\\&+68[B[A[B[B[B,A]]]]]+4[B[A[B[A[B,A]]]]]\big)\\&+\cdots,\end{aligned}\tag{16.15}$$

where

$A =$  , $B =$ 

and $[\cdot,\cdot]$ is the commutator.

Chapter 17
A Construction of the Kontsevich Invariant

Let $T_{w_0}^{w_1}$ be a framed q-tangle. Then $T_{w_0}^{w_1}$ is said to be a framed *q-tangle* on a 1-manifold X if the underlying tangle T is homeomorphic to X. For example, a framed knot is a framed q-tangle on $\mathbb{S}^1$, and if the q-tangle has an underlying tangle that is a diagram of a braid on three strands, then that q-tangle is on $I \cup I \cup I$. The relevance of the 1-manifold X here is that the value of the Kontsevich invariant of q-tangle on X will be a Jacobi diagram on X.

We shall describe the construction of the Kontsevich invariant in two steps. First we shall construct a function

$$Z\colon \{\text{framed}\, q\text{-tangles on}\, X\} \to \widehat{\mathscr{A}^c}(X) \tag{17.1}$$

that associates a formal power series of chord diagrams on X (or equivalently Jacobi diagrams on X, since $\widehat{\mathscr{A}^c}(X) \cong \widehat{\mathscr{A}}(X)$) with a q-tangle diagram on X. The construction of Z is similar in spirit to the construction of operator invariants as in Chap. 7; we associate an element of $\widehat{\mathscr{A}^c}(X)$ with basic q-tangles and extend Z to all tangles by defining $Z(T \circ S) := Z(T) \circ Z(S)$ and $Z(s_i(T)) := s_i(Z(T))$. However, the fact that we are using q-tangles, which are parenthesised, makes the definition of Z a little more involved than that of operator invariants. It also means that showing Z is well-defined (*i.e.*, independent of the choice of decomposition of a q-tangle into basic q-tangles) is a very involved process.

The issue with the function Z is that although it is (non-trivially) a function, it is *not* an invariant of q-tangles. The problem is that the function Z will give different values on the equivalent q-tangles such as

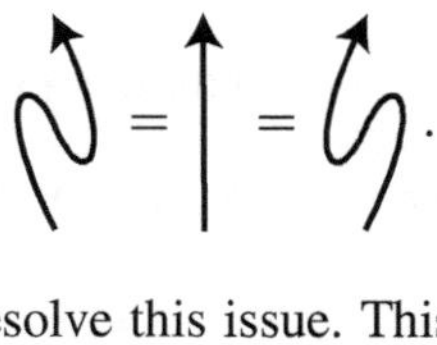

Fortunately, it is possible to resolve this issue. This is done by associating a copy of $\Big(Z\big($ 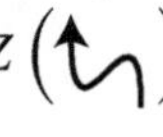 $\big)\Big)^{1/2} \in \widehat{\mathscr{A}^c}(I)$ with each critical point of the tangle diagram. (In

D. M. Jackson and I. Moffatt, *An Introduction to Quantum and Vassiliev Knot Invariants*, CMS Books in Mathematics, https://doi.org/10.1007/978-3-030-05213-3_17

more detail, for each critical point in a component of the q-tangle, one copy of $\left(Z\left(\uparrow\!\cap\right)\right)^{1/2}$ is connect summed to the corresponding component of the Jacobi diagram, as in Sect. 17.2.) Through this modification, Z gives rise to an invariant $\check{Z}$ of framed q-tangles:

$$\check{Z}: \frac{\{\text{framed } q\text{-tangles on } X\}}{\text{equivalence of framed } q\text{-tangles}} \longrightarrow \widehat{\mathscr{A}^c}(X). \tag{17.2}$$

We shall see in Chap. 18 the stunning results that this invariant contains all framed quantum and all Vassiliev knot invariants.

We emphasize that it is important to pay attention to whether or not Z has an accent or not since Z and $\check{Z}$ have differing properties and values.

17.1 The Function Z on q-Tangles

We approach the definition of the function Z on q-tangles by (i) defining it on basic q-tangles (see Definition 15.11); (ii) extending it to elementary q-tangles (see Definition 15.12) by defining $Z(s_a(T)) = s_a(Z(T))$, where the s_a are the operations from Definitions 15.5 and 16.10 that reverses the orientation of a component a of a q-tangle or Jacobi diagram; (iii) then extending Z to all q-tangles by setting $Z(T \circ T') = Z(T) \circ Z(T')$.

We start with the definition of the function Z on q-tangles that consist of a parenthesization of the trivial tangle $\uparrow\uparrow\cdots\uparrow$ on n components. The easiest instance of this occurs when the parenthesizations on the top and bottom of the trivial tangle are equal, in which case Z sends the q-tangle to the degree zero Jacobi diagram on $\cup_{i=1}^{n} I$:

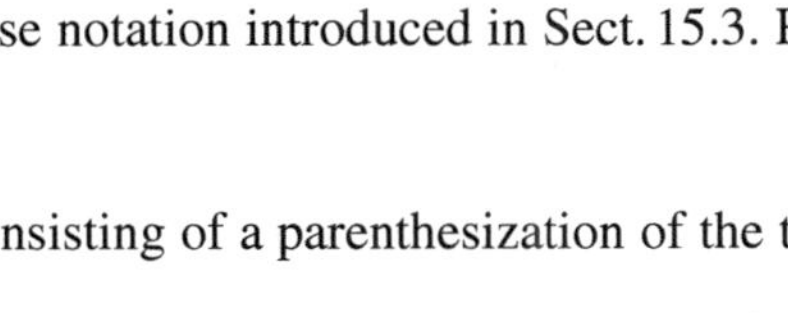

It is convenient to use notation introduced in Sect. 15.3. Recall that $\underset{(w)}{\overset{(w)}{\updownarrow}}$ denotes the q-tangle $\left(\updownarrow_n\right)_w^w$ consisting of a parenthesization of the trivial tangle $\uparrow\uparrow\cdots\uparrow$ with the same word w of length n on its top and bottom. Then we can write

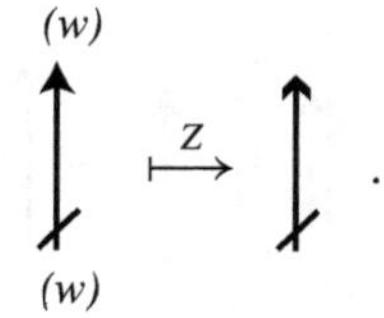

(17.3)

Of course, it may be that the parenthesizations on the top and bottom of a trivial tangle do not match. In this situation its value under Z will be complicated. To describe its values we first set

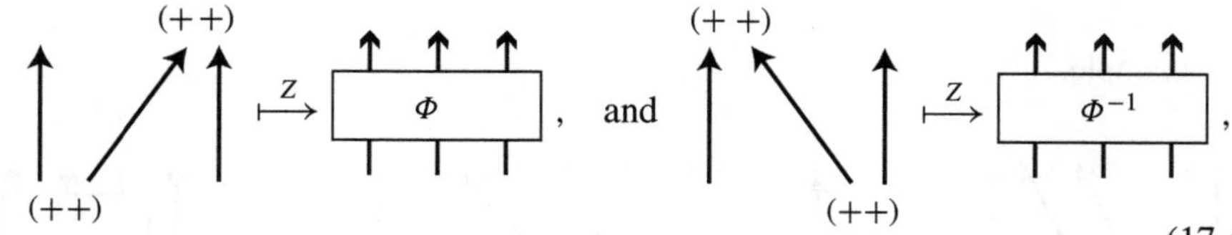

(17.4)

where 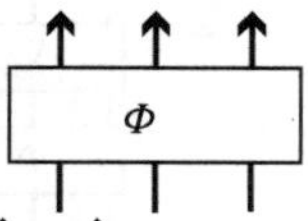

$\in \widehat{\mathscr{A}^c}(X_3)$ is an associator (see Definition 16.21), with inverse

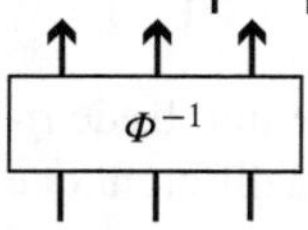

$\in \widehat{\mathscr{A}^c}(X_3)$.

More generally than in (17.4), a basic q-tangle can consist of more than three components and be of the form

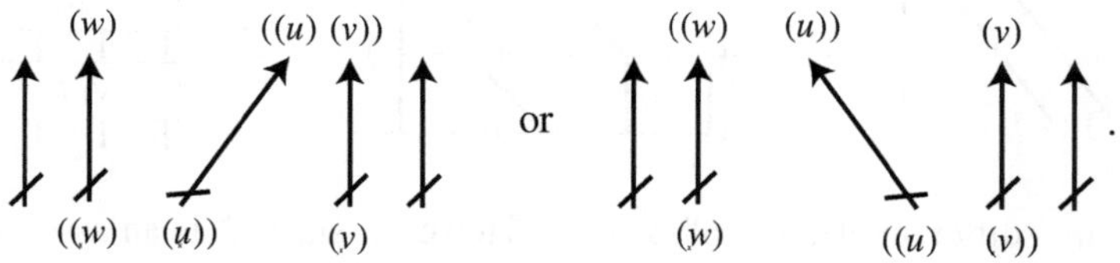

In this situation we use the operation ∇ of Sect. 16.3 to create an element in $\widehat{\mathscr{A}^c}(X_n)$ from an associator Φ or its inverse. We set

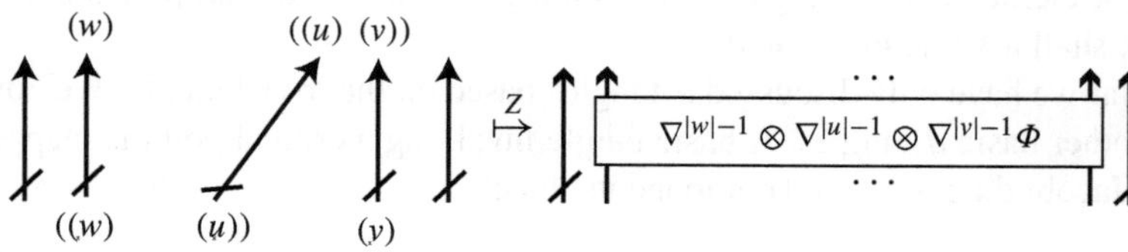

(17.5)

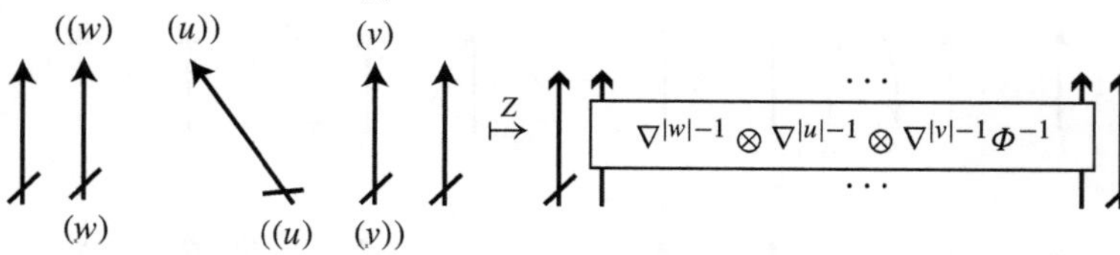

(17.6)

As an example,

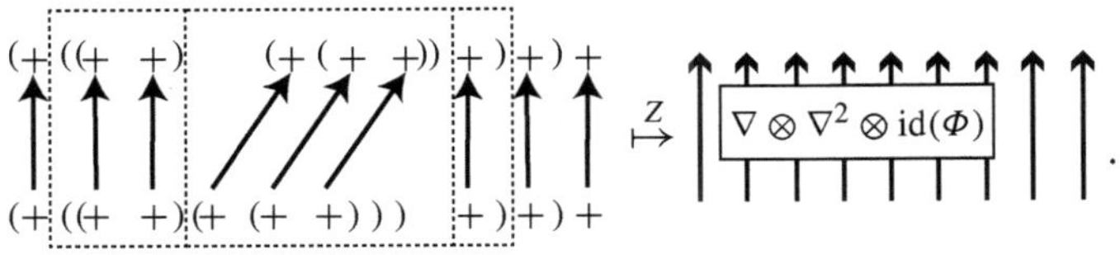

So far (17.3) and (17.5) allow us to compute Z of particular parenthesizations of a trivial tangle. We can extend this to all parenthesizations of it by defining

$$Z(T \circ T') := Z(T) \circ Z(T').$$

For example,

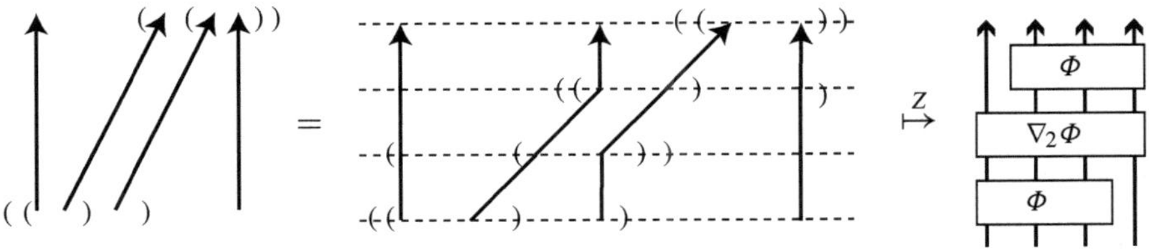

This computation involved a choice of decomposition of the q-tangle into basic q-tangles. Of course, this choice is not unique and we may have chosen a different one such as

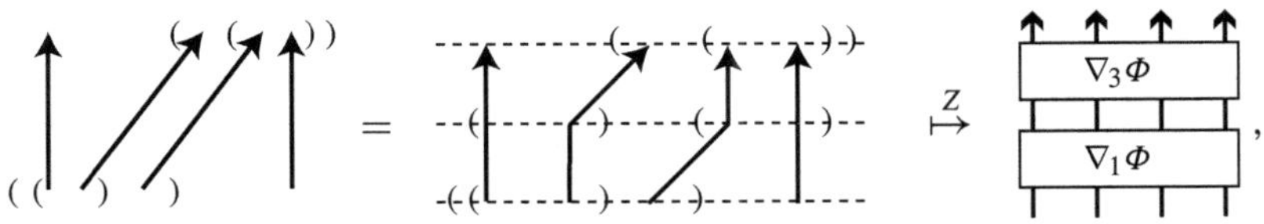

resulting in a different element of $\widehat{\mathscr{A}^c}(X_4)$. However, since Φ is an associator, the pentagon relation, as in Definition 16.21, holds in $\widehat{\mathscr{A}^c}(X_4)$ and hence the two computations in this case result in the same answer. In fact, our choice of Φ as an associator *always* means that the computation of Z is independent of the decomposition into basic (or elementary) q-tangles. This is a highly non-trivial fact [87, 88, 116] and one we shall assume in this text.

So far we have only discussed q-tangles based on the trivial tangle. We now look at the other basic q-tangles. A basic tangle involving a critical point is mapped to a trivial Jacobi diagram on its components thus:

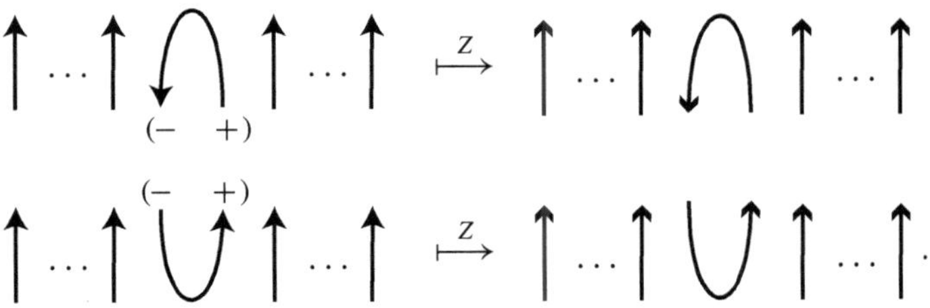

Recall that in an elementary q-tangle the two ends of a component with a critical point must be paired with each other in the parenthesization.

We can take advantage of our notation for basic q-tangles from Sect. 15.3 to express this formally as

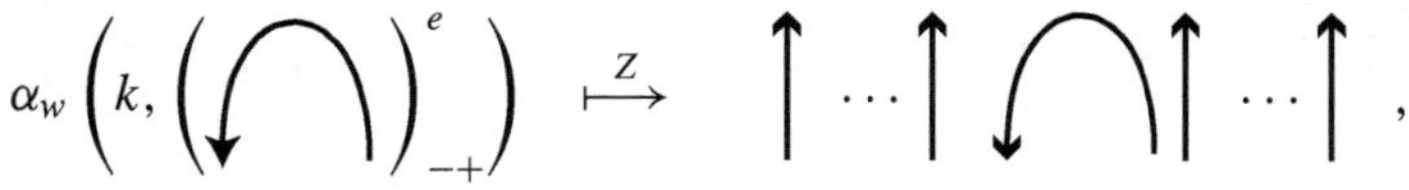
(17.7)

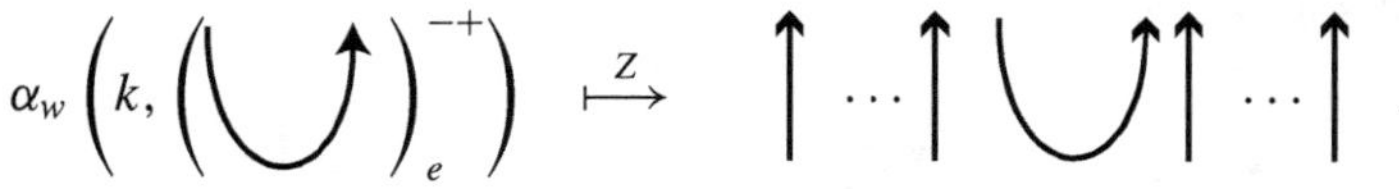
(17.8)

We can use Eqs. (17.3) and (17.5)–(17.8) to define Z on all crossing-free q-tangles by defining

$$Z(T \circ T') = Z(T) \circ Z(T') \quad \text{and} \quad Z(s_a(T)) = s_a(Z(T)),$$

where s_a denotes the operations from Definitions 15.5 and 16.10 that reverses the orientation of a component in a q-tangle or Jacobi diagram.

Example 17.1. 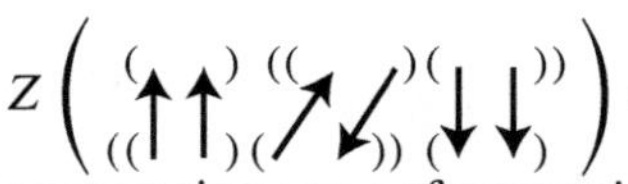 $= s_6 \circ s_5 \circ s_4(\nabla \otimes \nabla \otimes \nabla(\Phi))$.

Although there is not a unique way of expressing an element as a composition of s_i's and ∇_i's, the resulting elements are the same.

Example 17.2.

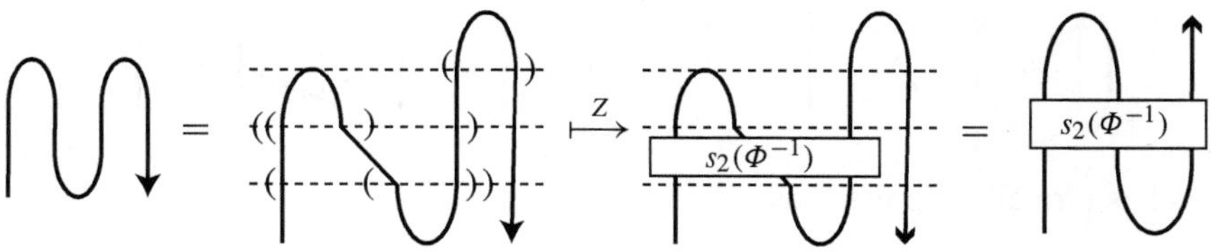

The next basic tangles we need to consider are those containing crossings. Recall that by convention all crossings in the basic tangle diagrams we consider here are directed upwards (although this is not the case in the elementary ones), and in a basic or elementary q-tangle a crossing can only happen when two strands are next to each other in the parenthesization.

Recall also the element $\mathsf{R} \in \widehat{\mathscr{A}^c}(X_2)$ given by

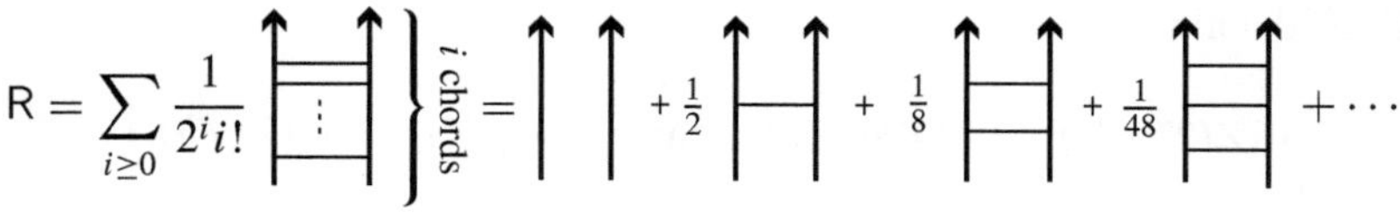

Since the power series defining R is that of an exponential, R is invertible with inverse given by

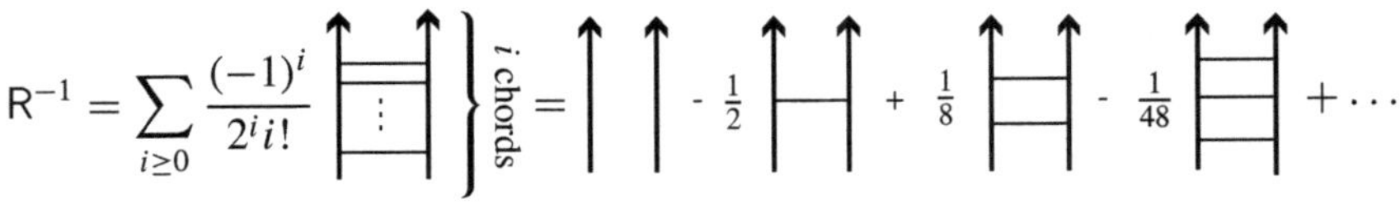

For a crossing we define

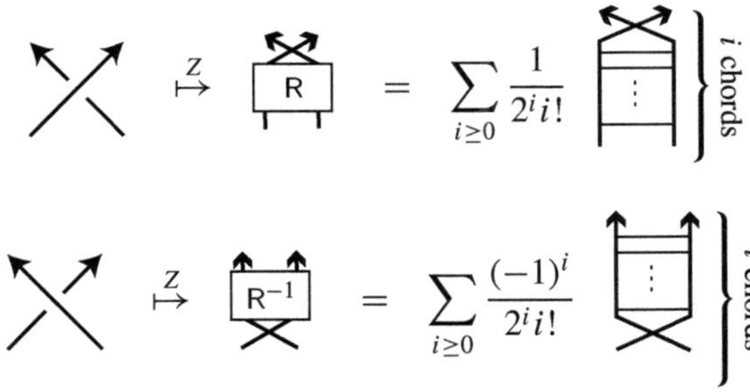

More generally, we define

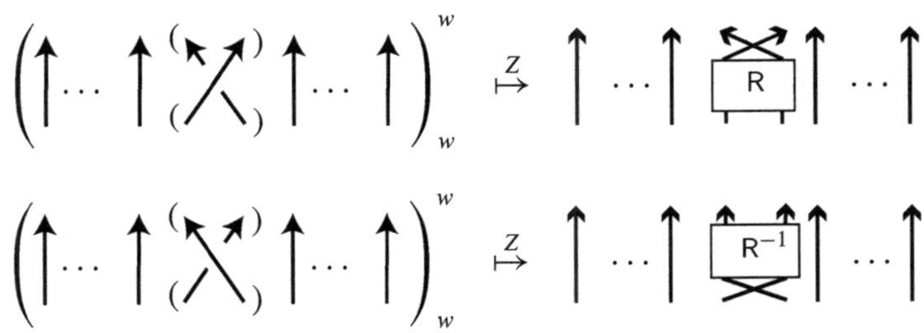

which is formally expressed as

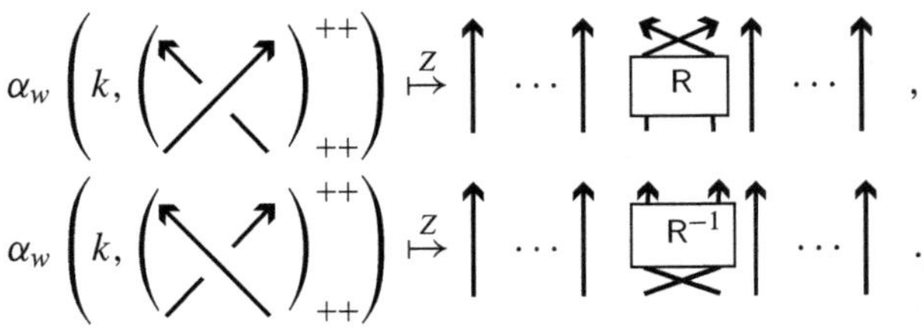

Upon defining

$$Z(T \circ T') := Z(T) \circ Z(T') \quad \text{and} \quad Z(s_a(T)) := s_a(Z(T)),$$

Z becomes defined on all q-tangles.

Example 17.3. An example of a computation of Z  is given below.

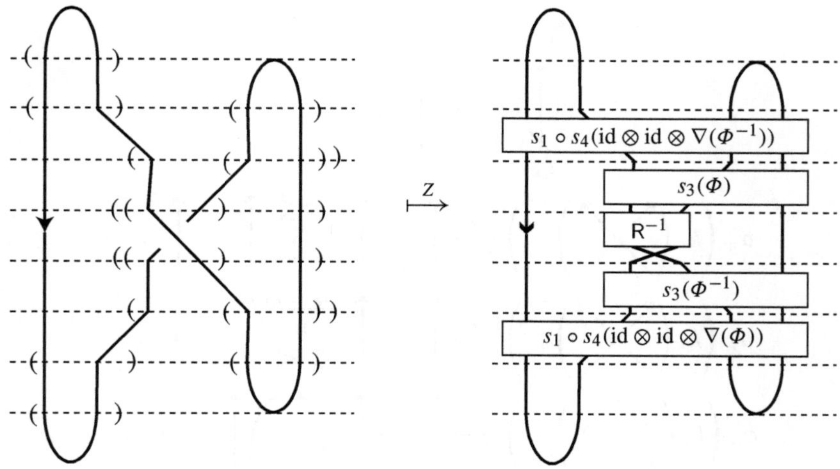

Example 17.4. As a second example we compute Z .

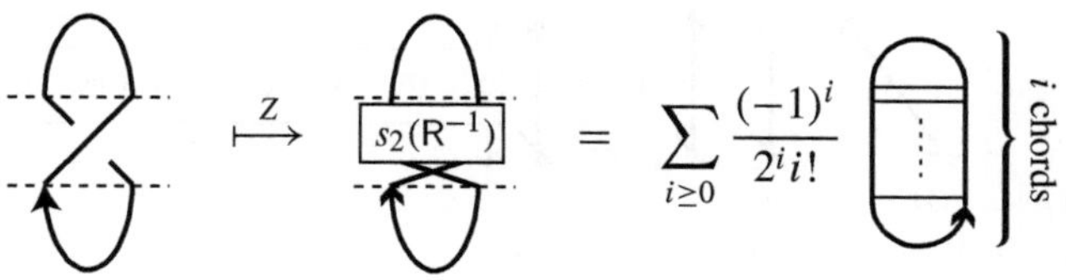

Note that the computation of Z  and of Z  from Example 17.3 appear (and indeed do) to give different values even though the knots are equivalent. This is because Z, unlike the later defined $\check{Z}$, is not a knot invariant. We discuss this further in the next section.

Definition 17.5 (Action of Z on q-tangles). Let u, v, and w be parenthesized words over the alphabet $\{+\}$, and let e denote the empty word. Then Z is defined on q-tangles by the following explicit element-wise actions,

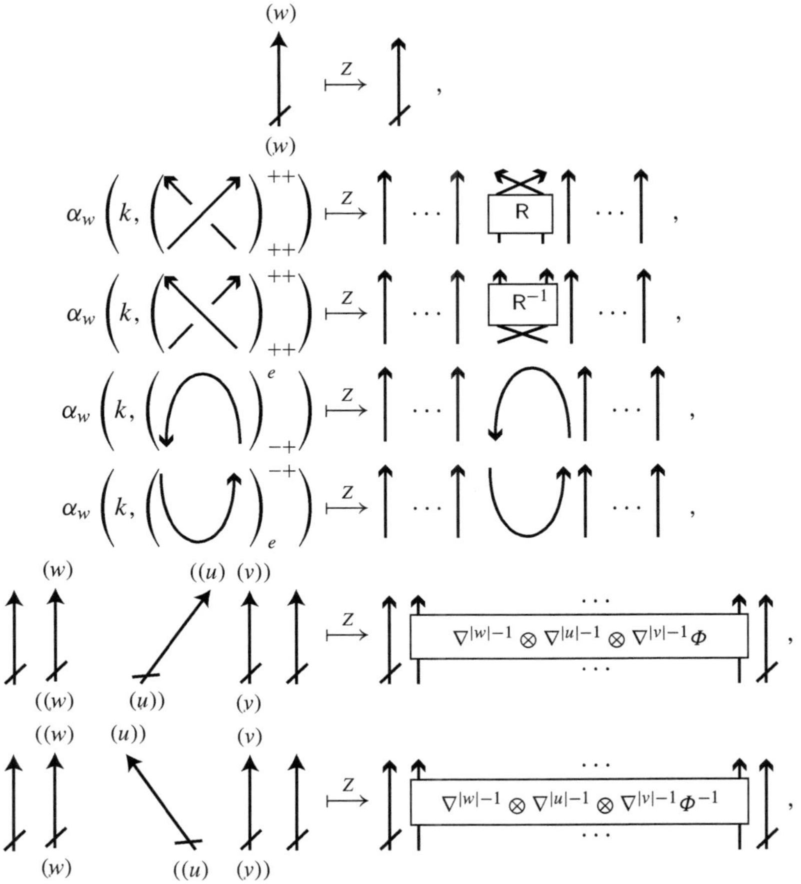

$$Z(s_a(T)) := s_a(Z(T)),$$

$$Z(T \circ T') := Z(T) \circ Z(T').$$

For a q-tangle diagram T, a first step in computing $Z(T)$ is expressing T as a composition of elementary q-tangles. There is not a unique way to do this, and it is necessary to show that the value of $Z(T)$ does not depend upon this choice. Proving that this is the case is an involved process, and we shall omit this argument. We refer the interested reader to [88] for details.

Theorem 17.6. *Let T be a fixed q-tangle diagram. Then $Z(T)$ is well-defined: that is $Z(T)$ is independent of the decomposition of T into elementary q-tangles.*

17.2 The Normalization Ž of Z

So far we have constructed a well-defined map

$$Z\colon \{q\text{-tangle diagrams on } X\} \to \widehat{\mathscr{A}^c}(X).$$

Recall that our aim is to obtain an *invariant* of framed q-tangles on X that evaluates in $\widehat{\mathscr{A}^c}(X)$, *i.e.*, we want a mapping whose values are unchanged under equivalence of framed q-tangles. The map Z is not an invariant of q-tangles, as is readily confirmed by the following example.

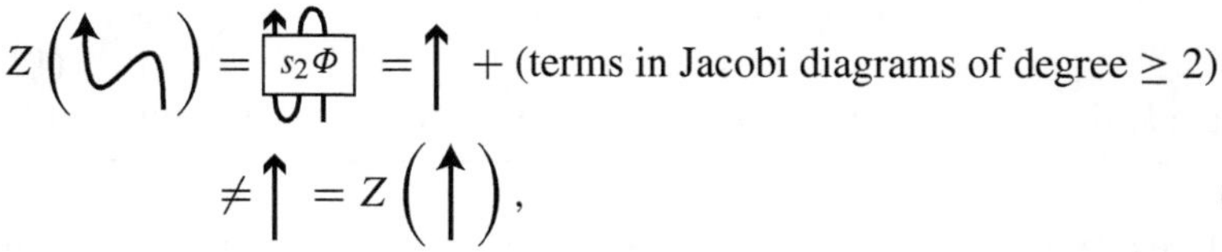

(17.9)

where we have used Lemma 16.22.

Fortunately, we can normalise Z to obtain a q-tangle invariant that is denoted by $\check{Z}$. This is the Kontsevich invariant. Essentially, the issue with removing maxima and minima displayed in (17.9) is all that stops Z from being a q-tangle invariant, and we can obtain a q-tangle invariant by cancelling each expression $s_2\Phi$ associated with maxima and minima, as we presently describe.

The fact from Lemma 16.12 that $\widehat{\mathscr{A}^c}$ (↑) is commutative, together with (17.9), shows that Z ( ) has an inverse. We shall use the notation

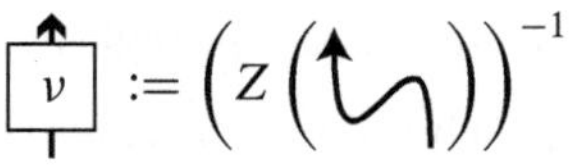

(17.10)

to describe this inverse. Since it has a square root (as it is a formal power series beginning with the identity), we can set $\nu^{1/2}$ to be the square root of ν starting with ↑.

Recall from Sect. 16.5 the connected sum operation $\#_i$ that acts by "splicing together" skeleton components of the Jacobi diagrams:

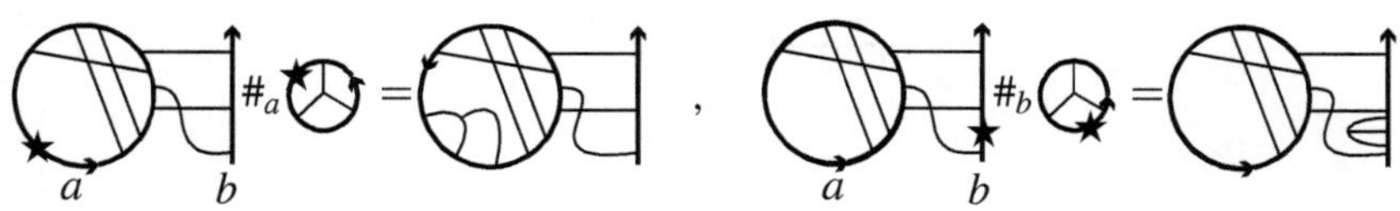

Of course, there is some choice of where on the skeletons we splice the Jacobi diagrams, but we saw in Theorem 16.19 that if D has a skeleton of exactly one component, then $E \#_a D$ is independent of this choice. A consequence of this is that the following definition is well-defined.

Definition 17.7 (Kontsevich invariant). Let T be a framed q-tangle on X with n components, and let m_i be the number of critical points in component i of T, for $i = 1, \ldots, n$. Then the *Kontsevich invariant* of T, denoted by $\check{Z}(T) \in \widehat{\mathscr{A}^c}(X)$, is defined to be

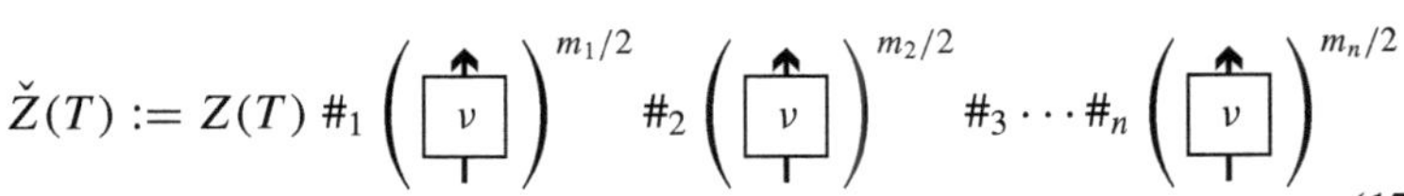

(17.11)

where $\#_i$ denotes connect sum at component i, and Z is as in Definition 17.1.

Before continuing with the exposition of $\check{Z}$ we address a point that may be worrying the reader. In the above we have normalised Z to get $\check{Z}$ using the value of . This gives prominence to the tangle 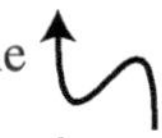 and the reader might question why we did not favour the tangles ,  or 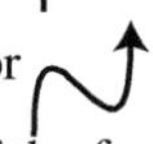 instead. The following lemma tells us that we could in fact have used any of the four tangles.

Lemma 17.8. *In* $\widehat{\mathscr{A}^c}(X_1)$

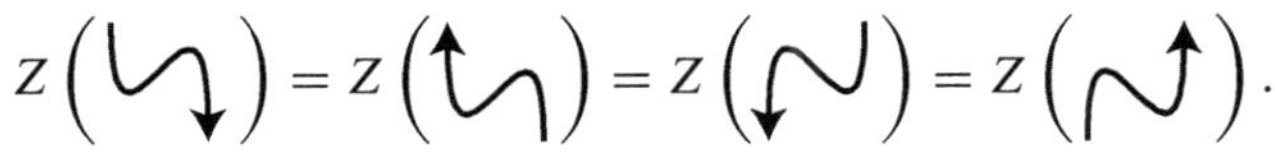

Proof. By Proposition 17.6, Z is well-defined so the q-tangles in the theorem statement may be decomposed into elementary q-tangles as we please.

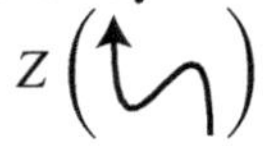 may be calculated by taking a q-tangle representation of it as follows:

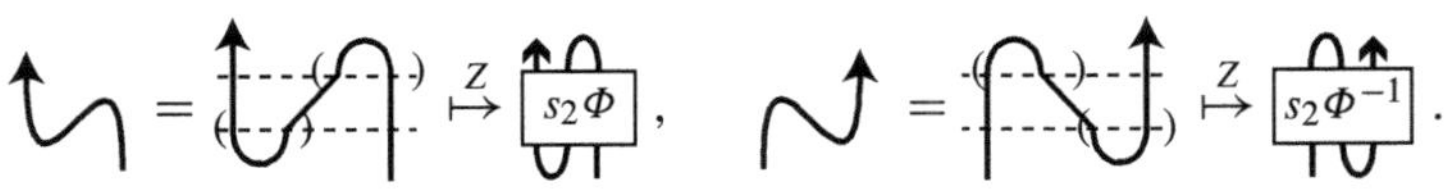

(17.12)

But from Definition 16.21(2)

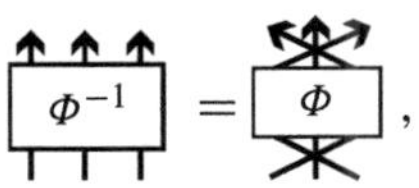

so

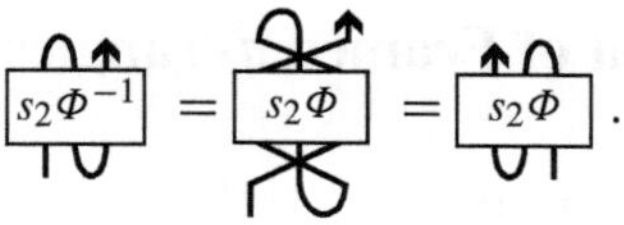

Thus Z  $= Z$ . Applying the map s from Definition 16.10 to the tangles in this identity gives Z  $= Z$ 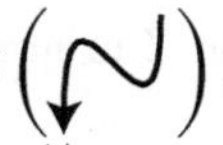.

To relate these two pairs, we determine Z 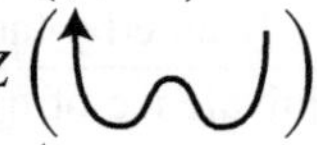 in two different ways. These are independent of the decomposition of  into elementary q-tangles, as Z is well-defined. For the first calculation,

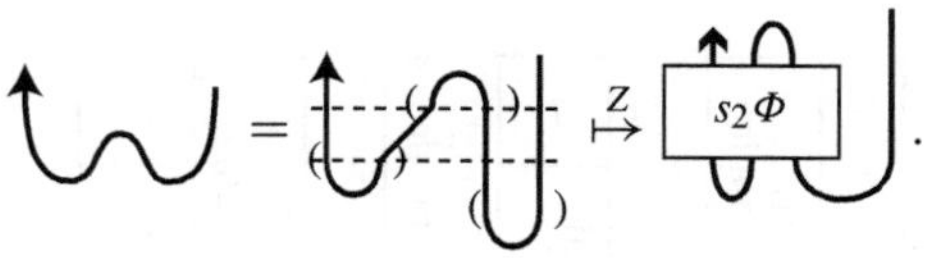 (17.13)

For the second calculation,

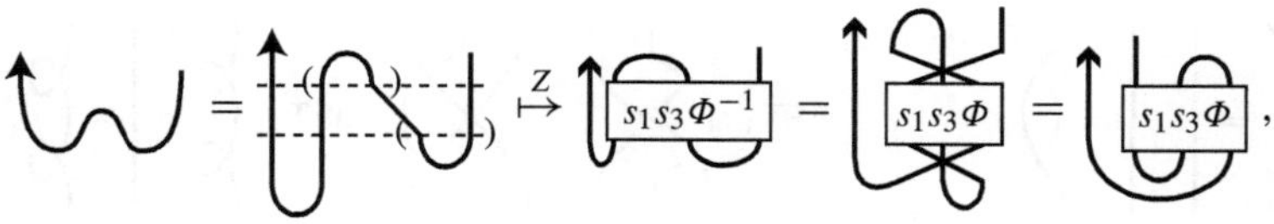 (17.14)

by Definition 16.21(2).

Therefore,

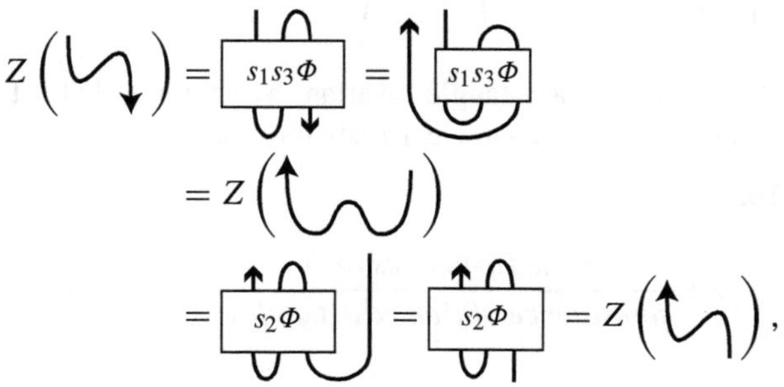

where the third equality is by (17.14) and the fourth equality is from (17.13) This completes the proof. □

Exercise 17.9. Let K and K' be oriented, framed knots. Show that

$\check{Z}(K \# K') =$ 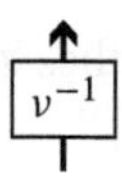 $\# \check{Z}(K) \# \check{Z}(K')$.

17.3 $\check{Z}$ is an Invariant of Framed q-Tangles

So far we have a well-defined map

$$\check{Z}\colon \{\text{framed } q\text{-tangles on } X\} \to \widehat{\mathscr{A}^c}(X).$$

Our goal now is to show that $\check{Z}$ is an invariant of framed q-tangles, *i.e.*, that

$$\check{Z}\colon \frac{\{\text{framed } q\text{-tangles on } X\}}{\text{equivalence of framed } q\text{-tangles}} \to \widehat{\mathscr{A}^c}(X).$$

We know from Chap. 15 that two framed q-tangles $S_{u_0}^{u_1}$ and $T_{w_0}^{w_1}$ are equivalent if and only if $u_1 = w_1$, $u_0 = w_0$ and their underlying tangles are related by the Turaev moves from Theorem 15.8, reproduced below for convenience.

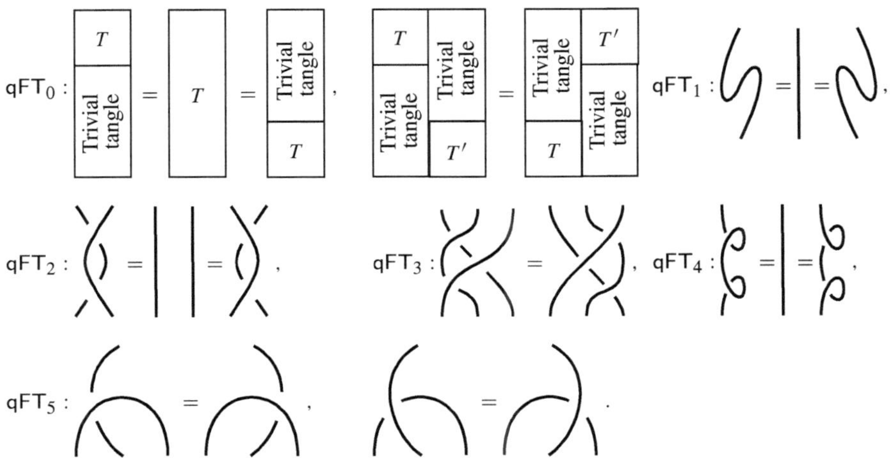

Thus we can show that $\check{Z}$ is a q-tangle invariant by showing that it takes the same value on each side of each of the above Turaev moves.

Theorem 17.10.

$$\check{Z}\colon \frac{\{\textit{framed } q\textit{-tangles on } X\}}{\textit{equivalence of framed } q\textit{-tangles}} \to \widehat{\mathscr{A}^c}(X).$$

That is, $\check{Z}$ is a framed q-tangle invariant.

Proof. We must show that $\check{Z}$ is invariant under the moves qFT_0–qFT_5. First observe that by the identity $Z(s_a(T)) = s_a(Z(T))$ we only need to show that $\check{Z}$ is invariant under one particular orientation of each qFT-move, and we are free to choose that orientation.

Since $\check{Z}$ is well-defined, we may parenthesize the diagrams as we wish. We can express any q-tangle $T_{w_0}^{w_1}$ as a composite $T_{w_0}^{w_1} = S_{u_1}^{w_1} \circ T_{u_0}^{u_1} \circ (S')_{w_0}^{u_0}$, where u_0 and u_1

are arbitrary parenthesizations of the same words as w_0 and w_1, S and S' are trivial tangles. With this we have

$$\check{Z}(T^{w_1}_{w_0}) = \check{Z}(S^{w_1}_{u_1}) \circ \check{Z}(T^{u_1}_{u_0}) \circ \check{Z}((S')^{u_0}_{w_0}).$$

This means that if we wanted to show that $\check{Z}(T^{w_1}_{w_0}) = \check{Z}(T'^{w_1}_{w_0})$ for some q-tangles T and T', we can instead prove that $\check{Z}(T^{u_1}_{u_0}) = \check{Z}(T'^{u_1}_{u_0})$ for any parenthesizations u_0 and u_1 that we choose. Thus when we show that $\check{Z}$ invariant under the moves qFT_0–qFT_5, without loss of generality, we can choose any parenthesisations of the tangles that we please (as long as the parenthesizations of two tangles in each move are the same).

Furthermore, this means that it is enough to show that $\check{Z}$ is unchanged only on the portion of the q-tangle that changes under each qFT-move. Thus we can consider the qFT-moves in isolation, without regard to the larger q-tangle they sit in, and with any parenthesization we please.

The qFT_0*-move*: Invariance of $\check{Z}$ under the first FT_0-move is trivial. The second follows trivially if we choose the parenthesizations of the form such that the tangles in the diagram equal $T \otimes T'$.

The qFT_1*-moves*: We have

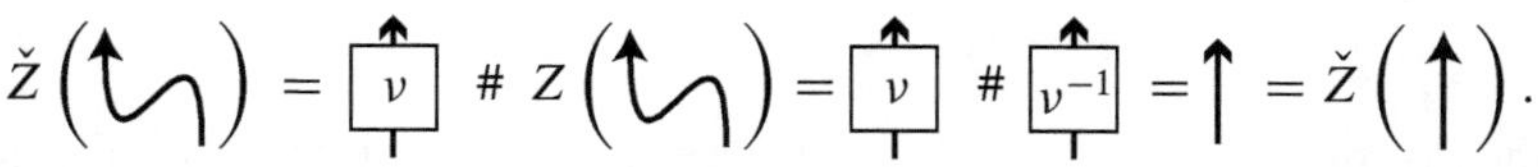

A similar argument, using Lemma 17.8, shows invariance under the second qFT_1-move.

The qFT_2*-move*: We have

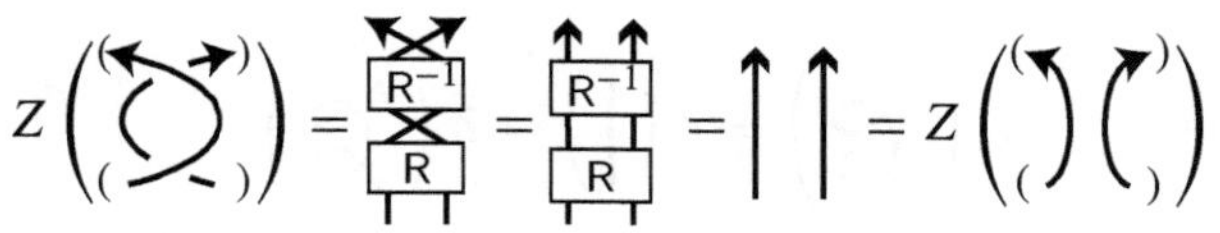

since 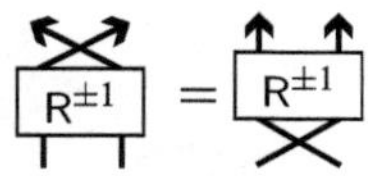. Thus, since there are no critical points and therefore no occurrences of the factor ν, we have $\check{Z}$ 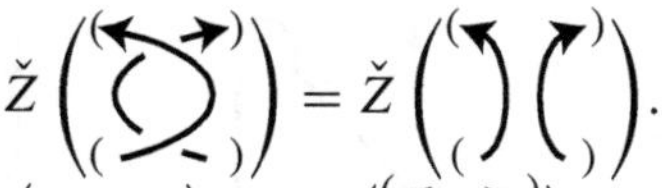.

A similar argument shows that $\check{Z}$ 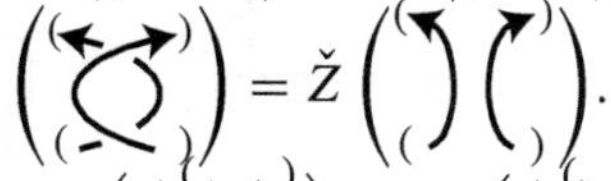.

The qFT_3*-move*: We must show that $\check{Z}$ 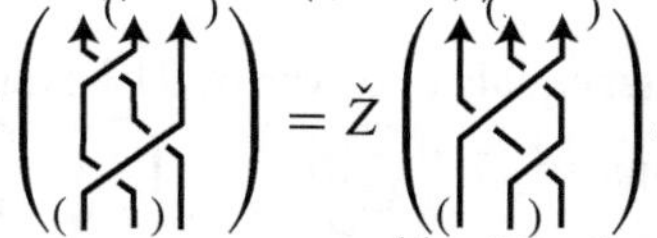. It is readily seen in (1) and (2) below, that the two diagrams are mapped by Z to the same image.

(1) 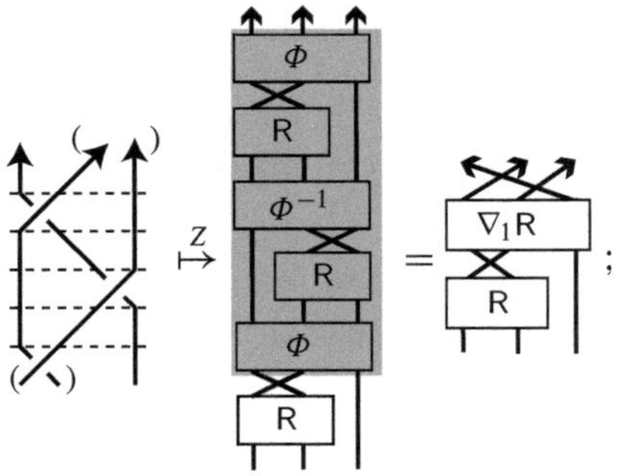

(2) 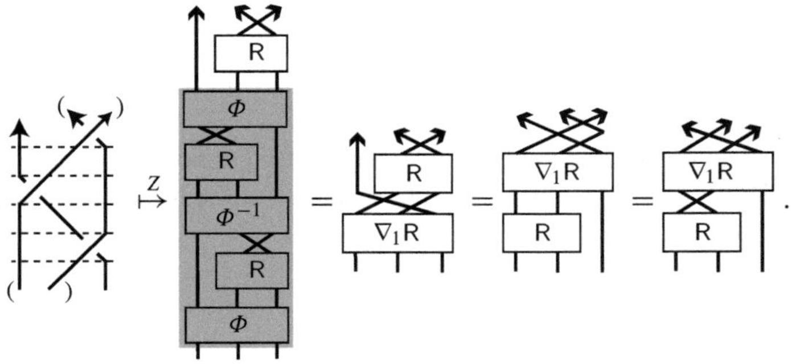

In the case of (1), the first equality is by the hexagon relation applied to the shaded box. In the case of (2), the first equality is again by the hexagon relation applied to the shaded box, the second equality is by commutation using Lemma 16.13 and the third equality is by the symmetry of $\nabla_1 R$.

<u>*The* qFT$_4$-*move*</u>: We must show that

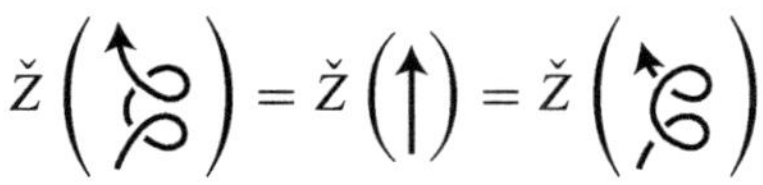

First, observe that

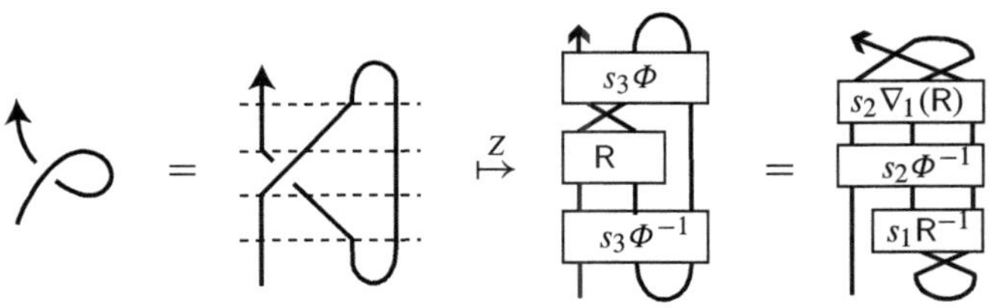

where the latter equality is given by the hexagon relation, given in Definition 16.21(4). But by Lemma 16.17, 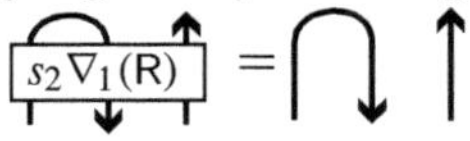, and, recall the effect of s_i on sign,

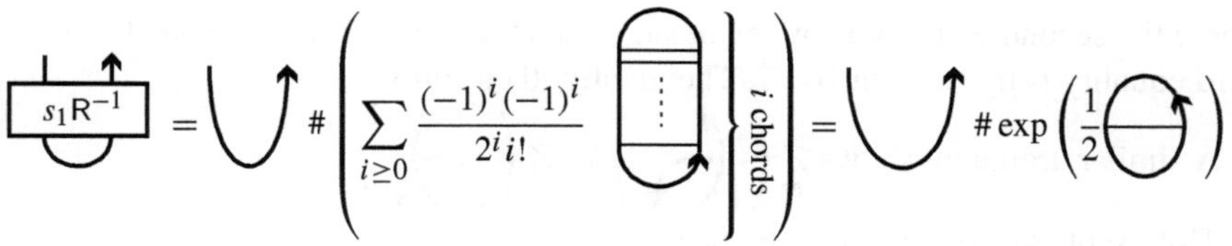

Continuing with the calculation we have

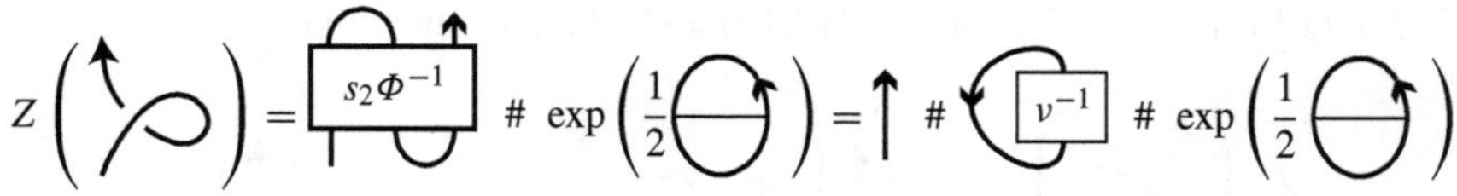

so

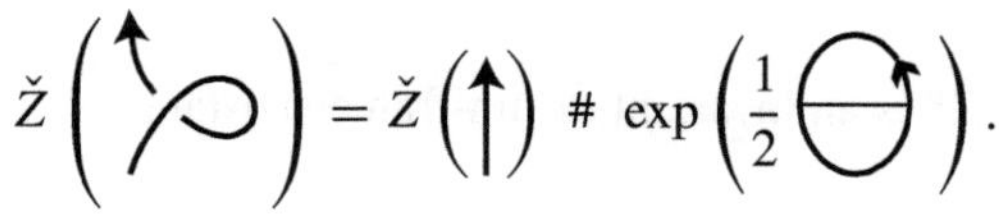 (17.15)

Similarly,

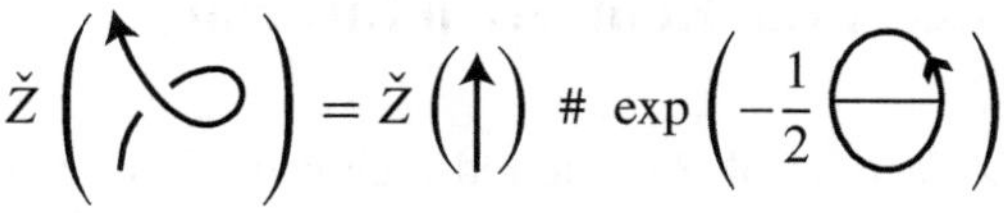 (17.16)

and, by combining these two expressions, we have immediately that

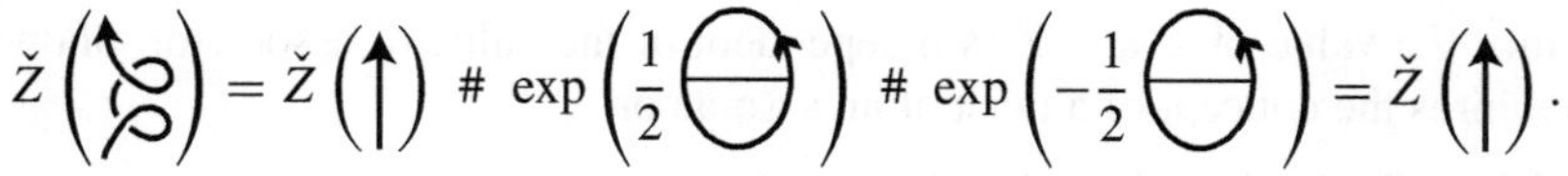

Similarly 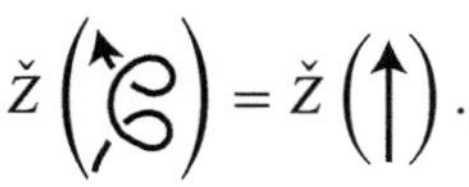

<u>*The* qFT_5*-move*</u>: We need to show that 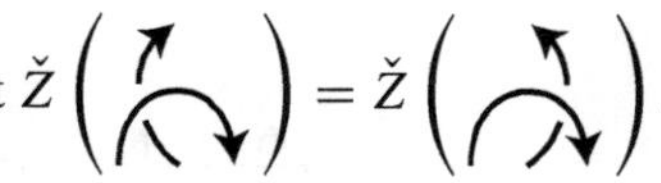

. Now

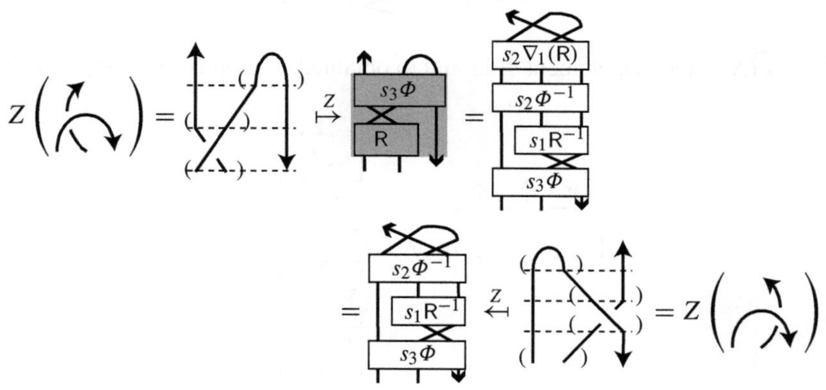

where the second equality is by the hexagon relation applied to the shaded box, the third equality is by Lemma 16.17. The identity then follows.

A similar argument shows 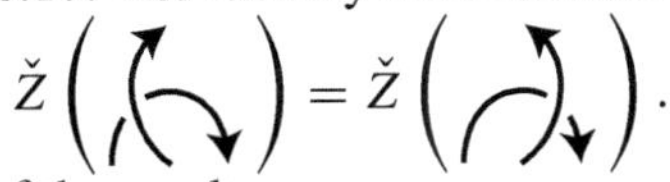

This completes the proof of the result. □

Exercise 17.11. Define $\check{Z}^u$ to be the mapping from q-tangle diagrams to $\widehat{\mathscr{A}^c}(X)/(1\text{T})$ given by $\check{Z}^u(T) = p \circ \check{Z}$, where p is projection from $\widehat{\mathscr{A}^c}(X)$ to $\widehat{\mathscr{A}^c}(X)/(1\text{T})$. By considering Eqs. (17.15) and (17.16), show that

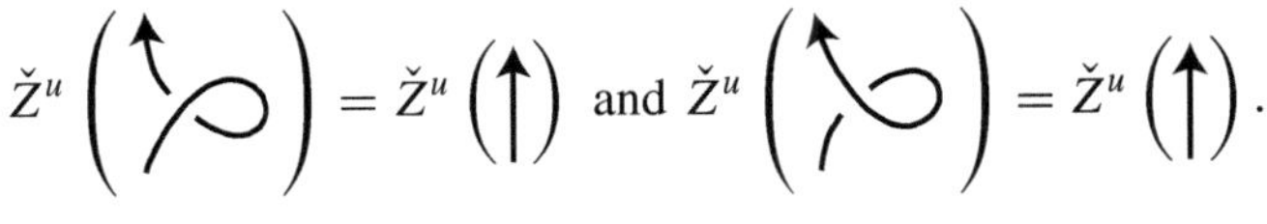

Hence deduce that $\check{Z}^u$ is an invariant of (un-framed) q-tangles.

17.4 Uniqueness of the Kontsevich Invariant

The definition of Z (and so of $\check{Z}$) required a choice of associator Φ. Since more than one associator exists in $\widehat{\mathscr{A}}$ this means that the value of Z on a given q-tangle can depend upon the choice of associator used in its definition. However, choosing a different associator has a predictable change in the values of Z. Moreover, for knots and links the value of Z and $\check{Z}$ is independent of the value of associator. Showing this requires the concept of a twist of an associator.

Definition 17.12 (Twisting). Let $F \in \widehat{\mathscr{A}}(X_2)$ be such that

$$\varepsilon_1(F) = \varepsilon_2(F) = \uparrow\uparrow, \tag{17.17}$$

and

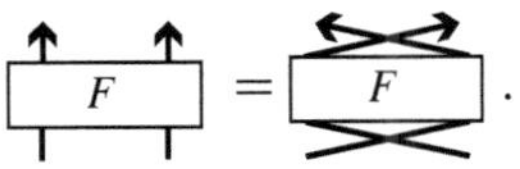

(17.18)

If $\Phi \in \widehat{\mathscr{A}}(X_3)$, then an element $\widetilde{\Phi}$ is said to obtained from Φ by *twisting* by F if

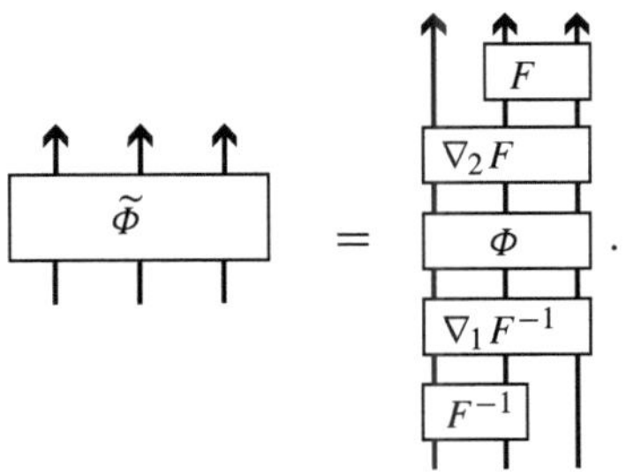

It should be noted that the first condition on F ensures that it is is invertible, since this condition implies that F has the form

$$F = \uparrow\uparrow + (\text{terms in Jacobi diagrams higher degree}),$$

which has an invertible term of lowest degree 0.

The significance of twisting lies in that any two associators are necessarily twists of each another. We will not provide a proof of this result, only a statement. The interested reader can find details of its proof in [114], which is a diagrammatic adaptation of a result for quasi-triangular Hopf algebras from [59].

Theorem 17.13. *If Φ and Φ' are associators, then Φ' is obtained from Φ by a twist by some $F \in \widehat{\mathscr{A}^c}(X_2)$ that satisfies* (17.17) *and* (17.18).

A consequence of Theorem 17.13 is that different choices of associator in the definition of the Kontsevich invariant result in closely related versions of the invariant. To describe the relation, let w be a parenthesized word on the alphabet $\{+\}$ and $F \in \widehat{\mathscr{A}}(X_2)$. Let $F_w \in \widehat{\mathscr{A}}(X_{|w|})$, where $|w|$ is the length of w, be an element inductively defined by

$$F_{(u)(v)} = \left(F_u \otimes \uparrow^{\otimes|v|}\right) \circ \left(\uparrow^{\otimes|u|} \otimes F_v\right) \circ \left((\nabla^{|v|-1} \otimes \nabla^{|u|-1})F\right), \tag{17.19}$$

with initial conditions $F_e = 1 \in \mathbb{C}$, $F_+ = \uparrow$ and $F_{++} = F$, where e is the empty word.

For example,

$$F_{(++)+} = \left(F_{++} \otimes \uparrow\right) \circ \left(\uparrow \otimes F_+\right) \circ ((\nabla \otimes \text{id})(F)) =$$ 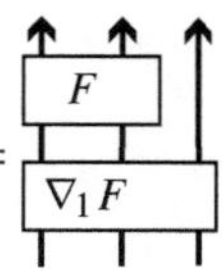

.

Theorem 17.14. *Let T_v^u be a framed q-tangle. Let Φ and $\widetilde{\Phi}$ be associators, with $\widetilde{\Phi}$ obtained from Φ by a twist by F. Let Z_Φ and $Z_{\widetilde{\Phi}}$ denote Kontsevich's knot invariant derived from Φ and $\widetilde{\Phi}$, respectively. Then*

$$Z_{\widetilde{\Phi}}(T_v^u) = F_v \circ Z_\Phi(T_v^u) \circ (F_u)^{-1},$$

and

$$\check{Z}_{\widetilde{\Phi}}(T_v^u) = F_v \circ \check{Z}_\Phi(T_v^u) \circ (F_u)^{-1},$$

A proof of this theorem can be found in [116].

An immediate corollary of Theorem 17.14 is that the Kontsevich invariant of a framed link does not depend upon the choice of associator used in its construction.

Corollary 17.15.

$$\check{Z}: \{\textit{framed } n \textit{ component links}\} \to \mathscr{A}\left(\sqcup_n \mathbb{S}^1\right)$$

is independent of the choice of associator in its construction.

Furthermore, since there exists an associator with rational coefficients (see (16.15)) this means that the Kontsevich invariant of a framed link has rational coefficients.

Chapter 18
Universality Properties of the Kontsevich Invariant

18.1 A Universal Vassiliev Invariant

The Kontsevich invariant (17.2) plays an important role in the theory of Vassiliev invariants. Recall from Chaps. 11 and 12 (specifically Lemmas 11.28 and 12.8) that there are well-defined injective linear maps

$$\overline{\alpha}_m\colon \frac{\mathscr{V}_m}{\mathscr{V}_{m-1}} \to \mathscr{W}_m \quad \text{and} \quad \overline{\alpha}_m\colon \frac{\mathscr{V}_m^f}{\mathscr{V}_{m-1}^f} \to \mathscr{W}_m^f.$$

Furthermore, we claimed in Theorems 11.29 and 12.9, but did not prove, that these maps define vector space isomorphisms and hence

$$\frac{\mathscr{V}_m}{\mathscr{V}_{m-1}} \cong \mathscr{W}_m \quad \text{and} \quad \frac{\mathscr{V}_m^f}{\mathscr{V}_{m-1}^f} \cong \mathscr{W}_m^f. \tag{18.1}$$

With the aid of the Kontsevich invariant, we may now prove these isomorphisms.

18.1.1 Ž as an Isomorphism

To prove the isomorphisms in (18.1), we first dualise the results. From Lemmas 11.38 and 12.11, we have that $\varphi_m^\star = \overline{\alpha}_m$ and so proving (18.1) is equivalent to proving that the maps

$$\varphi_m : \bar{\mathscr{A}}_m^c \to \mathscr{K}_m / \mathscr{K}_{m+1} \quad \text{and} \quad \varphi_m : \mathscr{A}_m^c \to \mathscr{K}_m^f / \mathscr{K}_{m+1}^f \tag{18.2}$$

are vector space isomorphisms. Here, we recall that φ_m is as in Definition 11.34 (and has the same action on chord diagrams in the framed and unframed cases). Its action on a chord diagram is given by the contraction operation:

D. M. Jackson and I. Moffatt, *An Introduction to Quantum and Vassiliev Knot Invariants*, CMS Books in Mathematics, https://doi.org/10.1007/978-3-030-05213-3_18

φ_m : 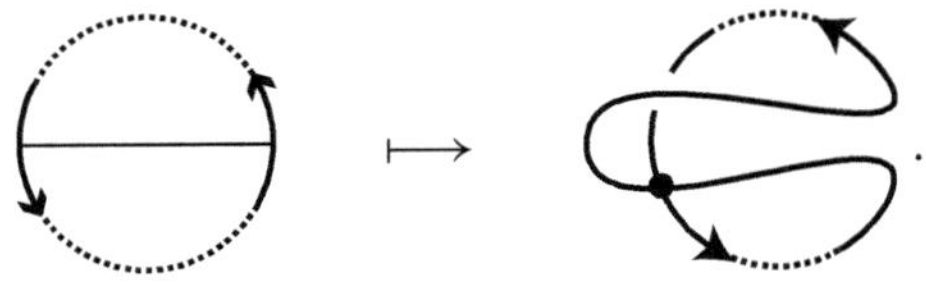 (18.3)

We have previously shown in Lemmas 11.39 and 12.11 that the maps φ_m are surjective linear maps, and hence, to prove that they are isomorphisms, we need to show they are injective. We shall do this by using the Kontsevich invariant to construct a map $\check{Z}_m$ that is a left inverse of φ_m, *i.e.* $\check{Z}_m \circ \varphi_m = \mathrm{id}$. Since a map is injective if and only if it has a left inverse, it will follow that φ_m is an isomorphism.

We shall prove the following theorem in this section.

Theorem 18.1. *Let $\check{Z}_m$ be the map defined by*

$$\check{Z}_m : \frac{\mathscr{K}_m^f}{\mathscr{K}_{m+1}^f} \to \mathscr{A}_m^c \colon [K] \mapsto p_m(\check{Z}(K))$$

where $p_m : \widehat{\mathscr{A}^c} \to \mathscr{A}_m^c$ is projection, and $[K]$ is the equivalence class of $K \in \mathscr{K}_m^f$. Then $\check{Z}_m$ is a vector space isomorphism. Moreover, $\check{Z}_m = \varphi_m^{-1}$ and hence

$$\varphi_m : \mathscr{A}_m^c \to \mathscr{K}_m^f / \mathscr{K}_{m+1}^f$$

is a vector space isomorphism.

This theorem has an unframed version.

Theorem 18.2. *Let $\check{Z}_m^u$ be the map defined by*

$$\check{Z}_m^u : \frac{\mathscr{K}_m}{\mathscr{K}_{m+1}} \to \bar{\mathscr{A}}_m^c \colon [K] \mapsto p_m(\check{Z}(K))$$

where $p_m : \widehat{\bar{\mathscr{A}}^c}/(1\mathrm{T}) \to \bar{\mathscr{A}}_m^c$ is projection. Then $\check{Z}_m^u$ is a vector space isomorphism. Moreover, $\check{Z}_m^u = \varphi_m^{-1}$ and hence

$$\varphi_m : \bar{\mathscr{A}}_m^c \to \mathscr{K}_m / \mathscr{K}_{m+1}$$

is a vector space isomorphism.

The proofs of Theorems 18.1 and 18.2 will make use of the following lemma.

Lemma 18.3. *Let $C \in \mathscr{A}_m^c$ be a degree m chord diagram. Then*

$$\check{Z}(\varphi_m(C)) = C + (termsofhigherdegree),$$

where the element-wise action of φ_m is given in (18.3).

Proof. For the proof, we represent $\varphi(C)$ as  where it is to be understood that we are representing $\varphi(C)$ by focussing on neighbourhoods of *all* of the singular points and operating upon *all* of them "simultaneously".

To determine the lowest degree term of $\check{Z}(\varphi(C))$, we consider

where the ends of the crossing segments are joined in the way specified by C. Now, from Vassiliev's relation and Definition 17.5,

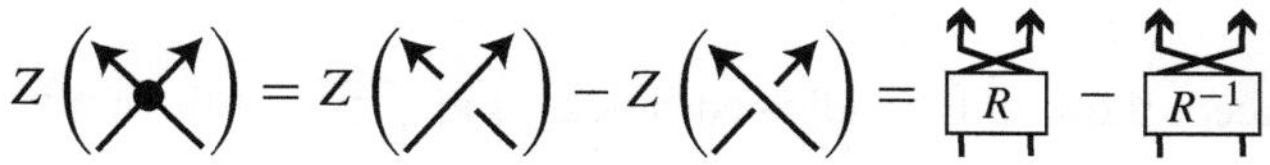

where the chord diagrams are identical except where shown and where R is given in (16.7). Moreover, we have used the symmetry of  to move the crossing to the top in . To calculate the lowest degree term in Z , we note that since, from Lemma 16.22, the associator has the form

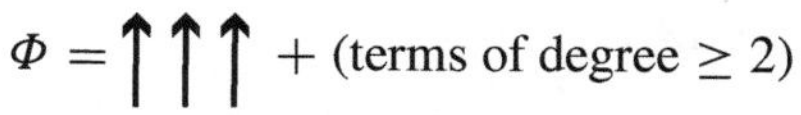

then any contribution of $\check{Z}$ arising from the associator will have a degree zero term with coefficient 1, and terms of degree ≥ 2. We have,

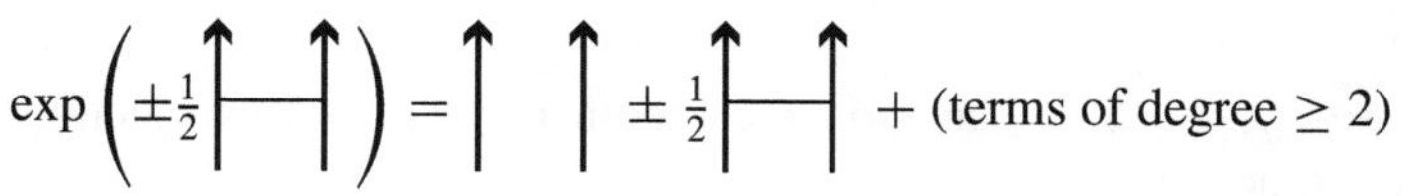

so,

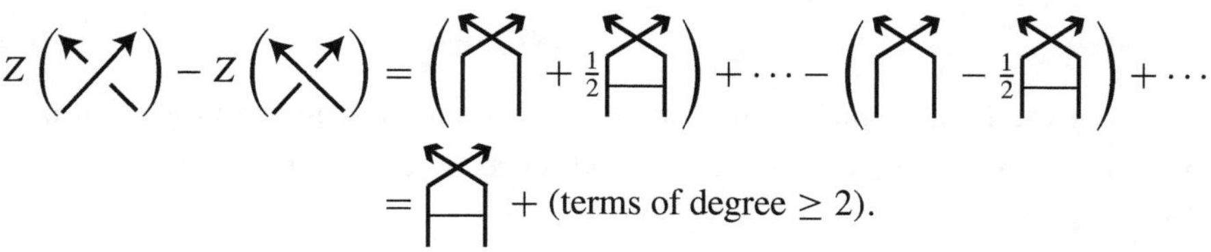

Let

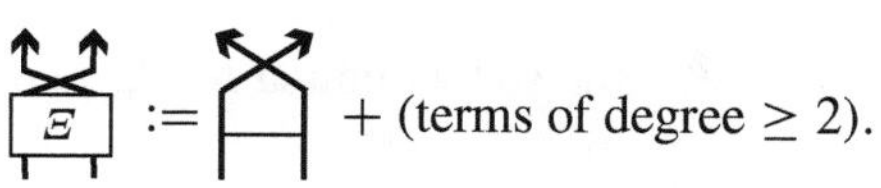

Then

$$Z(\varphi(C)) = \check{Z}$$ 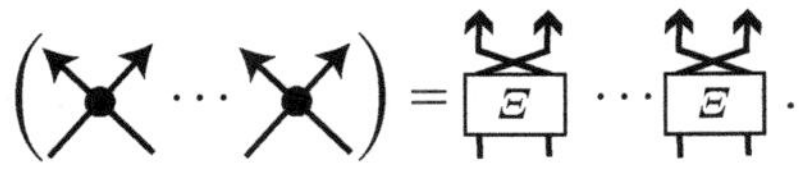

We now show that the rest of the knot contributes a multiplicative factor of 1. Since this part of the knot contains no singular crossings, the rest of the chord diagrams are constructed by taking products of applications of the operators s and ∇ to $\mathsf{R}^{\pm 1}$ and the associator Φ. Moreover, since both of these have constant terms equal to the identity, it follows that Z  $- Z$  is equal to the sum of  with no chords except those in the indicated blocks, and  with chords outside the indicated blocks. Thus

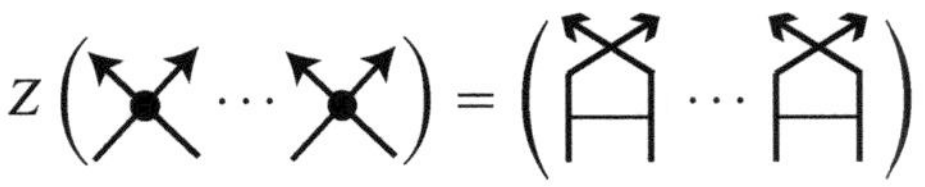 + (terms of degree ≥ 2)

$$= C + (\text{terms of degree} \geq 2)$$

by the construction of $\varphi(C)$. This completes the proof. □

We now prove Theorem 18.1.

Proof (*Theorem* 18.1). By Lemma 18.3, for each m, we have $\check{Z}(\mathcal{K}_m^f) \subseteq \widehat{\mathcal{A}^c}_{\geq m}$. Therefore, $\check{Z}$ induces a map

$$\check{Z}_{\geq m} \colon \mathcal{K}_m^f \to \widehat{\mathcal{A}^c}_{\geq m}.$$

We have $\widehat{\mathcal{A}^c}_{\geq m+1} \leq \widehat{\mathcal{A}^c}_{\geq m}$ and so we can compose $\check{Z}_{\geq m}$ with projection, π, onto the quotient space to get

$$\pi \circ \check{Z}_{\geq m} \colon \mathcal{K}_m^f \to \frac{\widehat{\mathcal{A}^c}_{\geq m}}{\widehat{\mathcal{A}^c}_{\geq m+1}}.$$

Applying the First Isomorphism Theorem (Theorem A.9) to the projection map $p_m : \widehat{\mathcal{A}^c}_{\geq m} \to \mathcal{A}_m^c$ gives $\frac{\widehat{\mathcal{A}^c}_{\geq m}}{\widehat{\mathcal{A}^c}_{\geq m+1}} \cong \mathcal{A}_m^c$, and hence, we can regard $\pi \circ \check{Z}_{\geq m}$ as giving rise to a map

$$p_m \circ \check{Z}_{\geq m} \colon \mathcal{K}_m^f \to \mathcal{A}_m^c.$$

Next, we note that $\mathcal{K}_{m+1}^f \leq \mathcal{K}_m^f$ and so by Lemma A.10, $p_m \circ \check{Z}_{\geq m}$ induces a well-defined linear map

$$p_m \circ \check{Z}_{\geq m} \colon \frac{\mathscr{K}_m^f}{\mathscr{K}_{m+1}^f} \to \mathscr{A}_m^c.$$

But this is precisely the map $\check{Z}_m$ from the statement of the theorem.

Having shown the existence of the linear map $\check{Z}_m$, the rest of the theorem follows easily. The fact that $Z_m \circ \varphi_m = \text{id}$ follows from Lemma 18.3. Since φ_m has a left inverse, it is injective. By Lemma 12.11, φ_m is surjective and hence is bijective. That $\check{Z}_m = \varphi_m^{-1}$ then follows since $Z_m \circ \varphi_m = \text{id}$. □

The proof of Theorem 18.2 is similar, with only small modifications, and is left as an exercise.

Exercise 18.4. Modify the proof of Theorem 18.1 to give a proof of Theorem 18.2.

We have just shown that Lemmas 11.38 and 12.11 hold. Hence, by duality, so do Lemmas 11.28 and 12.8.

Theorem 18.5. *The maps*

$$\overline{\alpha}_m \colon \frac{\mathscr{V}_m}{\mathscr{V}_{m-1}} \longrightarrow \mathscr{W}_m$$

and

$$\overline{\alpha}_m \colon \frac{\mathscr{V}_m^f}{\mathscr{V}_{m-1}^f} \longrightarrow \mathscr{W}_m^f$$

are vector space isomorphisms.

Proof. This result follows immediately from Theorems 18.1 and 18.2, and Lemmas 11.38 and 12.11. □

18.1.2 $\check{Z}$ as a Universal Vassiliev Invariant

In fact, the Kontsevich invariant of Theorem 18.1 is a universal Vassiliev invariant in the sense that it "contains" every Vassiliev invariant, or more specifically every Vassiliev invariant factors through it. This universality property, for the unframed case, may be summarised by the following commutative diagram, where θ is a degree m Vassiliev invariant.

$$\begin{array}{ccccc} \mathscr{K}^f & \xrightarrow{\check{Z}} & \widehat{\mathscr{A}^c} & \xrightarrow{p_{\leq m}} & \mathscr{A}^c_{\leq m} \\ & & & \searrow^{\theta} & \downarrow W \\ & & & & \mathbb{C} \end{array}$$

The universality property says for every given a degree m Vassiliev invariant θ, there exists a W that makes the diagram commute, and, conversely, for any W there exists a degree m Vassiliev invariant θ that makes the diagram commute. This is formalised in the following theorem of Kontsevich [102].

Theorem 18.6 (Universality). *Let $\check{Z}$ be the Kontsevich invariant.*

1. *If W is any linear map $W : \mathscr{A}^c_{\leq m} \to \mathbb{C}$ then the composite*

$$\theta_W : \mathscr{K}^f \xrightarrow{\check{Z}} \widehat{\mathscr{A}^c} \xrightarrow{p_{\leq m}} \mathscr{A}^c_{\leq m} \xrightarrow{W} \mathbb{C}$$

is a degree m Vassiliev invariant of framed knots.

2. *Any degree m Vassiliev invariant θ of framed knots can be written as a composite*

$$\theta : \mathscr{K}^f \xrightarrow{\check{Z}} \widehat{\mathscr{A}^c} \xrightarrow{p_{\leq m}} \mathscr{A}^c_{\leq m} \xrightarrow{W_\theta} \mathbb{C}.$$

for some $W_\theta : \mathscr{A}^c_{\leq m} \to \mathbb{C}$.

Proof. For the first part, by Lemma 18.3, $\check{Z}(\mathscr{K}^f_{m+1}) \subseteq \widehat{\mathscr{A}}_{\geq m+1}$. In particular, this means that $(p_{\leq m} \circ \check{Z})(\mathscr{K}^f_{m+1}) = 0$ and therefore $\theta_W = W \circ p_{\leq m} \circ \check{Z}$ is a degree m Vassiliev invariant.

For the second part, let $\theta \in \mathscr{V}^f_m$ be a degree m Vassiliev invariant. Since

$$\mathscr{V}^f_m \cong \frac{\mathscr{V}^f_m}{\mathscr{V}^f_{m-1}} \oplus \frac{\mathscr{V}^f_{m-1}}{\mathscr{V}^f_{m-2}} \oplus \cdots \oplus \frac{\mathscr{V}^f_1}{\mathscr{V}^f_0} \oplus \mathscr{V}^f_0,$$

we may write

$$\theta = \theta_m + \theta_{m-1} + \cdots + \theta_0,$$

where each $\theta_i \in \frac{\mathscr{V}^f_i}{\mathscr{V}^f_{i-1}}$ for $m \geq 1$, and $\theta_0 \in \mathscr{V}^f_0$. Using the isomorphism $\frac{\mathscr{V}^f_i}{\mathscr{V}^f_{i-1}} \cong \left(\frac{\mathscr{K}^f_i}{\mathscr{K}^f_{i+1}}\right)^*$, we may then define maps $W_i \in (\mathscr{A}^c_i)^*$ as the composite

$$W_i : \mathscr{A}^c_i \xrightarrow{\varphi_i} \frac{\mathscr{K}^f_i}{\mathscr{K}^f_{i+1}} \xrightarrow{\theta_i} \mathbb{C},$$

and then define $W_\theta \in (\mathscr{A}^c_{\leq m})^*$ as

$$W_\theta = W_m + W_{m-1} + \cdots + W_0.$$

We may write

$$p_{\leq m} \circ \check{Z} = \check{Z}_m + \check{Z}_{m-1} + \cdots + \check{Z}_0,$$

where $\check{Z}_i = p_m \circ \check{Z}$. Using, from Theorem 18.2, the fact that $\check{Z}_i = \varphi_i^{-1}$, and so $W_i \circ \check{Z}_i = \theta_i \circ \varphi_i \circ \check{Z}_i = \theta_i$, we then see that $W_\theta \circ p_{\leq m} \circ \check{Z}$ is equal to $\theta_m + \theta_{m-1} + \cdots + \theta_0 = \theta$. Thus, $W_\theta \circ p_{\leq m} \circ \check{Z}$ and θ are identical maps on $\frac{\mathscr{K}^f}{\mathscr{K}^f_{m+1}}$. Since, by Theorem A.23, $\mathscr{K}^f \cong \frac{\mathscr{K}^f}{\mathscr{K}^f_{m+1}} \oplus \mathscr{K}^f_{m+1}$ and $W_\theta \circ p_{\leq m} \circ \check{Z}$ and θ are both degree m Vassiliev invariants and hence vanish on $\mathscr{K}^f_{m+1}$, it follows that $W_\theta \circ p_{\leq m} \circ \check{Z}$ and θ take the same values on $\mathscr{K}^f$, as required. □

An unframed version of the universality property holds and is proved similarly.

Theorem 18.7 (Universality). *Let $\check{Z}^u$ be the unframed Kontsevich invariant.*

1. *If W is any linear map $W : \bar{\mathscr{A}}^c_{\leq m} \to \mathbb{C}$ then the composite*

$$\theta_W : \mathscr{K} \xrightarrow{\check{Z}^u} \widehat{\bar{\mathscr{A}}^c} \xrightarrow{p_{\leq m}} \bar{\mathscr{A}}^c_{\leq m} \xrightarrow{W} \mathbb{C}$$

is a degree m Vassiliev invariant.

2. *Any degree m Vassiliev invariant θ can be written as a composite*

$$\theta : \mathscr{K} \xrightarrow{\check{Z}^u} \widehat{\bar{\mathscr{A}}^c} \xrightarrow{p_{\leq m}} \bar{\mathscr{A}}^c_{\leq m} \xrightarrow{W_\theta} \mathbb{C}.$$

for some $W_\theta : \bar{\mathscr{A}}^c_{\leq m} \to \mathbb{C}$.

Exercise 18.8. Modify the proof of Theorem 18.6 to give a proof of Theorem 18.7.

18.2 A Universal Quantum Invariant

We have just seen that the Kontsevich invariant is a universal Vassiliev invariant; that is, it contains and unifies all of the invariants discussed in Part III of this book. In fact, it also contains and unifies the Reshetikhin–Turaev invariants from Part II. We shall prove this for the invariants arising from $\mathfrak{sl}_2$ and its standard two-dimensional representation that we described in Sect. 9.4.

For this, we need to modify a weight system coming from $\mathfrak{sl}_2$ so that it records the degree of a chord or Jacobi diagram by means of an indeterminate h. We therefore define $\widehat{W} : \widehat{\mathscr{A}} \to \mathbb{C}[[h]]$ by its action on a chord diagram or Jacobi diagram D by

$$\widehat{W}(D) := W_{\mathfrak{sl}_2,\rho}(D) \cdot h^{d(D)},$$

where $d(D)$ denotes the degree of D, and $W_{\mathfrak{sl}_2,\rho}$ is the Lie algebra weight systems associated with $\mathfrak{sl}_2$ and its standard two-dimensional representation, as described in Sect. 14.4.

We prove that the Reshetikhin–Turaev invariant $Q^{\mathfrak{sl}_2,\rho}$ factors through the Kontsevich invariant *via* the weight system $W_{\mathfrak{sl}_2,\rho}$. To do this, we need to be careful about codomains. The function $\widehat{W}$ is evaluated in $\mathbb{C}[[h,h^{-1}]]$. We need to modify the Reshetikhin–Turaev invariant $Q^{\mathfrak{sl}_2,\rho}$ so that it also is evaluated in this space. We shall do this by evaluating $Q^{\mathfrak{sl}_2,\rho}$ at $q=e^{h/2}$; then, since this is a map, evaluating this map at 1 gives an element in $\mathbb{C}[[h,h^{-1}]]$.

We shall prove the following.

Theorem 18.9. *Let K be a framed knot, then*

$$\widehat{W}\circ \check{Z}(K) = Q^{\mathfrak{sl}_2,\rho}(K)\big|_{q=e^{h/2}}(1),$$

where $\widehat{W}(D) := W_{\mathfrak{sl}_2,\rho}(D)\cdot h^{d(D)}$, $d(D)$ is the degree of D, $W_{\mathfrak{sl}_2,\rho}$ is the Lie algebra weight system associated with $\mathfrak{sl}_2$ and its standard two-dimensional representation, and $Q^{\mathfrak{sl}_2,\rho}(K)$ is the Reshetikhin–Turaev invariant associated with $\mathfrak{sl}_2$ and its standard two-dimensional representation.

The proof we give depends upon the specific details of the invariant $Q^{\mathfrak{sl}_2,\rho}$ and the weight systems $W_{\mathfrak{sl}_2,\rho}$. Writing $Q(K)$ for $Q^{\mathfrak{sl}_2,\rho}(K)\big|_{q=e^{h/2}}(1)$, we know by Theorem 9.5 that $Q^{\mathfrak{sl}_2,\rho}(K)\big|_{q=e^{h/2}}(1)$ is determined by the skein relations

(18.4)

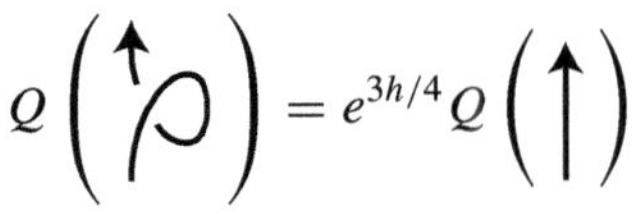
(18.5)

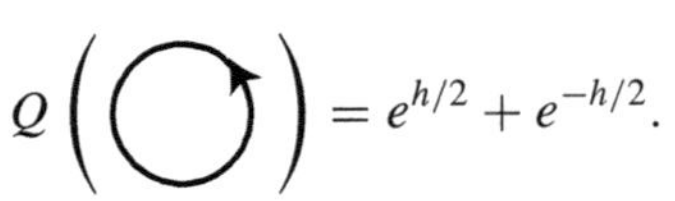
(18.6)

Our strategy is to show that the invariant $\widehat{W}\circ\check{Z}(K)$ satisfies the same skein relations. Then, since (18.4)–(18.6) define a unique invariant, $Q^{\mathfrak{sl}_2,\rho}(K)\big|_{q=e^{h/2}}(1)$ and $\widehat{W}\circ\check{Z}(K)$ must be equal.

To show that $\widehat{W}\circ\check{Z}(K)$ satisfies the skein relation, we take advantage of the combinatorics of $W_{\mathfrak{sl}_2,\rho}$. For simplicity, we shall denote the weight system $W_{\mathfrak{sl}_2,\rho}$ by W. By Theorem 14.24, we have that $W = W_{\mathfrak{sl}_2,\rho}$ is determined by the following skein relations:

(18.7)

(18.8)

We can rewrite (18.7) as

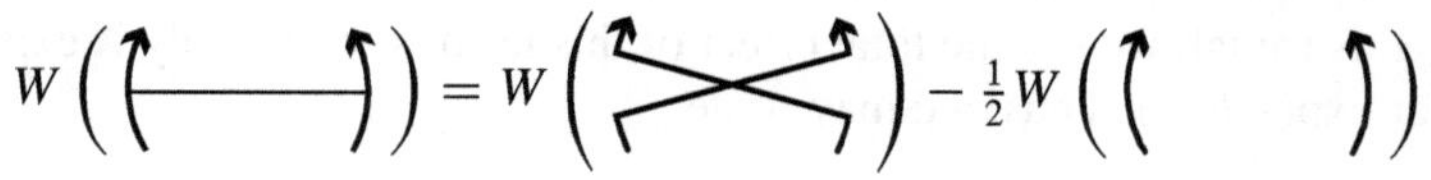

We use (18.8) and (18.7) to show that $\widehat{W} \circ \check{Z}$  satisfies the skein relations in (18.4)–(18.6). For this, we start by comparing the values of $\widehat{W} \circ \check{Z}$ , $\widehat{W} \circ \check{Z}$  and $\widehat{W} \circ \check{Z}$ 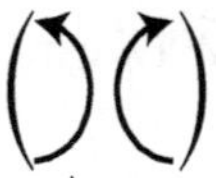. Since $\check{Z}$  $=$ 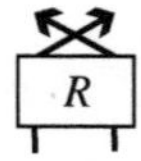, we need to compute the value of $\widehat{W}$ . (Technically, we have defined W only for chord diagrams on a circles, so the figures involving chord diagrams should be considered as being part of a larger circular skeleton.)

Let $\mathbf{X}$, $\mathbf{H}$ and $\mathbf{I}$ be the configurations given by

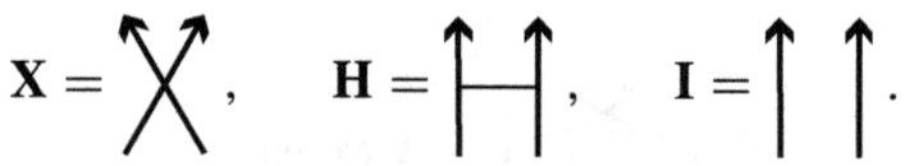

Then

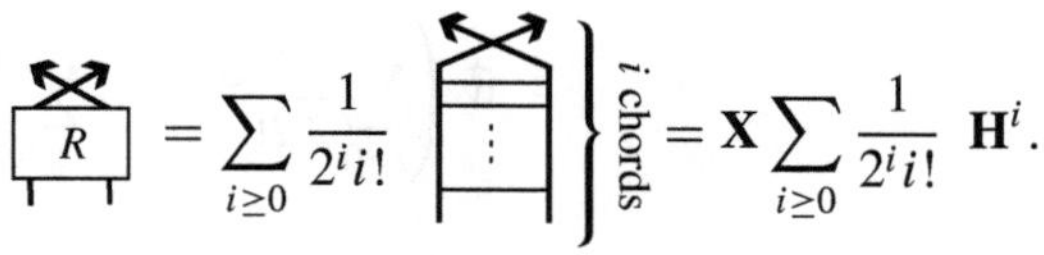

Turning to the computation of the weight system, we have

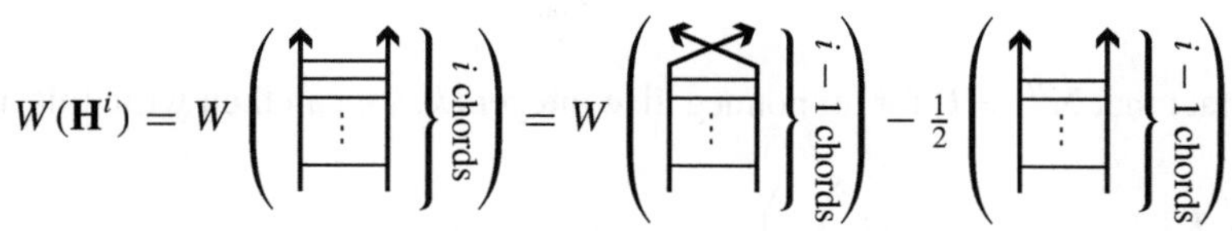

$$= W\left(\mathbf{X}\,\mathbf{H}^{i-1} - \tfrac{1}{2}\mathbf{H}^{i-1}\right) = W\left((\mathbf{X} - \tfrac{1}{2}\mathbf{I})\mathbf{H}^{i-1}\right) = \cdots = W\left((\mathbf{X} - \tfrac{1}{2}\mathbf{I})^i\right).$$

Then, bringing in the indeterminate h whose exponent records the degree of a Jacobi diagram,

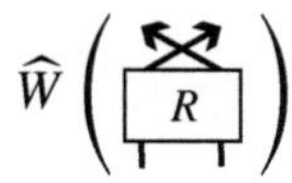

$$= \widehat{W}\left(\mathbf{X}\sum_{i\geq 0}\frac{h^i}{2^i i!}\mathbf{H}^i\right) = \widehat{W}\left(\mathbf{X}\sum_{i\geq 0}\frac{h^i}{2^i i!}\left(\mathbf{X} - \tfrac{h}{2}\mathbf{I}\right)^i\right)$$

$$= \widehat{W}\left(\mathbf{X}\exp(\tfrac{h}{2}(\mathbf{X} - \tfrac{h}{2}\mathbf{I}))\right) = \widehat{W}\left(\mathbf{X}\exp(h\mathbf{X}/2)\exp(-h\mathbf{I}/4)\right),$$

since $\mathbf{X}$ and $\mathbf{I}$ commute.

Examining the $\exp(-h\mathbf{I}/4)$ expression, since the relevant section of the cord diagram, $\mathbf{I}$, is trivial, we see the total effect of this term is to multiply $\mathbf{X}\exp(h\mathbf{X}/2)$ by a scalar $\exp(-h/4)$. Thus, we may write

$$\begin{aligned}\widehat{W}\left(\mathbf{X}\exp(h\mathbf{X}/2)\exp(-h\mathbf{I}/4)\right) &= \widehat{W}\left(\mathbf{X}\exp(h\mathbf{X}/2)e^{-h/4}\right)\\ &= e^{-h/4}\,\widehat{W}\left(\mathbf{X}\exp(h\mathbf{X}/2)\right).\end{aligned}$$

So we have shown

$e^{h/4}\,\widehat{W}\Big($  $\Big) = \widehat{W}\left(\mathbf{X}\exp(h\mathbf{X}/2)\right).$ (18.9)

Exercise 18.10. Show that

$e^{-h/4}\,\widehat{W}\Big($  $\Big) = \widehat{W}\left(\mathbf{X}\exp(-h\mathbf{X}/2)\right).$ (18.10)

Using Eqs. (18.9) and (18.10) and the linearity of W, we can then write

$e^{h/4}\,\widehat{W}\Big($  $\Big) - e^{-h/4}\,\widehat{W}\Big($  $\Big)$

$$\begin{aligned}&= \widehat{W}\left(\mathbf{X}\exp(h\mathbf{X}/2)\right) - \widehat{W}\left(\mathbf{X}\exp(-h\mathbf{X}/2)\right)\\ &= \widehat{W}\left(\mathbf{X}\exp(h\mathbf{X}/2) - \mathbf{X}\exp(-h\mathbf{X}/2)\right)\\ &= \widehat{W}\left(\sum_{i\geq 0}\frac{h^i}{2^i i!}\mathbf{X}^{i+1} - \sum_{i\geq 0}\frac{(-1)^i h^i}{2^i i!}\mathbf{X}^{i+1}\right)\\ &= \widehat{W}\left(\sum_{\substack{i\geq 0\\ i \text{ odd}}}\left(\frac{h^i}{2^i i!} + \frac{h^i}{2^i i!}\right)\mathbf{X}^{i+1}\right).\end{aligned}$$

Using the fact that $\mathbf{X}^{2k} = \mathbf{I}$, for all nonnegative integers k, we can then write this as

$$\begin{aligned}\widehat{W}\left(\left(\sum_{\substack{i\geq 0\\ i \text{ odd}}}\frac{h^i}{2^i i!} + \sum_{\substack{i\geq 0\\ i \text{ odd}}}\frac{h^i}{2^i i!}\right)\mathbf{I}\right) &= \widehat{W}\left(\left(\sum_{i\geq 0}\frac{h^i}{2^i i!} - \sum_{i\geq 0}\frac{(-1)^i h^i}{2^i i!}\right)\mathbf{I}\right)\\ &= \widehat{W}\left(\left(e^{h/2} - e^{-h/2}\right)\mathbf{I}\right)\\ &= \left(e^{h/2} - e^{-h/2}\right)\widehat{W}\left(\mathbf{I}\right).\end{aligned}$$

Thus, we have just shown that

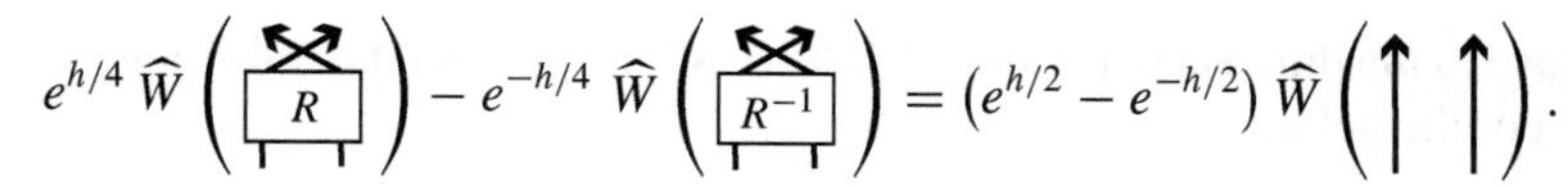 (18.11)

Observe that this equation holds if we apply it skein theoretically, in the sense that the 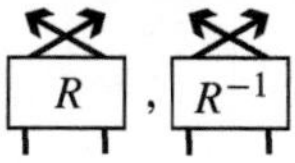 and  represent three elements of $\widehat{\mathscr{A}}$ that are identical everywhere except for in a small region where they differ as shown. Then, we have that $\widehat{W} \circ \check{Z}$ satisfies the skein relation

(18.12)

Having shown that $\widehat{W} \circ \check{Z}$ satisfies (18.4), we now show it satisfies (18.5). In showing this, we make use of the following exercise.

Exercise 18.11. Use Eqs. (18.7) and (18.8) to prove that

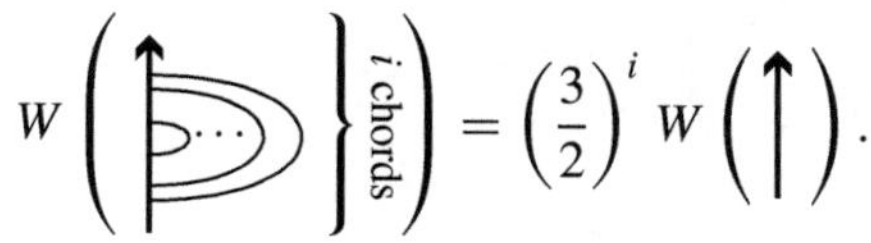

Hence, show that

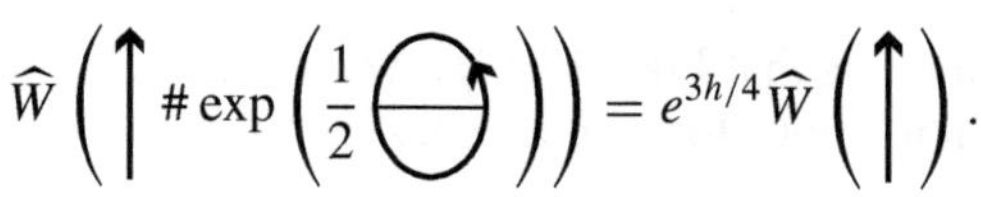 (18.13)

From the exercise, it follows easily that $\widehat{W} \circ \check{Z}$ satisfies (18.4). First observe that by Eq. (17.15),

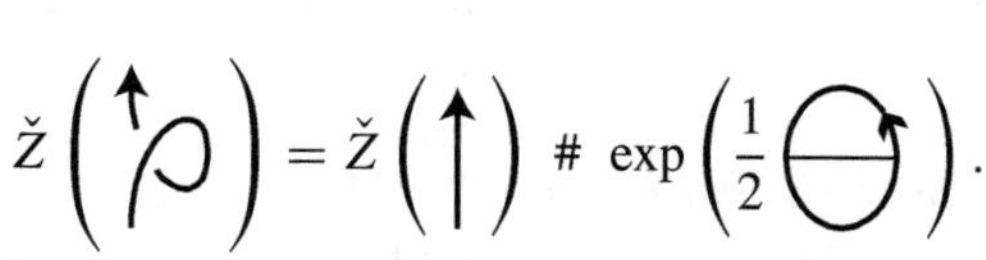

Then, apply Eq. (18.13) to obtain

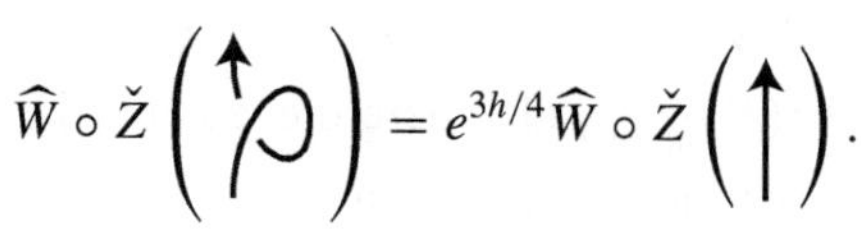 (18.14)

Exercise 18.12. Show that

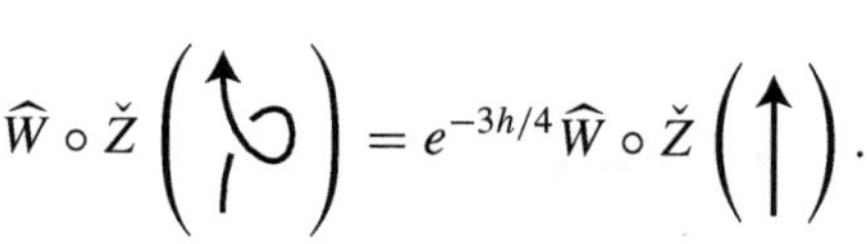 (18.15)

We move to the third skein relation (18.6). This follows from the following computation. For the unlink,

$(e^{h/2} - e^{-h/2})\Big(\widehat{W} \circ \check{Z}\Big($  $\Big) \cdot \widehat{W} \circ \check{Z}\Big($  $\Big)\Big)$

$= (e^{h/2} - e^{-h/2})\, \widehat{W} \circ \check{Z}\Big($  $\Big)$

$= e^{h/4}\, \widehat{W} \circ \check{Z}\Big($  $\Big) - e^{-h/4} \widehat{W} \circ \check{Z}\Big($  $\Big)$ (by (18.12))

$= e^{h}\, \widehat{W} \circ \check{Z}\Big($  $\Big) - e^{-h}\, \widehat{W} \circ \check{Z}\Big($  $\Big)$ (by (18.14) and (18.15))

$= (e^{h} - e^{-h})\, \widehat{W} \circ \check{Z}\Big($  $\Big).$

Solving for $\widehat{W} \circ \check{Z}\Big($  $\Big)$ then gives

$\widehat{W} \circ \check{Z}\Big($  $\Big) = e^{h/2} + e^{-h/2}.$ (18.16)

The key observation is that Eqs. (18.12), (18.14) and (18.16) together provide a set of defining skein relations for $\widehat{W} \circ \check{Z}(K)$.

Proof (*Theorem* 18.9). The defining skein relations, (18.4)–(18.6), for the knot invariant $Q^{\mathfrak{sl}_2,\rho}(K)\big|_{q=e^{h/2}}(1)$ coincide with the defining skein relations (18.12), (18.14) and (18.16) for $\widehat{W} \circ \check{Z}(K)$. Since these relations define a unique invariant, the values of $Q^{\mathfrak{sl}_2,\rho}(K)\big|_{q=e^{h/2}}(1)$ and $\widehat{W} \circ \check{Z}(K)$ coincide. □

Corollary 18.13. *The coefficient h^m in $Q^{\mathfrak{sl}_2,\rho}(K)\big|_{q=e^{h/2}}(1)$ is a degree m framed Vassiliev invariant. Its weight system is given by $W_{\mathfrak{sl}_2,\rho} \circ p_m$, where p_m is the projection of $\widehat{\mathscr{A}^c}$ onto $\mathscr{A}^c_m$.*

Proof. From Theorem 18.9 we know $\widehat{W} \circ \check{Z}(K) = Q^{\mathfrak{sl}_2,\rho}(K)\big|_{q=e^{h/2}}(1)$. The coefficient of h^m in $Q^{\mathfrak{sl}_2,\rho}(K)\big|_{q=e^{h/2}}(1)$ therefore equals the coefficient of h^m in $\widehat{W} \circ \check{Z}(K)$. Thus, if we define a weight system $W^{(m)}_{\mathfrak{sl}_2,\rho}$ by $W^{(m)}_{\mathfrak{sl}_2,\rho}(D) = W_{\mathfrak{sl}_2,\rho}(D)$ if D is of degree m and equals zero otherwise, then we see the coefficient of h^m in $Q^{\mathfrak{sl}_2,\rho}(K)\big|_{q=e^{h/2}}(1)$ is given by

$$\mathscr{K}^f \xrightarrow{\check{Z}} \widehat{\mathscr{A}^c} \xrightarrow{p_{\leq m}} \mathscr{A}^c_{\leq m} \xrightarrow{W^{(m)}_{\mathfrak{sl}_2,\rho}} \mathbb{C}$$

and so is therefore a degree m Vassiliev invariant, with weight system $W^{(m)}_{\mathfrak{sl}_2,\rho}$, by Theorem 18.6. □

Theorem 18.9 and Corollary 18.13 together give the stunning result that the Reshetikhin–Turaev invariants arising from the Lie algebra $\mathfrak{sl}_2$ and its standard representation are the same as the Vassiliev invariants arising from the Lie algebra $\mathfrak{sl}_2$ and its standard representation.

Theorem 18.9, however, does not only hold for the Lie algebras $\mathfrak{sl}_2$, but more generally for all semi-simple Lie algebras.

Every Reshetikhin–Turaev invariant coming from any semi-simple Lie algebra $\mathfrak{g}$ and any representation ρ of it factors through the Kontsevich invariant *via* the $(\mathfrak{g}, \rho)$-weight system:

$$W_{\mathfrak{g},\rho} \circ h^{\deg} \circ \check{Z}(K) = Q^{\mathfrak{g},\rho}(K)\big|_{q=e^{h/2}}(1). \tag{18.17}$$

The skein theoretic proof given here for Theorem 18.9 comes from [114]. (In fact, this reference proves the result, more generally, for the Lie algebra $\mathfrak{sl}_n$.) Proving (18.17) however, is much more difficult and beyond the scope of this text. An interested reader can find a proof in the source paper for the result [116], or in the book [90]. That the Reshetikhin–Turaev invariants give rise to Vassiliev invariants first appeared in [24, 124], and that their corresponding weight systems are the Lie algebra weight systems first appeared in [150].

As with Theorem 18.7, the result given in (18.17) is stunning. It says, somewhat informally, that

- $\check{Z}^u(K)$ contains and unifies all Reshetikhin–Turaev invariants.
- Understanding the Reshetikhin–Turaev invariants $Q^{\mathfrak{g},\rho}$ is equivalent to understanding the weight systems $W_{\mathfrak{g},\rho}$.

This should also be viewed together with Theorem 18.6 which gives the following.

Every Vassiliev invariant θ factors through the Kontsevich invariant:

$$W_\theta \circ p_{\leq m} \circ \check{Z} = \theta.$$

This means, somewhat informally, that

- $\check{Z}(K)$ contains and unifies all Vassiliev invariants.
- Understanding the Vassiliev invariants is equivalent to understanding weight systems.

A consequence of Theorem 18.6 is that each coefficient of $Q^{\mathfrak{g},\rho}(K)|_{q=e^{h/2}}(1)$ is a Vassiliev invariant. However, Vogel, in [187], showed that not all Vassiliev invariants (or equivalently weight systems) arise from semi-simple Lie algebras. Thus, tantalisingly, although Reshetikhin–Turaev invariants and Lie algebras play an important role in the theory of Vassiliev invariants, they do not form the whole story.

Appendix A
Background on Modules and Linear Algebra

In this chapter, we provide an overview of the various algebraic prerequisites for the text. Rather than trying to provide a comprehensive introduction to the relevant algebraic structures, our exposition here is tightly focussed on introducing specific structures we use in the way that we think about them. This means that our presentation is not in as general a setting as it could be. For example, we focus on finite dimensional vector spaces and free modules of finite rank. It also means that we favour constructive definitions over universal characterisations of algebraic objects. Additional background on modules can be found in, for example, [2, 27, 72, 97]. Our exposition is based on [27].

We shall let $\mathbb{K}$ denote a unital ring or a field. All rings in this text are assumed to have a unit, even when this is not explicitly stated. (Usually, we assume rings are commutative too.) Throughout the text, rings will generally be one of the following examples.

1. The field of complex numbers $\mathbb{C}$.
2. A ring of Laurent polynomials over $\mathbb{C}$, $\mathbb{C}[t, t^{-1}] = \{a_{-m}t^{-m} + \cdots + a_{-1}t^{-1} + a_0t^0 + a_1t^1 + \cdots + a_nt^n : m, n \in \mathbb{N}_0, a_i \in \mathbb{C}\}$.
3. A ring of formal power series over $\mathbb{C}$, $\mathbb{C}[[t]] = \{\sum_{i=0}^{\infty} a_it^i : a_i \in \mathbb{C}\}$.

The ring operations are the standard ones.

A.1 Vector Spaces

We assume a familiarity with basic linear algebra. However, we will make a few comments about the way that we think of vector spaces here, which is likely to be different from that encountered in a standard undergraduate class. The reader will be very familiar with the fact that a finite dimensional vector space V has a basis $\{e_1, e_2, \cdots, e_n\}$, and thus, we can represent every element of V as a linear combination $\lambda_1e_1 + \lambda_2e_2 + \ldots + \lambda_ne_n$. Here, rather than thinking of these linear

D. M. Jackson and I. Moffatt, *An Introduction to Quantum and Vassiliev Knot Invariants*, CMS Books in Mathematics, https://doi.org/10.1007/978-3-030-05213-3

combinations of descriptions of elements of a vector space, we will *think of the linear combinations as being the objects of the vector space*. This means that we treat vector spaces as the set of formal linear combinations of basis elements, $\lambda_1 e_1 + \lambda_2 e_2 + \ldots + \lambda_n e_n$, thinking of each e_i as a formal symbol (rather than thinking of it as an element of $\mathbb{R}^n$, say). Typically our bases will consist of indexed letters such as $\{e_1, e_2, \cdots, e_n\}$. However, sometimes they will consist of objects such as link diagrams. In either case, the basis elements are treated in the same way as formal symbols with no relations between them.

For example, an n-dimensional vector space over $\mathbb{C}$ with basis $\{e_1, e_2, \cdots, e_n\}$ has elements $\{\lambda_1 e_1 + \lambda_2 e_2 + \ldots + \lambda_n e_n : \lambda_i \in \mathbb{C}\}$, scalar multiplication defined by

$$\lambda(\lambda_1 e_1 + \ldots + \lambda_n e_n) := (\lambda\lambda_1)e_1 + \ldots + (\lambda\lambda_n)e_n, \tag{A.1}$$

and addition defined by

$$(\lambda_1 e_1 + \ldots + \lambda_n e_n) + (\lambda_1' e_1 + \ldots + \lambda_n' e_n) = (\lambda_1 + \lambda_1')e_1 + \ldots + (\lambda_n + \lambda_n')e_n. \tag{A.2}$$

There will also be situations where, rather than defining a vector space by specifying a basis for it, we instead define a vector space as consisting of finite formal linear combinations of some infinite set of objects (typically knots or diagrams). In this case, if $\{D_i : i \in \mathscr{I}\}$, where $\mathscr{I}$ is some indexing set, a typical element of the space is

$$\lambda_1 D_{i_1} + \lambda_2 D_{i_2} + \cdots + \lambda_p D_{i_p},$$

for some nonnegative integer p and scalars λ_i. Note that the length of the linear combination is not fixed between different elements. Addition and scalar multiplication are defined componentwise, as in (A.1) and (A.2) above.

A.2 Modules

As just mentioned, we view vector spaces as consisting of formal linear combinations of objects or symbols. We also need to be able to consider formal linear combinations of objects where the scalars are taken to be in a ring rather than a field. Thus, we are not able to work exclusively with vector spaces but have to consider modules. These can be thought of as vector spaces over rings rather than just fields. Although their definitions are very similar, modules, in general, can have very different properties from vector spaces, particularly where bases are concerned. However, for the knot theory considered here, we are generally able to work with a class of modules, namely, free modules of finite rank, to which key properties of vector space bases extend.

Definition A.1 (Module). A *module* over a unital ring $\mathbb{K}$ (or a $\mathbb{K}$-*module* or a *left* $\mathbb{K}$-*module*) consists of an abelian group $(V, +)$ together with a map from $\mathbb{K} \times V$ to V, with action given by $(a, x) \mapsto ax$, such that (i) $a(x + y) = ax + ay$, (ii) $(a + b)x = ax + bx$, (iii) $(ab)x = a(bx)$, (iv) $1x = x$, for all $a, b \in \mathbb{K}$ and $x, y \in V$.

Note that if $\mathbb{K}$ is a field, then a $\mathbb{K}$-module is precisely a vector space over $\mathbb{K}$.

Definition A.2 (Submodule). A module W is a *submodule* of V if its elements form a subset of V and its module operations are those of V. We use $W \leq V$ to denote that W is a submodule of V and write $W < V$ if $W \leq V$ and $W \neq V$.

Definition A.3 (Module morphism). If V and W are modules over the same ring $\mathbb{K}$, then $f : V \to W$ is a *morphism* (or $\mathbb{K}$-*morphism*) if (i) $f(x + y) = f(x) + f(y)$, and (ii) $f(\lambda x) = \lambda f(x)$, for all $x, y \in V$ and for all $\lambda \in \mathbb{K}$.

If the module morphism f is a bijection, then it is said to be an *isomorphism*, and V and W are said to be *isomorphic*, written $V \cong W$.

If V and W are vector spaces, then a morphism is traditionally called a *linear map* (or *linear transformation*).

We use the following notation throughout the text.

1. $\mathrm{Hom}(V, W)$ is the set of all morphisms from V to W.
2. $\mathrm{End}(V)$ is the set of all morphisms from V to V. Its elements are called *endomorphisms*.
3. $\mathrm{Aut}(V)$ is the set of all isomorphisms from V to V. Its elements are called *automorphisms*.

A.3 Generating Sets and Free Modules

Let S be a non-empty subset of elements of V. Then, the module *generated* by S is

$$\langle S \rangle := \{\lambda_1 x_1 + \ldots + \lambda_n x_n : \lambda_i \in \mathbb{K}, x_i \in S, n \in \mathbb{N}_0\}.$$

Thus, $\langle S \rangle$ consists of all finite linear combinations of the elements of S. If S is the empty set, then $\langle S \rangle$ is defined to be the trivial module $\{0\}$. We say that V is *generated* by S if $\langle S \rangle = V$. A module is *finitely generated* if it has a finite set of generators. In the case where V is a vector space, S is also said to *span* V.

Equivalently, the module $\langle S \rangle$ can be defined as the smallest submodule of V containing the set S, where smallest here means that it is a submodule of any module over $\mathbb{K}$ containing S.

For notational simplicity, when writing $\langle S \rangle$ we usually omit the braces for the set S. For example, writing $\langle x_1, x_2 \rangle$ rather than $\langle \{x_1, x_2\} \rangle$.

Definition A.4 (Linear independence). A non-empty subset S of elements of V is said to be *linearly independent* or *free* if, for every finite number of distinct elements $x_1, \ldots, x_n$ of S, we have

$$\lambda_1 x_1 + \ldots + \lambda_n x_n = 0 \implies \lambda_1 = \lambda_2 = \cdots = \lambda_n = 0.$$

Definition A.5 (Basis). A *basis* of V is a set of linearly independent elements of V that generate V.

The reader will undoubtably be familiar with the fact that in a vector space every one-element set is linearly independent. For modules in general, this is not true. For example, if we consider $\mathbb{Z}_6$ as a $\mathbb{Z}_6$-module, then $\{2\}$ is not linearly independent since $3 \cdot 2 \equiv 0 \mod 6$ and 3 is nonzero. The difficulty in this case is that the ring $\mathbb{Z}_6$ has zero-divisors (*i.e.* nonzero elements whose product is zero).

Definition A.6 (Free module). A module is said to be a *free module* if it has a basis.

A property of vector spaces that extends to free modules is that every element of a free module can be expressed as a unique linear combination of basis elements.

Unlike vector spaces, not all modules are free; *i.e.*, not every module has a basis. Furthermore, for modules that do have bases, not all bases necessarily have the same number of elements. Thus, we need to make the following definition.

Definition A.7 (Rank). A module is said to be a *free module of rank n* if it is free and every basis has exactly n elements.

At this point, we are now able to describe the ways in which modules appear throughout the text. We almost always define a module as a free module of rank n with a given basis. We will usually specify a basis $\{e_1, e_2, \ldots, e_n\}$, and every element of our module can be expressed as a unique linear combination of these elements. In fact, we consider the elements of the module exactly as being linear combinations of the basis elements.

At times, we will define a module or vector space structure by considering a set S of objects, such as knots or chord diagrams, and considering the $\mathbb{K}$-module consisting of all finite formal linear combinations of these elements. In these constructions, $\mathbb{K}$ will either be the field of complex numbers $\mathbb{C}$, a ring of Laurent polynomials over $\mathbb{C}$, or a ring of formal power series over $\mathbb{C}$. Since these rings do not have zero-divisors and since our objects are treated as formal symbols in the module (so there is no relation between them), we may assert that S forms a basis of the module.

Exactly as in the vector space case, if V and W are free modules of finite rank, $\{x_1, \ldots, x_n\}$ is a canonically ordered basis for V, $\{y_1, \ldots, y_m\}$ is a canonically ordered basis for W and $f \in \operatorname{Hom}(V, W)$, then f is completely defined by its action on the basis

$$f : x_i \mapsto \sum_{j=1}^{m} f_i^j y_j,$$

where this action is extended linearly to all elements of V, and so f can be described as a $m \times n$ matrix $\mathbf{F}$ whose (p, q)-entry is f_q^p.

A.4 Quotient Modules

We next record the definition of a quotient module. Before doing so, we emphasise that the fact that V/W forms a module in the way described is non-trivial and requires justification. We do not include this justification here, but it can be found in standard algebra texts or can be verified by the reader.

Definition A.8 (Quotient module). Let V be a module over $\mathbb{K}$, and let W be a subspace of V. Let $\sim$ be the equivalence relation defined on V by

$$x \sim y \quad \Longleftrightarrow \quad x - y \in W.$$

The set

$$V/W := \{[x] \colon x \in V\},$$

where $[x]$ is the equivalence class containing x (*i.e.* the set of all elements equivalent to x under $\sim$), is a module over $\mathbb{K}$ in which addition and scalar multiplication are defined by

$$[x] + [y] := [x + y] \qquad \text{for all } x, y \in V\ ,$$
$$\lambda\, [x] := [\lambda\, x] \qquad \text{for all } x \in V \text{ and } \lambda \in \mathbb{K}.$$

The module V/W is called a *quotient module*.

In practice, the way we think about and use quotient spaces here is as follows. We are given a module or vector space V and some nonzero elements $r_1, r_2, \ldots, r_n$. For our applications to knot theory, we will want to modify V in such a way that the elements $r_1, r_2, \ldots, r_n$ become zero. We do this by forming the quotient module $V/\langle r_1, r_2, \ldots, r_n\rangle$. This quotient module should be thought of as the space that arises from V by imposing the relations $r_1 = 0$, $r_2 = 0$, ..., $r_n = 0$ and all consequences of these relations.

Extending this way of thinking a little further, we will find that we want to modify V in a way that will result in equations $r_1 = s_1, r_2 = s_2, \ldots, r_n = s_n$ holding. To do this, we form the quotient $V/\langle r_1 - s_1, \ldots, r_n - s_n\rangle$, which we often denote by $V/\langle r_1 = s_1, \ldots, r_n = s_n\rangle$. At times, when $r_1 = s_1, r_2 = s_2, \ldots, r_n = s_n$ constitutes a named collection of relations (such as the "Turaev moves"), we denote $V/\langle r_1 = s_1, \ldots, r_n = s_n\rangle$ by "V/name of relations".

For example, let V be the free $\mathbb{Z}$-module of rank 3 with basis $\{e_1, e_2, e_3\}$, so $V = \{\lambda_1 e_1 + \lambda_2 e_2 + \lambda_3 e_3 : \lambda_1, \lambda_2, \lambda_3 \in \mathbb{Z}\}$. If we wished to impose the condition $e_1 + e_2 = e_3$, we could consider the module $V/\langle e_1 + e_2 - e_3\rangle$, which we would write as $V/\langle e_1 + e_2 = e_3\rangle$. The elements of this quotient module are equivalence classes. Two elements x and y of V are in the same class and hence represent the same element of the quotient module, if and only if there is a sequence of applications of the equation $e_1 + e_2 = e_3$, and all equations arising by taking linear combinations of this equation, that takes x to y. So, for example, $2e_1 - 3e_2 + e_3$ and $5e_1 - 2e_3$ represent the same element of this quotient space since $2e_1 - 3e_2 + e_3 = 2e_1 - 3(e_3 - e_1) + e_3 = 5e_1 - 2e_3$.

If $f \in \mathrm{Hom}(V, W)$, then $\mathrm{Im}(f)$ denotes the *image* of f and is defined by $\mathrm{Im}(f) := \{f(x) : x \in V\}$. The *kernel* of f, denoted by $\ker(f)$, is defined by $\ker(f) := \{x \in V : f(x) = 0\}$.

Theorem A.9 (First Isomorphism Theorem). *Let $f \in \mathrm{Hom}(V, W)$. Then $\mathrm{Im}(f)$ is a submodule of W, $\ker(f)$ is a submodule of V, and f induces an isomorphism*

$$\overline{f} : V/\ker(f) \to \mathrm{Im}(f) : [x] \mapsto f(x).$$

Important to us here is the question of when a morphism will induce a morphism on quotient spaces. For this let $f \in \mathrm{Hom}(V, W)$, I be a submodule of V, and J a submodule of W.

Then, f will always induce a morphism $\tilde{f} : V \to W/J$ *via* the action $f : x \mapsto [f(x)]_J$, where $[f(x)]_J$ denotes the equivalence class in W/J containing the element $f(x)$.

The question of when f induces a morphism $\overline{f} : V/I \to W$ is more involved. Using $[x]_I$ to denote the equivalence class of V/I containing x, we would like to define $\overline{f}$ by the action $[x]_I \mapsto f(x)$. The difficulty is that this map may not be well-defined since its evaluation on $[x]_I$ requires a choice of representative of the equivalence class, and its value could depend upon this choice of representative. Thus, we need to ensure that if $[x]_I = [y]_I$, then $f(x) = f(y)$, or equivalently if $y, z \in [x]_I$, then $f(y) = f(z)$. This can be shown to happen exactly when $f(I) := \{f(i) : i \in I\} = \{0\}$. We state this result, without proof, as the following theorem.

Lemma A.10 *Let V and W be $\mathbb{K}$-modules and let I be a submodule of V. Let $f \in \mathrm{Hom}(V, W)$. Then there exists a morphism*

$$\overline{f} \colon V/I \to W \colon [x] \mapsto f(x)$$

if and only if $f(I) = \{0\}$.

Many of our applications of Lemma A.10 use it in the following form that we state for vector spaces (since this is our application).

Lemma A.11 *Let I, V, and W be vector spaces and suppose that $I \leq V$. Then*

$$\mathrm{Hom}(V/I, W) \cong \{f \in \mathrm{Hom}(V, W) \colon I \leq \ker(f)\}.$$

Another way that we shall apply Lemma A.10 is through the observation it provides that if I is finitely generated by $x_1, \ldots, x_n$, to show that f induces a well-defined morphism from V/I to W we just need to check that $f(x_1) = 0, \cdots, f(x_n) = 0$. With this in mind, and for application, we record the following special case of Lemma A.11.

Lemma A.12 *Let U and V be finite dimensional vector spaces, and $r_1, \ldots, r_n \in V$ Let F be the vector space of all linear maps from U to V that satisfy the relations $r_1 = 0, \ldots, r_n = 0$, that is*

$$F = \{f \in \mathrm{Hom}(U, V) : f(r_1) = 0, \ldots, f(r_n) = 0\}.$$

Then

$$F \cong \mathrm{Hom}(U/\langle r_1, \ldots, r_n \rangle, V).$$

A.5 Dual Modules

The dual of a module V is the module obtained by considering the set of morphisms from V to the ground ring $\mathbb{K}$ and endowing it with the obvious addition and scalar multiplication.

Definition A.13 (Dual). Let V be a $\mathbb{K}$-module. Then, the *dual* of V, denoted by $V^\star$, is the $\mathbb{K}$-module $V^\star := \mathrm{Hom}(V, \mathbb{K})$ where for $\lambda \in \mathbb{K}$ and $f, g \in \mathrm{Hom}(V, \mathbb{K})$, the morphisms $\lambda \cdot f$ and $f + g$ are defined by

$$\lambda \cdot f : x \mapsto \lambda f(x) \quad \text{and} \quad f + g : x \mapsto f(x) + g(x).$$

It is readily verified that $V^\star$ forms a module. (It is perhaps worth noting that, unlike the vector space case, in general the above operations will not define a module structure on $\mathrm{Hom}(V, W)$ for modules V and W since the resulting maps need not be module morphisms.)

Recall the *Kronecker delta*

$$\delta_{i,j} = \begin{cases} 0 & \text{if } i \neq j, \\ 1 & \text{if } i = j. \end{cases}$$

At times it will be natural to write the Kronecker delta $\delta_{i,j}$ as δ_i^j or $\delta^{i,j}$.

Definition A.14 (Dual element e^i). Let V be a free $\mathbb{K}$-module with basis $\{e_i : i \in \mathscr{I}\}$, where $\mathscr{I}$ is an indexing set, then for each $i \in \mathscr{I}$ we define $e^i \in \mathrm{Hom}(V, \mathbb{K})$ by its action on the basis elements

$$e^i(e_j) = \delta^i_j,$$

where δ^i_j is the Kronecker delta.

The elements e^i can be used to form a basis of the dual space.

Theorem A.15 *Let V be a free $\mathbb{K}$-module with basis $\{e_i : i \in \mathscr{I}\}$, where $\mathscr{I}$ is an indexing set. Then, $\{e^i : i \in \mathscr{I}\}$ is a linearly independent subset of $V^\star$. Moreover, if $\{e_i : i \in \mathscr{I}\}$ is a finite set, then $\{e^i : i \in \mathscr{I}\}$ is a basis of $V^\star$.*

Note that a consequence of this theorem is that if V is a free module of rank n, then so is $V^\star$.

Definition A.16 (Dual basis). Let V be a free $\mathbb{K}$-module of finite rank with basis $B = \{e_1, e_2, \ldots, e_n\}$, then the basis *dual* to B is the basis $\{e^1, e^2, \ldots, e^n\}$ of $V^\star$.

We shall think of dual spaces in the following way in this text. We usually work with free module of rank n, V, given by a basis $\{e_1, e_2, \ldots, e_n\}$. We will think of its dual space $V^\star$ as the free module of rank n, $V^\star$, given by a basis $\{e^1, e^2, \ldots, e^n\}$. The elements of $V^\star$ are exactly formal linear combinations of $\{e^1, e^2, \ldots, e^n\}$.

At times it will be convenient to denote the dual element to e_i by $(e_i)^\star$, so for an element $v \in V$, we use $v^\star$ to denote the element $v^\star \in V^\star$ defined by $v^\star(v) = 1$ and $v^\star(x) = 0$ for all $x \in V$ with $x \neq v$. Note that $(e_i)^\star = e^i$.

Theorem A.17 *Let V be a free module with a finite basis. Then $V^{\star\star} \cong V$. More precisely, the map $\phi : V \to V^{\star\star}$ defined by $e_i \mapsto e_i^{\star\star} = (e^i)^\star$ is an isomorphism.*

We also need to consider *dual maps*. Since our applications only require the generality of free modules of finite rank, for simplicity we work in this setting.

Definition A.18 (Dual morphism). Let V and W be free modules of finite rank and $f : V \to W$ be a module morphism. Then, the *dual* of f denoted $f^\star$ is the map $f^\star : W^\star \to V^\star$ defined by

$$f^\star : g \mapsto (g \circ f),$$

where $g \in W^\star$.

We will make use of the following properties.

Theorem A.19 *Let V and W be free modules of finite rank and $f : V \to W$ be a module morphism. Then*

1. *$f^\star : W^\star \to V^\star$ is a module morphism.*
2. *f is injective (respectively, surjective) if and only if $f^\star$ is surjective (respectively, injective).*
3. *If f is an isomorphism, then $(f^{-1})^\star = (f^\star)^{-1}$.*

A.6 Direct Sum

We consider two ways to combine modules. The first of these is the direct sum.

Definition A.20 (Direct sum). A $\mathbb{K}$-module V is the *direct sum* of a family of submodules $\{V_i\}_{i\in\mathscr{I}}$, where $\mathscr{I}$ is an indexing set, if every $x \in V$ can be written in a unique way as $x = \sum_{i\in\mathscr{I}} x_i$ where each $x_i \in V_i$ and all but finitely many of the x_i are zero. We write $V = \bigoplus_{i\in\mathscr{I}} V_i$.

Note that the uniqueness condition in the definition means that no element in any given V_i can be obtained as a finite linear combination of elements in the other V_j.

We remark that there are two notions of direct sum in a module, the *internal direct sum* and the *external direct sum*. The definition above is the internal direct sum. The external direct sum is defined by considering Cartesian products, but we exclude its definition here since we always think of our direct sums as internal direct sums. Both forms of direct sum result in isomorphic modules and so it has become standard to refer to both constructions simply as the direct sum, to use the same symbol $\oplus$ for both, and to let the reader deduce from the context which type of direct sum is intended (which is usually clear).

We record two results about the direct sum in the generality we need them.

Theorem A.21 *Let V be a free $\mathbb{K}$-module of finite rank with basis $\{e_1, e_2, \dots, e_n\}$. Then*

$$V \cong \langle e_1\rangle \oplus \langle e_2\rangle \oplus \cdots \oplus \langle e_n\rangle.$$

Theorem A.22 *Let V be a free $\mathbb{K}$-module of finite rank, and suppose that $V = V_1 \oplus V_2$. Then*

$$(V_1 \oplus V_2)^\star \cong V_1^\star \oplus V_2^\star.$$

At times we will need to consider the *graded completion* of vector spaces, which essentially means allowing infinite sums in the definition of the direct sum. For this, we will be given a family of vector spaces $\{V_i\}_{i\in\mathbb{N}_0}$ indexed by the nonnegative integers. The *graded completion* $\widehat{V} = \bigoplus_{i=0}^{\infty} V_i$ of $\{V_i\}_{i\in\mathbb{N}_0}$ is defined by removing the condition that all but finitely many of the x_i are zero from Definition A.20.

We shall make use of the following direct sum decomposition of vector spaces.

Theorem A.23 *Let V be a vector space, and U be a subspace of V. Then*

$$V \cong (V/U) \oplus U.$$

In particular, if V is finite dimensional, then $\dim(V) = \dim(V/U) + \dim(U)$.

An important application of Theorem A.23 is that when we have a *filtration*

$$V_m \geq V_{m-1} \geq V_{m-2} \geq \cdots \geq V_1 \geq V_0$$

of a vector space V_m, we can write that vector space as a direct sum of quotient spaces:

$$V_m \cong (V_m/V_{m-1}) \oplus (V_{m-1}/V_{m-2}) \oplus \cdots \oplus (V_1/V_0) \oplus V_0.$$

This isomorphism enables us to examine V_m by examining the quotient spaces V_i/V_{i-1}, which may be more amenable to study.

We shall need to consider the direct sum of morphisms.

Definition A.24 (Direct sum of morphisms). Suppose that V and W are $\mathbb{K}$-modules such that $V = \bigoplus_{i\in\mathscr{I}} V_i$ and $W = \bigoplus_{i\in\mathscr{I}} W_i$. Suppose also that $\{f_i\}_{i\in\mathscr{I}}$ is a family of morphisms where $f_i : V_i \to W_i$ for each $i \in \mathscr{I}$. Then the *direct sum* of $\{f_i\}_{i\in\mathscr{I}}$ is the morphism $\bigoplus_{i\in\mathscr{I}} f_i : V \to W$ defined by, for each $x = \sum_{i\in\mathscr{I}} x_i \in V$, setting $\left(\bigoplus_{i\in\mathscr{I}} f_i\right)(x) := \sum_{i\in\mathscr{I}} f_i(x_i)$.

It is routine to check that $\bigoplus_{i\in\mathscr{I}} f_i$ is a module morphism and that $\bigoplus_{i\in\mathscr{I}} f_i$ is injective (or surjective) if and only if each f_i is.

A.7 Tensor Products

A discussion of the tensor product of modules in full generality would be considerably more involved than what is required in this text. Accordingly, in this section we work in the generality that we need for the applications in this book and define a tensor product of modules accordingly. Throughout, we shall let U and V denote free $\mathbb{K}$-modules of finite rank.

Definition A.25 (Tensor product). The *tensor product*, $U \otimes V$, of free $\mathbb{K}$-modules U and V of finite rank, is the quotient module $F(U, V)/I(U, V)$, where $F(U, V)$ is the free $\mathbb{K}$-module with basis $U \times V$ and $I(U, V)$ is the submodule generated by

1. $(u_1 + u_2, v) - (u_1, v) - (u_2, v)$,
2. $(u, v_1 + v_2) - (u, v_1) - (u, v_2)$,
3. $\lambda(u, v) - (\lambda u, v)$,
4. $\lambda(u, v) - (u, \lambda v)$,

for all $u, u_1, u_2 \in U$, $v, v_1, v_2 \in V$ and $\lambda \in \mathbb{K}$.

The image of (u, v) in $U \otimes V$ is denoted by $u \otimes v$, and we therefore have the relations

1. $(u_1 + u_2) \otimes v = u_1 \otimes v + u_2 \otimes v$.
2. $u \otimes (v_1 + v_2) = u \otimes v_1 + u \otimes v_2$.
3. $\lambda(u \otimes v) = (\lambda u) \otimes v = u \otimes (\lambda v)$.

In the current setting, the best way to think about the module $U \otimes V$ is not as a quotient module, but rather as the module consisting of all finite linear combinations of elements of $\{u \otimes v : u \in U, v \in V\}$, where $u \otimes v$ should be thought of as just a formal symbol, subject to the above relations.

As an example of a tensor product, let $\mathbb{Q}[x]$ and $\mathbb{Q}[y]$ be the vector space of polynomials in indeterminates x and y, respectively, over $\mathbb{Q}$. Then, $\mathbb{Q}[x] \otimes \mathbb{Q}[y] \cong \mathbb{Q}[x, y]$, where $\mathbb{Q}[x, y]$ is the vector space of bivariate polynomials in x and y. A natural isomorphism is given by the linear extension of the map $ax^i \otimes by^j \mapsto ab(x^i y^j)$.

Any element of $U \otimes V$ can be represented as a finite sum

$$\sum_{i,j=1}^{n} a_{i,j}\, u_i \otimes v_j .$$

It should be noted that it is *not* true that every element is of the form $\left(\sum_{i=1}^{n} a_i\, u_i\right) \otimes \left(\sum_{i=1}^{m} b_i\, v_i\right)$.

Note that in this book, our convention is to omit or include the comma separating indices (e.g. a_{ij} versus $a_{i,j}$ in the above) depending on which choice in a particular instance adds clarification.

Theorem A.26 *Let U, U_i, V and W be free modules of finite rank. Then*

1. *the map $u \otimes v \mapsto v \otimes u$ defines an isomorphism*

$$U \otimes V \cong V \otimes U;$$

2. *the map $(u \otimes v) \otimes w \mapsto u \otimes (v \otimes w)$ defines an isomorphism*

$$(U \otimes V) \otimes W \cong U \otimes (V \otimes W);$$

3. *the maps $\lambda \otimes v \mapsto \lambda v$ and $v \otimes \lambda \mapsto \lambda v$ define isomorphisms*

$$\mathbb{K} \otimes V \cong V \cong V \otimes \mathbb{K};$$

4. *the map $\left(\sum_{i \in I} u_i\right) \otimes v \mapsto \sum_{i \in I} (u_i \otimes v)$ defines an isomorphism*

$$\Big(\bigoplus_{i \in I} U_i\Big) \otimes V \cong \bigoplus_{i \in I} (U_i \otimes V) .$$

Definition A.27 (Twist map τ). The isomorphism $\tau : U \otimes V \to V \otimes U$ defined by $\tau : u \otimes v \mapsto v \otimes u$ is called the *twist* map.

Definition A.28 (Contraction map). The maps defined by $\lambda \otimes v \mapsto \lambda v$ or $v \otimes \lambda \mapsto \lambda v$ that provide the isomorphisms $\mathbb{K} \otimes V \cong V$ or $V \otimes \mathbb{K} \cong V$, respectively, are called *contraction* maps. They are denoted by κ. The inverses of the contraction maps, κ^{-1}, are denoted by ι. Their element-wise actions are defined by $v \mapsto 1 \otimes v$ or $v \mapsto v \otimes 1$.

Theorem A.29 *Let U and V be free $\mathbb{K}$-modules of finite rank with bases $\{u_1, u_2, \ldots, u_m\}$ and $\{v_1, v_2, \ldots, v_n\}$, respectively. Then*

$$\{u_i \otimes v_j : 1 \leq i \leq m, 1 \leq j \leq n\}$$

forms a basis for $U \otimes V$. Thus, $U \otimes V$ is a free module of rank mn.

The following can be proved by comparing the actions on the standard basis elements.

Proposition A.30 *Let U and V be free modules of finite rank. Then*

$$(U \otimes V)^\star \cong V^\star \otimes U^\star.$$

Furthermore, a natural isomorphism is defined on basis elements by $(u_i \otimes v_i)^* \mapsto v^i \otimes u^i$.

The following definition introduces a map called contraction and denoted by κ that is different from that appearing in Definition A.28. We use the same name and notation for the two types of map since it is standard in the literature, and since, in practice, it will cause no confusion.

Definition A.31 (Contraction map). Let V be a free $\mathbb{K}$-module of finite rank with basis $\{e_1, e_2, \ldots, e_n\}$, and let $\{e^1, e^2, \ldots, e^n\}$ be the dual basis of $V^\star$. Then, the *contraction* map κ refers to either the morphism defined by its action on basis elements by

$$\kappa : V \otimes V^\star \to \mathbb{K} : e_i \otimes e^j \mapsto e^j(e_i) = \delta_i^j,$$

or the morphism defined by its action on basis elements by

$$\kappa : V^\star \otimes V \to \mathbb{K} : e^i \otimes e_j \mapsto e^i(e_j) = \delta_j^i.$$

A.8 The Tensor Product of Morphisms

Definition A.32 (Tensor product of morphisms). Let $f : U \to U'$ and $g : V \to V'$ be morphisms between free modules of finite rank. The *tensor product* $f \otimes g$ of f and g is the module morphism

$$f \otimes g \colon U \otimes V \to U' \otimes V' \colon (f \otimes g)(u \otimes v) = f(u) \otimes g(v).$$

To facilitate the construction of the matrices representing tensor products of module morphisms, we specify a canonical ordering for the bases of the tensor product of vector spaces with ordered bases. If U and V are free modules of finite rank with ordered bases $\{u_1, u_2, \ldots, u_m\}$ and $\{v_1, v_2, \ldots, v_n\}$ and $f : U^{\otimes p} \to V^{\otimes q}$, then we index the coordinates according to

$$f(u_{i_1} \otimes \cdots \otimes u_{i_p}) = \sum_{j_1,\ldots j_m=1}^{n} f_{i_1,\ldots,i_p}^{j_1,\ldots,j_q}\, v_{j_1} \otimes \cdots \otimes v_{j_q}.$$

For example, let U be a free module of rank 2 with basis $\{u_1, u_2\}$. Then, the map

$$f\colon U \otimes U \to U \otimes U : u_i \otimes u_j \mapsto \sum_{k,l=1}^{2} f_{i,j}^{k,l} u_k \otimes u_l$$

is represented by the matrix

$$\begin{bmatrix} f_{1,1}^{1,1} & f_{1,2}^{1,1} & f_{2,1}^{1,1} & f_{2,2}^{1,1} \\ f_{1,1}^{1,2} & f_{1,2}^{1,2} & f_{2,1}^{1,2} & f_{2,2}^{1,2} \\ f_{1,1}^{2,1} & f_{1,2}^{2,1} & f_{2,1}^{2,1} & f_{2,2}^{2,1} \\ f_{1,1}^{2,2} & f_{1,2}^{2,2} & f_{2,1}^{2,2} & f_{2,2}^{2,2} \end{bmatrix}.$$

There is the following product for matrices that is related to the tensor product of the maps they represent.

Definition A.33 (Kronecker product of matrices). Let $\mathbf{M} = [m^i_j]$ and $\mathbf{N}$ be matrices, where m^i_j is the (i, j)-entry of $\mathbf{M}$. The *Kronecker product* $\mathbf{M} \otimes \mathbf{N}$ is the block matrix

$$\mathbf{M} \otimes \mathbf{N} = \left[m^i_j \mathbf{N}\right].$$

For example, the Kronecker products $\mathbf{R} \otimes \mathbf{I}_2$ and $\mathbf{I}_2 \otimes \mathbf{R}$ where

$$\mathbf{R} = \begin{bmatrix} a & 0 & 0 & 0 \\ 0 & b & c & 0 \\ 0 & d & e & 0 \\ 0 & 0 & 0 & f \end{bmatrix},$$

are, respectively, the 8×8 matrices

$$\begin{bmatrix} a & 0 & 0 & 0 & 0 & 0 & 0 & 0 \\ 0 & a & 0 & 0 & 0 & 0 & 0 & 0 \\ 0 & 0 & b & 0 & c & 0 & 0 & 0 \\ 0 & 0 & 0 & b & 0 & c & 0 & 0 \\ 0 & 0 & d & 0 & e & 0 & 0 & 0 \\ 0 & 0 & 0 & d & 0 & e & 0 & 0 \\ 0 & 0 & 0 & 0 & 0 & 0 & f & 0 \\ 0 & 0 & 0 & 0 & 0 & 0 & 0 & f \end{bmatrix} \quad \text{and} \quad \begin{bmatrix} a & 0 & 0 & 0 & 0 & 0 & 0 & 0 \\ 0 & b & c & 0 & 0 & 0 & 0 & 0 \\ 0 & d & e & 0 & 0 & 0 & 0 & 0 \\ 0 & 0 & 0 & f & 0 & 0 & 0 & 0 \\ 0 & 0 & 0 & 0 & a & 0 & 0 & 0 \\ 0 & 0 & 0 & 0 & 0 & b & c & 0 \\ 0 & 0 & 0 & 0 & 0 & d & e & 0 \\ 0 & 0 & 0 & 0 & 0 & 0 & 0 & f \end{bmatrix}. \tag{A.3}$$

Proposition A.34 *Let U, V, S and T be free modules of finite rank, and let $f : U \to V$ and $g : S \to T$ be module morphisms. If* **F** *and* **G** *are matrices of the module morphisms f and g, then the matrix of the module morphism $f \otimes g$ with respect to the canonical order on the bases of the tensor products is the Kronecker product* $\mathbf{F} \otimes \mathbf{G}$.

We shall also use the following result.

Proposition A.35 *Let U, V, S and T be free modules of finite rank, and let $f : U \to V$ and $g : S \to T$ be module morphisms. Then, if both are considered as maps from $(U \otimes S)^\star$ to $(V \otimes T)^\star$ (or from $U^\star \otimes S^\star$ to $V^\star \otimes T^\star$) we have $(f \otimes g)^\star = f^\star \otimes g^\star$.*

A.9 The Fundamental Morphism $\mathrm{Hom}(U, V) \cong U^\star \otimes V$

We will now describe a key result about tensor products: that $\mathrm{Hom}(U, V)$ and $U^\star \otimes V$ are isomorphic. Extensive use will be made of this since it enables us to represent morphisms by tensors.

Lemma A.36 *Let U and V be free modules of finite rank. Then*

$$\mathrm{Hom}(U, V) \cong U^\star \otimes V.$$

Moreover, isomorphisms between these spaces can be defined as follows. Let $\{u_1, \dots, u_m\}$ and $\{v_1, \dots, v_n\}$ be bases for U and V respectively, and let $\{u^1, \dots, u^m\}$ be the corresponding dual basis of U.

1. *Let $\psi : \mathrm{Hom}(U, V) \to U^\star \otimes V$ be defined by*

$$\psi : f \mapsto \sum_{i,j} f_i^j u^i \otimes v_j,$$

where $f \in \mathrm{Hom}(U, V)$ with its action on the basis given by $f(u_i) = \sum_{j=1}^n f_i^j v_j$, for $i = 1, \dots m$.

2. *Let $\phi : U^\star \otimes V \to \mathrm{Hom}(U, V)$ be defined by*

$$\phi : \sum_{i,j} f_i^j u^i \otimes v_j \mapsto f,$$

where $f \in \mathrm{Hom}(U, V)$ *is defined by, for* $x \in U$,

$$f(x) := \sum_{i,j} f_i^j u^i(x)\, v_j.$$

Then ψ *and* ϕ *are mutually inverse isomorphisms. (In the sums,* i *ranges from 1 to* m *and* j *ranges from 1 to* n*.)*

As an example, if U and V are free modules of rank 2 and

$$f : u_j \mapsto f_j^1 v_1 + f_j^2 v_2,$$

then f is identified with the tensor

$$f_1^1(u^1 \otimes v_1) + f_2^1(u^2 \otimes v_1) + f_1^2(u^1 \otimes v_2) + f_2^2(u^2 \otimes v_2).$$

Appendix B
Rewriting the Definition of Operator Invariants

In Sect. 7.3, the following rewriting of the definition of operator invariants was stated. Here, we provide the computations for that rewriting.

Theorem B.1 *Let V and W be free modules of the same finite rank. Then*

$$Q_{(V,W,\mathsf{R},\vec{n},\overleftarrow{n},\vec{u},\overleftarrow{u})}$$

is an operator invariant if and only if

1. *there exist isomorphisms* $\alpha\colon W^\star \to V$ *and* $\beta\colon V^\star \to W$ *such that*

$$\vec{u} = \alpha^{\cup}, \qquad \vec{n} = (\alpha^{-1})^{\cap}, \qquad \overleftarrow{u} = \beta^{\cup}, \qquad \overleftarrow{n} = (\beta^{-1})^{\cap};$$

2. R *is an invertible R-matrix;*
3. *If* $\mu := \beta^\star \circ \alpha^{-1}\colon V \to V$ *then*

    a. $\left(\tau \circ \mathsf{R}^{-1}\right)^{t_1} \circ (\mathrm{id}_{V^\star} \otimes \mu) \circ (\mathsf{R} \circ \tau)^{t_1} \circ (\mathrm{id}_{V^\star} \otimes \mu)^{-1} = \mathrm{id}_{V^\star \otimes V}$;
    b. $\mathrm{Tr}_2\left(\mathsf{R}^{\pm 1} \circ (\mathrm{id} \otimes \mu)\right) = \mathrm{id}_V$;
    c. $\mathsf{R} \circ (\mu \otimes \mu) = (\mu \otimes \mu) \circ \mathsf{R}$.

We make use of the following isomorphisms.

The $\cap$ *−map*

Lemma B.2 *There is a natural isomorphism*

$$(\cdot)^{\cap}\colon \mathrm{Hom}(X, Y) \to \mathrm{Hom}(X \otimes Y^\star, \mathbb{K})\colon f \mapsto (x \otimes g \mapsto g(f(x)))\,.$$

For a proof, see Exercise 7.22.

The $\cup$ *−map*

Lemma B.3 *There is a natural isomorphism*

$$(\cdot)^{\cup}\colon \mathrm{Hom}(X, Y) \to \mathrm{Hom}(\mathbb{K}, X^\star \otimes Y)\colon 1_{\mathbb{K}} \mapsto \sum_i x^i \otimes f(x_i).$$

D. M. Jackson and I. Moffatt, *An Introduction to Quantum and Vassiliev Knot Invariants*, CMS Books in Mathematics, https://doi.org/10.1007/978-3-030-05213-3

For a proof, see Exercise 7.24.

We shall use these maps to describe the four maps $\overrightarrow{u}, \overleftarrow{u}, \overrightarrow{n}, \overleftarrow{n}$ in terms of two isomorphisms

The maps $(\cdot)^{\cap}$ and $(\cdot)^{\cup}$ provide a relation between the spaces $\mathrm{Hom}(X, Y)$, $\mathrm{Hom}(X \otimes Y^{\star}, \mathbb{K})$ and $\mathrm{Hom}(\mathbb{K}, X^{\star} \otimes Y)$. Maps in spaces of the form $\mathrm{Hom}(X \otimes Y^{\star}, \mathbb{K})$ and $\mathrm{Hom}(\mathbb{K}, X^{\star} \otimes Y)$ arise in the definition of operator invariants (namely the maps $\overrightarrow{n}, \overleftarrow{n}, \overrightarrow{u}, \overleftarrow{u}$), and we shall see that the natural idea of expressing $\overrightarrow{n}, \overleftarrow{n}, \overrightarrow{u}, \overleftarrow{u}$ in terms of maps in $\mathrm{Hom}(X, Y)$ is a very good one. Let

$$\alpha\colon W^{\star} \xrightarrow{\cong} V,$$
$$\beta\colon V^{\star} \xrightarrow{\cong} W,$$

be isomorphisms. The cap- and cup-maps applied to α, β and their inverses have the following domain and codomain pairs:

$$\alpha^{\cup}\colon \mathbb{K} \to W \otimes V, \qquad (\alpha^{-1})^{\cap}\colon V \otimes W \to \mathbb{K},$$
$$\beta^{\cup}\colon \mathbb{K} \to V \otimes W, \qquad (\beta^{-1})^{\cap}\colon W \otimes V \to \mathbb{K}.$$

Determination of Element-Wise Actions of Compositions of Some ∪ and ∩ Maps

Explicit element-wise actions will be needed for the maps $\alpha^{\cup}$, $\beta^{\cup}$, $\left(\alpha^{-1}\right)^{\cap}$, and $\left(\beta^{-1}\right)^{\cap}$. These are gathered here in one place, together with the definitions of scalars for representing α, β, α^{-1} and β^{-1} with respect to the standard bases of V and W.

Let

$$\begin{aligned} &\alpha\colon W^{\star} \to V\colon w^i \mapsto \textstyle\sum_j \alpha^{ij} v_j, && \alpha^{-1}\colon V \to W^{\star}\colon v_i \mapsto \textstyle\sum_j \overline{\alpha}_{ij} w^j, \\ &\beta\colon V^{\star} \to W\colon v^i \mapsto \textstyle\sum_j \beta^{ij} w_j, && \beta^{-1}\colon W \to V^{\star}\colon w_i \mapsto \textstyle\sum_j \overline{\beta}_{ij} v^j, \\ &\mathsf{R}\colon V^{\otimes 2} \to V^{\otimes 2}\colon \textstyle\sum_{k,l} R_{ij}^{kl}\, v_k \otimes v_l, && \mathsf{R}^{-1}\colon V^{\otimes 2} \to V^{\otimes 2}\colon \textstyle\sum_{k,l} \overline{R}_{ij}^{kl}\, v_k \otimes v_l. \end{aligned} \tag{B.1}$$

Lemma B.4 *We have the following element-wise actions.*

1. $\alpha^{\star}\colon V^{\star} \to W\colon v^i \mapsto \sum_j \alpha^{ji} w_j$;
2. $\beta^{\star}\colon W^{\star} \to V\colon w^i \mapsto \sum_j \beta^{ji} v_j$;
3. $\left(\alpha^{-1}\right)^{\star}\colon W \to V^{\star}\colon w_i \mapsto \sum_j \overline{\alpha}_{ji} v^j$;
4. $\left(\beta^{-1}\right)^{\star}\colon V \to W^{\star}\colon v_i \mapsto \sum_j \overline{\beta}_{ji} w^j$;
5. $\alpha^{\cup}\colon \mathbb{K} \to W \otimes V\colon 1_{\mathbb{K}} \mapsto \sum_i w_i \otimes \alpha(w^i) = \sum_{i,j} \alpha^{ij}\, w_i \otimes v_j$;
6. $\beta^{\cup}\colon \mathbb{K} \to V \otimes W\colon 1_{\mathbb{K}} \mapsto \sum_i v_i \otimes \beta(v^i) = \sum_{i,j} \beta^{ij} v_i \otimes w_j$;
7. $(\alpha^{-1})^{\cap}\colon V \otimes W \to \mathbb{K}\colon v_i \otimes w_j \mapsto \left(\alpha^{-1}(v_i)\right)(w_j) = \overline{\alpha}_{ij}$;
8. $(\beta^{-1})^{\cap}\colon W \otimes V \to \mathbb{K}\colon w_i \otimes v_j \mapsto \left(\beta^{-1}(w_i)\right)(v_j) = \overline{\beta}_{ij}$.

Proof. We consider the items in turn.
Part 1. From the notation for α in (B.1),

$$\alpha \colon W^{\star} \to V \colon w^i \mapsto \sum_j \alpha^{ij}\, v_j.$$

Now $\alpha^{\star} \colon V^{\star} \to W^{\star\star}$, and $\{(w^i)^{\star}\}_i$ is the natural basis of $W^{\star\star}$, so let

$$\alpha^{\star}(v^j) = \sum_i c^{ij}\,(w^i)^{\star}.$$

To determine c^{ij}, consider the action of $\alpha^{\star}(v^j)$ on w^k. Then

$$\alpha^{\star}(v^j)(w^k) = \sum_i c^{ij}\,(w^i)^{\star}(w^k) = c^{kj}.$$

Thus

$$c^{kj} = (v^j \circ \alpha)(w^k) = v^j\left(\sum_m \alpha^{km}\, v_m\right) = \alpha^{kj}$$

and therefore $\alpha^{\star}(v^j) = \sum_i \alpha^{ij}\,(w^i)^{\star}$ so $\alpha^{\star} \colon v^j \mapsto \sum_i \alpha^{ij} w_i$.
Part 2. This follows similarly.
Part 3. From the notation for α^{-1} in (B.1), it follows that

$$\left(\alpha^{-1}\right)^{\star} \colon W^{\star\star} \to V^{\star} \colon (w^i)^{\star} \mapsto (w^i)^{\star} \circ \alpha^{-1}$$

where

$$(w^i)^{\star} \circ \alpha^{-1} \colon V \to \mathbb{K} \colon v_j \mapsto (w^i)^{\star}\left(\sum_k \overline{\alpha}_{jk}\, w^k\right) = \overline{\alpha}_{ji}.$$

Thus

$$\left(\alpha^{-1}\right)^{\star} \colon W^{\star\star} \to V^{\star} \colon (w^i)^{\star} \longmapsto \left(v_j \mapsto \overline{\alpha}_{ji}\right).$$

Rewriting this map in terms of the standard basis $\{v_j\}_j$ of V, we have

$$\left(\alpha^{-1}\right)^{\star} \colon \left(w^i\right)^{\star} \mapsto \sum_k \overline{\alpha}_{ki}\, v^k.$$

But

$$W^{\star\star} \xrightarrow{\cong} W \colon (w^i)^{\star} \mapsto w_i$$

so, under this isomorphism $(\alpha^{-1})^{\star}$ may be regarded as a map

$$\left(\alpha^{-1}\right)^{\star} : W \to V^{\star} : w_i \mapsto \sum_k \overline{\alpha}_{ki}\, v^k .$$

Part 4. This follows similarly.
Part 5 – 8. These follow easily from the element-wise action of $(\cdot)^{\cap}$ and $(\cdot)^{\cup}$ described in Lemmas B.2 and B.3. □

Simplification of the Conditions for the T_1*- and* T_2*-Moves*

We next show that the existence of the four maps $\overrightarrow{u}$, $\overleftarrow{u}$, $\overrightarrow{n}$, $\overleftarrow{n}$ in the definition of an operator invariant is equivalent to the existence of two isomorphisms $\alpha\colon W^{\star} \to V$ and $\beta\colon V^{\star} \to W$. This equivalence will allow us to rewrite the remaining conditions in the definition of an operator invariant in a simpler form.

Lemma B.5 *Let $\alpha\colon W^{\star} \to V$ and $\beta\colon V^{\star} \to W$ be isomorphisms, and let $\overrightarrow{u}$, $\overrightarrow{n}$, $\overleftarrow{u}$ and $\overleftarrow{n}$ be maps defined in terms of these by*

$$\overrightarrow{u} := \alpha^{\cup}, \qquad \overrightarrow{n} := (\alpha^{-1})^{\cap}, \qquad \overleftarrow{u} := \beta^{\cup}, \qquad \overleftarrow{n} := (\beta^{-1})^{\cap}.$$

Then, the Turaev Conditions (7.8) and (7.9) for moves T_1 *and* T_2*, respectively, are satisfied.*

Proof. We shall prove the result only for one T_1-move, since the result for the other moves follow similarly.

By considering the left-hand side of the Turaev Condition (7.8) and using the actions given in Lemma B.4, we have

$$\begin{aligned}
(\overrightarrow{n} \otimes \mathrm{id}_V) \circ (\mathrm{id}_V \otimes \overrightarrow{u}) \circ \iota_2(v_i) &= \left((\alpha^{-1})^{\cap} \otimes \mathrm{id}_V\right) \circ (\mathrm{id}_V \otimes \alpha^{\cup})(v_i \otimes 1_{\mathbb{K}}) \\
&= \left((\alpha^{-1})^{\cap} \otimes \mathrm{id}_V\right) \left(v_i \otimes \sum_{j,k} \alpha^{jk}\, w_j \otimes v_k\right) \\
&= \sum_{j,k} \left((\alpha^{-1})^{\cap} \otimes \mathrm{id}_V\right) \left(\alpha^{jk}\, v_i \otimes w_j \otimes v_k\right) \\
&= \sum_{j,k} \alpha^{jk} \left((\alpha^{-1})^{\cap}(v_i \otimes w_j)\right) \otimes v_k \\
&= \sum_{j,k} \overline{\alpha}_{ij}\, \alpha^{jk}\, (1_{\mathbb{K}} \otimes v_k) \quad (\text{action of } (\alpha^{-1})^{\cap}) \\
&= \sum_{k} \delta_{i,k}\, (1_{\mathbb{K}} \otimes v_k) \quad (\text{since } \alpha^{-1} \circ \alpha = \mathrm{id}_{W^{\star}}) \\
&= 1_{\mathbb{K}} \otimes v_i \overset{\kappa_1}{\longmapsto} v_i .
\end{aligned}$$

Thus, we have shown that $(\overrightarrow{n} \otimes \mathrm{id}_V) \circ (\mathrm{id}_V \otimes \overrightarrow{u}) = \mathrm{id}_V$ if $\overrightarrow{u} = \alpha^{\cup}$ and $\overrightarrow{n} = (\alpha^{-1})^{\cap}$.

The results for the T_2-move and the remaining T_1-move follow similarly. □

The next result, which is the converse of Lemma B.5, shows that if the Turaev Conditions (7.8) and (7.9) hold, then α and β exist and are unique.

Lemma B.6 *Let $\overrightarrow{n}$, $\overleftarrow{n}$, $\overrightarrow{u}$ and $\overleftarrow{u}$, with domains and codomains as indicated in (7.3), satisfy Turaev Conditions (7.8) and (7.9) for the T_1- and T_2-Moves, respectively. Then there exist unique isomorphisms*

$$\alpha\colon W^\star \xrightarrow{\cong} V, \qquad \textit{and} \qquad \beta\colon V^\star \xrightarrow{\cong} W$$

such that $\quad \overrightarrow{u} = \alpha^{\cup}, \quad \overrightarrow{n} = (\alpha^{-1})^{\cap}, \quad \overleftarrow{u} = \beta^{\cup}, \quad \overleftarrow{n} = (\beta^{-1})^{\cap}.$

Proof. Let the element-wise actions of $\overrightarrow{n}$ and $\overrightarrow{u}$ be

$$\overrightarrow{n}: V \otimes W \to \mathbb{K}\colon v_i \otimes w_j \mapsto \overrightarrow{n}_{ij}, \qquad \overrightarrow{u}: \mathbb{K} \to W \otimes V: 1_{\mathbb{K}} \mapsto \sum_{i,j} \overrightarrow{u}^{ij}\; w_i \otimes v_j.$$

where $\overrightarrow{n}_{ij}, \overrightarrow{u}^{ij} \in \mathbb{K}$. We know that

$$(\overrightarrow{n} \otimes \mathrm{id}_V) \circ (\mathrm{id}_V \otimes \overrightarrow{u}) = \mathrm{id}_V \qquad \text{(left-hand side of (7.8))} \qquad \text{(B.2)}$$

$$(\mathrm{id}_W \otimes \overrightarrow{n}) \circ (\overrightarrow{u} \otimes \mathrm{id}_W) = \mathrm{id}_W \qquad \text{(right-hand side of (7.9))}. \qquad \text{(B.3)}$$

Let α be the map

$$\alpha\colon W^\star \to V : w^i \mapsto \sum_j \overrightarrow{u}^{ij}\; v_j.$$

Then, it is easily checked that $\alpha^{\cup} = \overrightarrow{u}$, and α is the unique map with this property since the $\cup$-map is invertible. Now let γ be the map defined by

$$\gamma\colon V \to W^\star : v_i \mapsto \sum_j \overrightarrow{n}_{ij}\; w^j.$$

Then, by the action of the $\cap$-map of γ we have $\gamma^{\cap}\colon v_i \otimes w_j \mapsto \overrightarrow{n}_{ij}$, so $\gamma^{\cap} = \overrightarrow{n}$ and γ is the unique map with this property since the $\cap$-map is invertible. It remains to show that $\alpha^{-1} = \gamma$ which may be done by showing that $\gamma \circ \alpha = \mathrm{id}_{W^\star}$ and $\alpha \circ \gamma = \mathrm{id}_V$.

First consider $\gamma \circ \alpha$. The element-wise action of this map is

$$(\gamma \circ \alpha)(w^i) = \gamma\left(\sum_j \overrightarrow{u}^{ij}\; v_j\right) = \sum_{j,k} \overrightarrow{u}^{ij}\, \overrightarrow{n}_{jk}\; w^k,$$

so $\gamma \circ \alpha = \mathrm{id}_{W^\star}$ if and only if $\sum_j \vec{u}^{ij} \vec{n}_{jk} = \delta_{jk}$. Now, from (B.3), and using Convention 7.14, we have

$$\kappa \circ \left(\mathrm{id}_W \otimes \vec{n}\right) \circ \left(\vec{u} \otimes \mathrm{id}_W\right) \circ \iota = \mathrm{id}_W$$

so

$$\begin{aligned} w_i = \mathrm{id}_W(w_i) &= \kappa \circ \left(\mathrm{id}_W \otimes \vec{n}\right) \circ \left(\vec{u} \otimes \mathrm{id}_W\right) \circ \iota(w_1) \\ &= \kappa \circ \left(\mathrm{id}_W \otimes \vec{n}\right) \circ \left(\vec{u} \otimes \mathrm{id}_W\right)(1_{\mathbb{K}} \otimes w_i) \\ &= \kappa \circ \left(\mathrm{id}_W \otimes \vec{n}\right) \left(\sum_{j,k} \vec{u}^{jk} \, w_j \otimes v_k \otimes w_i\right) \\ &= \kappa \left(\sum_{j,k} \vec{u}^{jk} \vec{n}_{ki} \, w_j \otimes 1_{\mathbb{K}}\right) = \sum_{j,k} \vec{u}^{jk} \vec{n}_{ki} \, w_j. \end{aligned}$$

Therefore, $\sum_k \vec{u}^{jk} \vec{n}_{ki} = \delta_{i,j}$ whence $\gamma \circ \alpha = \mathrm{id}_{W^\star}$.

The relation (B.3) implies that $\alpha \circ \gamma = \mathrm{id}_V$ in a similar way, and therefore, there exists a map $\alpha \colon W^\star \xrightarrow{\cong} V$ such that $\vec{u} = \alpha^\cup$ and $\vec{n} = (\alpha^{-1})^\cap$.

The construction for $\beta \colon V^\star \to W$ follows in a similar way, completing the proof. □

Simplification of the Conditions for the T_3*-Move*

Lemma B.7 *Let* $(V, W, \mathsf{R}, \vec{u}, \overleftarrow{u}, \vec{n}, \overleftarrow{n})$ *define an operator invariant. Then the Turaev Condition (7.10) is equivalent to*

$$\mathsf{R} \circ (\mu \otimes \mu) = (\mu \otimes \mu) \circ \mathsf{R} \tag{B.4}$$

where

$$\mu = \beta^\star \circ \alpha^{-1} \tag{B.5}$$

and α *and* β *are the unique maps (from Lemma B.6) such that*

$$\vec{u} = \alpha^\cup, \quad \overleftarrow{u} = \beta^\cup, \quad \vec{n} = (\alpha^{-1})^\cap, \quad \overleftarrow{n} = (\beta^{-1})^\cap.$$

Proof. Consider the T_3-move for the positive crossing. We prove this lemma as follows. In Stage 1, we calculate the element-wise actions of the maps on the left- and right-hand sides of (7.10). In Stage 2, we show that (7.10) is equivalent to

$$(\alpha^\star \otimes \alpha^\star)^{-1} \circ (\beta \otimes \beta) \circ \mathsf{R}^\star = \mathsf{R}^\star \circ \left((\alpha^{-1})^\star \otimes (\alpha^{-1})^\star\right) \otimes (\beta^{-1} \otimes \beta^{-1})^{-1}$$

by calculating the element-wise actions of the left- and right-hand sides of this equation. The lemma, in effect, then follows in Stage 3 as the dual of this identity.

Stage 1: From (7.10), the Turaev Condition for T_3 is

$$a_5 \circ \cdots \circ a_1 = b_5 \circ \cdots \circ b_1, \tag{B.6}$$

where

$$\begin{aligned} a_5 &:= \mathrm{id}_W^{\otimes 2} \otimes \vec{n}, & a_4 &:= \mathrm{id}_W^{\otimes 2} \otimes \mathrm{id}_V \otimes \vec{n} \otimes \mathrm{id}_W, \\ a_3 &:= \mathrm{id}_W^{\otimes 2} \otimes \mathsf{R} \otimes \mathrm{id}_W^{\otimes 2}, & a_2 &:= \mathrm{id}_W \otimes \vec{u} \otimes \mathrm{id}_V \otimes \mathrm{id}_W^{\otimes 2}, \\ a_1 &:= \vec{u} \otimes \mathrm{id}_W^{\otimes 2}, \end{aligned} \tag{B.7}$$

and

$$\begin{aligned} b_5 &:= \overleftarrow{n} \otimes \mathrm{id}_W^{\otimes 2}, & b_4 &:= \mathrm{id}_W \otimes \overleftarrow{n} \otimes \mathrm{id}_V \otimes \mathrm{id}_W^{\otimes 2}, \\ b_3 &:= \mathrm{id}_W^{\otimes 2} \otimes \mathsf{R} \otimes \mathrm{id}_W^{\otimes 2}, & b_2 &:= \mathrm{id}_W^{\otimes 2} \otimes \mathrm{id}_V \otimes \overleftarrow{u} \otimes \mathrm{id}_W, \\ b_1 &:= \mathrm{id}_W^{\otimes 2} \otimes \overleftarrow{u}\ . \end{aligned} \tag{B.8}$$

Applying $a_5 \circ \cdots \circ a_1$, making use of ι, to the element $w_i \otimes w_j$ gives

$$\begin{aligned} w_i \otimes w_j \stackrel{\iota}{\longmapsto}\ & 1_{\mathbb{K}} \otimes w_i \otimes w_j \\ \stackrel{a_1}{\longmapsto}\ & \sum_k w_k \otimes \alpha(w^k) \otimes w_i \otimes w_j \\ \stackrel{\iota}{\longmapsto}\ & \sum_k w_k \otimes 1_{\mathbb{K}} \otimes \alpha(w^k) \otimes w_i \otimes w_j \\ \stackrel{a_2}{\longmapsto}\ & \sum_{k,l} w_k \otimes w_l \otimes \big(\alpha(w^l) \otimes \alpha(w^k)\big) \otimes w_i \otimes w_j \\ =\ & \sum_{k,l,m,n} \alpha^{lm}\alpha^{kn} w_k \otimes w_l \otimes (v_m \otimes v_n) \otimes w_i \otimes w_j \\ \stackrel{a_3}{\longmapsto}\ & \sum_{k,l,m,n,p,q} \alpha^{lm}\alpha^{kn} R^{pq}_{mn}\, w_k \otimes w_l \otimes v_p \otimes \big(v_q \otimes w_i\big) \otimes w_j \\ \stackrel{a_4}{\longmapsto}\ & \sum_{k,l,m,n,p,q} \alpha^{lm}\alpha^{kn}\overline{\alpha}_{qi} R^{pq}_{mn}\, w_k \otimes w_l \otimes \big(v_p \otimes w_j\big) \\ \stackrel{a_5}{\longmapsto}\ & \sum_{k,l,m,n,p,q} \alpha^{lm}\alpha^{kn}\overline{\alpha}_{qi}\overline{\alpha}_{pj}\, R^{pq}_{mn}\, w_k \otimes w_l. \end{aligned}$$

Thus

$$(a_5 \circ \cdots \circ a_1)(w_i \otimes w_j) = \sum_{k,l,m,n,p,q} \alpha^{lm}\alpha^{kn}\overline{\alpha}_{qi}\overline{\alpha}_{pj}\, R^{pq}_{mn}\, w_k \otimes w_l. \tag{B.9}$$

Similarly, applying $b_5 \circ \cdots \circ b_1$ to the element $w_i \otimes w_j$ gives

$$\begin{aligned} w_i \otimes w_j &\overset{\iota}{\longmapsto} w_i \otimes w_j \otimes 1_{\mathbb{K}} \\ &\overset{b_1}{\longmapsto} \sum_{k,l} \beta^{kl} w_i \otimes w_j \otimes v_k \otimes w_l \\ &\overset{b_2 \circ \iota}{\longmapsto} \sum_{k,l,m,n} \beta^{kl}\beta^{mn} w_i \otimes w_j \otimes (v_k \otimes v_m) \otimes w_n \otimes w_l \\ &\overset{b_3}{\longmapsto} \sum_{k,l,m,n} \beta^{kl}\beta^{mn}\, R^{pq}_{km}\, w_i \otimes (w_j \otimes v_p) \otimes v_q \otimes w_n \otimes w_l \\ &\overset{b_4}{\longmapsto} \sum_{k,l,m,n,p,q} \beta^{kl}\beta^{mn}\overline{\beta}_{jp}\, R^{pq}_{km} \left(w_i \otimes v_q\right) \otimes w_n \otimes w_l \\ &\overset{b_5}{\longmapsto} \sum_{k,l,m,n,p,q} \beta^{kl}\beta^{mn}\overline{\beta}_{jp}\overline{\beta}_{iq}\, R^{pq}_{km}\, w_n \otimes w_l. \end{aligned}$$

Thus

$$(b_5 \circ \cdots \circ b_1)(w_i \otimes w_j) = \sum_{k,l,m,n,p,q} \beta^{kl}\beta^{mn}\overline{\beta}_{jp}\,\overline{\beta}_{iq} R^{pq}_{km}\, w_n \otimes w_l. \tag{B.10}$$

It follows from (B.6), together with the evaluations given in (B.9) and (B.10), that Turaev's Condition for the T_3-move is equivalent to

$$\sum_{k,l,m,n,p,q} \alpha^{lm}\alpha^{kn}\overline{\alpha}_{qi}\overline{\alpha}_{pj}\, R^{pq}_{mn}\, w_k \otimes w_l = \sum_{k,l,m,n,p,q} \beta^{kl}\beta^{mn}\overline{\beta}_{jp}\,\overline{\beta}_{iq} R^{pq}_{km}\, w_n \otimes w_l. \tag{B.11}$$

Stage 2: The next stage is the simplification of (B.11), beginning with its right-hand side. It is convenient to consider the map

$$\tau \circ (\beta \otimes \beta) \circ \mathsf{R}^\star \circ (\beta^{-1} \otimes \beta^{-1}) \circ \tau \tag{B.12}$$

where τ is the twist map. Now

$$w_i \otimes w_j \overset{\tau}{\longmapsto} w_j \otimes w_i \overset{\beta^{-1}\otimes\beta^{-1}}{\longmapsto} \sum_{k,l} \overline{\beta}_{jk}\,\overline{\beta}_{il}\, v^k \otimes v^l. \tag{B.13}$$

But

$$\mathsf{R}^\star \colon (V \otimes V)^\star \to (V \otimes V)^\star \colon (v_i \otimes v_j)^\star \mapsto (v_i \otimes v_j)^\star \circ \mathsf{R}$$

and therefore, element-wise,

$$(v_i \otimes v_j)^\star \circ \mathsf{R} \colon v_k \otimes v_l \mapsto (v_i \otimes v_j)^\star \sum_{m,n} \mathsf{R}^{mn}_{kl} v_m \otimes v_n = \mathsf{R}^{ij}_{kl}$$

so

$$\mathsf{R}^\star\left((v_i \otimes v_j)^\star\right) = (v_i \otimes v_j)^\star \circ \mathsf{R} = \sum_{m,n} \mathsf{R}^{ij}_{mn}(v_m \otimes v_n)^\star = \sum_{m,n} \mathsf{R}^{ij}_{mn} v^m \otimes v^n.$$

Continuing with the application of the map (B.12) from the point reached in (B.13), we have

$$\begin{aligned}
\sum_{k,l} \overline{\beta}_{jk}\,\overline{\beta}_{il}\, v^k \otimes v^l &\overset{\mathsf{R}^\star}{\longmapsto} \sum_{k,l,m,n} \overline{\beta}_{jk}\,\overline{\beta}_{il}\, R^{kl}_{mn}\, v^m \otimes v^n \\
&\overset{\beta\otimes\beta}{\longmapsto} \sum_{k,l,m,n,p,q} \overline{\beta}_{jk}\,\overline{\beta}_{il}\, \beta^{mp}\, \beta^{nq}\, R^{kl}_{mn}\, w_p \otimes w_q \\
&\overset{\tau}{\longmapsto} \sum_{k,l,m,n,p,q} \overline{\beta}_{jk}\,\overline{\beta}_{il}\, \beta^{mp}\, \beta^{nq}\, R^{kl}_{mn}\, w_q \otimes w_p \\
&= \sum_{k,l,m,n,p,q} \overline{\beta}_{iq}\,\overline{\beta}_{jp}\, \beta^{mn}\, \beta^{kl}\, R^{pq}_{km} w_n \otimes w_l
\end{aligned}$$

where the latter is under reindexing the summation through $m \mapsto k, k \mapsto p, l \mapsto q, q \mapsto n, p \mapsto l, n \mapsto m$. But this expression is precisely the expression on the right-hand side of (B.11) so

$$a_1 \circ \cdots \circ a_5 = \tau \circ (\beta \otimes \beta) \circ \mathsf{R}^\star \circ (\beta^{-1} \otimes \beta^{-1}) \circ \tau. \tag{B.14}$$

The left-hand side of (B.11) may be simplified in a similar way by considering the map

$$\tau \circ (\alpha^\star \otimes \alpha^\star) \circ \mathsf{R}^\star \circ \left((\alpha^{-1})^\star \otimes (\alpha^{-1})^\star\right) \circ \tau. \tag{B.15}$$

Now

$$\begin{aligned}
w_i \otimes w_j &\overset{\tau}{\longmapsto} w_j \otimes w_i \\
&\overset{(\alpha^{-1})^\star\otimes(\alpha^{-1})^\star}{\longmapsto} \sum_{k,l} \overline{\alpha}_{kj}\,\overline{\alpha}_{li}\, v^k \otimes v^l \\
&\overset{\mathsf{R}^\star}{\longmapsto} \sum_{k,l,m,n} \overline{\alpha}_{kj}\,\overline{\alpha}_{li}\, R^{kl}_{mn}\, v^m \otimes v^n \\
&\overset{\alpha^\star\otimes\alpha^\star}{\longmapsto} \sum_{k,l,m,n} \overline{\alpha}_{kj}\,\overline{\alpha}_{li}\, \alpha^{pm}\, \alpha^{qn}\, R^{kl}_{mn}\, w_p \otimes w_q \\
&\overset{\tau}{\longmapsto} \sum_{k,l,m,n,p,q} \overline{\alpha}_{kj}\,\overline{\alpha}_{li}\, \alpha^{pm}\, \alpha^{qn}\, R^{kl}_{mn}\, w_q \otimes w_p \\
&= \sum_{k,l,m,n,p,q} \alpha^{lm}\, \alpha^{kn}\, \overline{\alpha}_{pj}\,\overline{\alpha}_{qi}\, R^{pq}_{mn}\, w_k \otimes w_l
\end{aligned}$$

where the latter is obtained by reindexing through $q \mapsto k$, $p \mapsto l$, $k \mapsto p$, $l \mapsto q$. This expression is precisely the expression on the left-hand side of (B.11) so

$$b_1 \circ \cdots \circ b_5 = \tau \circ (\alpha^\star \otimes \alpha^\star) \circ \mathsf{R}^\star \circ \left((\alpha^{-1})^\star \otimes (\alpha^{-1})^\star\right) \circ \tau. \tag{B.16}$$

Thus, combining (B.14) and (B.16), it follows that (B.4) is equivalent to

$$\tau \circ (\beta \otimes \beta) \circ \mathsf{R}^\star \circ (\beta^{-1} \otimes \beta^{-1}) \circ \tau = \tau \circ (\alpha^\star \otimes \alpha^\star) \circ \mathsf{R}^\star \circ \left((\alpha^{-1})^\star \otimes (\alpha^{-1})^\star\right) \circ \tau$$

and therefore, since τ is invertible, (B.4) is equivalent to

$$(\beta \otimes \beta) \circ \mathsf{R}^\star \circ (\beta^{-1} \otimes \beta^{-1}) = (\alpha^\star \otimes \alpha^\star) \circ \mathsf{R}^\star \circ \left((\alpha^{-1})^\star \otimes (\alpha^{-1})^\star\right). \tag{B.17}$$

Stage 3: Condition (B.17) is equivalent to

$$(\alpha^\star \otimes \alpha^\star)^{-1} \circ (\beta \otimes \beta) \circ \mathsf{R}^\star = \mathsf{R}^\star \circ \left((\alpha^{-1})^\star \otimes (\alpha^{-1})^\star\right) \circ (\beta^{-1} \otimes \beta^{-1})^{-1}$$

which is equivalent to

$$\left((\alpha^{-1})^\star \circ \beta\right) \otimes \left((\alpha^{-1})^\star \circ \beta\right) \circ \mathsf{R}^\star = \mathsf{R}^\star \circ \left((\alpha^{-1})^\star \circ \beta^{-1}\right) \otimes \left((\alpha^{-1})^\star \circ \beta^{-1}\right)$$

By Theorem A.19, $\left(f^{-1}\right)^\star = (f^\star)^{-1}$, so this is equivalent to

$$(\mu \otimes \mu)^\star \circ \mathsf{R}^\star = \mathsf{R}^\star \circ (\mu \otimes \mu)^\star$$

since, by Proposition A.35, $(f \otimes g)^\star = f^\star \otimes g^\star$. Thus $\mathsf{R} \circ (\mu \otimes \mu) = (\mu \otimes \mu) \circ \mathsf{R}$.

A similar analysis for the T_3-move with the negative crossing gives that Condition (7.10) is equivalent to $\mathsf{R}^{-1} \circ (\mu \otimes \mu) = (\mu \otimes \mu) \circ \mathsf{R}^{-1}$. But, by multiplying on the left and right by R, we see that this holds if and only if $\mathsf{R} \circ (\mu \otimes \mu) = (\mu \otimes \mu) \circ \mathsf{R}$. completing the proof. □

Simplification of the Conditions for the T_6*-Move*

Lemma B.8 *Let* $(V, W, \mathsf{R}, \overrightarrow{u}, \overleftarrow{u}, \overrightarrow{n}, \overleftarrow{n})$ *define an operator invariant. Then, the Turaev Condition (7.16) is equivalent to*

$$\mathrm{Tr}_2\left(\mathsf{R}^{\pm 1} \circ (\mathrm{id} \otimes \mu)\right) = \mathrm{id}_V \tag{B.18}$$

where μ *is as in Lemma B.7.*

Proof. From (7.13), the Turaev Condition for the T_6-move is

$$\left(\mathrm{id}_V \otimes \overrightarrow{n}\right) \circ \left(\mathsf{R}^{\pm 1} \otimes \mathrm{id}_W\right) \circ \left(\mathrm{id}_V \otimes \overleftarrow{u}\right) = \mathrm{id}_V. \tag{B.19}$$

We show that the element-wise actions of the left-hand sides of this and (B.18) are equal. For the left-hand side of (B.19), we have

$$\begin{aligned}
v_i &\overset{\iota}{\longmapsto} v_i \otimes 1_{\mathbb{K}} \\
&\overset{\mathrm{id}\otimes\overleftarrow{u}}{\longmapsto} \sum_{j,k} \beta^{jk}\, v_i \otimes v_j \otimes w_k && \text{(since } \overleftarrow{u} = \beta^{\cup}) \\
&\overset{\mathsf{R}\otimes\mathrm{id}_W}{\longmapsto} \sum_{j,k,l,m} \beta^{jk}\, R_{ij}^{lm}\, v_l \otimes v_m \otimes w_k \\
&\overset{\mathrm{id}_V\otimes\overrightarrow{n}}{\longmapsto} \sum_{j,k,l,m} \overline{\alpha}_{mk}\, \beta^{jk}\, R_{ij}^{lm}\, v_l && \text{(since } \overleftarrow{n} = (\alpha^{-1})^{\cap}).
\end{aligned}$$

For the left-hand side of (B.18),

$$\begin{aligned}
v_i \otimes v_j &\overset{\mathrm{id}\otimes\mu}{\longmapsto} v_i \otimes \mu(v_j) \\
&= v_i \otimes \beta^{\star}\left(\alpha^{-1}(v_j)\right) && \text{(since } \mu = \beta^{\star} \circ \alpha^{-1}) \\
&= v_i \otimes \beta^{\star}\left(\sum_k \overline{\alpha}_{jk}\, w^k\right) \\
&= \sum_{k,l} \overline{\alpha}_{jk}\, \beta^{lk}\, v_i \otimes v_l && \text{(Lem. B.4 (2))} \\
&\overset{\mathsf{R}}{\longmapsto} \sum_{k,l,m,n} \overline{\alpha}_{jk}\, \beta^{lk}\, R_{il}^{mn}\, v_m \otimes v_n && \text{(from (B.1))} \\
&= \sum_{m,n} A_{ij}^{mn}\, v_m \otimes v_n && \text{(where } A_{ij}^{mn} := \sum_{k,l} \overline{\alpha}_{jk}\, \beta^{lk}\, R_{il}^{mn}).
\end{aligned}$$

Having found the action of the map $\mathsf{R} \circ (\mathrm{id}_V \otimes \mu)$, we may determine the action of the map $\mathrm{Tr}_2\,(\mathsf{R} \circ (\mathrm{id}_V \otimes \mu))$. Recalling from Exercise 6.4 that if

$$A\colon v_i \otimes v_j \mapsto \sum_{m,n} A_{ij}^{mn} v_m \otimes v_n \qquad \text{then} \qquad \mathrm{Tr}_2 A : v_i \mapsto \sum_{m,j} A_{ij}^{mj}\, v_m,$$

so

$$v_i \overset{\mathrm{Tr}_2(\mathsf{R}\circ(\mathrm{id}_V\otimes\mu))}{\longmapsto} \sum_{k,l,m,j} \overline{\alpha}_{jk}\, \beta^{lk}\, R_{il}^{mj}\, v_m = \sum_{j,k,l,m} \overline{\alpha}_{mk}\, \beta^{jk}\, R_{ij}^{lm}\, v_l$$

where the latter equality is by reindexing through $m \mapsto l$, $j \mapsto m$, $l \mapsto j$.

We have shown that the actions of

$$\left(\mathrm{id}_V \otimes \overrightarrow{n}\right) \circ (\mathsf{R} \otimes \mathrm{id}_W) \circ \left(\mathrm{id}_V \otimes \overleftarrow{u}\right) \qquad \text{and} \qquad \mathrm{Tr}_2\left(\mathsf{R} \circ (\mathrm{id} \otimes \mu)\right)$$

on $v_i \otimes v_j$ are identical. The argument for R^{-1} is similar, and the result therefore follows. □

The $(\cdot)^{t_1}$- *and* $(\cdot)^{t_2}$-*Maps*

To simplify Condition (7.14) for the T_7-move, two further maps, the $(\cdot)^{t_1}$-map and the $(\cdot)^{t_2}$-map, are introduced together with a technical result concerning them.

Lemma B.9 *There is a natural isomorphism*

$$\begin{aligned} (\cdot)^{t_1}: \quad & \mathrm{Hom}(V \otimes W, X \otimes Y) \quad \rightarrow \quad \mathrm{Hom}(X^\star \otimes W, V^\star \otimes Y) \\ : & \left(v_i \otimes w_j \mapsto \textstyle\sum_{k,l} f_{ij}^{kl}\, x_k \otimes y_l\right) \mapsto \left(x^k \otimes w_j \mapsto \textstyle\sum_{i,l} f_{ij}^{kl}\, v^i \otimes y_l\right). \end{aligned}$$

Proof. Consider the space $\mathrm{Hom}(V \otimes W, X \otimes Y)$. There is a composition of natural isomorphisms such that

$$\begin{aligned} \mathrm{Hom}(V \otimes W, X \otimes Y) &\cong (V \otimes W)^\star \otimes X \otimes Y && \text{(Lem. A.36)} \\ &\cong V^\star \otimes W^\star \otimes X \otimes Y \\ &\cong (X \otimes W^\star) \otimes (V^\star \otimes Y) \\ &\cong (X^\star \otimes W)^\star \otimes (V^\star \otimes Y) \\ &\cong \mathrm{Hom}(X^\star \otimes W,\ V^\star \otimes Y) && \text{(Lem. A.36)} \end{aligned}$$

so let

$$(\cdot)^{t_1}: \mathrm{Hom}(V \otimes W, X \otimes Y) \xrightarrow{\cong} \mathrm{Hom}(X^\star \otimes W, V^\star \otimes Y).$$

The element-wise action of the isomorphism $(\cdot)^{t_1}$ is obtained in the familiar way, by following the above chain of natural isomorphisms. Thus

$$\begin{aligned} (\cdot)^{t_1}: f := &\left(v_i \otimes w_j \mapsto \sum_{k,l} f_{ij}^{kl}\, x_k \otimes y_l\right) \\ &\longmapsto \sum_{i,j,k,l} f_{ij}^{kl}\, (v_i \otimes w_j)^\star \otimes x_k \otimes y_l && \text{(Lem. A.36)} \\ &\longmapsto \sum_{i,j,k,l} f_{ij}^{kl} v^i \otimes w^j \otimes x_k \otimes y_l \end{aligned}$$

$$\longmapsto \sum_{i,j,k,l} f_{ij}^{kl}\,(x_k \otimes w^j) \otimes (v^i \otimes y_l)$$

$$\longmapsto \sum_{i,j,k,l} f_{ij}^{kl}\,(x^k \otimes w_j)^\star \otimes (v^i \otimes y_l)$$

$$\longmapsto \left(x^k \otimes w_j \mapsto \sum_{i,l} f_{ij}^{kl}\, v^i \otimes y_l\right) \qquad \text{(Lem. A.36)}$$

and therefore $(f)^{t_1} : x^k \otimes w_j \mapsto \sum_{i,l} f_{ij}^{kl}\, v^i \otimes y_l$. The result now follows. □

Lemma B.10 *There is a natural isomorphism*

$$\begin{array}{rccc} (\cdot)^{t_2}: & \mathrm{Hom}(V \otimes W, X \otimes Y) & \xrightarrow{\cong} & \mathrm{Hom}(V \otimes Y^\star, X \otimes W^\star) \\ : & \left(v_i \otimes w_j \mapsto \sum_{k,l} f_{ij}^{kl}\, x_k \otimes y_l\right) & \longmapsto & \left(v_i \otimes y^l \mapsto \sum_{j,k} f_{ij}^{kl}\, x_k \otimes w^j\right). \end{array}$$

Proof. There is a chain of natural isomorphisms through which

$$(\cdot)^{t_2} : \mathrm{Hom}(V \otimes W, X \otimes Y) \xrightarrow{\cong} \mathrm{Hom}(V \otimes Y^\star, X \otimes W^\star)$$

may be defined, and the element-wise action of this isomorphism is constructed in a similar way:

$$(\cdot)^{t_2} : f := \left(v_i \otimes w_j \mapsto \sum_{k,l} f_{ij}^{kl}\, x_k \otimes y_l\right)$$

$$\longmapsto \sum_{i,j,k,l} f_{ij}^{kl}\, v^i \otimes w^j \otimes x_k \otimes y_l \qquad \text{(Lem. A.36)}$$

$$\longmapsto \sum_{i,j,k,l} f_{ij}^{kl}\,(v^i \otimes y_l) \otimes (x_k \otimes w^j)$$

$$\longmapsto \sum_{i,j,k,l} f_{ij}^{kl}\,(v_i \otimes y^l)^\star \otimes (x_k \otimes w^j)$$

$$\longmapsto \left(v_i \otimes y^l \mapsto \sum_{j,k} f_{ij}^{kl}\, x_k \otimes w^j\right) \qquad \text{(Lem. A.36)}$$

and therefore $(f)^{t_2} : v_i \otimes y^l \mapsto \sum_{j,k} f_{ij}^{kl}\, x_k \otimes w^j$. The result now follows. □

The isomorphisms $(\cdot)^{t_1}$ and $(\cdot)^{t_2}$ have the following properties that are stated in the next three lemmas.

Lemma B.11 *The maps* $(\cdot)^{t_1}$ *and* $(\cdot)^{t_2}$ *are involutory.*

Proof. We show that $(\cdot)^{t_1}$ is involutory and omit the proof for $(\cdot)^{t_2}$ since the proof is similar. Now

$$\begin{aligned}\mathrm{Hom}(V \otimes W, X \otimes Y) &\cong \mathrm{Hom}(X^\star \otimes W, V^\star \otimes Y)\\ &\cong \mathrm{Hom}(V^{\star\star} \otimes W, X^{\star\star} \otimes Y)\\ &\cong \mathrm{Hom}(V \otimes W, X \otimes Y).\end{aligned}$$

The result follows immediately by computing the element-wise actions. □

Let $f^{t_1 t_2}$ denote $\left((f)^{t_1}\right)^{t_2}$, and $f^{t_2 t_1}$ denote $\left((f)^{t_2}\right)^{t_1}$.

Lemma B.12 *We have,* $f^{t_1 t_2} = f^\star = f^{t_2 t_1}$.

Proof. Applying t_1 to f using Lemma B.9, we have

$$f^{t_1} : X^\star \otimes W \to V^\star \otimes Y : x^k \otimes w_j \mapsto \sum_{i,l} f_{ij}^{kl}\, v^i \otimes y_l,$$

and then applying t_2 to this gives

$$f^{t_1 t_2} : X^\star \otimes Y^\star \to V^\star \otimes W^\star : x^k \otimes y^l \mapsto \sum_{i,j} f_{ij}^{kl}\, v^i \otimes w^j.$$

But $f^\star : (X \otimes Y)^\star \to (V \otimes W)^\star$ with the element-wise action

$$f^\star : (x_k \otimes y_l)^\star \longmapsto \left((x_k \otimes y_l)^\star \circ f : v_i \otimes w_j \mapsto \sum_{p,q} f_{ij}^{pq}\, (x_k \otimes y_l)^\star (x_p \otimes y_q) \right).$$

However,

$$\sum_{p,q} f_{ij}^{pq}\, (x_k \otimes y_l)^\star (x_p \otimes y_q) = \sum_{p,q} f_{ij}^{pq}\, \delta_{kp}\, \delta_{lq} = f_{ij}^{kl},$$

so, in terms of standard bases,

$$f^\star : (x_k \otimes y_l)^\star \mapsto \sum_{i,j} f_{ij}^{kl}\, (v_i \otimes w_j)^\star,$$

and then

$$f^\star : x^k \otimes y^l \mapsto \sum_{i,j} f_{ij}^{kl}\, v^i \otimes w^j.$$

Thus, the actions of $f^{t_1 t_2}$ and $f^\star$ on a basis element $x^k \otimes y^l$ of $X^\star \otimes Y^\star$ are equal so

$$f^{t_1 t_2} = f^\star.$$

Similarly, $f^{t_2 t_1} = f^\star$, completing the proof. □

Exercise B.13 Complete the proof of Lemma B.12 by showing that $f^{t_2\,t_1} = f^{\star}$.

The final property of the maps t_1 and t_2 that is need is the following.

Lemma B.14 *In the above notation,* $f^{t_2\,t_1\,t_2} = f^{t_1}$ *and* $f^{t_1\,t_2\,t_1} = f^{t_2}$.

Proof. From Lemma B.12, $(f^{t_2\,t_1})^{t_2} = (f^{\star})^{t_2} = (f^{t_1\,t_2})^{t_2} = (f^{t_1})^{t_2\,t_2} = f^{t_1}$ since $(\cdot)^{t_2}$ is an involution by Lemma B.11. Thus $f^{t_2\,t_1\,t_2} = f^{t_1}$. Similarly, $f^{t_1\,t_2\,t_1} = f^{t_2}$. □

The Simplification of the T_7 *Condition*

With the introduction of the maps t_1 and t_2 and their properties, we may now proceed with the simplification of Turaev's Condition for the T_7-move.

Lemma B.15 *Suppose that conditions* T_1 *and* T_2 *hold. Let*

$$\begin{aligned} \mathsf{Y} &:= \left(\mathrm{id}_V \otimes \mathrm{id}_V \otimes \overrightarrow{n}\right) \circ (\mathrm{id}_W \otimes \mathsf{R} \otimes \mathrm{id}_W) \circ \left(\overrightarrow{u} \otimes \mathrm{id}_V \otimes \mathrm{id}_W\right), \\ \mathsf{T} &:= \left(\overleftarrow{n} \otimes \mathrm{id}_V \otimes \mathrm{id}_W\right) \circ \left(\mathrm{id}_W \otimes \mathsf{R}^{-1} \otimes \mathrm{id}_W\right) \circ \left(\mathrm{id}_W \otimes \mathrm{id}_V \otimes \overleftarrow{u}\right). \end{aligned} \tag{B.20}$$

Then, under the conditions of Lemma B.7, the relations

$$\mathsf{Y} \circ \mathsf{T} = \mathrm{id}_W \otimes \mathrm{id}_V \quad \textit{and} \quad \mathsf{T} \circ \mathsf{Y} = \mathrm{id}_V \otimes \mathrm{id}_W$$

hold if and only if

$$\left(\tau \circ \mathsf{R}^{-1}\right)^{t_1} \circ (\mathrm{id} \otimes \mu) \circ (\mathsf{R} \circ \tau)^{t_1} \circ (\mathrm{id} \otimes \mu)^{-1} = \mathrm{id}_{V^\star} \otimes \mathrm{id}_V.$$

Proof. The result is proved in Part 1 of the proof and is subject to two claims, which are proved in Part 2 of the proof.
Part 1: From Lemma B.6,

$$\overrightarrow{u} = \alpha^{\cup}, \qquad \overleftarrow{u} = \beta^{\cup}, \qquad \overrightarrow{n}\ (\alpha^{-1})^{\cap}, \qquad \overleftarrow{n} = (\beta^{-1})^{\cap}.$$

Claim 1. $\mathsf{Y} = (\alpha^{\star} \otimes \mathrm{id}) \circ \tau \circ (\mathsf{R} \circ \tau)^{t_2} \circ (\mathrm{id} \otimes \alpha^{-1})^{\star}$.
Claim 2. $\mathsf{T} = (\mathrm{id} \otimes \beta) \circ (\tau \circ \mathsf{R}^{-1})^{t_2} \circ \tau \circ (\beta^{-1} \otimes \mathrm{id})$.

Assume that these to Claims are true. Then

$$\mathsf{Y} \circ \mathsf{T} = \mathrm{id}_W \otimes \mathrm{id}_V$$

if and only if

$$\begin{aligned} &(\alpha^{\star} \otimes \mathrm{id}) \circ \tau \circ (\mathsf{R} \circ \tau)^{t_2} \circ (\mathrm{id} \otimes \alpha^{-1})^{\star} \\ &\qquad \circ (\mathrm{id} \otimes \beta) \circ (\tau \circ \mathsf{R}^{-1})^{t_2} \circ \tau \circ (\beta^{-1} \otimes \mathrm{id}) = \mathrm{id}_W \otimes \mathrm{id}_V. \end{aligned}$$

Setting $\mu = \beta^\star \circ \alpha^{-1}$, then

$$(\mathrm{id} \otimes \alpha^{-1})^\star \circ (\mathrm{id} \otimes \beta) = \mathrm{id} \otimes \mu^\star,$$

so $\mathsf{Y} \circ \mathsf{T} = \mathrm{id}_W \otimes \mathrm{id}_V$ if and only if

$$(\alpha^\star \otimes \mathrm{id}) \circ \tau \circ (\mathsf{R} \circ \tau)^{t_2} \circ (\mathrm{id} \otimes \mu^\star) \circ (\tau \circ \mathsf{R}^{-1})^{t_2} \circ \tau \circ (\beta^{-1} \otimes \mathrm{id}) = \mathrm{id}_W \otimes \mathrm{id}_V.$$

Conjugating both sides with $\tau \circ (\beta^{-1} \otimes \mathrm{id})$, it follows that $\mathsf{Y} \circ \mathsf{T} = \mathrm{id}_W \otimes \mathrm{id}_V$ if and only if

$$\tau \circ (\beta^{-1} \otimes \mathrm{id}) \circ (\alpha^\star \otimes \mathrm{id}) \circ \tau \circ (\mathsf{R} \circ \tau)^{t_2} \circ (\mathrm{id} \otimes \mu^\star) \circ (\tau \circ \mathsf{R}^{-1})^{t_2} = \mathrm{id}_V \otimes \mathrm{id}_{V^\star}. \tag{B.21}$$

The composition of the first four factors can be simplified by first observing that

$$\tau \circ (\beta^{-1} \otimes \mathrm{id}) \circ (\alpha^\star \otimes \mathrm{id}) \circ \tau = \tau \circ \big((\beta^{-1} \circ \alpha^\star) \otimes \mathrm{id}\big) \circ \tau \tag{B.22}$$

and then that $\beta^{-1} \circ \alpha^\star = (\mu^\star)^{-1}$. Therefore, from (B.22),

$$\tau \circ (\beta^{-1} \otimes \mathrm{id}) \circ (\alpha^\star \otimes \mathrm{id}) \circ \tau = \tau \circ \big((\mu^\star)^{-1} \otimes \mathrm{id}\big) \circ \tau. \tag{B.23}$$

Now

$$\begin{aligned} v_i \otimes v_j \overset{\tau}{\longmapsto} v_j \otimes v_i &\overset{(\mu^\star)^{-1}\otimes \mathrm{id}}{\longmapsto} \big((\mu^\star)^{-1}(v_j)\big) \otimes v_i \\ &\overset{\tau}{\longmapsto} v_i \otimes \big((\mu^\star)^{-1}(v_j)\big) = \big(\mathrm{id}_V \otimes (\mu^\star)^{-1}\big)(v_i \otimes v_j). \end{aligned}$$

Then

$$\tau \circ \big((\mu^\star)^{-1} \otimes \mathrm{id}\big) \circ \tau = \mathrm{id} \otimes (\mu^\star)^{-1}$$

and so, from (B.23),

$$\tau \circ (\beta^{-1} \otimes \mathrm{id}) \circ (\alpha^\star \otimes \mathrm{id}) \circ \tau = \big(\mathrm{id} \otimes \mu^\star\big)^{-1}.$$

Substituting this into (B.21) gives the condition, $\mathsf{Y} \circ \mathsf{T} = \mathrm{id}_W \otimes \mathrm{id}_V$ if and only if

$$\big(\mathrm{id} \otimes \mu^\star\big)^{-1} \circ (\mathsf{R} \circ \tau)^{t_2} \circ (\mathrm{id} \otimes \mu^\star) \circ (\tau \circ \mathsf{R}^{-1})^{t_2} = \mathrm{id}_V \otimes \mathrm{id}_{V^\star}. \tag{B.24}$$

Taking the dual of this condition, we obtain

$$\big((\tau \circ \mathsf{R}^{-1})^{t_2}\big)^\star \circ (\mathrm{id} \otimes \mu^\star)^\star \circ \big((\mathsf{R} \circ \tau)^{t_2}\big)^\star \circ \Big(\big(\mathrm{id} \otimes \mu^\star\big)^{-1}\Big)^\star = \mathrm{id}_{V^\star} \otimes \mathrm{id}_V$$

and then, using Lemma B.12, gives

$$(\tau \circ \mathsf{R}^{-1})^{t_2 t_1 t_2} \circ (\mathrm{id} \otimes \mu) \circ (\mathsf{R} \circ \tau)^{t_2 t_1 t_2} \circ (\mathrm{id} \otimes \mu)^{-1} = \mathrm{id}_{V^\star} \circ \mathrm{id}_V.$$

We therefore conclude from Lemma B.14 that $\mathsf{Y} \circ \mathsf{T} = \mathrm{id}_W \otimes \mathrm{id}_V$ if and only if

$$(\tau \circ \mathsf{R}^{-1})^{t_1} \circ (\mathrm{id} \otimes \mu) \circ (\mathsf{R} \circ \tau)^{t_1} \circ (\mathrm{id} \otimes \mu)^{-1} = \mathrm{id}_{V^\star} \circ \mathrm{id}_V.$$

A similar analysis shows that $\mathsf{T} \circ \mathsf{Y} = \mathrm{id}_V \otimes \mathrm{id}_W$ if and only if

$$(\tau \circ \mathsf{R}^{-1})^{t_1} \circ (\mathrm{id} \otimes \mu) \circ (\mathsf{R} \circ \tau)^{t_1} \circ (\mathrm{id} \otimes \mu)^{-1} = \mathrm{id}_{V^\star} \circ \mathrm{id}_V.$$

This establishes the result, conditional on the truth of Claims 1 and 2, which is proved next.

<u>*Part 2*</u>:

Proof of Claim 1. By definition,

$$\mathsf{Y} := \left(\mathrm{id}_V \otimes \mathrm{id}_W \otimes (\alpha^{-1})^{\cap}\right) \circ (\mathrm{id}_W \otimes \mathsf{R} \otimes \mathrm{id}_W) \circ \left(\alpha^{\cup} \otimes \mathrm{id}_V \otimes \mathrm{id}_W\right).$$

Determining the element-wise action of Y on $v_i \otimes w_j$, we have

$$\begin{aligned}
v_i \otimes v_j &\overset{\alpha^{\cup}\otimes \mathrm{id}_V \otimes \mathrm{id}_W}{\longmapsto} \sum_{k,l} \alpha^{kl}\, w_k \otimes v_l \otimes v_i \otimes w_j && \text{(Lem. B.4(5))}\\
&\overset{\mathrm{id}_W \otimes \mathsf{R} \otimes \mathrm{id}_W}{\longmapsto} \sum_{k,l,m,n} \alpha^{kl}\, R_{li}^{mn}\, w_k \otimes v_m \otimes v_n \otimes w_j && \text{(from (B.1))}\\
&\overset{\mathrm{id}_V \otimes \mathrm{id}_W \otimes (\alpha^{-1})^{\cap}}{\longmapsto} \sum_{k,l,m,n} \alpha^{kl}\, R_{li}^{mn}\, \overline{\alpha}_{nj}\, w_k \otimes v_m && \text{(Lem. B.4(7))}.
\end{aligned}$$

To determine the element-wise action of $(\alpha^\star \otimes \mathrm{id}) \circ \tau \circ (\mathsf{R} \circ \tau)^{t_2} \circ (\mathrm{id} \otimes \alpha^{-1})^\star$, the right-hand side of Claim 1, it is first necessary to determine the element-wise action of $(\mathsf{R} \circ \tau)^{t_2}$. This is straightforward since, from (B.1),

$$(\mathsf{R} \circ \tau)(v_i \otimes v_j) = \mathsf{R}(v_j \otimes v_i) = \sum_{k,l} R_{ji}^{kl}\, v_k \otimes v_l,$$

so, from Lemma B.10,

$$(\mathsf{R} \circ \tau)^{t_2} : v_i \otimes v^l \mapsto \sum_{k,j} R_{ji}^{kl}\, v_k \otimes v^j.$$

Thus

$$
\begin{aligned}
v_i \otimes w_j &\overset{(\mathrm{id}_V \otimes \alpha^{-1})^\star}{\longmapsto} \sum_k \overline{\alpha}_{kj}\, v_i \otimes v^k \\
&\overset{(\mathsf{R}\circ\tau)^{t_2}}{\longmapsto} \sum_{k,l,m} \overline{\alpha}_{kj}\, R^{mk}_{li}\, v_m \otimes v^l && \text{(Lem. B.10)} \\
&\overset{\tau}{\longmapsto} \sum_{k,l,m} \overline{\alpha}_{kj}\, R^{mk}_{li}\, v^l \otimes v_m \\
&\overset{\alpha^\star \otimes \mathrm{id}_V}{\longmapsto} \sum_{k,l,m,n} \overline{\alpha}_{kj}\, R^{mk}_{li}\, \alpha^{nl}\, w_n \otimes v_m && \text{(Lem. B.4(1))} \\
&= \sum_{k,l,m,n} \overline{\alpha}_{nj}\, R^{mn}_{li}\, \alpha^{ki}\, w_k \otimes v_m,
\end{aligned}
$$

where the last step is by reindexing through $k \mapsto n$, $n \mapsto k$. Thus, the two maps have the same element-wise actions and are therefore equal. This completes the proof of Claim 1.

Proof of Claim 2. Since the proof of Claim 2 is very similar to the proof of Claim 1, we leave it as an exercise. The completion of the exercise completes the proof of the Lemma. □

Exercise B.16 Complete the proof of Lemma B.15 by proving Claim 2 in its proof.

The Proof of Theorem 7.19

Proof (Theorem 7.19*).* In view of the previous three lemmas, all that remains to be shown is that

$$\mathsf{R} \circ \mathsf{R}^{-1} = \mathrm{id}_V \otimes \mathrm{id}_V = \mathsf{R}^{-1} \circ \mathsf{R}, \tag{B.25}$$

which comes from the T_4-move, and that

$$(\mathrm{id}_V \otimes \mathsf{R}) \circ (\mathsf{R} \otimes \mathrm{id}) \circ (\mathrm{id} \otimes \mathsf{R}) = (\mathsf{R} \otimes \mathrm{id}) \circ (\mathrm{id} \otimes \mathsf{R}) \circ (\mathsf{R} \circ \mathrm{id}), \tag{B.26}$$

which comes from the T_5-move and is the Yang–Baxter Equation (see Definition 5.4). But (B.25) holds since R is invertible and (B.26) holds by definition since R is an R-matrix. □

Appendix C
Computations in Quasi-triangular Hopf Algebras

This Appendix contains proofs of some results stated in Chap. 8. The proofs serve as good practice for working with algebras and, in particular, ribbon Hopf algebras.

The Following Result Appears Earlier in the Text as Proposition 8.31

Proposition C.1 *Let* $(\mathfrak{A}, m, \Delta, \varepsilon, \eta, S, \mathsf{R})$ *be a quasi-triangular Hopf algebra. Then*

1. $(\varepsilon \otimes \mathrm{id}_{\mathfrak{A}})(\mathsf{R}) = 1_{\mathfrak{A}} = (\mathrm{id}_{\mathfrak{A}} \otimes \varepsilon)(\mathsf{R})$,
2. $(S \otimes \mathrm{id}_{\mathfrak{A}})(\mathsf{R}) = \mathsf{R}^{-1} = (\mathrm{id}_{\mathfrak{A}} \otimes S^{-1})(\mathsf{R})$,
3. $(S \otimes S)(\mathsf{R}) = \mathsf{R}$.

Proof. Part 1 was proved after the statement of Proposition 8.31.

For Part 2, since $\mathfrak{A}$ is a Hopf algebra, it follows from Definition 8.23 that

$$m \circ (S \otimes \mathrm{id}_{\mathfrak{A}}) \circ \Delta = \eta \circ \varepsilon = m \circ (\mathrm{id}_{\mathfrak{A}} \otimes S) \circ \Delta.$$

Using Part 1,

$$(m \otimes \mathrm{id}_{\mathfrak{A}}) \circ (S \otimes \mathrm{id}_{\mathfrak{A}} \otimes \mathrm{id}_{\mathfrak{A}}) \circ (\Delta \otimes \mathrm{id}_{\mathfrak{A}})(\mathsf{R}) = ((\eta \circ \varepsilon) \otimes \mathrm{id}_{\mathfrak{A}})(\mathsf{R}) = 1_{\mathfrak{A}\otimes\mathfrak{A}}.$$

Then, by Definition 8.29(2),

$$(m \otimes \mathrm{id}_{\mathfrak{A}}) \circ (S \otimes \mathrm{id}_{\mathfrak{A}} \otimes \mathrm{id}_{\mathfrak{A}})(\mathsf{R}_{13} \cdot \mathsf{R}_{23}) = 1_{\mathfrak{A}\otimes\mathfrak{A}}.$$

Applying the operators on the left-hand side to $\mathsf{R}_{13} \cdot \mathsf{R}_{23}$, we have

D. M. Jackson and I. Moffatt, *An Introduction to Quantum and Vassiliev Knot Invariants*, CMS Books in Mathematics, https://doi.org/10.1007/978-3-030-05213-3

$$\begin{aligned}
\mathsf{R}_{13} \cdot \mathsf{R}_{23} &= \sum_{i,j} \alpha_i \otimes \alpha_j \otimes (\beta_i \cdot \beta_j) \overset{S \otimes \mathrm{id}_{\mathfrak{A}} \otimes \mathrm{id}_{\mathfrak{A}}}{\longmapsto} \sum_{i,j} S(\alpha_i) \otimes \alpha_j \otimes (\beta_i \cdot \beta_j) \\
&\overset{m \otimes \mathrm{id}_{\mathfrak{A}}}{\longmapsto} \sum_{i,j} S(\alpha_i) \cdot \alpha_j \otimes (\beta_i \cdot \beta_j) = \left(\sum_i S(\alpha_i) \otimes \beta_i \right) \cdot \left(\sum_j \alpha_j \otimes \beta_j \right) \\
&= (S \otimes \mathrm{id}_{\mathfrak{A}})(\mathsf{R}) \cdot \mathsf{R},
\end{aligned}$$

so

$$1_{\mathfrak{A} \otimes \mathfrak{A}} = (m \otimes \mathrm{id}_{\mathfrak{A}}) \circ (S \otimes \mathrm{id}_{\mathfrak{A}} \otimes \mathrm{id}_{\mathfrak{A}}) \circ (\Delta \otimes \mathrm{id}_{\mathfrak{A}})(\mathsf{R}) = (S \otimes \mathrm{id}_{\mathfrak{A}})(\mathsf{R}) \cdot \mathsf{R}.$$

Therefore $(S \otimes \mathrm{id}_{\mathfrak{A}})(\mathsf{R}) = \mathsf{R}^{-1}$, giving the left-hand side of Part (2).

For the right-hand side of Part 2, we use the opposite algebra from Exercise 8.27,

$$\mathfrak{A}^{\mathrm{op}} := (\mathfrak{A}, m^{\mathrm{op}}, \Delta, \varepsilon, \eta, S^{-1}).$$

Since $(\mathfrak{A}, \mathsf{R})$ is a quasi-triangular Hopf algebra, then so is $(\mathfrak{A}^{\mathrm{op}}, \mathsf{R}^{21})$ where $\mathsf{R}^{21} = \tau(\mathsf{R})$. Moreover, since we have proved that $(S \otimes \mathrm{id}_{\mathfrak{A}})(\mathsf{R}) = \mathsf{R}^{-1}$ in $(\mathfrak{A}, \mathsf{R})$, it follows, by applying the identity to $\mathfrak{A}^{\mathrm{op}}$, that

$$m^{\mathrm{op}} \left((S^{-1} \otimes \mathrm{id}_{\mathfrak{A}^{\mathrm{op}}})(\mathsf{R}^{21}), \mathsf{R}^{21} \right) = 1_{\mathfrak{A}^{\mathrm{op}} \otimes \mathfrak{A}^{\mathrm{op}}}.$$

Then

$$\begin{aligned}
m^{\mathrm{op}} \left((S^{-1} \otimes \mathrm{id}_{\mathfrak{A}^{\mathrm{op}}})(\mathsf{R}^{21}), \mathsf{R}^{21} \right) &= m \left(\mathsf{R}^{21}, (S^{-1} \otimes \mathrm{id}_{\mathfrak{A}^{\mathrm{op}}})(\mathsf{R}^{21}) \right) \\
&= m \left(\tau(\mathsf{R}), (S^{-1} \otimes \mathrm{id}_{\mathfrak{A}^{\mathrm{op}}}) \circ \tau(\mathsf{R}) \right) \\
&= \left(\sum_i \beta_i \otimes \alpha_i \right) \cdot \left(\sum_j S^{-1}(\beta_j) \otimes \alpha_j \right) \\
&= \sum_{i,j} \left(\beta_i \cdot S^{-1}(\beta_j) \right) \otimes (\alpha_i \cdot \alpha_j).
\end{aligned}$$

But

$$\begin{aligned}
\sum_{i,j} \left(\beta_i \cdot S^{-1}(\beta_j) \right) \otimes (\alpha_i \cdot \alpha_j) &= \tau \left(\sum_{i,j} (\alpha_i \cdot \alpha_j) \otimes \left(\beta_i \cdot S^{-1}(\beta_j) \right) \right) \\
&= \tau \left(\sum_{i,j} (\alpha_i \otimes \beta_i) \cdot \left(\alpha_j \otimes S^{-1}(\beta_j) \right) \right)
\end{aligned}$$

$$= \tau\left(\left(\sum_i \alpha_i \otimes \beta_i\right) \cdot \left(\sum_j \alpha_j \otimes S^{-1}(\beta_j)\right)\right)$$
$$= \tau\left(\mathsf{R} \cdot (\mathrm{id}_{\mathfrak{A}^{\mathrm{op}}} \otimes S^{-1})(\mathsf{R})\right).$$

Therefore,

$$\tau\left(\mathsf{R} \cdot (\mathrm{id}_{\mathfrak{A}^{\mathrm{op}}} \otimes S^{-1})(\mathsf{R})\right) = 1_{\mathfrak{A}^{\mathrm{op}} \otimes \mathfrak{A}^{\mathrm{op}}}.$$

Applying τ to this identity gives

$$\mathsf{R} \cdot (\mathrm{id}_{\mathfrak{A}^{\mathrm{op}}} \otimes S^{-1})(\mathsf{R}) = \tau(1_{\mathfrak{A}^{\mathrm{op}} \otimes \mathfrak{A}^{\mathrm{op}}}) = \tau(1_{\mathfrak{A}^{\mathrm{op}}} \otimes 1_{\mathfrak{A}^{\mathrm{op}}}) = 1_{\mathfrak{A}^{\mathrm{op}}} \otimes 1_{\mathfrak{A}^{\mathrm{op}}} = 1_{\mathfrak{A}^{\mathrm{op}} \otimes \mathfrak{A}^{\mathrm{op}}}.$$

By Exercise 8.3, $1_{\mathfrak{A}^{\mathrm{op}}} = 1_{\mathfrak{A}}$, and Part 2 follows. |

For Part 3, from Part 2 we have

$$(S \otimes \mathrm{id}_{\mathfrak{A}})(\mathsf{R}) = (\mathrm{id}_{\mathfrak{A}} \otimes S^{-1})(\mathsf{R}).$$

Then $(\mathrm{id}_{\mathfrak{A}} \otimes S) \circ (S \otimes \mathrm{id}_{\mathfrak{A}})(\mathsf{R}) = \mathsf{R}$ so $(S \otimes S)(\mathsf{R}) = \mathsf{R}$, giving Part (3). |

The result therefore follows. □

The Following Result Appears Earlier in the Text as Lemma 8.35

Recall that in a quasi-triangular Hopf algebra the element $\mathbf{u}$ is defined by

$$\mathbf{u} := \sum_i S(\beta_i) \cdot \alpha_i$$

where $\sum_i \alpha_i \otimes \beta_i = \mathsf{R}$.

Lemma C.2 (Almost-centrality). *Let $\mathfrak{A}$ be a quasi-triangular Hopf algebra. Then, the element* $\mathbf{u}$ *is* almost central. *That is,*

$$\mathbf{u} \cdot x = S^2(x) \cdot \mathbf{u}, \qquad \text{for all } x \in \mathfrak{A}.$$

Proof. For an arbitrary $x \in \mathfrak{A}$, let

$$\Delta(x) := \sum_i y_i \otimes z_i$$

and

$$(\Delta \otimes \mathrm{id}_{\mathfrak{A}}) \circ \Delta(x) := \sum_k a_k \otimes b_k \otimes c_k. \tag{C.1}$$

Since $\mathfrak{A}$ is a quasi-triangular Hopf algebra, for all $a \in \mathfrak{A}$ we have, from Definition 8.29(1),

$$(\tau \circ \Delta)(a) = \mathsf{R} \cdot \Delta(a) \cdot \mathsf{R}^{-1},$$

so

$$\mathsf{R} \cdot \Delta(a) = (\tau \circ \Delta(a)) \cdot \mathsf{R}. \tag{C.2}$$

Now

$$\begin{aligned}
(\mathsf{R} \otimes 1_{\mathfrak{A}}) \cdot (((\Delta \otimes \mathrm{id}_{\mathfrak{A}}) \circ \Delta)(x)) &= (\mathsf{R} \otimes 1_{\mathfrak{A}}) \cdot \left((\Delta \otimes \mathrm{id}_{\mathfrak{A}}) \left(\sum_i y_i \otimes z_i \right) \right) \\
&= (\mathsf{R} \otimes 1_{\mathfrak{A}}) \cdot \left(\sum_i \Delta(y_i) \otimes z_i \right) \\
&= \sum_i (\mathsf{R} \cdot \Delta(y_i)) \otimes z_i \\
&= \sum_i ((\tau \circ \Delta(y_i)) \cdot \mathsf{R}) \otimes z_i \qquad \text{(from (C.2))} \\
&= \sum_i ((\tau \circ \Delta(y_i)) \otimes z_i) \cdot (\mathsf{R} \otimes 1_{\mathfrak{A}}) \\
&= \left((\tau \otimes \mathrm{id}_{\mathfrak{A}}) \left(\sum_i \Delta(y_i) \otimes z_i \right) \right) \cdot (\mathsf{R} \otimes 1_{\mathfrak{A}}) \\
&= (\tau \otimes \mathrm{id}_{\mathfrak{A}}) \left((\Delta \otimes \mathrm{id}_{\mathfrak{A}}) \circ \Delta(x) \right) \cdot (\mathsf{R} \otimes 1_{\mathfrak{A}})
\end{aligned}$$

so

$$(\mathsf{R} \otimes 1_{\mathfrak{A}}) \cdot (((\Delta \otimes \mathrm{id}_{\mathfrak{A}}) \circ \Delta)(x)) = (\tau \otimes \mathrm{id}_{\mathfrak{A}}) \left((\Delta \otimes \mathrm{id}_{\mathfrak{A}}) \circ \Delta(x) \right) \cdot (\mathsf{R} \otimes 1_{\mathfrak{A}}).$$

Thus, using (C.1),

$$\sum_{i,k} (\alpha_i \cdot a_k) \otimes (\beta_i \cdot b_k) \otimes c_k = \sum_{i,k} (b_k \cdot \alpha_i) \otimes (a_k \cdot \beta_i) \otimes c_k.$$

By applying

$$m \circ (m \otimes \mathrm{id}_{\mathfrak{A}}) \circ \left(S^2 \otimes S \otimes \mathrm{id}_{\mathfrak{A}} \right) \circ \tau_{13},$$

where τ_{13} switches the first and third tensor factors, to each side, we obtain

$$\sum_{i,k} S^2(c_k)\cdot S(\beta_i\cdot b_k)\cdot(\alpha_i\cdot a_k)=\sum_{i,k} S^2(c_k)\cdot S(a_k\cdot\beta_i)\cdot(b_k\cdot\alpha_i)$$

so rewriting and remembering that S is an anti-homomorphism,

$$\sum_{i,k} S\left(b_k S(c_k)\right)\cdot S(\beta_i)\cdot(\alpha_i\cdot a_k)=\sum_{i,k} S^2(c_k)\cdot S(\beta_i)\cdot S(a_k)\cdot(b_k\cdot\alpha_i) \tag{C.3}$$

We now show that the left- and right-hand sides of (C.3) equal $\mathbf{u}\cdot x$ and $S^2(x)\cdot\mathbf{u}$, respectively.
For the right-hand side of (C.3) From the definition of the antipode S of the Hopf algebra $(\mathfrak{A}, m, \Delta, \varepsilon, \eta, S)$ we have

$$m\circ(S\otimes \mathrm{id}_{\mathfrak{A}})\circ\Delta=\eta\circ\varepsilon=m\circ(\mathrm{id}_{\mathfrak{A}}\otimes S)\circ\Delta. \tag{C.4}$$

Therefore, tensoring on the right of each side with $\mathrm{id}_{\mathfrak{A}}$ and applying them to $\Delta(x)$ gives

$$(m\otimes \mathrm{id}_{\mathfrak{A}})\circ(S\otimes \mathrm{id}_{\mathfrak{A}}\otimes \mathrm{id}_{\mathfrak{A}})\otimes(\Delta\otimes \mathrm{id}_{\mathfrak{A}})(\Delta(x))=((\eta\circ\varepsilon)\otimes \mathrm{id}_{\mathfrak{A}})(\Delta(x)). \tag{C.5}$$

It is immediate by direct calculation that, in the notation of (C.1), the left-hand side of (C.5) is equal to

$$\sum_k (S(a_k)b_k)\otimes c_k.$$

Since ε is a counit,

$$(\varepsilon\otimes \mathrm{id}_{\mathfrak{A}})(\Delta(x))=1_{\mathbb{K}}\otimes x\quad\text{and}\quad(\mathrm{id}_{\mathfrak{A}}\otimes\varepsilon)(\Delta(x))=x\otimes 1_{\mathbb{K}}, \tag{C.6}$$

so the right-hand side of (C.5) is

$$((\eta\circ\varepsilon)\otimes \mathrm{id}_{\mathfrak{A}})\Delta(x)=(\eta\otimes \mathrm{id}_{\mathfrak{A}})((\varepsilon\otimes \mathrm{id}_{\mathfrak{A}})(\Delta(x)))\;=(\eta\otimes \mathrm{id}_{\mathfrak{A}})(1_{\mathbb{K}}\otimes x)=1_{\mathfrak{A}}\otimes x.$$

Thus, from (C.5),

$$\sum_k (S(a_k)b_k)\otimes c_k=1_{\mathfrak{A}}\otimes x$$

and, on applying $(S^2\otimes \mathrm{id}_{\mathfrak{A}})\circ\tau$ to each side, we obtain

$$\sum_k S^2(c_k)\otimes S(a_k)b_k=S^2(x)\otimes 1_{\mathfrak{A}}.$$

Then, multiplying on the right throughout by $\sum_i S(\beta_i) \otimes \alpha_i$ gives

$$\sum_{k,i} \left(S^2(c_k)\, S(\beta_i)\right) \otimes (S(a_k) b_k \alpha_i) = \sum_i \left(S^2(x)\, S(\beta_i)\right) \otimes \alpha_i .$$

Applying m to each side gives

$$\sum_{k,i} \left(S^2(c_k)\, S(\beta_i)\right) \cdot (S(a_k) b_k \alpha_i) = \sum_i S^2(x)\, S(\beta_i)\, \alpha_i = S^2(x) \cdot \mathbf{u}.$$

Therefore, the right-hand side of (C.3) is equal to $S^2(x) \cdot \mathbf{u}$.
For the left-hand side of (C.3) Tensoring the property (C.4) of the antipode S, on the left by $\mathrm{id}_{\mathfrak{A}}$, and then applying each side to $\Delta(x)$, we have

$$\begin{aligned}((\mathrm{id}_{\mathfrak{A}} \otimes m) \circ (\mathrm{id}_{\mathfrak{A}} \otimes \mathrm{id}_{\mathfrak{A}} \otimes S) \circ (\mathrm{id}_{\mathfrak{A}} \otimes \Delta))\,(\Delta(x)) &= (\mathrm{id}_{\mathfrak{A}} \otimes (\eta \circ \varepsilon))(\Delta(x)) \\ &= ((\mathrm{id}_{\mathfrak{A}} \otimes \eta) \circ (\mathrm{id}_{\mathfrak{A}} \otimes \varepsilon))\,(\Delta(x)) \\ &= (\mathrm{id}_{\mathfrak{A}} \otimes \eta)(x \otimes 1_{\mathbb{K}}) \quad \text{(by (C.6))} \\ &= x \otimes 1_{\mathfrak{A}} .\end{aligned}$$

But, by direct computation with (C.1), we have

$$((\mathrm{id}_{\mathfrak{A}} \otimes m) \circ (\mathrm{id}_{\mathfrak{A}} \otimes \mathrm{id}_{\mathfrak{A}} \otimes S) \circ (\mathrm{id}_{\mathfrak{A}} \otimes \Delta))\,(\Delta(x)) = \sum_k a_k \otimes (b_k \cdot S(c_k)) ,$$

and therefore

$$\sum_k a_k \otimes (b_k \cdot S(c_k)) = x \otimes 1_{\mathfrak{A}}$$

so applying $(S \otimes \mathrm{id}_{\mathfrak{A}}) \circ \tau$ to both sides gives

$$\sum_k S(b_k\, S(c_k)) \otimes a_k = S(1_{\mathfrak{A}}) \otimes x = 1_{\mathfrak{A}} \otimes x ,$$

where we have used that since S is an anti-homomorphism, $S(1_{\mathfrak{A}}) = 1_{\mathfrak{A}}$. Multiplying on the left throughout by $1_{\mathfrak{A}} \otimes \mathbf{u}$, we have

$$\sum_k S(b_k\, S(c_k)) \otimes \mathbf{u}\, a_k = 1_{\mathfrak{A}} \otimes (\mathbf{u} \cdot x)$$

and then applying m to each side and expanding $\mathbf{u}$, we obtain

$$\sum_{i,k} S(b_k S(c_k))\, S(\beta_i)\, \alpha_i\, a_k = \mathbf{u} \cdot x .$$

Therefore the left-hand side of (C.3) is equal to $\mathbf{u} \cdot x$. We therefore conclude from (C.3) that $S^2(x) \cdot \mathbf{u} = \mathbf{u} \cdot x$, completing the proof. □

The Following Result Appears Earlier in the Text as Proposition 8.37

Proposition C.3 *Let $\mathfrak{A}$ be a quasi-triangular Hopf algebra, and let $\mathbf{u}$ be as given in Definition 8.34. Then*

1. $S^2(\mathbf{u}) = \mathbf{u}$,
2. $\mathbf{u} \cdot S(\mathbf{u}) = S(\mathbf{u}) \cdot \mathbf{u}$,
3. $\mathbf{u}\, S(\mathbf{u})$ *is in the centre of* $\mathfrak{A}$,
4. $\Delta(\mathbf{u}) = (\tau(\mathsf{R}) \cdot \mathsf{R})^{-1} \cdot (\mathbf{u} \otimes \mathbf{u}) = (\mathbf{u} \otimes \mathbf{u}) \cdot (\tau(\mathsf{R}) \cdot \mathsf{R})^{-1}$,
5. $\mathbf{u}^{-1} = \sum_i \beta_i\, S^2(\alpha_i)$,
6. $\Delta(S(\mathbf{u})) = (\tau(\mathsf{R}) \cdot \mathsf{R})^{-1} \cdot (S(\mathbf{u}) \otimes S(\mathbf{u})) = (S(\mathbf{u}) \otimes S(\mathbf{u})) \cdot (\tau(\mathsf{R}) \cdot \mathsf{R})^{-1}$,
7. $\varepsilon(\mathbf{u}) = 1_{\mathbb{K}}$.

Proof. We consider the items in turn.

For Part 1. From Lemma 8.35, we know that $\mathbf{u} \cdot x = S^2(x) \cdot \mathbf{u}$ for all $x \in \mathfrak{A}$. But $\mathbf{u}$ is invertible from Lemma 8.36, so the result follows by setting $x = \mathbf{u}$. |

For Part 2. From Lemma 8.35, we know that $\mathbf{u} \cdot x = S^2(x) \cdot \mathbf{u}$ for all $x \in \mathfrak{A}$. Then, applying S and recalling that S is an anti-homomorphism gives

$$S(\mathbf{u}) \cdot S^3(x) = S(x) \cdot S(\mathbf{u}). \tag{C.7}$$

The result follows from Part (1) by setting $x = S^{-1}(\mathbf{u})$. |

For Part 3. Let $a \in \mathfrak{A}$. Then, setting $x = S^{-1}(a)$ in (C.7), we have

$$S(\mathbf{u}) \cdot S^2(a) = a \cdot S(\mathbf{u}).$$

But $S^2(a) = \mathbf{u}\, a\, \mathbf{u}^{-1}$ from Lemma 8.36 so $S(\mathbf{u}) \cdot \mathbf{u}\, a\, \mathbf{u}^{-1} = a \cdot S(\mathbf{u})$ and then $(S(\mathbf{u})\, \mathbf{u}) \cdot a = a \cdot (S(\mathbf{u})\, \mathbf{u})$ for all $a \in \mathfrak{A}$. The result follows. |

For Part 4. Our proof follows that from [90]. We have

$$\begin{aligned}
\Delta \otimes \Delta(\mathsf{R}) &= (\Delta \otimes \mathrm{id}_{\mathfrak{A}} \otimes \mathrm{id}_{\mathfrak{A}})(\mathrm{id}_{\mathfrak{A}} \otimes \Delta(\mathsf{R})) \\
&= (\Delta \otimes \mathrm{id}_{\mathfrak{A}} \otimes \mathrm{id}_{\mathfrak{A}})\mathsf{R}_{13}\mathsf{R}_{12} && \text{(Def. 8.29(3))} \\
&= (\Delta \otimes \mathrm{id}_{\mathfrak{A}} \otimes \mathrm{id}_{\mathfrak{A}})(\mathsf{R}_{13}) \cdot (\Delta \otimes \mathrm{id}_{\mathfrak{A}} \otimes \mathrm{id}_{\mathfrak{A}})(\mathsf{R}_{12}) \\
&= (\mathsf{R}_{14}\mathsf{R}_{24})(\mathsf{R}_{13}\mathsf{R}_{23}) && \text{(Def. 8.29(2))},
\end{aligned}$$

Then, writing $\mathsf{R} = \sum_i \alpha_i \otimes \beta_i$ gives

$$\sum_i \Delta(\alpha_i) \otimes \Delta(\beta_i) = \mathsf{R}_{14}\mathsf{R}_{24}\mathsf{R}_{13}\mathsf{R}_{23}$$
$$= \sum_{j,k,l,m} (\alpha_j\alpha_l \otimes \alpha_k\alpha_m) \otimes (\beta_l\beta_m \otimes \beta_j\beta_k). \qquad \text{(C.8)}$$

Next, writing R_{21} for $\tau(\mathsf{R})$,

$$\begin{aligned}\Delta(\mathbf{u})\tau(\mathsf{R})\mathsf{R} &= \Delta(\mathbf{u})\mathsf{R}_{21}\mathsf{R}\\ &= \sum_i \Delta\left(S(\beta_i)\alpha_i\right) \cdot \mathsf{R}_{21} \cdot \mathsf{R}\\ &= \sum_i \Delta\left(S(\beta_i)\right) \cdot \Delta\left(\alpha_i\right) \cdot \mathsf{R}_{21} \cdot \mathsf{R}\\ &= \sum_i ((S \otimes S) \circ \tau(\Delta(\beta_i))) \cdot \Delta\left(\alpha_i\right) \cdot \mathsf{R}_{21} \cdot \mathsf{R} \qquad \text{(Eq. (8.9)).}\end{aligned}$$

Using Definition 8.29(1) we have, for all $a \in \mathfrak{A}$, that $\tau \circ \Delta(a) = \mathsf{R} \cdot \Delta(a) \cdot \mathsf{R}^{-1}$ and so,

$$\Delta(a) = \mathsf{R}_{21} \cdot (\tau \circ \Delta(a)) \cdot \mathsf{R}_{21}^{-1}.$$

Thus, from the above,

$$\begin{aligned}\Delta(\mathbf{u})\tau(\mathsf{R})\mathsf{R} &= \sum_i ((S \otimes S) \circ \tau \circ \Delta)(\beta_i) \cdot \mathsf{R}_{21} \cdot \tau(\Delta(\alpha_i)) \cdot \mathsf{R}\\ &= \sum_i ((S \otimes S) \circ \tau \circ \Delta)(\beta_i) \cdot \mathsf{R}_{21} \cdot \mathsf{R} \cdot \Delta(\alpha_i) \qquad \text{(Def. 8.29(1))}\\ &= \sum_{j,k,l,m,p} \big((S \otimes S)(\beta_j\beta_k \otimes \beta_l\beta_m)\big) \cdot \mathsf{R}_{21} \cdot \big(\alpha_p\alpha_j\alpha_l \otimes \beta_p\alpha_k\alpha_m\big) \quad \text{(Eq. (C.8)).}\end{aligned}$$

We can define an action of $\mathfrak{A}^{\otimes 4}$ on $\mathfrak{A} \otimes \mathfrak{A}$ through an operation $*$ by

$$(x \otimes y) * (X \otimes Y) := (S \otimes S)(Y) \cdot (x \otimes y) \cdot X.$$

Observe that, since S is an anti-homomorphism,

$$(x \otimes y) * (X \otimes Y) * (A \otimes B) = (x \otimes y) * (XA \otimes YB).$$

By expanding the R's in terms of α's and β's, and then applying Lemma 8.32, we see that

$$\begin{aligned}
\Delta(\mathbf{u})\tau(\mathsf{R})\mathsf{R} &= \sum_{j,k,l,m,p} \big((S\otimes S)(\beta_j\beta_k\otimes\beta_l\beta_m)\big)\cdot \mathsf{R}_{21}\cdot\big(\alpha_p\alpha_j\alpha_l\otimes\beta_p\alpha_k\alpha_m\big)\\
&= \mathsf{R}_{21} * (\mathsf{R}_{12}\mathsf{R}_{13}\mathsf{R}_{23}\mathsf{R}_{14}\mathsf{R}_{24})\\
&= \mathsf{R}_{21} * (\mathsf{R}_{23}\mathsf{R}_{13}\mathsf{R}_{12}\mathsf{R}_{14}\mathsf{R}_{24}).
\end{aligned}$$

Now,

$$\begin{aligned}
\mathsf{R}_{21} * \mathsf{R}_{23} &= \sum_{ij} S(\beta_j)\beta_i\otimes\alpha_i\alpha_j\\
&= (S\otimes \mathrm{id}_{\mathfrak{A}})\left(\sum_{ij} S^{-1}(\beta_i)\beta_j\otimes\alpha_i\alpha_j\right)\\
&= (S\otimes \mathrm{id}_{\mathfrak{A}})\left(\mathsf{R}_{21}^{-1}R_{21}\right) && \text{(Prop. 8.31(2))}\\
&= (S\otimes \mathrm{id}_{\mathfrak{A}})(1_{\mathfrak{A}}\otimes 1_{\mathfrak{A}})\\
&= 1_{\mathfrak{A}}\otimes 1_{\mathfrak{A}}.
\end{aligned}$$

Then

$$\mathsf{R}_{21} * (\mathsf{R}_{23}\mathsf{R}_{13}) = (\mathsf{R}_{21} * \mathsf{R}_{23}) * \mathsf{R}_{13} = (1_{\mathfrak{A}}\otimes 1_{\mathfrak{A}}) * \mathsf{R}_{13} = \sum_i S(\beta_i)\alpha_i\otimes 1_{\mathfrak{A}} = \mathbf{u}\otimes 1_{\mathfrak{A}},$$

and

$$\mathsf{R}_{21} * (\mathsf{R}_{23}\mathsf{R}_{13}\mathsf{R}_{12}) = (\mathbf{u}\otimes 1_{\mathfrak{A}}) * \mathsf{R}_{12} = (\mathbf{u}\otimes 1_{\mathfrak{A}})\cdot \mathsf{R}.$$

Then

$$\begin{aligned}
\mathsf{R}_{21} * (\mathsf{R}_{23}\mathsf{R}_{13}\mathsf{R}_{12}\mathsf{R}_{14}) &= ((\mathbf{u}\otimes 1_{\mathfrak{A}})\cdot\mathsf{R}) * \mathsf{R}_{14}\\
&= \sum_i ((S\otimes S)(1_{\mathfrak{A}}\otimes\beta_i))\cdot(\mathbf{u}\otimes 1_{\mathfrak{A}})\cdot\mathsf{R}\cdot(\alpha_i\otimes 1_{\mathfrak{A}})\\
&= \sum_{i,j} ((1_{\mathfrak{A}}\otimes S(\beta_i)\cdot(\mathbf{u}\otimes 1_{\mathfrak{A}})\cdot(\alpha_j\otimes\beta_j)\cdot(\alpha_i\otimes 1_{\mathfrak{A}})\\
&= (\mathbf{u}\otimes 1_{\mathfrak{A}})\cdot\left(\sum_{i,j}(\alpha_j\alpha_i\otimes S(\beta_i)\beta_j)\right)\\
&= (\mathbf{u}\otimes 1_{\mathfrak{A}})\cdot(\mathrm{id}_{\mathfrak{A}}\otimes S)\left(\sum_{i,j}\alpha_j\alpha_i\otimes S^{-1}(\beta_j)\beta_i\right)\\
&= (\mathbf{u}\otimes 1_{\mathfrak{A}})\cdot(\mathrm{id}_{\mathfrak{A}}\otimes S)(\mathsf{R}^{-1}\mathsf{R})) && \text{(Prop. 8.31(2))}\\
&= (\mathbf{u}\otimes 1_{\mathfrak{A}}).
\end{aligned}$$

Finally,

$$\begin{aligned}\mathsf{R}_{21} * (\mathsf{R}_{23}\mathsf{R}_{13}\mathsf{R}_{12}\mathsf{R}_{14}\mathsf{R}_{24}) &= (\mathbf{u} \otimes 1_{\mathfrak{A}}) * \mathsf{R}_{24} \\ &= \sum_i (1_{\mathfrak{A}} \otimes S(\beta_i)) \cdot (\mathbf{u} \otimes 1_{\mathfrak{A}}) \cdot (1_{\mathfrak{A}} \otimes \alpha_i) \\ &= (\mathbf{u} \otimes 1_{\mathfrak{A}}) \cdot (1_{\mathfrak{A}} \otimes \mathbf{u}) \\ &= \mathbf{u} \otimes \mathbf{u},\end{aligned}$$

showing that $\Delta(\mathbf{u}) \cdot \tau(\mathsf{R}) \cdot \mathsf{R} = \mathbf{u} \otimes \mathbf{u}$, giving one of the identities of the claim.

For the other identity, Definition 8.29(1) gives

$$\tau \circ \Delta(\mathbf{u}) = \mathsf{R} \cdot \Delta(\mathbf{u}) \cdot \mathsf{R}^{-1}.$$

So

$$\Delta(\mathbf{u}) = \mathsf{R}_{21} \cdot (\tau \circ \Delta(\mathbf{u})) \cdot (\mathsf{R}_{21}^{-1}).$$

Then

$$\Delta(\mathbf{u}) = \mathsf{R}_{21} \cdot \mathsf{R} \cdot (\Delta(\mathbf{u})) \cdot \mathsf{R}^{-1} \cdot (\mathsf{R}_{21}^{-1}),$$

and, since $\Delta(\mathbf{u}) = (\mathbf{u} \otimes \mathbf{u}) \cdot (\tau(\mathsf{R}) \cdot \mathsf{R})^{-1}$, it follows that

$$(\mathbf{u} \otimes \mathbf{u}) = \tau(\mathsf{R}) \cdot \mathsf{R} \cdot \Delta(\mathbf{u}),$$

completing the proof of this item. |

For Part 5. We have

$$\begin{aligned}\mathbf{u}^{-1} &= \sum_j S^{-1}(\overline{\beta}_j) \cdot \overline{\alpha}_j && \text{(Prop. 8.36)} \\ &= m \circ \tau \circ (\mathrm{id}_{\mathfrak{A}} \otimes S^{-1})(\mathsf{R}^{-1}) \\ &= m \circ \tau \circ (\mathrm{id}_{\mathfrak{A}} \otimes S^{-1}) \circ (\mathrm{id}_{\mathfrak{A}} \otimes S^{-1})(\mathsf{R}) && \text{(Prop. 8.31(2))} \\ &= \sum_i S^{-2}(\beta_i) \cdot \alpha_i\end{aligned}$$

Using that $S^2(\mathbf{u}) = \mathbf{u}$, the first item of this proposition, and that S is an anti-homomorphism,

$$1_{\mathfrak{A}} = S(1_{\mathfrak{A}}) = S^2(1_{\mathfrak{A}}) = S^2(\mathbf{u}\mathbf{u}^{-1}) = S^2(\mathbf{u}) \cdot S^2(\mathbf{u}^{-1}) = \mathbf{u} \cdot S^2(\mathbf{u}^{-1}).$$

Since $\mathbf{u}$ is invertible, it follows that $\mathbf{u}^{-1} = S^2(\mathbf{u}^{-1})$. Then,

$$\mathbf{u}^{-1} = S^2(\mathbf{u}^{-1}) = S^2\left(\sum_i S^{-2}(\beta_i) \cdot \alpha_i\right) = \sum_i \beta_i\, S^2(\alpha_i),$$

as required. |

For Part 6. By Eq. (8.9), we have $(S \otimes S) \circ \Delta = \tau \circ \Delta \circ S$. We also know from Part (4) that $\Delta(\mathbf{u}) = (\tau(\mathsf{R}) \cdot \mathsf{R})^{-1} \cdot (\mathbf{u} \otimes \mathbf{u}) = (\mathbf{u} \otimes \mathbf{u}) \cdot (\tau(\mathsf{R}) \cdot \mathsf{R})^{-1}$. Then

$$\begin{aligned}
\tau(\mathsf{R}) \cdot \mathsf{R} \cdot \Delta(S(\mathbf{u})) &= \tau(\mathsf{R}) \cdot \mathsf{R} \cdot (\tau \circ (S \otimes S))(\Delta(\mathbf{u})) \\
&= \tau(\mathsf{R} \cdot \tau(\mathsf{R})) \cdot (\tau \circ (S \otimes S))(\Delta(\mathbf{u})) \\
&= \tau\left(\mathsf{R} \cdot \tau(\mathsf{R}) \cdot (S \otimes S)(\Delta(\mathbf{u}))\right) \\
&= \tau\left((S \otimes S)(\mathsf{R}) \cdot \tau((S \otimes S)(\mathsf{R})) \cdot (S \otimes S)(\Delta(\mathbf{u}))\right) \quad \text{(Prop. 8.31(3))} \\
&= \tau\left((S \otimes S)(\mathsf{R}) \cdot (S \otimes S)(\tau(\mathsf{R})) \cdot (S \otimes S)(\Delta(\mathbf{u}))\right) \\
&= \tau\left((S \otimes S)(\Delta(\mathbf{u}) \cdot \tau(\mathsf{R}) \cdot \mathsf{R})\right) \quad (S \text{ is an anti-hom}) \\
&= \tau\left((S \otimes S)(\mathbf{u} \otimes \mathbf{u})\right) \quad \text{(Part (4))} \\
&= S(\mathbf{u}) \otimes S(\mathbf{u})
\end{aligned}$$

The other equality follows similarly

For Part 7. By Proposition 8.31,

$$1_{\mathfrak{A}} \otimes 1_{\mathfrak{A}} = 1_{\mathfrak{A} \otimes \mathfrak{A}} = ((\eta \circ \varepsilon) \otimes \mathrm{id}_{\mathfrak{A}})(\mathsf{R}) = \sum_i \varepsilon(\alpha_i) 1_{\mathfrak{A}} \otimes \beta_i.$$

Applying the product m to this expression gives $1_{\mathfrak{A}} = \sum_i \varepsilon(\alpha_i)\beta_i$. Since S is an anti-homomorphism and $\varepsilon(\alpha_i) \in \mathbb{K}$, $1_{\mathfrak{A}} = S(1_{\mathfrak{A}}) = \sum_i \varepsilon(\alpha_i) S(\beta_i)$. Then since ε is an algebra morphism, $1_{\mathbb{K}} = \varepsilon(1_{\mathfrak{A}}) = \sum_i \varepsilon(\alpha_i)\varepsilon(S(\beta_i))$. However,

$$\varepsilon(\mathbf{u}) = \varepsilon\left(\sum_i S(\beta_i) \cdot \alpha_i\right) = \sum_i \varepsilon(S(\beta_i))\varepsilon(\alpha_i) = \sum_i \varepsilon(\alpha_i)\varepsilon(S(\beta_i)),$$

where we have used the fact that $\varepsilon(\alpha_i) \in \mathbb{K}$ in the last equality. It follows that $\varepsilon(\mathbf{u}) = 1_{\mathbb{K}}$, as required.

This completes the proof of the Proposition. □

Appendix D
A Proof of the Invariance of the Reshetikhin–Turaev Invariants

In this appendix, we complete the proof of Theorem 9.3, which is restated below for convenience.

Theorem D.1. *The Reshetikhin–Turaev invariant, $Q^{\mathfrak{A}}$, is an invariant of framed oriented coloured tangles.*

The following technical lemma will be needed in subsequent calculations.

Lemma D.2. *Let $\{e_i\}_i$ and $\{e^i\}_i$ be bases of V and $V^\star$, where $e^i(e_j) = \delta_{i,j}$. Let $f, g \in \mathrm{End}(V)$. Then*

1. $\sum_j f(e_j) \cdot e^j\,(g(e_i)) = (f \circ g)(e_i)$,
2. $\sum_k e^i\,(f(e_k)) \cdot e^k\,(g(e_m)) = e^i\,(f \circ g(e_m))$.

Proof. Let $f(e_j) = \sum_p f_p^j\, e_p$ and $g(e_i) = \sum_q g_q^i\, e_q$.

For Part 1,

$$\sum_j f(e_j) \cdot e^j\,(g(e_i)) = \sum_{j,p} f_p^j\, e_p \cdot e^j \left(\sum_q g_q^i\, e_q \right) = \sum_{j,p} f_p^j\, e_p \cdot g_j^i$$
$$= \sum_{j,p} g_j^i\ f_p^j\, e_p = f \circ g(e_i),$$

completing the proof of Part 1.

Part 2 follows by applying e^i to each side of the first identity, completing the proof. □

Proof (Theorem 9.3 *(and Theorem* D.1*)).* We must show that $Q^{\mathfrak{A}}$ is invariant under each of the framed, oriented Turaev moves $\mathsf{FT}_0, \ldots, \mathsf{FT}_7$ given in Theorem 7.33. For each such move, the appropriate tangle diagram is sliced into elementary pieces in the usual way. To clarify certain steps in the proof, we have occasionally indicated with an underbrace sub-expressions that are scalar.

D. M. Jackson and I. Moffatt, *An Introduction to Quantum and Vassiliev Knot Invariants*, CMS Books in Mathematics, https://doi.org/10.1007/978-3-030-05213-3

For FT_0: This is trivially satisfied since $\mathrm{id}^{\otimes n} \otimes Q^{\mathfrak{A}}(T) = Q^{\mathfrak{A}}(T) = Q^{\mathfrak{A}}(T) \otimes \mathrm{id}^{\otimes n}$ and $(Q^{\mathfrak{A}}(T') \otimes \mathrm{id}^{\otimes m}) \circ (\mathrm{id}^{\otimes n} \otimes Q^{\mathfrak{A}}(T)) = Q^{\mathfrak{A}}(T') \otimes Q^{\mathfrak{A}}(T) = (\mathrm{id}^{\otimes n} \otimes Q^{\mathfrak{A}}(T)) \circ (Q^{\mathfrak{A}}(T') \otimes \mathrm{id}^{\otimes m})$.

For FT_1: The proof of invariance under FT_1 was given in Sect. 9.2 after the statement of Theorem 9.3.

For FT_2: Since the diagram  has only one component, there is only one colour. Let V be the module associated with this colour. By slicing, we have $Q^{\mathfrak{A}}\Big($  $\Big) =$ $Q^{\mathfrak{A}}\Big($  $\Big) \circ Q^{\mathfrak{A}}\Big($  $\Big)$. Then

$e^i \overset{\iota}{\longmapsto} e^i \otimes 1_{\mathbb{K}}$ $\overset{Q^{\mathfrak{A}}(\;)}{\longmapsto}$  $\sum_j e^i \otimes e_j \otimes e^j$ $\overset{Q^{\mathfrak{A}}(\;)}{\longmapsto}$  $\sum_j e^i(e_j) \otimes e^j = 1_{\mathbb{K}} \otimes e^i \overset{\kappa}{\longmapsto} e^i,$

so $Q^{\mathfrak{A}}\Big($  $\Big) = \mathrm{id}_V$.

Similarly $Q^{\mathfrak{A}}\Big($  $\Big) = Q^{\mathfrak{A}}\Big($  $\Big) \circ Q^{\mathfrak{A}}\Big($  $\Big)$. Then

$e^i \overset{\iota}{\longmapsto} 1_{\mathbb{K}} \otimes e^i$ $\overset{Q^{\mathfrak{A}}(\;)}{\longmapsto}$  $\sum_j e^j \otimes \rho^V_{\mathbf{u}^{-1}\mathbf{v}}(e_j) \otimes e^i$

$\overset{Q^{\mathfrak{A}}(\;)}{\longmapsto}$ 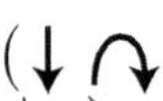 $\sum_j e^j \otimes e^i \left(\rho^V_{\mathbf{v}^{-1}\mathbf{u}} \circ \rho^V_{\mathbf{u}^{-1}\mathbf{v}}(e_j)\right)$

$$= \sum_j e^j \otimes e^i(e_j) = e^i \otimes 1_{\mathbb{K}} \overset{\kappa}{\longmapsto} e^i,$$

so $Q^{\mathfrak{A}}\Big($  $\Big) = \mathrm{id}_V$.

Thus $Q^{\mathfrak{A}}\Big($  $\Big) = Q^{\mathfrak{A}}\Big($  $\Big) = Q^{\mathfrak{A}}\Big($ 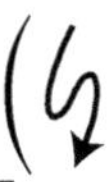 $\Big)$ so $Q^{\mathfrak{A}}$ is invariant under FT_2.

For FT_3: The proof of invariance under FT_3 was given in Sect. 9.2 after the statement of Theorem 9.3.

For FT_4: The tangle 

has two components that are coloured as shown using the representations ρ^V and ρ^W. By slicing, we have

$$Q^{\mathfrak{A}}\Big($$  $$\Big) = Q^{\mathfrak{A}}\Big($$  $$\Big) \circ Q^{\mathfrak{A}}\Big($$  $$\Big) : V \otimes W \to V \otimes W.$$

The colouring of the strands of both  and  is induced by the colouring of . Then, writing $\mathsf{R}^{-1} = \sum_i \overline{\alpha}_i \otimes \overline{\beta}_i$,

$$e_i \otimes w_j \overset{Q^{\mathfrak{A}}(\cdot)}{\longmapsto} \sum_k \rho^W_{\beta_k}(w_j) \otimes \rho^V_{\alpha_k}(e_i)$$

$$\overset{Q^{\mathfrak{A}}(\cdot)}{\longmapsto} \Big(\sum_l \rho^V_{\overline{\alpha}_l} \otimes \rho^W_{\overline{\beta}_l}\Big) \circ \tau \Big(\sum_k \rho^W_{\beta_k}(w_j) \otimes \rho^V_{\alpha_k}(e_i)\Big)$$

$$= \sum_{k,l} \big(\rho^V_{\overline{\alpha}_l} \circ \rho^V_{\alpha_k}(e_i)\big) \otimes \big(\rho^W_{\overline{\beta}_l} \circ \rho^W_{\beta_k}(w_j)\big)$$

$$= \sum_{k,l} \Big(\big(\rho^V_{\overline{\alpha}_l} \circ \rho^V_{\alpha_k}\big) \otimes \big(\rho^W_{\overline{\beta}_l} \circ \rho^W_{\beta_k}\big)\Big)(e_i \otimes w_j)$$

$$= \Big(\sum_{k,l} \big(\rho^V_{\overline{\alpha}_l} \otimes \rho^W_{\overline{\beta}_l}\big) \circ \big(\rho^V_{\alpha_k} \otimes \rho^W_{\beta_k}\big)\Big)(e_i \otimes w_j)$$

$$= \sigma^{V\otimes W}_{\sum_l \overline{\alpha}_l \otimes \overline{\beta}_l} \circ \sigma^{V\otimes W}_{\sum_k \alpha_k \otimes \beta_k}(e_i \otimes w_j),$$

where $\sigma^{V\otimes W}$ is a representation of $\mathfrak{A} \otimes \mathfrak{A}$ on $V \otimes W$. The final term of this equals

$$\sigma^{V\otimes W}_{\mathsf{R}^{-1}} \circ \sigma^{V\otimes W}_{\mathsf{R}}(e_i \otimes w_j) = \sigma^{V\otimes W}_{\mathsf{R}^{-1}\mathsf{R}}(e_i \otimes w_j) = \sigma^{V\otimes W}_{1_{\mathfrak{A}}}(e_i \otimes w_j) = e_i \otimes w_j,$$

so $Q^{\mathfrak{A}}\Big($  $\Big) = Q^{\mathfrak{A}}\Big($  $\Big)$.

The argument for the other FT_4-move is similar and left as an exercise. With this exercise, it follows that $Q^{\mathfrak{A}}$ is invariant under FT_4.

Exercise D.3. Show that $Q^{\mathfrak{A}}\Big($  $\Big) = Q^{\mathfrak{A}}\Big($ 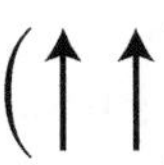 $\Big)$ to complete the proof of invariance under FT_4.

For FT_5: We consider the tangle 

and 

, with their three components coloured as shown. By slicing these, we have, using the induced colouring for the slices,

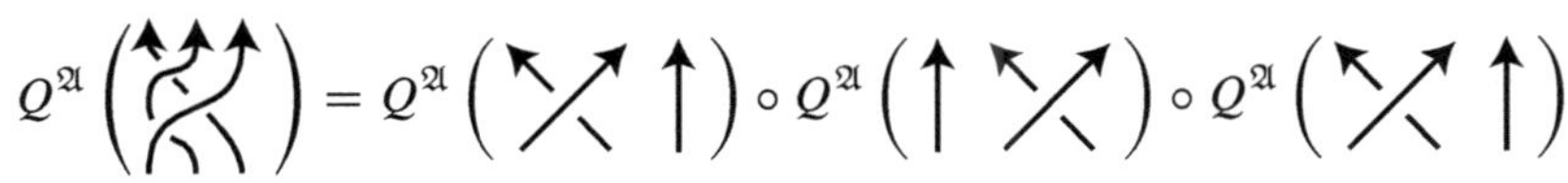

$$= \left(\tau \circ \left((\rho^V \otimes \rho^W)(\mathsf{R})\right) \otimes \mathrm{id}_U\right) \circ \left(\mathrm{id}_V \otimes \left(\tau \circ (\rho^U \otimes \rho^W)(\mathsf{R})\right)\right)$$
$$\circ \left(\tau \circ \left((\rho^U \otimes \rho^V)(\mathsf{R})\right) \otimes \mathrm{id}_W\right)$$

and

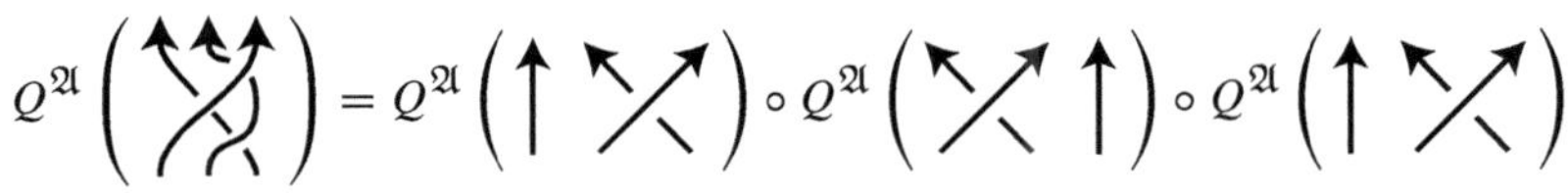

$$= \left(\mathrm{id}_W \otimes \left(\tau \circ (\rho^U \otimes \rho^V)(\mathsf{R})\right)\right) \circ \left(\tau \circ \left((\rho^U \otimes \rho^W)(\mathsf{R})\right) \otimes \mathrm{id}_V\right)$$
$$\circ \left(\mathrm{id}_U \otimes \left(\tau \circ (\rho^V \otimes \rho^W)(\mathsf{R})\right)\right).$$

The proof that these two expressions are equal is similar to the proof of Proposition 8.33 and is left as an exercise.

Exercise D.4. By considering the element-wise actions of the maps, show that 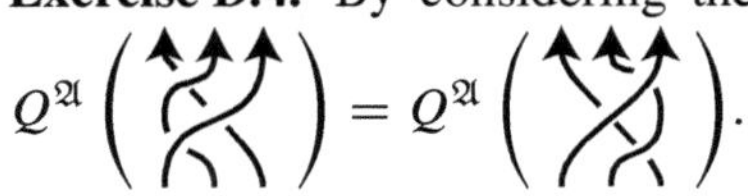

For FT$_6$: The tangle  has one component, so we can assume that it is coloured using the representation ρ^V. Regarding this tangle as a composition of 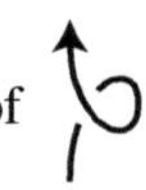 and , with the same colouring, we have

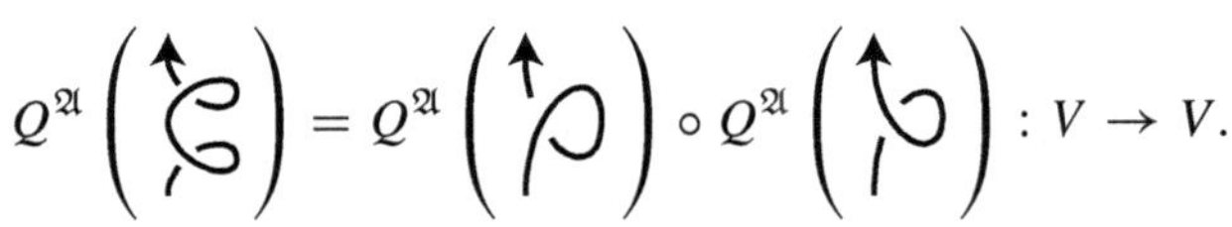 (D.1)

We consider 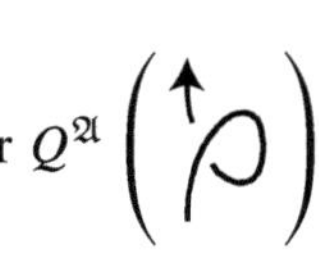 first. Slicing this gives

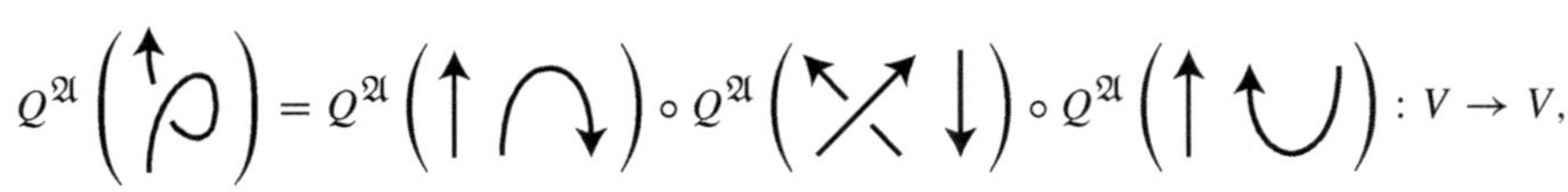

where each elementary piece is coloured using V. Then

$$e_i \overset{\iota}{\longmapsto} e_i \otimes 1_{\mathbb{K}}$$

$Q^{\mathfrak{A}}$  $\longmapsto \sum_j e_i \otimes e_j \otimes e^j$

$Q^{\mathfrak{A}}$  $\longmapsto \sum_{j,k} \rho^V_{\beta_k}(e_j) \otimes \rho^V_{\alpha_k}(e_i) \otimes e^j$

$Q^{\mathfrak{A}}$ 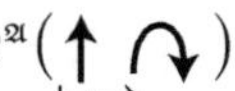 $\longmapsto \sum_{j,k} \rho^V_{\beta_k}(e_j) \otimes \underbrace{\left(e^j\left(\rho^V_{\mathbf{v}^{-1}\mathbf{u}\,\alpha_k}(e_i)\right)\right)}_{\in\mathbb{K}}$

$$\begin{aligned} &\overset{\kappa}{\longmapsto} && \sum_{j,k} \rho^V_{\beta_k}(e_j) \cdot \left(e^j\left(\rho^V_{\mathbf{v}^{-1}\mathbf{u}\,\alpha_k}(e_i)\right)\right) \\ &= && \sum_k \rho^V_{\beta_k\,\mathbf{v}^{-1}\mathbf{u}\,\alpha_k}(e_i) && \text{(Lem. D.2)} \\ &= && \rho^V_{\sum_k \beta_k\,\mathbf{v}^{-1}\mathbf{u}\,\alpha_k}(e_i) && (\rho^V \text{ is a rep. of } \mathfrak{A}.) \end{aligned}$$

But

$$\begin{aligned} \sum_k \beta_k\,\mathbf{v}^{-1}\mathbf{u}\,\alpha_k &= \sum_k \beta_k \mathbf{u}\,\alpha_k\,\mathbf{v}^{-1} && \text{(}\mathbf{v}\text{ is central: Def. 8.38(1))} \\ &= \sum_k \left(\beta_k\, S^2(\alpha_k)\right)\,\mathbf{u}\,\mathbf{v}^{-1} && \text{(Lem. 8.35)} \\ &= \mathbf{u}^{-1}\mathbf{u}\mathbf{v}^{-1} && \text{(Prop. 8.37(5))} \\ &= \mathbf{v}^{-1}. \end{aligned}$$

Thus

$Q^{\mathfrak{A}}$ 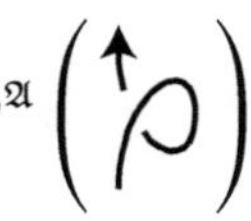 $= \rho^V_{\mathbf{v}^{-1}}.$ (D.2)

Similarly, slicing $Q^{\mathfrak{A}}$  gives

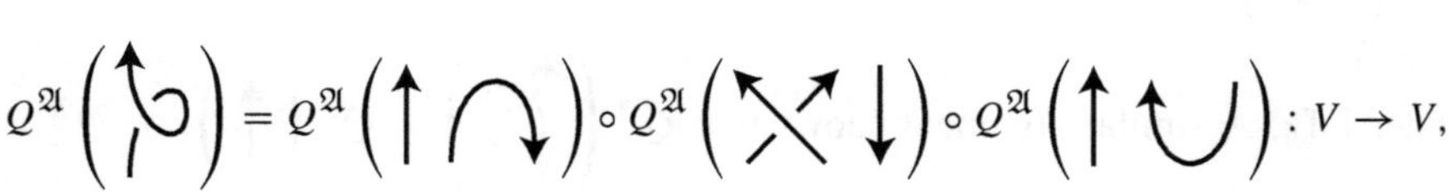

where each elementary piece is coloured using V. Then

$$e_i \overset{\iota}{\longmapsto} e_i \otimes 1_{\mathbb{K}}$$

$$\sum_j e_i \otimes e_j \otimes e^j$$

$$\sum_{j,k} \rho^V_{S(\alpha_k)}(e_j) \otimes \rho^V_{\beta_k}(e_i) \otimes e^j$$

$$\sum_{j,k} \rho^V_{S(\alpha_k)}(e_j) \otimes \underbrace{\left(e^j\left(\rho^V_{\mathbf{v}^{-1}\mathbf{u}\beta_k}(e_i)\right)\right)}_{\in\mathbb{K}}$$

$$\begin{aligned} &\overset{\kappa}{\longmapsto} \sum_{j,k} \rho^V_{S(\alpha_k)}(e_j) \cdot \left(e^j\left(\rho^V_{\mathbf{v}^{-1}\mathbf{u}\beta_k}(e_i)\right)\right) \\ &= \sum_k \left(\rho^V_{S(\alpha_k)} \circ \rho^V_{\mathbf{v}^{-1}\mathbf{u}\beta_k}\right)(e_i) && \text{(Lem. D.2)} \\ &= \rho^V_{\sum_k S(\alpha_k)\mathbf{v}^{-1}\mathbf{u}\beta_k}(e_i). \end{aligned}$$

But

$$\begin{aligned} \sum_k S(\alpha_k)\mathbf{v}^{-1} \cdot \mathbf{u} \cdot \beta_k &= \mathbf{v}^{-1} \sum_k S(\alpha_k) \cdot \mathbf{u} \cdot \beta_k && (\mathbf{v} \text{ is central: Def. 8.38(1)}) \\ &= \mathbf{v}^{-1} \sum_k S(\alpha_k) \cdot S^2(\beta_k) \cdot \mathbf{u} && \text{(Lem. 8.36(2))} \\ &= \mathbf{v}^{-1} S\left(\sum_k S(\beta_k) \cdot \alpha_k\right) \cdot \mathbf{u} && (S \text{ is anti-homom.}) \\ &= \mathbf{v}^{-1} \cdot S(\mathbf{u}) \cdot \mathbf{u} && \text{(Def. 8.34)} \\ &= \mathbf{v} && \text{(Def. 8.38(2)).} \end{aligned}$$

Thus $Q^{\mathfrak{A}}$  $= \rho^V_{\mathbf{v}}$. Then from this, together with (D.1) and (D.2), we have $Q^{\mathfrak{A}}$  $= \rho^V_{\mathbf{v}^{-1}} \circ \rho^V_{\mathbf{v}} = \rho^V_{\mathbf{v}^{-1}\mathbf{v}} = \rho^V_{1_{\mathfrak{A}}} = \mathrm{id}_V = Q^{\mathfrak{A}}(\uparrow)$ so $Q^{\mathfrak{A}}$ is invariant under FT_6. A similar argument shows that $Q^{\mathfrak{A}}$  $= Q^{\mathfrak{A}}(\uparrow)$, so $Q^{\mathfrak{A}}$ is also invariant under FT_6.

<u>*For*</u> FT_7 : The tangle  has two components so let its colouring be ,

and let the colouring of  be .

Let

$A :=$  and $B :=$ ,

so that

 $= B \circ A$ and  $= A \circ B.$

Let

$\mathsf{Y} = Q^{\mathfrak{A}}(A) = Q^{\mathfrak{A}}\Big($ 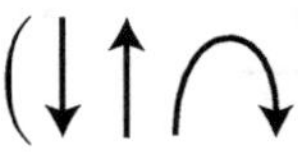 $\Big) \circ Q^{\mathfrak{A}}\Big($  $\Big) \circ Q^{\mathfrak{A}}\Big($  $\Big)$

and

$\mathsf{T} = Q^{\mathfrak{A}}(B) = Q^{\mathfrak{A}}\Big($  $\Big) \circ Q^{\mathfrak{A}}\Big($  $\Big) \circ Q^{\mathfrak{A}}\Big($  $\Big).$

To prove FT_7 we need to show that

$$\mathsf{Y} \circ \mathsf{T} = \mathrm{id}_{V^\star} \otimes \mathrm{id}_W \quad \text{and} \quad \mathsf{T} \circ \mathsf{Y} = \mathrm{id}_W \otimes \mathrm{id}_{V^\star}.$$

The action of T is given by

$$e^i \otimes w_j \quad \overset{\iota}{\longmapsto} \quad e^i \otimes w_j \otimes 1_{\mathbb{K}}$$

$Q^{\mathfrak{A}}\big($  $\big) \longmapsto \displaystyle\sum_k e^i \otimes w_j \otimes e_k \otimes e^k$

$Q^{\mathfrak{A}}\big($  $\big) \longmapsto \displaystyle\sum_{k,l} e^i \otimes \rho^V_{S(\alpha_l)}(e_k) \otimes \rho^W_{\beta_l}(w_j) \otimes e^k$

$Q^{\mathfrak{A}}\big($ 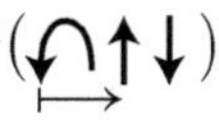 $\big) \longmapsto \displaystyle\sum_{k,l} \underbrace{e^i \left(\rho^V_{S(\alpha_l)}(e_k)\right)}_{\in \mathbb{K}} \otimes \rho^W_{\beta_l}(w_j) \otimes e^k$

$$\overset{\kappa}{\longmapsto} \quad \sum_{k,l} \underbrace{e^i \left(\rho^V_{S(\alpha_l)}(e_k)\right)}_{\in \mathbb{K}} \cdot \rho^W_{\beta_l}(w_j) \otimes e^k.$$

The action of Y is given by

$$w_p \otimes e^q \overset{\iota}{\longmapsto} 1_{\mathbb{K}} \otimes w_p \otimes e^q$$

$Q^{\mathfrak{A}}$(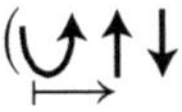)

$$\longmapsto \sum_m e^m \otimes \rho^V_{\mathbf{u}^{-1}\mathbf{v}}(e_m) \otimes w_p \otimes e^q$$

$Q^{\mathfrak{A}}$()

$$\longmapsto \sum_{m,n} e^m \otimes \rho^W_{\beta_n}(w_p) \otimes \rho^V_{\alpha_n \mathbf{u}^{-1}\mathbf{v}}(e_m) \otimes e^q$$

$Q^{\mathfrak{A}}$(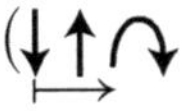)

$$\longmapsto \sum_{m,n} e^m \otimes \rho^W_{\beta_n}(w_p) \otimes \underbrace{e^q\left(\rho^V_{\mathbf{v}^{-1}\mathbf{u}\alpha_n\mathbf{u}^{-1}\mathbf{v}}(e_m)\right)}_{\in \mathbb{K}}$$

$$\overset{\kappa}{\longmapsto} \sum_{m,n} \underbrace{e^q\left(\rho^V_{\mathbf{v}^{-1}\mathbf{u}\alpha_n\mathbf{u}^{-1}\mathbf{v}}(e_m)\right)}_{\in \mathbb{K}} \cdot e^m \otimes \rho^W_{\beta_n}(w_p)$$

$$= \sum_{m,n} \underbrace{e^q\left(\rho^V_{S^2(\alpha_n)}(e_m)\right)}_{\in \mathbb{K}} \cdot e^m \otimes \rho^W_{\beta_n}(w_p),$$

where for the last equality we have used that $\mathbf{v}$ is central and Lemma 8.35 to write

$$\rho^V_{\mathbf{v}^{-1}\mathbf{u}\alpha_n\mathbf{u}^{-1}\mathbf{v}} = \rho^V_{\mathbf{u}\alpha_n\mathbf{u}^{-1}} = \rho^V_{S^2(\alpha_n)}.$$

Then $Q^{\mathfrak{A}}$() $= \mathsf{Y} \circ \mathsf{T}$ and so acts by

$Q^{\mathfrak{A}}$()

$$e^i \otimes w_j \longmapsto \sum_{k,l,m,n} \underbrace{e^i\left(\rho^V_{S(\alpha_l)}(e_k)\right)}_{\in \mathbb{K}} \cdot \underbrace{e^k\left(\rho^V_{S^2(\alpha_n)}(e_m)\right)}_{\in \mathbb{K}} \cdot e^m \otimes \rho^W_{\beta_n\beta_l}(w_j)$$

$$= \sum_{l,m,n} \underbrace{e^i\left(\rho^V_{S(\alpha_l)} \circ \rho^V_{S^2(\alpha_n)}(e_m)\right)}_{\in \mathbb{K}} \cdot e^m \otimes \rho^W_{\beta_n\beta_l}(w_j)$$

where the last equality was by Lemma D.2(2). Thus

$Q^{\mathfrak{A}}$(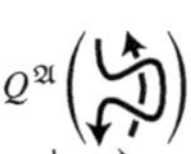)

$$e^i \otimes w_j \longmapsto \sum_{l,m,n} \underbrace{e^i\left(\rho^V_{S(\alpha_l)\cdot S^2(\alpha_n)}(e_m)\right)}_{\in \mathbb{K}} \cdot e^m \otimes \rho^W_{\beta_n\beta_l}(w_j). \tag{D.3}$$

But

$$\begin{aligned}\sum_{a,b}\left(S(\alpha_b)\cdot S^2(\alpha_a)\right)\otimes(\beta_a\cdot\beta_b) &= (S\otimes \mathrm{id}_{\mathfrak{A}})\sum_{a,b}(S(\alpha_a)\cdot(\alpha_b))\otimes(\beta_a\cdot\beta_b)\\ &= (S\otimes \mathrm{id}_{\mathfrak{A}})\left(\left(\sum_a S(\alpha_a)\otimes\beta_a\right)\cdot \mathsf{R}\right)\\ &= (S\otimes \mathrm{id}_{\mathfrak{A}})\left(((S\otimes \mathrm{id}_{\mathfrak{A}})(\mathsf{R}))\cdot \mathsf{R}\right)\\ &= (S\otimes \mathrm{id}_{\mathfrak{A}})(\mathsf{R}^{-1}\cdot \mathsf{R}) \qquad \text{(Prop. 8.31(2))}\\ &= (S\otimes \mathrm{id}_{\mathfrak{A}})(1_{\mathfrak{A}\otimes\mathfrak{A}})\\ &= 1_{\mathfrak{A}\otimes\mathfrak{A}}.\end{aligned}$$

Thus

$$\begin{aligned}\sum_{l,n}\rho^V_{S(\alpha_l)\cdot S^2(\alpha_n)}(e_m)\otimes\rho^W_{\beta_n\beta_l}(w_j) &= \left(\rho^V\otimes\rho^W\right)\left(\sum_{l,n} S(\alpha_l)S^2(\alpha_n)\otimes\beta_n\beta_l\right)(e_m\otimes w_j)\\ &= (\rho^V_{1_{\mathfrak{A}}}\otimes\rho^W_{1_{\mathfrak{A}}})(e_m\otimes w_j)\\ &= e_m\otimes w_j.\end{aligned}$$

This gives

$$\sum_{l,n}\rho^V_{S(\alpha_l)\cdot S^2(\alpha_n)}(e_m)\otimes\rho^W_{\beta_n\beta_l}(w_j) = e_m\otimes w_j.$$

Applying the map $e^i\otimes \mathrm{id}_{\mathfrak{A}}$ to each side gives

$$\sum_{l,n} e^i\left(\rho^V_{S(\alpha_l)\cdot S^2(\alpha_n)}(e_m)\right)\otimes\rho^W_{\beta_n\beta_l}(w_j) = e^i(e_m)\otimes w_j.$$

Multiplying by $e^m\otimes 1_{\mathfrak{A}}$ gives

$$\sum_{l,n} e^i\left(\rho^V_{S(\alpha_l)\cdot S^2(\alpha_n)}(e_m)\right)\cdot e^m\otimes\rho^W_{\beta_n\beta_l}(w_j) = e^i(e_m)\cdot e^m\otimes w_j.$$

Then, summing over m gives

$$\sum_{l,n,m} e^i\left(\rho^V_{S(\alpha_l)\cdot S^2(\alpha_n)}(e_m)\right)\cdot e^m\otimes\rho^W_{\beta_n\beta_l}(w_j) = \sum_m e^i(e_m)\cdot e^m\otimes w_j = e^i\otimes w_j.$$

and therefore, from (D.3),

$e^i \otimes w_j \xrightarrow{Q^{\mathfrak{A}}\left(\right.}$  $\left.\right)$ $e^i \otimes w_j.$

Thus

$Q^{\mathfrak{A}}\left(\right.$  $\left.\right) = \mathrm{id}_{V^\star} \otimes \mathrm{id}_W = Q^{\mathfrak{A}}\left(\downarrow\uparrow\right).$

Similarly,

$Q^{\mathfrak{A}}\left(\right.$  $\left.\right) = \mathsf{T} \circ \mathsf{Y},$

and so acts by

$w_p \otimes e^q \xrightarrow{Q^{\mathfrak{A}}\left(\right.}$  $\left.\right)$

$$\begin{aligned} & \sum_{k,l,m,n} \rho^W_{\beta_l\beta_n}(w_p) \otimes \left(e^q\left(\rho^V_{S^2(\alpha_n)}(e_m)\right)\right) \cdot \left(e^m\left(\rho^V_{S(\alpha_l)}(e_k)\right)\right) e^k \\ = & \sum_{k,l,n} \rho^W_{\beta_l\beta_n}(w_p) \otimes \left(e^q\left(\rho^V_{S^2(\alpha_n)S(\alpha_l)}(e_k)\right)\right) e^k \end{aligned}$$

Next,

$$\begin{aligned} \sum_{a,b} S^2(\alpha_b)S(\alpha_a) \otimes \beta_a\beta_b &= (S \otimes \mathrm{id})\left(\sum_{a,b} \alpha_a S(\alpha_b) \otimes \beta_a\beta_b\right) \\ &= (S \otimes \mathrm{id})\left(\left(\sum_a \alpha_a \otimes \beta_a\right) \cdot \left(\sum_b S(\alpha_b) \otimes \beta_b\right)\right) \\ &= (S \otimes \mathrm{id})(\mathsf{R} \cdot \mathsf{R}^{-1}) \qquad \text{(Prop. 8.31(2))} \\ &= 1_{\mathfrak{A}\otimes\mathfrak{A}} \end{aligned}$$

Thus

$Q^{\mathfrak{A}}\left(\right.$  $\left.\right) = \mathsf{T} \circ \mathsf{Y} : w_p \otimes e^q \mapsto w_p \otimes e^q,$

so

$Q^{\mathfrak{A}}\Bigg($  $\Bigg) = \mathrm{id}_W \otimes \mathrm{id}_{V^*} = Q^{\mathfrak{A}}\left(\uparrow\downarrow\right),$

so $Q^{\mathfrak{A}}$ is invariant under the FT_7-move.

This completes the proof of the theorem. □

Hints for the Exercises

Exercise 0.1: From [148],

Exercise 1.6: Consider how the strands at the top and bottom of each "box" are connected to each other when p_i is odd and when it is even. Deduce that it is enough to consider each p_i modulo 2.

Exercise 1.15: It is the figure-of-eight knot.

Exercise 1.18: Move the rightmost vertical strand out to the right.

Exercise 1.19: Choose a base point on the knot, and start travelling round it from this base point. Change each crossing (if needed) so that the first time you meet it you pass under the other strand. Then, you can obtain a knot in $\mathbb{R}^3$ from this diagram that has the property that starting from the base point you travel up until you return to the base point, then travel straight down to "close the knot". Thus the diagram corresponds to a closed curve in $\mathbb{R}^3$ that has exactly one maximum and exactly one minimum and hence must be an unknot.

Exercise 1.24: RII and RI.

Exercise 1.27: Which Reidemeister moves can you apply to the diagram?

Exercise 1.32: Straightforward.

Exercise 2.3: Zero.

Exercise 2.5: For the first part, 0. For the second, 0, 0, 1, 1.

Exercise 2.9: Adapt the method of Fig. 2.1.

D. M. Jackson and I. Moffatt, *An Introduction to Quantum and Vassiliev Knot Invariants*,
CMS Books in Mathematics, https://doi.org/10.1007/978-3-030-05213-3

Exercise 2.11: Try colouring the diagram for 4_1 in Fig. 1.1. How much choice do you have?

Exercise 2.12: Adapt the proof of Theorem 2.8.

Exercise 2.22: Replace an unlink component with L in the computation in the proof of Lemma 2.21.

Exercise 2.25: Just compute.

Exercise 2.26: Reflecting the link in the plane $z = 0$ has the effect of reversing all of the crossings in the diagram. How do the skein relations applied to L and $\overline{L}$ compare?

Exercise 2.29: Adapt Examples 2.23 and 2.24.

Exercise 2.31: Adapt Examples 2.23 and 2.24, or evaluate the HOMFLY polynomial at $x = 1$ and $y = z$.

Exercise 2.32:

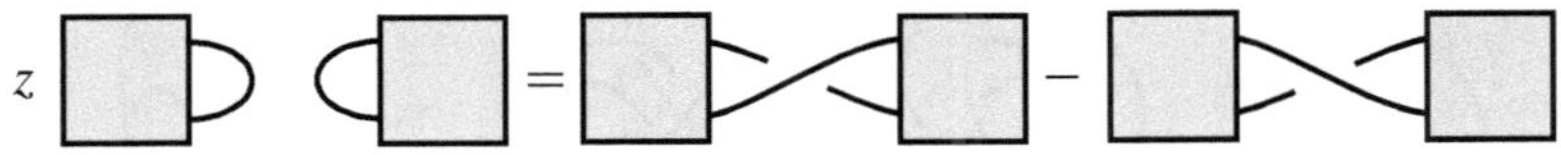

Exercise 3.4:

Exercise 3.6: If the framed links are equivalent, consider their projections onto the plane. Consider the link diagrams formed by one boundary component of each annulus. There exists a sequence of Reidemeister moves between them. Consider the framed links formed by thickening these diagrams. What happens to the twists throughout this sequence of Reidemeister moves?

Exercise 3.10: Adapt the proof of Theorem 2.17.

Exercise 3.13: In the sequence of Reidemeister moves between the two diagrams, whenever a "kink" is added to the diagram through an RI-move it must be balanced with a "kink" of the opposite sign. (This result is from [177], where a full proof can be found.)

Exercise 3.18: Just compute.

Exercise 3.19: Identify each state with a terminal (without crossings) form in the computation of $\langle\cdot\rangle$. Alternatively, show the state sum satisfies the defining relations for $\langle\cdot\rangle$.

Exercise 3.25: The state sum in Exercise 3.19 is obviously well-defined.

Exercise 4.6: What are the restrictions on the Δ-moves?

Exercise 5.1: Theorem A.29 and Proposition A.34 may help.

Exercise 5.5: $(\mathrm{id}\otimes\lambda\mathsf{R}) = \lambda(\mathrm{id}\otimes\mathsf{R})$, $(\mathrm{id}\otimes\mathsf{R}^{-1}) = (\mathrm{id}\otimes\mathsf{R})^{-1}$, etc. Also $\tau\circ\mathsf{R}\circ\tau : e_i\otimes e_j \mapsto \sum_{k,l} R^{kl}_{ji} e_l \otimes e_k$.

Exercise 5.9: Adapt the example.

Exercise 6.1: The isomorphism is given by identifying $f : u_i \mapsto \sum_j f_i^j v_j \in \mathrm{Hom}(U, V)$ with $\sum_{i,j} f_i^j u^i \otimes v_j \in U^\star \otimes V$. See Lemma A.36.

Exercise 6.2: $g\circ f = \sum_{i,j,k,l} f_i^j g_k^l u^i \otimes v^k(v_j)\otimes w_l$.

Exercise 6.3: If $f = \sum_{i,j} f_i^j u^i \otimes w_j$ and $g = \sum_{k,l} g_k^l w^k \otimes u_l$, then $g\circ f = \sum_{i,j,k,l} f_i^j g_k^l u^i \otimes w^k(w_j)\otimes u_l$, and so $\mathrm{Tr}(g\circ f) = \sum_{i,j,k,l} f_i^j g_k^l \cdot u^i(u_l)\cdot w^k(w_j) = \sum_{i,j} f_i^j g_j^i$. Now calculate $\mathrm{Tr}(f\circ g)$ in a similar way.

Exercise 6.4: $h = \sum_{i,j,k,l} h_{i,j}^{k,l} e^i \otimes e^j \otimes e_k \otimes e_l$, so $\mathrm{Tr}_2(h) = \sum_{i,j,k,l} h_{i,j}^{k,l} e^i \otimes e^j(e_l)\otimes e_k = \sum_{i,j,k} h_{i,j}^{k,j} e^i \otimes e_l$.

Exercise 6.5: Write f as an element of $(V^*)^{\otimes n}\otimes V^{\otimes n}$. Then $\mathrm{Tr}(f)$ is obtained by contracting the i-th V^* with the $(n+i)$-th V, for each $i = 1,\ldots,n$. Then $\mathrm{Tr}_n(f)$ is obtained by contracting the n-th V^* with the $2n$-th V, for each $i = 1,\ldots,n$, and then to get $\mathrm{Tr}(\mathrm{Tr}_n(f))$ we contract the i-th V^* with the $(n+i)$-th V, for each $i = 1,\ldots,n-1$. Overall the result is the same.

Exercise 6.6: For the first item, $f\circ(g\otimes\mathrm{id}_V) = \sum g_{i_1,\ldots,i_n}^{j_1,\ldots,j_n} f_{j_1,\ldots,j_{n+1}}^{k_1,\ldots,k_{n+1}} e^{i_1}\otimes\cdots e^{i_{n+1}}\otimes e_{k_1}\otimes\cdots\otimes e_{k_{n+1}}$, so $\mathrm{Tr}_{n+1}(f\circ(g\otimes\mathrm{id}_V)) = \sum g_{i_1,\ldots,i_n}^{j_1,\ldots,j_n} f_{j_1,\ldots,j_n,l}^{k_1,\ldots,k_n,l} e^{i_1}\otimes\cdots e^{i_n}\otimes e_{k_1}\otimes\cdots\otimes e_{k_n}$. On the other hand, $\mathrm{Tr}_{n+1}(f) = f_{j_1,\ldots,j_n,l}^{k_1,\ldots,k_n,l} e^{j_1}\otimes\cdots e^{j_n}\otimes e_{k_1}\otimes\cdots\otimes e_{k_n}$, so $\mathrm{Tr}_{n+1}(f)\circ g = \sum g_{i_1,\ldots,i_n}^{j_1,\ldots,j_n} f_{j_1,\ldots,j_n,l}^{k_1,\ldots,k_n,l} e^{i_1}\otimes\cdots e^{i_n}\otimes e_{k_1}\otimes\cdots\otimes e_{k_n}$, giving $\mathrm{Tr}_{n+1}(f\circ(g\otimes\mathrm{id}_V)) = \mathrm{Tr}_{n+1}(f)\circ g$.

The other identities follow from similar calculations.

Exercise 6.7:

$$\begin{bmatrix} 1 & 0 & 0 & 0 \\ 0 & 0 & c & 0 \\ 0 & c^{-1} & 0 & 0 \\ 0 & 0 & 0 & 1 \end{bmatrix}$$

Exercise 6.11: Invariance under MI follows from Lemma 6.8. For MII follow the proof of Lemma 6.9 and observe that $\rho_{\mathsf{R},\mu}\left((\sigma\otimes 1)\circ\sigma_n^{\pm1}\right) = \alpha^{\pm1}\beta\rho_{\mathsf{R},\mu}(\sigma)$, and deduce that $\rho_{(\mathsf{R},\,\mu,\,\alpha,\,\beta)}$ is invariant under MII. Invariance under the braid moves follows since ω, n, ρ_{R} are.

Exercise 6.14: Make use of Proposition A.34 and work with the matrices.

Exercise 6.15: Use matrices.

Exercise 6.18: For the skein relation part, see Exercise 2.26. Using $\rho_{\mathsf{R},\mu}$, observe that R^{-1} is obtained by changing $t \leftrightarrow t^{-1}$ in R and reflecting in the antidiagonal, and μ is invariant under this change. When you take the trace, the overall change is the exchange $t \leftrightarrow t^{-1}$.

Exercise 6.19: Take the trace of μ.

Exercise 6.20: Examine at the element-wise actions of the maps. Use that R^{-1} : $e_i \otimes e_j \mapsto \sum_{k,l}(R^{-1})^{k,l}_{i,j} e_k \otimes e_l$, where

$$(R^{-1})^{k,l}_{i,j} = \begin{cases} q^{m-1} & \text{if } i = j = k = l, \\ -q^m & \text{if } i = l \neq k = j, \\ q^m(q^{-1} - q) & \text{if } i = k > l = j, \\ 0 & \textit{otherwise}. \end{cases}$$

Exercise 6.22: With $(R^{-1})^{k,l}_{i,j}$ as above, check the identity holds in each position of the matrices.

Exercise 6.24: A straightforward adaptation.

Exercise 7.18: As matrices,

$$\overrightarrow{n} = \begin{bmatrix} -t^{\frac{1}{2}} \; 0 \; 0 \; -t^{-\frac{1}{2}} \end{bmatrix}, \quad \overleftarrow{n} = \begin{bmatrix} 1 \; 0 \; 0 \; 1 \end{bmatrix}, \quad \overrightarrow{u} = \begin{bmatrix} -t^{-\frac{1}{2}} \\ 0 \\ 0 \\ -t^{\frac{1}{2}} \end{bmatrix}, \quad \overleftarrow{u} = \begin{bmatrix} 1 \\ 0 \\ 0 \\ 1 \end{bmatrix}.$$

Make use of these to check the identities and find the value of $Q(T)$ using a computer package.

Exercise 7.22: Tracking through the actions on a basis, $\left(f : u_i \mapsto \sum_j f_i^j v_j\right) \mapsto$
$\left(\sum_{i,j} f_i^j u^i \otimes v_j\right) \mapsto \left(\sum_{i,j} f_i^j (u_i \otimes v^j)^\star\right) \mapsto \left(\sum_{i,j} f_i^j (u_i \otimes v^j)^\star \otimes 1_{\mathbb{K}}\right) \mapsto$
$\left(u_i \otimes v^j \mapsto f_i^j \otimes 1_{\mathbb{K}}\right)$, but $v^j(f(u_i)) = v^j\left(\sum_k f_i^k v_k\right) = f_i^j$.

Exercise 7.24: $\left(f : u_i \mapsto \sum_j f_i^j v_j\right) \mapsto \left(\sum_{i,j} f_i^j u^i \otimes v_j\right) \mapsto \left(\sum_{i,j} 1_{\mathbb{K}} \otimes f_i^j u^i \otimes v_j\right)$
$\mapsto \left(\sum_{i,j} 1_{\mathbb{K}}^\star \otimes f_i^j u^i \otimes v_j\right) \mapsto \left(1_{\mathbb{K}} \mapsto \sum_{i,j} f_i^j u^i \otimes v_j\right) = \left(1_{\mathbb{K}} \mapsto \sum_i u^i \otimes f(u_i)\right).$

Exercise 7.27: See the proof of Lemma B.9.

Exercise 7.28: Make use of Lemma B.4.

Exercise 7.38: By Lemma B.8, Q^f  $= c\, Q^f$ 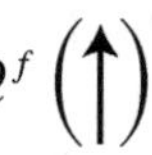 and Q^f 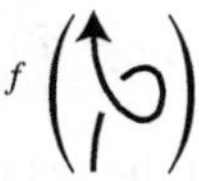 $=$ $c^{-1}\, Q^f$ .

Exercise 8.1: Routine.

Exercise 8.2: Routine.

Exercise 8.3: $m^{\mathrm{op}} \circ (m^{\mathrm{op}} \otimes \mathrm{id})(x \otimes y \otimes z) = z(yx) = (zy)x = m^{\mathrm{op}} \circ (\mathrm{id} \otimes m^{\mathrm{op}})$ $(x \otimes y \otimes z)$ and $m^{\mathrm{op}}(1 \otimes x) = m(x \otimes 1) = x = m(1 \otimes x) = m^{\mathrm{op}}(x \otimes 1)$.

Exercise 8.4: First suppose that η is a unit in the functional sense of (8.3). Prove that the element $\eta(1_{\mathbb{K}})$, where $1_{\mathbb{K}}$ is the unit of $\mathbb{K}$, satisfies (8.2).

Conversely, suppose that $1_{\mathfrak{A}}$ is a unit in the sense of (8.2). Show that the map defined by $f : 1_{\mathbb{K}} \mapsto 1_{\mathfrak{A}}$ defines a unit in the sense of (8.3).

Exercise 8.15: If $(\Delta \otimes \mathrm{id}) \circ \Delta(x) = \sum \Delta(x_{(1)}) \otimes x_{(2)} = \sum y_{(1)} \otimes y_{(2)} \otimes x_{(2)}$, and $(\mathrm{id} \otimes \Delta) \circ \Delta(x) = \sum x_{(1)} \otimes \Delta(x_{(2)}) = \sum x_{(1)} \otimes z_{(1)} \otimes z_{(2)}$, then $(\Delta^{\mathrm{op}} \otimes \mathrm{id}) \circ \Delta^{\mathrm{op}}(x) = \sum \Delta^{\mathrm{op}}(x_{(2)}) \otimes x_{(1)} = \sum z_{(2)} \otimes z_{(1)} \otimes x_{(1)}$, and $(\mathrm{id} \otimes \Delta^{\mathrm{op}}) \circ \Delta^{\mathrm{op}}(x) = \sum x_{(2)} \otimes \Delta^{\mathrm{op}}(x_{(1)}) = \sum x_{(2)} \otimes y_{(2)} \otimes y_{(1)}$. Then use that $(\mathrm{id} \otimes \Delta) \circ \Delta = (\Delta \otimes \mathrm{id}) \circ \Delta$ and permute tensor factors.

Exercise 8.17: Write down the actions of $((\Delta^\star \circ \alpha) \otimes \mathrm{id}) \circ (\Delta^\star \circ \alpha)$ and $(\mathrm{id} \otimes (\Delta^\star \circ \alpha)) \circ (\Delta^\star \circ \alpha)$ on $f \otimes g \otimes h$ and use the cocommutativity of Δ, or adapt the approach of Example 8.16.

Exercise 8.18: Write out $(f * g) * h$ and $f * (g * h)$ in full.

Exercise 8.20: Write down the commutative diagrams for the two sets of conditions.

Exercise 8.21: Make use of Exercises 8.3 and 8.15.

Exercise 8.22: Make use of Example 8.16 and Exercise 8.17.

Exercise 8.26: $S^{-1} \circ S(m^{\mathrm{op}} \circ (S^{-1} \otimes \mathrm{id}) \circ \Delta(x)) = S^{-1} \circ S\left(\sum x_{(2)} S^{-1}(x_{(1)})\right) = S^{-1}\left(\sum x_{(1)} S(x_{(2)})\right) = S^{-1}((m \circ (\mathrm{id} \otimes S) \circ \Delta)(x)) = S^{-1}(\eta \circ \varepsilon(x)) = \eta \circ \varepsilon(x)$.

Exercise 8.27: Take the dual of (8.8).

Exercise 8.30: Write $\mathsf{R} =: \sum \alpha_i \otimes \beta_i$, $\mathsf{R}^{-1} =: \sum \bar{\alpha}_j \otimes \bar{\beta}_j$ and $\Delta(a) = \sum a_{(1)} \otimes a_{(2)}$, then do an element-wise calculation.

Exercise 8.41: Expand $0 = [x + y, x + y]$.

Exercise 8.43: Routine.

Exercise 8.45: Routine.

Exercise 8.48: Just check the axioms.

Exercise 8.50: Check the matrices satisfy the defining relations for $U(\mathfrak{sl}_2)$.

Exercise 8.51: For the second item, $(q - \overline{q}\,)^2 [x][y] = \left(q^{x+y} + \overline{q}^{\,x+y}\right) - \left(q^{y-x} + \overline{q}^{\,y-x}\right)$, by expanding the left-hand side using this twice,

$$(q-\overline{q})^2\left([x][y]-[x-a][y+a]\right)=q^{y-x}\left(q^{2a}-1\right)+\overline{q}^{\,y-x}\left(\overline{q}^{\,2a}-1\right)$$

and the result follows. The third item follows from the second by replacing with x with $-x$.

Exercise 8.53**:** Write K in terms of H.

Exercise 8.54**:** Check the Hopf algebra axioms hold for each generator.

Exercise 8.58**:** Check the Hopf algebra axioms hold for each generator.

Exercise 8.62**:** Adapt the argument for $\rho(KX)=\rho(qXK)$.

Exercise 9.6**:** Deduce from FRI that $Q^{\mathfrak{sl}_2}$  $= q^{-\frac{3}{2}}\ Q^{\mathfrak{sl}_2}$ 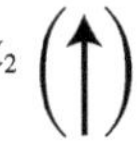. Deframing then gives that $\widetilde{Q}(L) := q^{-\frac{3}{2}\omega(L)}\ Q^{\mathfrak{sl}_2}(L)$ is an invariant of oriented links. For the skein relation, observe $\omega(L_+)=\omega(L_0)+1$ and $\omega(L_-)=\omega(L_0)-1$. Using this,

$$\begin{aligned}
q^2\widetilde{Q}(L_+)-q^{-2}\widetilde{Q}(L_-) &= q^2q^{-\frac{3}{2}\omega(L_+)}\ Q^{\mathfrak{sl}_2}(L_+)-q^{-2}q^{-\frac{3}{2}\omega(L_-)}\ Q^{\mathfrak{sl}_2}(L_-)\\
&= q^{-\frac{3}{2}\omega(L_0)}\left(q^{\frac{1}{2}}\ Q^{\mathfrak{sl}_2}(L_+)-q^{-\frac{1}{2}}\ Q^{\mathfrak{sl}_2}(L_-)\right)\\
&= q^{-\frac{3}{2}\omega(L_0)}(q-q^{-1})\ Q^{\mathfrak{sl}_2}(L_0)\\
&= (q-q^{-1})\widetilde{Q}(L_0)
\end{aligned}$$

Also, $\widetilde{Q}$  $= q+q^{-1}$.

By comparing skein relations, $\widetilde{Q}(L)=(q+q^{-1})\ J(L)|_{-t^{-\frac{1}{2}}=q}$.

Exercise 9.7**:** Adapt the argument given for the two-dimensional case.

Exercise 10.8**:** Choose a crossing c, then partition the sum into terms in which c is resolved positively and terms in which it is resolved negatively.

Exercise 10.13**:** This follows from the filtration (10.3).

Exercise 10.14**:**

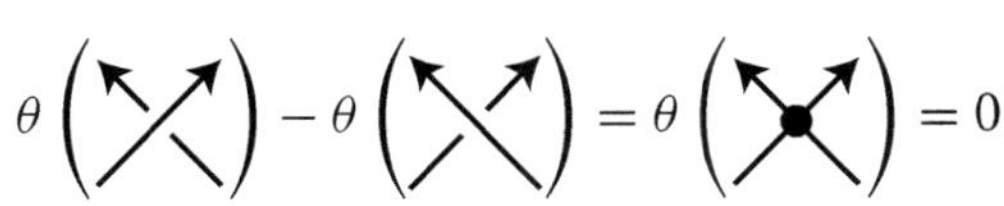

Exercise 10.17**:** Take $n=2$, or $t^{\frac{1}{2}}=-q^{-1}=-e^{-\frac{h}{2}}$.

Exercise 10.19: For $K \in \mathscr{K}_{m+1}$, use that $\theta(K) = \theta'(K) = 0$.

Exercise 11.15: Compute the Vassiliev resolutions of the two singular knots, simplify the resulting knots, then use Example 2.24 and Exercise 2.25.

Exercise 11.36: Look at the proofs of Lemmas 11.22 and 11.24.

Exercise 12.13: Just work through carefully.

Exercise 13.3: The degree of a vertex is the number of half-edges it meets, so counting the total degree by summing over edges gives that the sum of all degree equals twice the number of edges. Then, since every vertex has odd degree, there must be an even number of vertices.

Exercise 13.12: Since the STU-relation can be used to resolve all trivalent vertices, $\mathscr{A}_3$ is spanned by chord diagrams of degree 3. These can be obtained by adding one chord to the degree two cord diagrams in (13.4). Eliminating repeats, this gives a spanning set of five chord diagrams:

Applying the STU-relation to eliminate the trivalent vertex in two different ways in each of

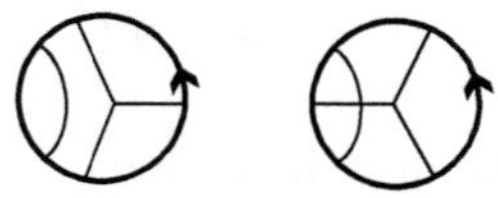

shows that two of these cord diagrams belong to the span of three of the others.

Exercise 13.17: Applying the AS relation to the second and third terms of the Jacobi form gives

We can spot the IHX-relation thus:

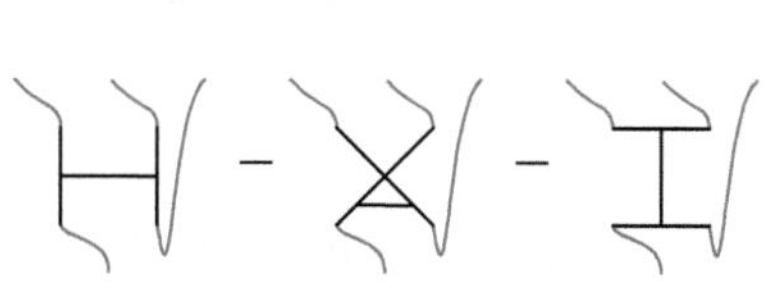

$= 0.$

Similarly, applying the AS relation to the second and third terms of the symmetric form gives

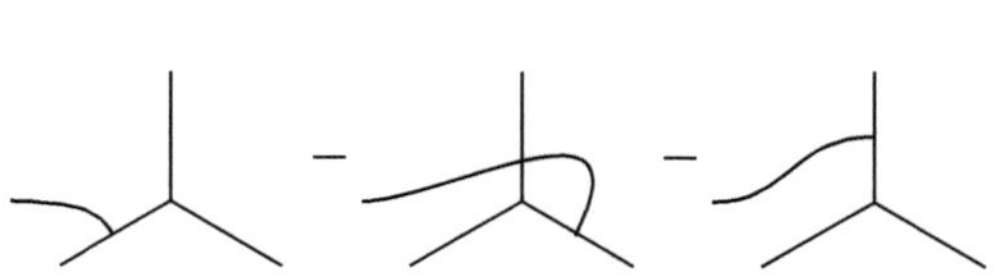

We can spot the IHX-relation thus:

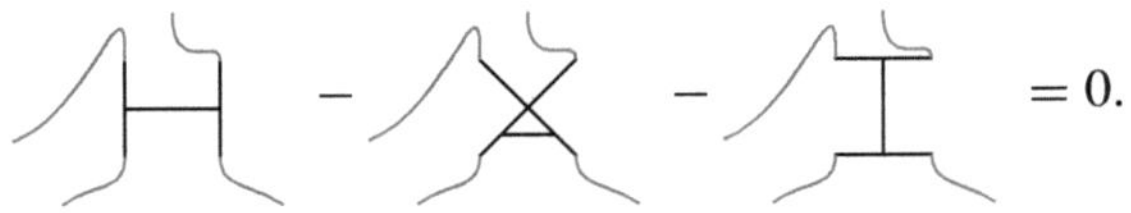

Exercise 13.18:

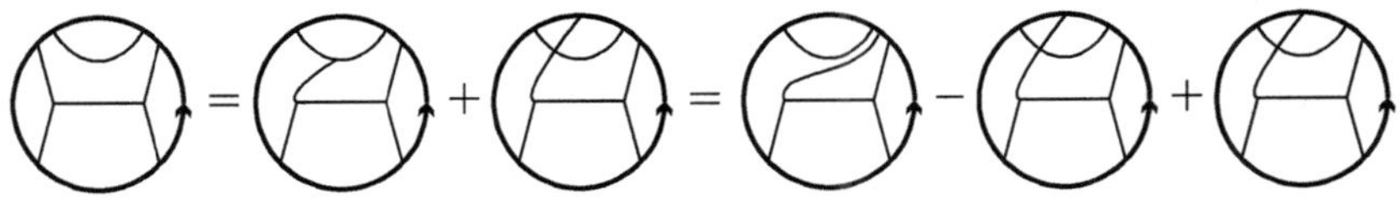

Exercise 13.23:

$$(\mathrm{id} \otimes \Delta(D)) \circ \Delta(D) = \sum_{\substack{(A,B,C)\text{ a} \\ \text{partition of } Y}} D_A \otimes D_B \otimes D_C = (\Delta(D) \otimes \mathrm{id}) \circ \Delta(D).$$

Exercise 14.1: $\langle \cdot, e_i \rangle : e_j \mapsto h_{i,j}$, so $\langle \cdot, e_i \rangle = \sum_k h_{i,k} e^k$.

Exercise 14.2: $\langle \cdot, \cdot \rangle^{-1} \circ \langle \cdot, \cdot \rangle : e_p \mapsto \sum_q h_{pq} h^{qr} e_r$, but $\langle \cdot, \cdot \rangle^{-1} \circ \langle \cdot, \cdot \rangle = \mathrm{id}$, so $h_{pq} h^{qr} = \delta^r_p$.

Exercise 14.4: Symmetry and bilinearity follow from the standard properties of the trace. For non-degeneracy, $\langle \cdot, \cdot \rangle : e_i \mapsto \sum_j \langle \cdot, e_i \rangle e^j$. Thus $\langle \cdot, \cdot \rangle$ sends X to $Y^\star$, Y to $X^\star$, and H to $2H^\star$. Its easy to write down an inverse of this map, and so $\langle \cdot, \cdot \rangle$ is non-degenerate. For ad-invariance, check the action on the bases using the properties of $[\cdot, \cdot]$ and $\langle \cdot, \cdot \rangle$ to reduce the number of cases.

Exercise 14.5: $E_{i_1} E_{i_2} \cdots E_{i_n} \longleftrightarrow E_{i_1} \otimes E_{i_2} \otimes \cdots \otimes E_{i_n}$.

Exercise 14.8: With the labelling of the vertices on the skeleton in the order $i_1, \ldots, i_4$, the two tensors are

$$\sum h^{i_1,i_3} h^{i_2,i_4}\, e_{i_1} \otimes e_{i_2} \otimes e_{i_3} \otimes e_{i_4},$$

and

$$\sum h^{i_1,i_5} h_{i_5,i_6} h^{i_6,i_7} h_{i_7,i_8} h^{i_8,i_3} h^{i_2,i_9} h_{i_9,i_{10}} h^{i_{10},i_4}\, e_{i_1} \otimes e_{i_2} \otimes e_{i_3} \otimes e_{i_4}.$$

Exercise 14.9: In $U(\mathfrak{g})$, $[e_i, e_j] = e_i \otimes e_j - e_j \otimes e_i$, so $\sum_c \gamma_{i,j}{}^c e_c = e_i \otimes e_j - e_j \otimes e_i$.

Exercise 14.21:

$$\overset{T_{\mathfrak{sl}_2}}{\longmapsto} \sum_{i,j,k,l\in\{X,Y,H\}} h^{i,l}h^{j,k}\, i\otimes j\otimes k\otimes l$$

$$\begin{aligned} = \; & X\otimes X\otimes Y\otimes Y + X\otimes Y\otimes X\otimes Y + \tfrac{1}{2}X\otimes H\otimes H\otimes Y + \\ & Y\otimes X\otimes Y\otimes X + Y\otimes Y\otimes X\otimes X + \tfrac{1}{2}Y\otimes H\otimes H\otimes X + \\ & \tfrac{1}{2}H\otimes X\otimes Y\otimes H + \tfrac{1}{2}H\otimes Y\otimes X\otimes H + \tfrac{1}{4}H\otimes H\otimes H\otimes H \end{aligned}$$

$$\overset{\mathrm{Tr}\circ\rho}{\longmapsto} (0+1+\tfrac{1}{2}) + (1+0+\tfrac{1}{2}) + (\tfrac{1}{2}+\tfrac{1}{2}+\tfrac{1}{2})$$

so $W_{\mathfrak{sl}_2,\rho}\Big($  $\Big) = \frac{9}{2}$.

Exercise 14.26: Consider the sum of the four chord diagrams in the 4T-relation. Apply (14.32) to the two distinguished chords in it to get a sum of 16 diagrams. Spot that the terms cancel to give zero.

Exercise 14.27: Choose a chord c and partition the sum into terms in which c is assigned a 1, and terms in which c is assigned a $-\frac{1}{2}$. An alternative way to see the result is to identify each summand in (14.39) with a terminal form of resulting from the application of (14.32) to C.

Exercise 14.28: If C_n is the chord diagram of the question, then Theorem 14.24 gives $W_{\mathfrak{sl}_2,\rho}(C_n) = 2W_{\mathfrak{sl}_2,\rho}(C_{n-1}) - \frac{1}{2}W_{\mathfrak{sl}_2,\rho}(C_{n-1})$, and so $W_{\mathfrak{sl}_2,\rho}(C_n) = 2(\frac{3}{2})^n$. It follows that there are framed Vassiliev invariants of each degree.

Exercise 14.29: Follow the approach in Sect. 14.4. You can make the exercise easier by just looking at the special case of $\mathfrak{gl}_2$.

Exercise 14.30: Follow the approach in Sect. 14.4.

Exercise 15.9: Forget about the tangles. It is enough to prove that it is possible to move between any two parenthesisations *via* $(\cdot(\cdot\cdot)) \leftrightarrow ((\cdot\cdot)\cdot)$, which can be done using induction on the number of pairs of brackets.

Exercise 16.11: Each expression distributes the ends in all ways over three copies of the skeleton component.

Exercise 16.12: Use the symmetry of R.

Exercise 16.20: Corollary 13.29 gives independence of the linearisation, and Lemma 16.12 independence of the location of the "splicing".

Exercise 16.24: If Φ had a degree one summand, then the diagram in that summand is a single chord between two of the stands. Applying ε to the other strand would contradict Condition (1) of Definition 16.21.

Exercise 17.9: "Cut open" K and K' to get $(1, 1)$-tangles T and T', so that K and K' are recovered by taking the closures T and T'. Then, $K\#K'$ is the closure of

$T \circ T'$. Now, $\check{Z}(K)$ can be obtained by computing $\check{Z}(T)$, closing up the skeleton components of the chord diagrams, and then connect summing a copy of ν. $\check{Z}(K')$ and $\check{Z}(K \# K')$ can be computed similarly but starting with T' and $T \circ T'$. The result is not hard to see when the values of $\check{Z}$ are computed in this way.

Exercise 17.11**:** Straightforward.

Exercise 18.4**:** Modify the proof of Theorem 18.1.

Exercise 18.8**:** Modify the proof of Theorem 18.6.

Exercise 18.10**:** Adapt the argument given for (18.9).

Exercise 18.11**:** Apply (18.7) i times working from the innermost chord outwards.

Then, use that 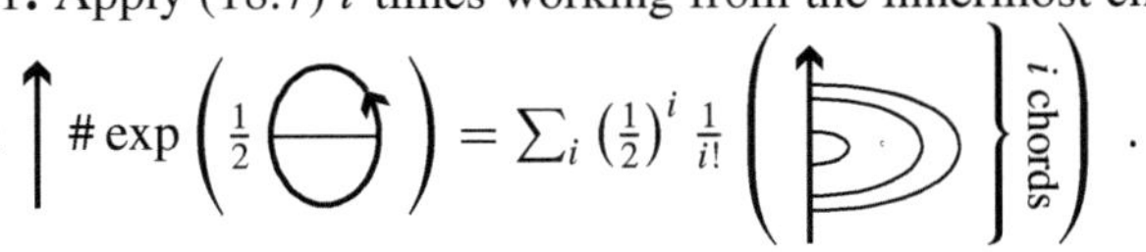

Exercise 18.12**:** Use (17.16), and modify (18.2) accordingly.

Exercise B.13**:** Adapt the argument for $f^{t_1 t_2} = f^{\star}$.

Exercise B.16**:** Adapt the argument for Claim 1.

Exercise D.3**:** Modify the argument given for the other case.

Exercise D.4**:** Adapt the proof of Proposition 8.33.

References

1. E. Abe, *Hopf Algebras*. Cambridge Tracts in Mathematics, vol. 74 (Cambridge University Press, Cambridge, 1980). Translated from the Japanese by Hisae Kinoshita and Hiroko Tanaka
2. W.A. Adkins, S.H. Weintraub, *Algebra*. Graduate Texts in Mathematics, vol. 136 (Springer, New York, 1992). An approach via module theory
3. A. Alekseev, E. Meinrenken, The non-commutative Weil algebra. Invent. Math. **139**(1), 135–172 (2000)
4. J.W. Alexander, Topological invariants of knots and links. Trans. Am. Math. Soc. **30**(2), 275–306 (1928)
5. D. Altschüler, A. Coste, Quasi-quantum groups, knots, three-manifolds, and topological field theory. Commun. Math. Phys. **150**(1), 83–107 (1992)
6. D. Altschüler, L. Freidel, On universal Vassiliev invariants. Commun. Math. Phys. **170**(1), 41–62 (1995)
7. M. Andler, A. Dvorsky, S. Sahi, Kontsevich quantization and invariant distributions on Lie groups. Ann. Sci. École Norm. Sup. (4) **35**(3), 371–390 (2002)
8. V. I. Arnol'd, The Vassiliev theory of discriminants and knots, in *First European Congress of Mathematics, Vol. I (Paris, 1992)*. Progress in Mathematics. vol. 119 (Birkhäuser, Basel, 1994), pp. 3–29
9. J.C. Baez, Link invariants of finite type and perturbation theory. Lett. Math. Phys. **26**(1), 43–51 (1992)
10. B. Bakalov, A. Kirillov Jr., *Lectures on Tensor Categories and Modular Functors*. University Lecture Series, vol. 21 (American Mathematical Society, Providence, 2001)
11. D. Bar-Natan, *Perturbative aspects of the Chern-Simons topological quantum field theory*. Ph.D. thesis, Princeton University (1991)
12. D. Bar-Natan, Weights of Feynman diagrams and the Vassiliev knot invariants. Preprint (1991)
13. D. Bar-Natan, On the Vassiliev knot invariants. Topology **34**(2), 423–472 (1995)
14. D. Bar-Natan, Vassiliev homotopy string link invariants. J. Knot Theory Ramif. **4**(1), 13–32 (1995)
15. D. Bar-Natan, Lie algebras and the four color theorem. Combinatorica **17**(1), 43–52 (1997)
16. D. Bar-Natan, Non-associative tangles, in *Geometric topology (Athens, GA, 1993)*. AMS/IP Studies in Advanced Mathematics, vol. 2 (American Mathematical Society, Providence, 1997), pp. 139–183
17. D. Bar-Natan, S. Garoufalidis, On the Melvin-Morton-Rozansky conjecture. Invent. Math. **125**(1), 103–133 (1996)
18. D. Bar-Natan, S. Garoufalidis, L. Rozansky, D.P. Thurston, Wheels, wheeling, and the Kontsevich integral of the unknot. Israel J. Math. **119**, 217–237 (2000)

D. M. Jackson and I. Moffatt, *An Introduction to Quantum and Vassiliev Knot Invariants*,
CMS Books in Mathematics, https://doi.org/10.1007/978-3-030-05213-3

19. D. Bar-Natan, T.T.Q. Le, D.P. Thurston, Two applications of elementary knot theory to Lie algebras and Vassiliev invariants. Geom. Topol. **7**, 1–31 (electronic) (2003)
20. D. Bar-Natan, A. Stoimenow, The fundamental theorem of Vassiliev invariants, in *Geometry and Physics (Aarhus, 1995)*. Lecture Notes in Pure and Applied Mathematics, vol. 184 (Dekker, New York, 1997), pp. 101–134
21. R.J. Baxter, Partition function of the eight-vertex lattice model. Ann. Phys. **70**(1), 193–228 (1972)
22. J.S. Birman, *Braids, Links, and Mapping Class Groups*. Annals of Mathematics Studies, vol. 82 (Princeton University Press, Princeton; University of Tokyo Press, Tokyo, 1974)
23. J.S. Birman, New Points of View in Knot Theory. Bull. Am. Math. Soc. (N.S.), **28**(2), 253–287 (1993)
24. J.S. Birman, X.-S. Lin, Knot polynomials and Vassiliev's invariants. Invent. Math. **111**(2), 225–270 (1993)
25. S.A. Bleiler, A note on unknotting number. Math. Proc. Camb. Philos. Soc. **96**(3), 469–471 (1984)
26. T.S. Blyth, *Categories (Longman* (Harlow; Wiley Inc, New York, 1986)
27. T.S. Blyth, *Module Theory*, 2nd edn. (Oxford Science Publications. The Clarendon Press, Oxford University Press, New York, 1990)
28. B. Bollobás, O. Riordan, Linearized chord diagrams and an upper bound for Vassiliev invariants. J. Knot Theory Ramif. **9**(7), 847–853 (2000)
29. L. Brand, *Differential and Difference Equations* (Wiley, New York, 1966)
30. D. Bump, *Lie Groups*. Graduate Texts in Mathematics, vol. 225 (Springer, New York, 2004)
31. G. Burde, H. Zieschang, *Knots*. de Gruyter Studies in Mathematics, vol. 5, 2nd edn. (Walter de Gruyter & Co., Berlin, 2003)
32. P. Cartier, Construction combinatoire des invariants de Vassiliev-Kontsevich des nœuds. C. R. Acad. Sci. Paris Sér. I Math. **316**(11), 1205–1210 (1993)
33. V. Chari, A. Pressley, *A Guide to Quantum Groups* (Cambridge University Press, Cambridge, 1995). Corrected reprint of the 1994 original
34. R. Cheng, D.M. Jackson, G. Stanley, Combinatorial aspects of the quantized universal enveloping algebra of $\mathfrak{sl}_{n+1}$. Ann. Combin. **22**(4), 681–710 (2018)
35. S. Chmutov, A proof of the Melvin-Morton conjecture and Feynman diagrams. J. Knot Theory Ramif. **7**(1), 23–40 (1998)
36. S. Chmutov, S. Duzhin, A lower bound for the number of Vassiliev knot invariants. Topol. Appl. **92**(3), 201–223 (1999)
37. S. Chmutov, S. Duzhin, J. Mostovoy, *Introduction to Vassiliev Knot Invariants* (Cambridge University Press, Cambridge, 2012)
38. S.V. Chmutov, S.V. Duzhin, An upper bound for the number of Vassiliev knot invariants. J. Knot Theory Ramif. **3**(2), 141–151 (1994)
39. S.V. Chmutov, S.V. Duzhin, S.K. Lando, Vassiliev knot invariants. I. Introduction, in *Singularities and Bifurcations*. Advances in Soviet Mathematics, vol. 21 (American Mathematical Society, Providence, 1994), pp. 117–126
40. S.V. Chmutov, S.V. Duzhin, S.K. Lando, Vassiliev knot invariants. II. Intersection graph conjecture for trees, in *Singularities and Bifurcations*. Advances in Soviet Mathematics, vol. 21. (American Mathematical Society, Providence, 1994), pp. 127–134
41. S.V. Chmutov, S.V. Duzhin, S.K. Lando, Vassiliev knot invariants. III. Forest algebra and weighted graphs, in *Singularities and Bifurcations*. Advances in Soviet Mathematics, vol. 21 (American Mathematical Society, Providence, 1994), pp. 135–145
42. S.V. Chmutov, S.K. Lando, Mutant knots and intersection graphs. Algebr. Geom. Topol. **7**, 1579–1598 (2007)
43. S.V. Chmutov, A.N. Varchenko, Remarks on the Vassiliev knot invariants coming from sl_2. Topology **36**(1), 153–178 (1997)
44. J.H. Conway, An enumeration of knots and links, and some of their algebraic properties, in *Computational Problems in Abstract Algebra (Proc. Conf., Oxford, 1967)* (Pergamon, Oxford, 1970), pp. 329–358

45. P.R. Cromwell, *Knots and Links* (Cambridge University Press, Cambridge, 2004)
46. R.H. Crowell, R.H. Fox, Introduction to Knot Theory (Springer, New York, *Reprint of the 1963 original* (Graduate Texts in Mathematics, No, 1977), p. 57
47. O.T. Dasbach, On the combinatorial structure of primitive Vassiliev invariants. II. J. Combin. Theory Ser. A **81**(2), 127–139 (1998)
48. O.T. Dasbach, On the combinatorial structure of primitive Vassiliev invariants. III. A lower bound. Commun. Contemp. Math. **2**(4) 579–590 (2000)
49. J. Dean, Many classical knot invariants are not Vassiliev invariants. J. Knot Theory Ramif. **3**(1), 7–10 (1994)
50. P. Di Francesco, Meander determinants. Commun. Math. Phys. **191**(3), 543–583 (1998)
51. P. Di Francesco. Folding and coloring problems in mathematics and physics. *Bull. Amer. Math. Soc. (N.S.)*, 37(3):251–307, 2000
52. P. Di Francesco, O. Golinelli, E. Guitter, Meanders and the Temperley-Lieb algebra. Commun. Math. Phys. **186**(1), 1–59 (1997)
53. J. Dixmier, *Enveloping algebras*. Graduate Studies in Mathematics, vol. 11 (American Mathematical Society, Providence, 1996). Revised reprint of the 1977 translation
54. V.G. Drinfel'd, Hopf algebras and the quantum Yang-Baxter equation. Dokl. Akad. Nauk SSSR **283**(5), 1060–1064 (1985)
55. V.G. Drinfel'd, Quantum groups, in *Proceedings of the International Congress of Mathematicians, Vols. 1, 2 (Berkeley, Calif., 1986)* (American Mathematical Society, Providence, 1987), pp. 798–820
56. V.G. Drinfel'd, Almost cocommutative Hopf algebras. Algebra i Analiz **1**(2), 30–46 (1989)
57. V.G. Drinfel'd, Quasi-Hopf algebras. Algebra i Analiz **1**(6), 114–148 (1989)
58. V.G. Drinfel'd, Quasi-Hopf algebras and Knizhnik-Zamolodchikov equations, in *Problems of modern quantum field theory (Alushta, 1989)*, Res. Rep. Phys., pp. 1–13. Springer, Berlin, 1989
59. V.G. Drinfel'd, On quasitriangular quasi-Hopf algebras and on a group that is closely connected with $\mathrm{Gal}(\overline{Q}/Q)$. Algebra i Analiz **2**(4), 149–181 (1990)
60. S. Eliahou, L.H. Kauffman, M.B. Thistlethwaite, Infinite families of links with trivial Jones polynomial. Topology **42**(1), 155–169 (2003)
61. K. Erdmann, M.J. Wildon, *Introduction to Lie Algebras, Springer Undergraduate Mathematics Series* (Springer, London Ltd, London, 2006)
62. J.M. Figueroa-O'Farrill, T. Kimura, A. Vaintrob, The universal Vassiliev invariant for the Lie superalgebra $\mathrm{gl}(1|1)$. Commun. Math. Phys. **185**(1), 93–127 (1997)
63. P. Freyd, D. Yetter, J. Hoste, W.B.R. Lickorish, K. Millett, A. Ocneanu, A new polynomial invariant of knots and links. Bull. Am. Math. Soc. (N.S.) **12**(2):239–246 (1985)
64. P. Freyd, D.N. Yetter, Coherence theorems via knot theory. J. Pure Appl. Algebra **78**(1), 49–76 (1992)
65. P.J. Freyd, D.N. Yetter, Braided compact closed categories with applications to low-dimensional topology. Adv. Math. **77**(2), 156–182 (1989)
66. L. Geissinger, Hopf algebras of symmetric functions and class functions, in *Combinatoire et représentation du groupe symétrique (Actes Table Ronde C.N.R.S., Universite Louis-Pasteur Strasbourg, Strasbourg, 1976)*. Lecture Notes in Mathematics, vol. 579 (Springer, Berlin, 1977), pp. 168–181
67. L. Goeritz, Bemerkungen zur knotentheorie. Abh. Math. Sem. Univ. Hamburg **10**(1), 201–210 (1934)
68. F.M. Goodman, P. de la Harpe, V.F.R. Jones, *Coxeter Graphs and Towers of Algebras*, vol. 14 (Mathematical Sciences Research Institute Publications (Springer, New York, 1989)
69. M. Gusarov, A new form of the Conway-Jones polynomial of oriented links, in *Topology of Manifolds and Varieties*. Advances in Soviet Mathematics, vol. 18, pp. 167–172 (American Mathematical Society, Providence, 1994)
70. M. Gusarov, On n-equivalence of knots and invariants of finite degree, in *Topology of manifolds and varieties*. Advances in Soviet Mathematics, vol. 18 (American Mathematical Society, Providence, 1994), pp. 173–192

71. B. Hall, *Lie Groups, Lie Algebras, and Representations*. Graduate Texts in Mathematics, vol. 222, 2nd edn. (Springer, Cham, 2015)
72. B. Hartley, T.O. Hawkes, *Rings, Modules and Linear Algebra* (Chapman & Hall, London, 1980)
73. J. Hass, J.C. Lagarias, The number of Reidemeister moves needed for unknotting. J. Am. Math. Soc. **14**(2), 399–428 (electronic) (2001)
74. J. Hass, J.C. Lagarias, N. Pippenger, The computational complexity of knot and link problems. J. ACM **46**(2), 185–211 (1999)
75. C. Hayashi, A lower bound for the number of Reidemeister moves for unknotting. J. Knot Theory Ramif. **15**(3), 313–325 (2006)
76. V. Hinich, A. Vaintrob, Cyclic operads and algebra of chord diagrams. Selecta Math. (N.S.) **8**(2), 237–282 (2002)
77. J. Hoste, A polynomial invariant of knots and links. Pac. J. Math. **124**(2), 295–320 (1986)
78. J.E. Humphreys, *Introduction to Lie Algebras and Representation Theory*. Graduate Texts in Mathematics, vol. 9 (Springer, New York, 1978). Second printing, revised
79. N. Jacobson, *Basic Algebra. I* (W. H. Freeman and Co., San Francisco, 1974)
80. N. Jacobson, *Lie Algebras* (Dover Publications, Inc., New York, 1979). Republication of the 1962 original
81. N. Jacobson, *Basic Algebra. II* (W. H. Freeman and Co., San Francisco, 1980)
82. F. Jaeger, D.L. Vertigan, D.J.A. Welsh, On the computational complexity of the Jones and Tutte polynomials. Math. Proc. Camb. Philos. Soc. **108**(1), 35–53 (1990)
83. M. Jimbo, A q-difference analogue of $U(\mathfrak{g})$ and the Yang-Baxter equation. Lett. Math. Phys. **10**(1), 63–69 (1985)
84. V.F.R. Jones, A polynomial invariant for knots via von Neumann algebras. Bull. Am. Math. Soc. (N.S.) **12**(1), 103–111 (1985)
85. V.F.R. Jones, Hecke algebra representations of braid groups and link polynomials. Ann. Math. (2) **126**(2), 335–388 (1987)
86. V.F.R. Jones, On knot invariants related to some statistical mechanical models. Pac. J. Math. **137**(2), 311–334 (1989)
87. A. Joyal, R. Street, The geometry of tensor calculus. I. Adv. Math. **88**(1), 55–112 (1991)
88. A. Joyal, R. Street, Braided tensor categories. Adv. Math. **102**(1), 20–78 (1993)
89. T. Kanenobu, Infinitely many knots with the same polynomial invariant. Proc. Am. Math. Soc. **97**(1), 158–162 (1986)
90. C. Kassel, *Quantum Groups*. Graduate Texts in Mathematics, vol. 155 (Springer, New York, 1995)
91. C. Kassel, M. Rosso, V. Turaev, *Quantum Groups and Knot Invariants*. Panoramas et Synthèses, vol. 5 (Société Mathématique de France, Paris, 1997)
92. L.H. Kauffman, *Formal Knot Theory*. Mathematical Notes, vol. 30 (Princeton University Press, Princeton, 1983)
93. L.H. Kauffman, *On Knots*. Annals of Mathematics Studies, vol. 115 (Princeton University Press, Princeton, 1987)
94. L.H. Kauffman, *Knots and Physics*. Series on Knots and Everything, vol. 53, 4th edn. (World Scientific Publishing Co. Pte. Ltd., Hackensack, 2013)
95. L.H. Kauffman, S. Lambropoulou, Hard unknots and collapsing tangles, in *Introductory Lectures on Knot Theory*. Ser, Knots Everything, vol. 46 (World Scientific Publishing, Hackensack, 2012), pp. 187–247
96. L.H. Kauffman, H. Saleur, Free fermions and the Alexander-Conway polynomial. Commun. Math. Phys. **141**(2), 293–327 (1991)
97. M.E. Keating, *A First Course in Module Theory* (Imperial College Press, London, 1998)
98. R. Kirby, P. Melvin, The 3-manifold invariants of Witten and Reshetikhin-Turaev for sl$(2, C)$. Invent. Math. **105**(3), 473–545 (1991)
99. A.N. Kirillov, N.Y. Reshetikhin, Representations of the algebra $U_q(\mathrm{sl}(2))$, q-orthogonal polynomials and invariants of links, in *Infinite-dimensional Lie algebras and groups (Luminy-Marseille, 1988)*. Advances Series in Mathematical Physics, vol. 7 (World Scientific Publishing, Teaneck, 1989), pp. 285–339

100. J.A. Kneissler, The number of primitive Vassiliev invariants up to degree 12, arxiv.org/abs/q-alg/9706022
101. The knot atlas, http://katlas.org/
102. M. Kontsevich, Vassiliev's knot invariants, in *I. M. Gel'fand Seminar*. Advances in Soviet Mathematics, vol. 16 (American Mathematical Society, Providence, 1993), pp. 137–150
103. M. Kontsevich, Feynman diagrams and low-dimensional topology, in *First European Congress of Mathematics, Vol. II (Paris, 1992)*. Progress in Mathematics, vol. 120 (Birkhäuser, Basel, 1994), pp. 97–121
104. M. Kontsevich, Deformation quantization of Poisson manifolds. Lett. Math. Phys. **66**(3), 157–216 (2003)
105. A. Kricker, Alexander-Conway limits of many Vassiliev weight systems. J. Knot Theory Ramif. **6**(5), 687–714 (1997)
106. A. Kricker, Differential operators and the wheels power series. Algebr. Geom. Topol. **11**(2), 1107–1162 (2011)
107. A. Kricker, Noncommutative Chern-Weil theory and the combinatorics of wheeling. Duke Math. J. **157**(2), 223–281 (2011)
108. A. Kricker, B. Spence, I. Aitchison, Cabling the Vassiliev invariants. J. Knot Theory Ramif. **6**(3), 327–358 (1997)
109. P.P. Kulish, N.J. Reshetikhin, Quantum linear problem for the sine-Gordon equation and higher representations. Zap. Nauchn. Sem. Leningrad. Otdel. Mat. Inst. Steklov. (LOMI) **101**, 101–110, 207 (1981)
110. B.I. Kurpita, K. Murasugi, A graphical approach to the Melvin-Morton conjecture. I. Topol. Appl. **82**(1–3), 297–316 (1998)
111. S.K. Lando, On a Hopf algebra in graph theory. J. Combin. Theory Ser. B **80**(1), 104–121 (2000)
112. R.J. Lawrence, *A universal link invariant using quantum groups, in Differential Geometric Methods in Theoretical Physics (Chester, 1988)* (World Scientific Publishing, Teaneck, 1989), pp. 55–63
113. R.J. Lawrence, A universal link invariant, in *The interface of mathematics and particle physics (Oxford, 1988)*. Inst. Math. Appl. Conf. Ser. New Ser. vol. 24 (Oxford University Press, New York, 1990), pp. 151–156
114. T.Q.T. Le, J. Murakami, Kontsevich's integral for the Homfly polynomial and relations between values of multiple zeta functions. Topol. Appl. **62**(2), 193–206 (1995)
115. T.Q.T. Le, J. Murakami, Representation of the category of tangles by Kontsevich's iterated integral. Commun. Math. Phys. **168**(3), 535–562 (1995)
116. T.Q.T. Le, J. Murakami, The universal Vassiliev-Kontsevich invariant for framed oriented links. Compos. Math. **102**(1), 41–64 (1996)
117. T.T.Q. Le, J. Murakami, Kontsevich's integral for the Kauffman polynomial. Nagoya Math. J. **142**, 39–65 (1996)
118. T.T.Q. Le, J. Murakami, Parallel version of the universal Vassiliev-Kontsevich invariant. J. Pure Appl. Algebra **121**(3), 271–291 (1997)
119. S.Z. Levendorskiĭ, Y.S. Soĭbel'man, Some applications of the quantum Weyl groups. J. Geom. Phys. **7**(2), 241–254 (1990)
120. W.B.R. Lickorish, *An Introduction to Knot Theory*. Graduate Texts in Mathematics, vol. 175 (Springer, New York, 1997)
121. J. Lieberum, On Vassiliev invariants not coming from semisimple Lie algebras. J. Knot Theory Ramif. **8**(5), 659–666 (1999)
122. J. Lieberum, Chromatic weight systems and the corresponding knot invariants. Math. Ann. **317**(3), 459–482 (2000)
123. J. Lieberum, The number of independent Vassiliev invariants in the Homfly and Kauffman polynomials. Doc. Math. **5**, 275–299 (electronic) 2000
124. X.-S. Lin, Vertex modles, quantum groups and Vassiliev's knot invariants. Preprint (1991)
125. G. Lusztig, Quantum deformations of certain simple modules over enveloping algebras. Adv. Math. **70**(2), 237–249 (1988)

126. S. Mac Lane, Categories for the Working Mathematician. Graduate Texts in Mathematics, vol. 5, 2nd edn. (Springer, New York, 1998)
127. S. Majid, *A Quantum Groups Primer*. London Mathematical Society Lecture Note Series, vol. 292 (Cambridge University Press, Cambridge, 2002)
128. V. Manturov, *Knot theory* (Chapman & Hall/CRC, Boca Raton, 2004)
129. W.S. Massey, *A Basic Course in Algebraic Topology*. Graduate Texts in Mathematics, vol. 127 (Springer, New York, 1991)
130. P.M. Melvin, H.R. Morton, The coloured Jones function. Commun. Math. Phys. **169**(3), 501–520 (1995)
131. J.W. Milnor, J.C. Moore, On the structure of Hopf algebras. Ann. Math. **2**(81), 211–264 (1965)
132. T. Mochizuki, On the morphism of Duflo-Kirillov type. J. Geom. Phys. **41**(1–2), 73–113 (2002)
133. H.R. Morton, Invariants of links and 3-manifolds from skein theory and from quantum groups, in *Topics in Knot Theory (Erzurum, 1992)*. NATO Adv. Sci. Inst. Ser. C Math. Phys. Sci. (Kluwer Academic Publishers, Dordrecht, 1993), pp. 107–155
134. H.R. Morton, Quantum invariants given by evaluation of knot polynomials. J. Knot Theory Ramif. **2**(2), 195–209 (1993)
135. H.R. Morton, The coloured Jones function and Alexander polynomial for torus knots. Math. Proc. Camb. Philos. Soc. **117**(1), 129–135 (1995)
136. H.R. Morton, P.R. Cromwell, Distinguishing mutants by knot polynomials. J. Knot Theory Ramif. **5**(2), 225–238 (1996)
137. H.R. Morton, H.B. Short, Calculating the 2-variable polynomial for knots presented as closed braids. J. Algorithms **11**(1), 117–131 (1990)
138. H.R. Morton, P. Strickland, Jones polynomial invariants for knots and satellites. Math. Proc. Camb. Philos. Soc. **109**(1), 83–103 (1991)
139. H.R. Morton, P.M. Strickland, *Satellites and surgery invariants, in Knots 90 (Osaka, 1990)* (de Gruyter, Berlin, 1992), pp. 47–66
140. J. Murakami, The parallel version of polynomial invariants of links. Osaka J. Math. **26**(1), 1–55 (1989)
141. K.Y. Ng, Groups of ribbon knots. Topology **37**(2), 441–458 (1998)
142. K.Y. Ng, T. Stanford, On Gusarov's groups of knots. Math. Proc. Camb. Philos. Soc. **126**(1), 63–76 (1999)
143. T. Ohtsuki, Colored ribbon Hopf algebras and universal invariants of framed links. J. Knot Theory Ramif. **2**(2), 211–232 (1993)
144. T. Ohtsuki, *Quantum Invariants*. Series on Knots and Everything, vol. 29 (World Scientific Publishing Co., Inc., River Edge, 2002)
145. M. Ozawa, Knots and surfaces. Sūgaku **67**(4), 403–423 (2015)
146. R. Penrose, Applications of negative dimensional tensors, in *Combinatorial Mathematics and its Applications (Proc. Conf., Oxford, 1969)* (Academic, London, 1971), pp. 221–244
147. R. Penrose, *The Road to Reality* (Alfred A. Knopf, Inc., New York, 2005)
148. K.A. Perko, On the classification of knots. Proc. Am. Math. Soc. **45**, 262–266 (1974)
149. S. Piunikhin, Combinatorial expression for universal Vassiliev link invariant. Commun. Math. Phys. **168**(1), 1–22 (1995)
150. S. Piunikhin, Weights of Feynman diagrams, link polynomials and Vassiliev knot invariants. J. Knot Theory Ramif. **4**(1), 163–188 (1995)
151. M. Polyak, Minimal generating sets of Reidemeister moves. Quantum Topol. **1**(4), 399–411 (2010)
152. V.V. Prasolov, A.B. Sossinsky, *Knots, Links, Braids and 3-manifolds*. Translations of Mathematical Monographs, vol. 154 (American Mathematical Society, Providence, 1997)
153. J.H. Przytycki, History of knot theory. arXiv:math/0703096
154. J.H. Przytycki, P. Traczyk, Conway algebras and skein equivalence of links. Proc. Am. Math. Soc. **100**(4), 744–748 (1987)
155. N. Reshetikhin, V.G. Turaev, Invariants of 3-manifolds via link polynomials and quantum groups. Invent. Math. **103**(3), 547–597 (1991)

156. N.Y. Reshetikhin, Quasitriangular Hopf algebras and invariants of links. Algebra i Analiz **1**(2), 169–188 (1989)
157. N.Y. Reshetikhin, V.G. Turaev, Ribbon graphs and their invariants derived from quantum groups. Commun. Math. Phys. **127**(1), 1–26 (1990)
158. D. Rolfsen, *Knots and Links*. Mathematics Lecture Series, vol. 7 (Publish or Perish, Inc., Berkeley, 1976)
159. M. Rosso, Finite-dimensional representations of the quantum analog of the enveloping algebra of a complex simple Lie algebra. Commun. Math. Phys. **117**(4), 581–593 (1988)
160. M. Rosso, An analogue of P.B.W. theorem and the universal R-matrix for $U_h\mathrm{sl}(N+1)$. Commun. Math. Phys. **124**(2), 307–318 (1989)
161. M. Rosso, V. Jones, On the invariants of torus knots derived from quantum groups. J. Knot Theory Ramif. **2**(1), 97–112 (1993)
162. C.P. Rourke, B.J. Sanderson, *Introduction to piecewise-linear topology* (Springer, Berlin, 1982)
163. L. Rozansky, A contribution of the trivial connection to the Jones polynomial and Witten's invariant of 3d manifolds. I. Commun. Math. Phys. **175**(2), 275–296 (1996)
164. L. Rozansky, A contribution of the trivial connection to the Jones polynomial and Witten's invariant of 3d manifolds. II. Commun. Math. Phys. **175**(2), 297–318 (1996)
165. L. Rozansky, Higher order terms in the Melvin-Morton expansion of the colored Jones polynomial. Commun. Math. Phys. **183**(2), 291–306 (1997)
166. L. Rozansky, The universal R-matrix, Burau representation, and the Melvin-Morton expansion of the colored Jones polynomial. Adv. Math. **134**(1), 1–31 (1998)
167. H. Samelson, *Notes on Lie algebras*. Universitext, 2nd edn. (Springer, New York, 1990)
168. J. Sawada, A fast algorithm for generating nonisomorphic chord diagrams. SIAM J. Discret. Math. **15**(4), 546–561 (2002)
169. R.G. Scharein, Knotplot http://www.knotplot.com/
170. H. Schubert, Die eindeutige Zerlegbarkeit eines Knotens in Primknoten. S.-B. Heidelberger Akad. Wiss. Math.-Nat. Kl. **1949**(3), 57–104 (1949)
171. T. Stanford, Braid commutators and Vassiliev invariants. Pac. J. Math. **174**(1), 269–276 (1996)
172. A. Stoimenow, Enumeration of chord diagrams and an upper bound for Vassiliev invariants. J. Knot Theory Ramif. **7**(1), 93–114 (1998)
173. A. Stoimenow, On the number of chord diagrams. Discret. Math. **218**(1–3), 209–233 (2000)
174. M.C. Sullivan, Knot factoring. Am. Math. Monthly **107**(4), 297–315 (2000)
175. M. Thistlethwaite, Links with trivial Jones polynomial. J. Knot Theory Ramif. **10**(4), 641–643 (2001)
176. D. P. Thurston. *Wheeling: A diagrammatic analogue of the Duflo isomorphism.* Thesis (Ph.D.)–University of California, Berkeley (2000)
177. B. Trace, On the Reidemeister moves of a classical knot. Proc. Am. Math. Soc. **89**(4), 722–724 (1983)
178. R. Trapp, Twist sequences and Vassiliev invariants. J. Knot Theory Ramif. **3**(3), 391–405 (1994)
179. H.F. Trotter, Non-invertible knots exist. Topology **2**, 275–280 (1963)
180. V.G. Turaev, The Yang-Baxter equation and invariants of links. Invent. Math. **92**(3), 527–553 (1988)
181. V.G. Turaev, Operator invariants of tangles, and R-matrices. Izv. Akad. Nauk SSSR Ser. Mat. **53**(5), 1073–1107, 1135 (1989)
182. V.G. Turaev, *Quantum Invariants of Knots and 3-manifolds*. de Gruyter Studies in Mathematics, vol. 18 (Walter de Gruyter & Co., Berlin, 1994)
183. A. Vaintrob, Universal weight systems and the Melvin-Morton expansion of the colored Jones knot invariant. J. Math. Sci. **82**(1), 3240–3254 (1996). Algebraic geometry, 5
184. A. Vaintrob, Melvin-Morton conjecture and primitive Feynman diagrams. Int. J. Math. **8**(4), 537–553 (1997)
185. V.A. Vassiliev, Cohomology of knot spaces, in *Theory of Singularities and Its Applications*. Advances in Soviet Mathematics, vol. 1 (American Mathematical Society, Providence, 1990), pp. 23–69

186. V.A. Vassiliev, *Complements of Discriminants of Smooth Maps: Topology and Applications*. Translations of Mathematical Monographs, vol. 98 (American Mathematical Society, Providence, 1992)
187. P. Vogel, Algebraic structures on modules of diagrams. J. Pure Appl. Algebra **215**(6), 1292–1339 (2011)
188. L. Watson, Any tangle extends to non-mutant knots with the same Jones polynomial. J. Knot Theory Ramif. **15**(9), 1153–1162 (2006)
189. D.J.A. Welsh, *Complexity: Knots, Colourings and Counting*. London Mathematical Society Lecture Note Series, vol. 186 (Cambridge University Press, Cambridge, 1993)
190. F.Y. Wu, Knot theory and statistical mechanics. Rev. Modern Phys. **64**(4), 1099–1131 (1992)
191. S. Yamada, An operator on regular isotopy invariants of link diagrams. Topology **28**(3), 369–377 (1989)
192. C.N. Yang, Some exact results for the many-body problem in one dimension with repulsive delta-function interaction. Phys. Rev. Lett. **19**, 1312–1315 (1967)
193. D.N. Yetter, *Functorial Knot Theory*. Series on Knots and Everything, vol. 26 (World Scientific Publishing Co., Inc., River Edge, 2001)
194. D. Zagier, Vassiliev invariants and a strange identity related to the Dedekind eta-function. Topology **40**(5), 945–960 (2001)

Index

D. M. Jackson and I. Moffatt, *An Introduction to Quantum and Vassiliev Knot Invariants*, CMS Books in Mathematics, https://doi.org/10.1007/978-3-030-05213-3

MIX
Papier aus verantwortungsvollen Quellen
Paper from responsible sources
FSC® C105338

If you have any concerns about our products,
you can contact us on
ProductSafety@springernature.com

In case Publisher is established outside the EU,
the EU authorized representative is:
Springer Nature Customer Service Center GmbH
Europaplatz 3, 69115 Heidelberg, Germany

Printed by Libri Plureos GmbH
in Hamburg, Germany